WHAT OUR READERS/REVIEWERS SAY ABOUT VOLUME I?

In the last few years, several books have been published in Concurrent Engineering. The book *Concurrent Engineering Fundamentals* is the first comprehensive text book, which balances coverage of fundamental concepts, original research results, industrial applications and practical experiences. It deals with all major issues involved in CE ranging from information technology to life cycle management. *Concurrent Engineering Fundamentals* is essential reading for engineers, managers and academics who are working in the field of concurrent engineering.

...It is an excellent text book for senior undergraduate students and graduate students in the field of manufacturing engineering, production engineering, industrial engineering and business schools.

Peihua Gu, Ph.D. and P.Eng.
Professor and NSERC/AECL Chair
Dept. of Mechanical Engineering
University of Saskatchewan, Saskatoon, SK, Canada

I found *Concurrent Engineering Fundamentals* to be an easy-to-read introduction to an area that has intrigued me for several years. The book is rich in illustrations and tables, and this abundance of visual material helped me make sense of the concepts and jargon introduced in the book. Furthermore, I found that I could skip around the book to topics of particular interest without too much trouble. That is, the book permits the reader to select topics of interest without having to read it in its entirety.

...The *Concurrent Engineering Fundamentals* book will be eminently useful both to students taking a course in Concurrent Engineering and to engineers seeking to update their skills on their own.

Raphael (Rafi) T. Haftka, Ph.D.
Professor, University of Florida
Department of Aerospace Engineering Mechanics and Engineering Science
Gainesville, Florida

This book *Concurrent Engineering Fundamentals*—although it has Fundamentals in its title—is a book not only for the newcomers to the field, but also for the experts, too. What distinguishes this book from others is that it really embodies concurrent engineering in its writing. Concurrency is well maintained throughout—among the concepts, methodologies including discussions of the social and technical backgrounds. Further, these concepts and methodologies are so well illustrated that newcomers will not find any difficulty in understanding them. The well indexed technical terms help a great deal for the newcomers to understand. Experts will also find *Concurrent Engineering Fundamentals* very informative because there are so many descriptions and comparisons of different cultures. There are also many descriptions about Japan. Even to a Japanese like me, I found that the book contains many new findings about our Japanese industrial backgrounds that I did not know before.

...I would like to recommend *Concurrent Engineering Fundamentals* for all who have interests in CE, newcomers and experts as well.

Shuichi Fukuda, Ph.D.
Chair Professor and Department Chair
Department of Production, Information and Systems Engineering
Tokyo Metropolitan Institute of Technology
Asahigaoka, Hino, Tokyo, JAPAN

A long needed book …*Concurrent Engineering Fundamentals* is the first comprehensive text in the rapidly developing area of CE that covers the very fundamentals….

Apart from its merits of *high quality and timely content*, the book is very *well organized* editorially. The book *will appeal* to both the engineering and management practitioner, as well as the academic community, where it can serve as a textbook.

Dr. Marek B. Zaremba
Professor, Dept. of Computer Science
University of Quebec, CANADA

The cost and time it takes to do product and process engineering has been escalating over the last few decades due to several reasons among them: increasing customer satisfaction, increasing government regulations, and increasing design alternatives due to material and process innovations. *Concurrent Engineering Fundamentals* rightly sets forth the philosophy and methodology necessary to conduct a modern concurrent engineering process.

…*Concurrent Engineering Fundamentals* will be a very welcomed addition to the literature in this important growing field.

Mounir M. Kamal, Ph.D.
Executive Director (Retired)
General Motors Research Laboratory
Warren, Michigan

Concurrent Engineering Fundamentals is a very comprehensive, thorough, and visionary analysis of the concurrent engineering process. It serves a wide range of needs from an engineering textbook to a highly useful reference. In this day of exploding knowledge, intensifying global competition, and more demanding customers, it is imperative to significantly improve the engineering process. No longer is an undisciplined and often ad hoc process good enough. The entire process must be managed following a very disciplined approach.

Concurrent Engineering Fundamentals is a book that brings both breadth and depth to the issue. It is a coherent work integrating the "alphabet soup" of current thinking and new techniques related to the overall product development process and should help individuals, work teams, and companies improve their effectiveness. Key performance factors, including quality, time to market, and cost are given appropriate attention as is the important issue of continuous improvement and re-engineering.

…I recommend *Concurrent Engineering Fundamentals* to all who are faced with challenges of improving the effectiveness and efficiency of their engineering process.

David E. Cole, Ph.D.
Director, Transportation Research Institute
Office for the Study of Automotive Transportation
The University of Michigan
Ann Arbor, Michigan

Concurrent Engineering Fundamentals offers a *lot of new* information … *new material* and focused. Frankly speaking, *no book exits* in the market to this—CE Fundamentals book….

Nanua Singh, Ph.D.
Professor, Department of Industrial and Manufacturing Engineering
Wayne State University, Detroit, MI

CONCURRENT ENGINEERING FUNDAMENTALS

PRENTICE HALL INTERNATIONAL SERIES
IN INDUSTRIAL AND SYSTEMS ENGINEERING

W. J. Fabrycky and J. H. Mize, Editors

AMOS AND SARCHET   *Management for Engineers*
AMRINE, RITCHEY, MOODIE AND KMEC   *Manufacturing Organization and Management, 6/E*
ASFAHL   *Industrial Safety and Health Management, 3/E*
BABCOCK   *Managing Engineering and Technology, 2/E*
BADIRU   *Comprehensive Project Management*
BADIRU   *Expert Systems Applications in Engineering and Manufacturing*
BANKS, CARSON AND NELSON   *Discrete Event System Simulation, 2/E*
BLANCHARD   *Logistics Engineering and Management, 4/E*
BLANCHARD AND FABRYCKY   *Systems Engineering and Analysis, 2/E*
BROWN   *Technimanagement: The Human Side of the Technical Organization*
BURTON AND MORAN   *The Future Focused Organization*
BUSSEY AND ESCHENBACH   *The Economic Analysis of Industrial Projects, 2/E*
BUZACOTT AND SHANTHIKUMAR   *Stochastic Models of Manufacturing Systems*
CANADA AND SULLIVAN   *Economic and Multi-Attribute Evaluation of Advanced Manufacturing Systems*
CANADA, SULLIVAN AND WHITE   *Capital Investment Analysis for Engineering and Management, 2/E*
CHANG AND WYSK   *An Introduction to Automated Process Planning Systems*
CHANG, WYSK AND WANG   *Computer Aided Manufacturing*
DELAVIGNE AND ROBERTSON   *Deming's Profound Changes*
EAGER   *The Information Payoff: The Manager's Concise Guide to Making PC Communications Work*
EBERTS   *User Interface Design*
EBERTS AND EBERTS   *Myths of Japanese Quality*
ELSAYED AND BOUCHER   *Analysis and Control of Production Systems, 2/E*
FABRYCKY AND BLANCHARD   *Life-Cycle Cost and Economic Analysis*
FABRYCKY AND THUESEN   *Economic Decision Analysis*
FISHWICK   *Simulation Model Design and Execution: Building Digital Worlds*
FRANCIS, MCGINNIS AND WHITE   *Facility Layout and Location: An Analytical Approach, 2/E*
GIBSON   *Modern Management of the High-Technology Enterprise*
GORDON   *Systematic Training Program Design*
GRAEDEL AND ALLENBY   *Industrial Ecology*
HALL   *Queuing Methods: For Services and Manufacturing*
HANSEN   *Automating Business Process Reengineering*
HAMMER   *Occupational Safety Management and Engineering, 4/E*
HAZELRIGG   *Systems Engineering*
HUTCHINSON   *An Integrated Approach to Logistics Management*
IGNIZIO   *Linear Programming in Single- and Multiple-Objective Systems*
IGNIZIO AND CAVALIER   *Linear Programming*
KROEMER, KROEMER AND KROEMER-ELBERT   *Ergonomics: How to Design for Ease and Efficiency*
KUSIAK   *Intelligent Manufacturing Systems*
LAMB   *Availability Engineering and Management for Manufacuring Plant Performance*
LANDERS, BROWN, FANT, MALSTROM AND SCHMITT   *Electronic Manufacturing Processes*
LEEMIS   *Reliability: Probabilistic Models and Statistical Methods*
MICHAELS   *Technical Risk Management*
MUNDEL AND DANNER   *Motion and Time Study: Improving Productivity, 7/E*
OSTWALD   *Engineering Cost Estimating, 3/E*
PINEDO   *Scheduling: Theory, Algorithms, and Systems*
PRASAD   *Concurrent Engineering Fundamentals, Vol. I: Integrated Product and Process Organization*
PRASAD   *Concurrent Engineering Fundamentals, Vol. II: Integrated Product Development*
PULAT   *Fundamentals of Industrial Ergonomics*
SHTUB, BARD AND GLOBERSON   *Project Management: Engineering Technology and Implementation*
TAHA   *Simulation Modeling and SIMNET*
THUESEN AND FABRYCKY   *Engineering Economy, 8/E*
TURNER, MIZE, CASE AND NAZEMETZ   *Introduction to Industrial and Systems Engineering, 3/E*
TURTLE   *Implementing Concurrent Project Management*
VON BRAUN   *The Innovation War*
WALESH   *Engineering Your Future*
WOLFF   *Stochastic Modeling and the Theory of Queues*

CONCURRENT ENGINEERING FUNDAMENTALS
VOLUME II
Integrated Product Development

Biren Prasad

PRENTICE HALL INTERNATIONAL SERIES
IN INDUSTRIAL AND SYSTEMS ENGINEERING

To join a Prentice Hall PTR internet mailing list, point to:
http://www.prenhall.com/register

Prentice Hall PTR
Upper Saddle River, New Jersey 07458
http://www.prenhall.com

Library of Congress Cataloging-in-Publication Data

Prasad, Biren
    Concurrent engineering fundamentals: integrated product
development / Biren Prasad.
        p.   cm. — (Prentice-Hall international series in industrial
and systems engineering)
    Includes bibliographical references and index.
    ISBN 0–13–396946–0
    1. Production engineering.   2. Concurrent engineering.   3. Design,
Industrial.   I. Title.   II. Series.
    TS176.P694   1996
    670.4—dc20   95–43132
                                                            CIP

Acquisitions editor: Bernard Goodwin
Cover designer: Design Source
Cover design director: Jerry Votta
Manufacturing buyer: Alexis R. Heydt
Compositor/Production services: Pine Tree Composition, Inc.

 © 1997 by Prentice Hall PTR
Prentice-Hall, Inc.
A Simon & Schuster Company
Upper Saddle River, New Jersey 07458

The publisher offers discounts on this book when ordered in
bulk quantities. For more information contact:

Corporate Sales Department
Prentice Hall PTR
One Lake Street
Upper Saddle River, New Jersey 07458

Phone: 800–382–3419
Fax: 201–236–7141
email: corpsales@prenhall.com

Printed in the United States of America

10 9 8 7 6 5 4 3 2 1

ISBN: 0-13-396946-0

Prentice-Hall International (UK) Limited, *London*
Prentice-Hall of Australia Pty. Limited, *Sydney*
Prentice-Hall Canada, Inc., *Toronto*
Prentice-Hall Hispanoamericana, S.A., *Mexico*
Prentice-Hall of India Private Limited, *New Delhi*
Prentice-Hall of Japan, Inc., *Tokyo*
Simon & Schuster Asia Pte. Ltd., *Singapore*
Editora Prentice-Hall do Brasil, Ltda., *Rio de Janeiro*

To Pushpa, Rosalie, Gunjan, and Palak,
for your patience and support

Trademarks (TM)

Pro/Engineer[TM]: Parametric Technology Corp., Waltham, MA.

I-DEAS Master Series[TM]: SDRC, Milford, OH.

CADDS 5[TM]: Computervision Corp., Bedford, MA.

Anvil 5000[TM]: Manufacturing & Consulting Series, Scottsdale, AZ.

Catia[TM] Solutions: Dassault Systems, North Hollywood, CA.

Unigraphics[TM]: EDS Unigraphics, Maryland Heights, MO.

HP PE/SolidDesigner[TM]: Hewlett-Packard, Ft. Collins, CO.

ICAD[TM]: Concentra Corporation, Burlington, MA.

I/EMS[TM]: Intergraph Corp., Huntsville, AL.

CONTENTS

9   Life-cycle Mechanization   405

10   IPD Deployment Methodology   447

Index   481

ACRONYMS

3Ps	Policy, Practices, and Procedures
4GL	Fourth Generation Language
4Ms	Models, Methods, Metrics, and Measures
6Ms	Six Resource Elements (Materials, Manpower, Methods, Management, Money, and Machine)—(Figure 3.8/Volume I)
7Ts	Talents, Tasks, Teams, Techniques, Technology, Time, and Tools (Figure 4.1/Volume I)
7Cs	Collaboration, Commitment, Communications, Compromise, Consensus, Continuous Improvement, and Coordination
8Ws	Eight Waste Components (Figure 3.4/Volume I)
AFNOR	French Association for Standardization
AMICE	European CIM Architecture—in reverse
ANSI	American National Standards Institute
API	Application Programming Interface
ATIS	A Tools' Integration Standards
BOMs	Bill-of-materials
BSI	British Standard Institution
C4	CAD/CAM/CAE/CIM
CA	Computational Architecture
CAD	Computer-Aided Design
CAE	Computer-aided Engineering
CALS	Computer-aided Acquisition and Logistics Support (old)
CALS	Computer-aided Acquisition and Life-cycle Support (new)
CAM	Computer-aided Manufacturing

CAPP	Computer-aided Process Planning
CASA	Computer-aided Society of Manufacturing Engineers
CASE	Computer-aided Process Engineering or Computer-aided Simultaneous Engineering
CEC	Commission of the European Communities
CERA	Concurrent Engineering: Research and Applications
CFD	Concurrent Function Deployment
CFI	CAD Framework Initiative
CIM	Computer-integrated Manufacturing
CIMOSA	Open System Architecture for CIM
COe	Consistent Office environment
CORBA	Common Object Request Broker Architecture
CPU	Central Processing Unit
CWCe	Consistent Work group Computing environment
DARPA	Defense Agency for Research Projects
DBMS	Data Base Management Systems
DDL	Dynamic Data Linking
DECnet	Digital Electronic Computer (DEC) Network
DFM	Design for Manufacturability
DFm	Distributed File management
DICE	DARPA Initiative in Concurrent Engineering
DIN	German Industrial Standards Institute
DNS	Distributed Name Service
E-mail	Electronic-mail
EAL	Engineering Analysis Language
EC ESPRINT	European Strategic Program for Research and Development in Information Technology
EDI	Electronic Data Interchange
ESPRIT	European Strategic Planning for Research in Information Technology
EWS	Engineering Workstations
FEA	Finite Element Analysis
FEM	Finite Element Modeling
FMEA	Failure Mode and Effects Analysis
GUI	Graphics User Interface
GKS	Graphics Kernel System
ICAM	Integrated Computer-Aided Manufacturing
IDL	ICAD Design Language
IEEE	Institute of Electricals and Electronics Engineers
IGES	Initial Graphics Exchange Specification
IOS	Input/Output Sub-systems
IPPO	Integrated Product and Process Organization
IPD	Integrated Product Development
IPPD	Integrated Product and Process Development
IPR	Interactive Photorealistic Rendering

ISO	International Standard Organization
ISO/IGES	International Standard Organization/Initial Graphics Exchange Specification
ISPE	International Society for Productivity Enhancement
JISC	Japanese Industrial Standards Committee
LAN	Local Area Network
MAP	Manufacturing Automation Protocol
MCAE	Mechanical Computer-aided Engineering
MFLOPS	Million Instructions Per Second
MIS	Mainframe Information System
MRP	Manufacturing Resource Planning
NAS	Network Application Services
NC	Numerical Control
NCS	Network Computing System
NFS	Network File System
NIST	National Institute of Standards and Technology
NURBS	Non-Uniform Rational B-Splines
OSF	Open System Foundation
OSI/MAP	Open System Institute/Manufacturing Automation Protocol
PC	Personal Computer
PD^3	Product, Design, Development and Delivery
PDES	Product Data Exchange using STEP
PDES/Express	A Language developed using PDES
PDMS	Product Database Management System
PHIGS	Programmers' Hierarchical Interactive Graphic Standard
PIM	Product Information Management
PPO	Product, Process, and Organization
PsBS	Process Breakdown Structure
PtBS	Product Breakdown Structure
QC	Quality Control
QFD	Quality Function Deployment
RDBMS	Relational Data Base Management System
RISC	Reduced Instruction Set Computing
RPC	Remote Procedure Call
SARA	Systems Automation: Research and Applications
SBU	Strategic Business Unit
SLA	Stereolithography Apparatus
SME	Society of Manufacturing Engineers
SNA	Systems Network Architecture
SPEC	System Performance Evaluation Cooperative
SQL	Structured Query Language
SSD	Secondary Storage Device
STEP	Standard for the Exchange of Product Model Data
TaskBroker	(HP) TaskBroker program

TCP/IP	Transmission Control Protocol/Internet Protocol
TQM	Total Quality Management
TVM	Total Values Management
UI	User Interface
V-c	Video-conferenceing
WAN	Wide Area Network
WBS	Work Breakdown Structure
WC	Work-group Computing

PREFACE

As the name implies, the book describes the fundamentals of Concurrent Engineering (CE) and explains the basic principles on which this very subject is founded. Most of the material in this book is either original ideas or their extension to CE. Most is never reported elsewhere and is based on the author's successes while practicing CE on the job. They encompass decades of his research and learning while working with electronic, automotive, aerospace, computer, and railroad industries including Ford, General Motors, Electronic Data Systems, Association of American Railroads, NASA, and numerous other places. Concurrent Engineering approach to product design and development has two major themes. The first theme is establishing an *integrated product and process organization (PPO)*. This is referred herein as *process taxonomy*. The second theme is applying this process taxonomy (or a set of methodologies) to design and develop a total product system. This is referred to as *integrated product development (IPD)*. Each theme is divided into several essential parts forming major chapters of this book.

The first volume called *product and process organization (PPO)* had nine chapters. The second volume sub-titled integrated product development has ten chapters. The materials in these two volumes have been brought together to balance the interests of both the customers and the companies. The contents of "Volume I" were *Manufacturing competitiveness, Life-cycle Management: Process Re-engineering, Concurrent Engineering Techniques, Cooperative Work groups, System Engineering, Information Modeling, The Whole System*, and *Product Realization Taxonomy*.

The contents of "Volume II" are *Concurrent Function Deployment, CE Metrics and Measures, Total Value Management, Product Development Methodology, Frameworks and Architectures, Capturing Life-Cycle Values, Decision Support Systems, Intelligent Information System, Life-Cycle Mechanization*, and *IPD Deployment Methodology*.

In Concurrent Engineering (CE) system, each modification of the product represents a taxonomical relationship between specifications (inputs, requirements, and constraints), outputs, and the concept it (the modification) represents. At the beginning of the design process, the specifications are generally in abstract forms. As more and more of the specifications are satisfied, the product begins to take shape—begins to evolve into a physical form. To illustrate how a full CE system will work, and to show the inner-working of its elements, author defines this CE system as a set of two synchronized wheels. The representation is analogous to *a set of synchronized wheels of a bicycle*. Figure P1 shows this CE wheel set:

CONCURRENT ENGINEERING WHEELS

The first CE wheel represents the *integrated product and process organization (PPO)*. The second CE wheel accomplishes the *integrated product development (IPD)*. The two wheels together harmonize the interests of the customers and the CE organization (also frequently referred as an enterprise). The contents of first wheel were described in volume I and contents of second wheel are described in volume II of the *CE fundamental* books. Three concentric rings represent the three essential elements of a wheel. The innermost ring of the wheels constitutes the hubs of the wheels. A hub represents four supporting "M" elements: *models, methods, metrics and measures*. The chapters from the two volumes that contribute to "M" elements are contained in the following table.

Innermost Ring (Hub)	Volume I—PPO	Volume II—IPD
Models	Information Modeling (Chap. 7)	—
Methods	Product Realization Taxonomy (Chap. 9)	IPD Deployment Methodology (Chap. 10)
Metrics & Measures	—	CE Metrics and Measures (Chap. 2)

Life-cycle mechanization and IPD deployment methodology constitute the middle ring of the IPD wheel. The two are discussed in Chapters 9 and 10 of volume II, respectively. Each sector in the outer ring represents a chapter of this book. The sectors for the first wheel are discussed in volume I. Volume I explains how the CE design process (called herein CE process taxonomy) provides a stable, repeatable process through which increased accuracy is achieved. The sectors of the second wheel are discussed in volume II. Volume II explains how a product can be designed, developed and delivered using a process-based taxonomy of volume I. A separate chapter in the books is dedicated to discussing each part of the two CE wheels.

First CE Wheel: Integrated Product and Process Organization

The innermost ring of the first CE wheel is a hub. The layout of hub is the same for both wheels. The hub represents four supporting "M" elements: *models, methods, metrics and measures*. Models refer to *information modeling*. Methods refer to *product realization taxonomy*. They are discussed in Chapters 7 and 9 of volume I, respectively. *CE Metrics and Measures* are discussed in Chapter 2 of volume II. The middle ring represents the CE work groups, which drives the customer and the enterprise like how a human drives a bike. The work groups are divided into four quadrants representing the four so called CE teams. These teams are: *personnel team, technology team, logical team and the virtual team*. They are discussed in Chapter 5 of volume I. The outer ring for each wheel is divided into eight parts. *Volume I* starts with an introductory chapter on manufacturing competitiveness reviewing the history and emerging trends. The remaining chapters of the book (volume I) describe CE design techniques, explain how concurrent design process can create a competitive advantage, describe CE process taxonomy, and address a number of major issues related to *product and process organization*. The complexity of the product design, development, and delivery (PD^3) process differs depending upon the

1. Types of information and sources
2. Complexity of tasks
3. Degree of their incompleteness or ambiguity

Other dimensions encountered during this PD^3 process that cannot be easily accommodated using traditional process (such as serial engineering) are:

4. Timing of decision making
5. Order of decision making
6. Communication mechanism

The elements of the first CE wheel define a set of systems and processes that have the ability to handle all of the above six dimensions. In the following some salient points of the volume I chapters are briefly highlighted:

• *Manufacturing Competitiveness:* Price of the product is dictated by world economy and not by one's own economy or a company's market edge alone. Those companies that can quickly change to world changing market place can position themselves to complete globally. This chapter outlines what is required to become a market leader and compete globally. Successful companies have been the ones who have gained a better focus on eliminating waste, normally sneaked into their products, by understanding what drives product and process costs and, how can value be added. They have focused on product and process delivery-system—how to transform process innovations into technical success and how to leverage the implementation know-how into big commercial success. Many have chosen to emphasize high-quality flexible or agile production in product delivery rather than high-volume (mass) production.

• *Life-cycle Management:* Today, most companies are under extreme pressure to develop products within time periods that are rapidly shrinking. As the market changes so do the requirements. This has chilling effect in managing the complexity of such continuously varying product specifications and handling the changes

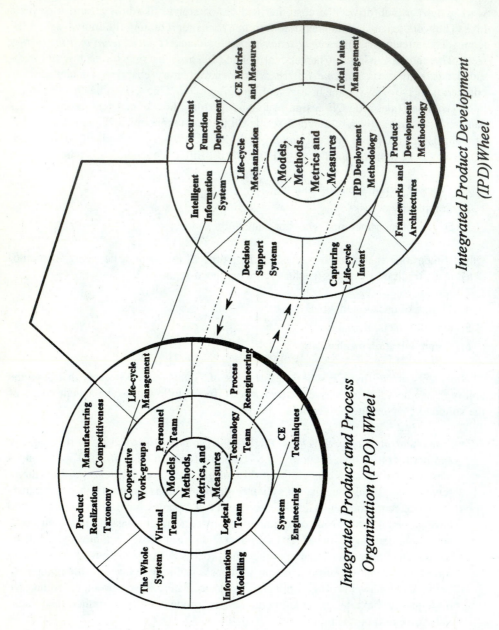

Integrated Product and Process Organization (PPO) Wheel

Integrated Product Development (IPD) Wheel

FIGURE P1 A synchronized set of CE Wheels

xx

within this shrinking time period. The ongoing success of an organization lies in its ability to: continue to evolve; quickly react to changing requirements; reinvent itself on a regular basis; and keep up with ever changing technology and innovation. Many companies are stepping up the pace of new product introduction, and are constantly learning new ways of engineering products more correctly the first time, and more often thereafter. This chapter outlines life-cycle management techniques, such as change management, and process improvement to remain globally competitive.

• *Process Reengineering:* The global marketplace of 1990s has shown no sympathy to tradition. The reality is that if the products manufactured do not meet the market needs, demand declines and profits dwindle. Many companies are finding that true increase in productivity and efficiency begins with such factors as clean and efficient process, good communication infrastructure, teamwork, a constancy of shared vision and purpose. The challenge is simply not to crank up the speed of the machines so that it outputs (per unit of time) are increased or doubled, but to change the basic machinery or process that produces the outputs. To accomplish the latter goals, this chapter describes several techniques to achieve competitive superiority such as benchmarking, CPI, organizational restructuring, renovation, process reengineering, etc.

• *CE Techniques:* The changing market conditions and international competitiveness are making the time-to-market a fast shrinking target. Over the same period, diversity and complexity of the products have increased multi-folds. Concurrency is the major force of Concurrent Engineering. Paralleling describes a "time overlap" of one or more work groups, activities, tasks, etc. This chapter describes seven CE principles to aim at: *Parallel work-group; Parallel Product Decomposition; Concurrent Resource Scheduling; Concurrent Processing; Minimize Interfaces; Transparent Communication; and Quick Processing;* This chapter also describes the seven forces that influence the domain of CE (called here as enabling agents or 7Ts) namely: *talents, tasks, teams, techniques, technology, time and tools.*

• *Cooperative Work Groups:* It has been the challenge for the design and manufacturing engineers to work together as teams to improve quality while reducing costs, weight, and lead-time. A single person, or a team of persons, is not enough to provide all the links between: human knowledge and skills; logical organization; technology; and a set of 7Cs coordination features. A number of supporting teams is required, some either virtual or at least virtually collocated. For the waltz of CE synthesis to succeed, CE teams need clear choreography. This chapter describes for the first time the four collaborative teams that are essential for managing a CE organization. Examples of collaborative features include capabilities of electronic meeting such as message-posting and interactions through voice, text, graphics and pictures.

• *System Engineering:* Most groups diligently work to optimize their subsystems, but due to lack of incentives they tend to work independently of each other. This results in a product, which is often suboptimized at each decomposed level. System engineering requires that product realization problem is viewed as a "system-centered"

problem as opposed to "component-centered." *Systems Engineering* does not dis-
agree with the idea of compartments or divisions of works, but it emphasizes that
the interface requirements between the divisions (inter-divisional) and across the
levels should be adequately covered. That way, when the time comes to modernize
other components of the system, an enterprise has the assurance that previously introduced technologies and processes will work logically in a fully integrated fashion, thereby increasing the net efficiency and profitability.

• *Information Modeling:* A successful integrated product development (IPD) requires
a sufficient understanding of the product and process behaviors. One way to achieve
this understanding is to use a series of reliable information models for planning, designing, optimizing and controlling each unit of an IPD process. The demands go
beyond the 3-D CAD geometric modeling. The demands require schemes that can
model all phases of a product's life-cycle from cradle to grave. The different aspects
of product design (planning, feasibility, design, process-planning), process design
(process-execution, production, manufacturing, product support), the human behavior in teamwork, and the organization or environment in which it will operate, all
have to be taken into account. Five major classes of modeling schemata are discussed in this chapter. They are:
    1. Product representation schemes and tools for capturing and describing the
       product development process and design of various interfaces, such as design-
       manufacturing interface
    2. Schemes for modeling physical processes, including simulation, as well as mod-
       els useful for product assessments, such as DFA/DFX, manufacturability evalua-
       tion of in-progress designs
    3. Schemes for capturing (product, process, and organization structure) require-
       ments or characteristics for setting strategic and business goals
    4. Schemes to model enterprise activities (data and work flow) in order to deter-
       mine what types of functions best fit the desired profitability, responsiveness,
       quality and productivity goals
    5. Schemes to model team behavior, because most effective manufacturing environ-
       ments involve a carefully orchestrated interplay between teams and machines.

• *The Whole System:* Often while designing an artifact, work groups forget that the
product is a system. It consists of a number of subassemblies, each fulfilling a different but distinct function. A product is far more than the collection of components.
Without a structure or some "constancy-of-purpose" there is no system. The central
difference between a CE transformation system and any other manufacturing system,
such as serial engineering, is the manner in which the tasks' distribution is stated and
requirements are accomplished. In a CE transformation system, the purpose of every
process step of a manufacturing system is not just to achieve a transformation but to
accomplish this in an optimal and concurrent way. This chapter proposes a system-
based taxonomy, which is founded on parallel scheduling of tasks. This chapter also
proposes a set of breakdown structures for product, process and work to realize a drastic reduction in time and cost in product and process realizations.

- *Product Realization Taxonomy:* This constitutes a "state of series of evolution or transformation" leading to a complete design maturity. *Product Realization Taxonomy* involves items related to design incompleteness, product development practices, readiness feasibility, and assessing goodness. In addition, CE requires these taxonomies to have a unified "product realization base." The enterprise integration metrics of the CE model should be well characterized and the modeling methodologies and/or associated ontology for developing them should be adequate for describing and integrating enterprise functions. The methodologies should have built-in product and service accelerators. Taxonomy comprises of the product, process descriptions, classification techniques, information concepts, representation, and transformation tasks (inputs, requirements, constraints and outputs). Specifications, describing the transformation model for product realization. They are included as part of the taxonomy descriptions.

Second CE Wheel: Integrated Product Development

The second CE wheel defines the integrated product development (IPD). This is discussed in this book (Volume II). IPD in this context does not imply a step-by-step serial process. Indeed, the beauty of this IPD wheel is that it offers a framework for a concurrent PD^3 process. A framework within which, a CE team has flexibility to move about, fitting together bits of the jigsaw as they come together. A CE team has an opportunity to apply a variety of techniques contained in this volume (such as: *Concurrent Function Deployment, Total Value Management, Metrics and Measures,* etc.) And through their use, teams could avail the opportunity to achieve steady overall progress towards a finished product.

- *Concurrent Function Deployment:* The role of the organization and engineers is changing today, as is the method of doing business. Competition has driven organization to consider concepts such as time compression (fast-to-market), Concurrent Engineering, Design for X-ability, and Tools and Technology (such as Taguchi, Value Engineering) while designing and developing an artifact. Quality Function Deployment (QFD) addresses major aspects of "quality" with reference to the functions it performs but this is one of the many functions that need to be deployed. With conventional deployment, it is difficult, however, to address all aspects of Total Values Management (TVM) such as *X-ability, Cost, Tools and Technology, Responsiveness and Organization* issues. It is not enough to deploy just the "*Quality*" into the product and expect the outcome to be the *World Class*. TVM efforts are vital in maintaining a competitive edge in today's world marketplace.

- *CE Merits and Measures:* Metrics are the basis of monitoring and measuring process improvement methodology and managing their effectiveness. Metric information assists in monitoring team progress, measuring quality of products produced, managing the effectiveness of the improved process, and providing related feedback. Individual assurances of DFX specifications (one at a time) do not capture the most important aspect of Concurrent Engineering—the system perspectives, or the trade-off across the different DFX principles. While satisfying these DFX principles in this isolated manner, only those which are not in conflict are usually met. Concurrent engineering views the design and evaluates the artifact as a system, which has a wider impact than just suboptimizing the subsystems within each domain.

- *Total Value Management:* The most acclaimed slogan for introducing a quality program in early corporate days simply was to provide the *most value for the lowest cost.* This changed as the competitiveness became more fierce. For example, during the introduction of traditional TQM program in 1990 "*getting a quality product to market for a fair price*" was the name of the game. The new paradigm for CE now is total value management (TVM). TVM mission is "*to provide the total value for the lowest cost in the least amount of time, which satisfies the customers the most and lets the company make a fair profit.*" Here use of value is not just limited to *quality.* To provide long lasting added value, companies must change their philosophy towards things like x-ability, *responsiveness, functionality, tools and technology, cost, architecture*, etc.

- *Product Development Methodology:* A systematic methodology is essential in order to be able to integrate:
  1. Teamwork
  2. Information modeling
  3. Product realization taxonomy
  4. Measures of merits (called CE metrics), and quantitatively assess the effectiveness of the transformation.

  This may involve identification of performance metrics for measuring the product and process behaviors. Integrated product development methodology is geared to take advantage of the product realization taxonomy.

- *Frameworks & Architectures:* In order to adequately support the CE and the 4Ms (namely: modeling, methods, metrics and measurements), it is necessary to have a flexible architecture. An architecture that is openly accessible across different CE teams, information systems, platforms, and networks. Architecture consists of information contents, integrated data structures, knowledge bases, behavior and rules. An architecture not only provides an information base for easy storage, retrieval, and tracking version control, but can also be accessed by different users simultaneously, under ramp-up scheduling of parallel tasks, and in synchronization. We also need a product management system containing work *HOW* management capabilities integrated with the database. This is essential because in CE there exists a large degree of flexibility for parallelism that must be carefully managed in conjunction with other routine file and data management tasks.

- *Capturing Life-cycle Intent:* Most CAD/CAM tools are not really *capture* tools. In static representation of CAD geometry, configuration changes cannot be handled easily, particularly when parts and dimensions are linked. This has resulted in loss of configuration control, proliferation of changes to fix the errors caused by other changes, and sometimes ambiguous designs. By capturing "design intent" as opposed to "static geometry," configuration changes could be made and controlled more effectively using the power of language constructs than through traditional CAD attributes (such as lines and surfaces). The power of a "capture" tool comes from the methods used in capturing the "design intent" initially so that the required changes can be made easily and quickly if needed. "Life-cycle capture" refers to the definition of the physical object and its environment in some generic form. "Life-cycle intent" means representing the life-cycle capture in a form, which can be modified and iterated until all the life-cycle specifications for the product are fully satisfied.

- *Decision Support System:* In CE, cooperation is required between CE teams, management, suppliers, and customers. A knowledge based support system will help the participating teams in decision making and to reflect balanced views. Tradeoffs between conflicting requirements can be made on the basis of information obtained from sensitivity, multi-criterion objectives, simulation, or feedback. The taxonomy can be made a part of decision support system (DSS) in supporting decisions about product decomposition. By keeping track of what specifications are satisfied, teams can ensure common visibility in the state of product realization, including dispatching and monitoring of tasks, structure, corporate design histories, etc.

- *Intelligent Information System* (IIS): Another major goal of CE is to handle information intelligently in multi-media—audio, video, text, graphics. Since IIS equals CIM plus CE, with IIS, many relevant CE demands can be addressed and quickly processed. Examples include:
  1. Over local or wide area networks, such as SQL, which connects remote, multiple databases and multimedia repositories
  2. Any needed information, such as recorded product designers' design notes, figures, decisions, etc. They can be made available on demand at the right place at the right time
  3. Any team can retrieve information in the right format and distribute it promptly to other members of the CE teams.

- *Life-cycle Mechanization:* Life-cycle mechanization equals CIM + Automation + CE. Life-cycle mechanization is arranged under a familiar acronym: CAE, for CIM, Automation, and CE. Since CAE also equals IIS plus automation, the major benefits of mechanization in CAE come from removing or breaking barriers. The three common barriers are:
  a. Integration (this is a term taken from CIM)
  b. Automation
  c. Cooperation (which is a term taken from CE).

- CE provides the decision support element, and CIM provides the framework & architecture plus the information management elements. *Life-cycle Mechanization* refers to the automation of life-cycle functions or creation of computerized modules that are built from one another and share the information from one another. This includes integration and seamless transfer of data between commercial computer-based engineering tools and product-specific in-house applications. This tends to reduce the dependency of many CE teams on communication links and product realization strategies, such as decomposition and concatenation.

- *IPD Deployment Methodology:* The purpose of this chapter is to offer an implementation guideline for product redesign and development through its life-cycle functions. IPD implementation is a multi-track methodology. The tracks overlap, but still provide a structured approach to organizing product ideas and measures for concurrently performing the associated tasks. Concurrency is built in a number of ways (similar to what was discussed in volume I), depending upon the complexity of the process or the system involved. This chapter proposes a set of *"Ten Commandments,"* that serves to guide the product and process iterative aspects of IPD rather than just the work group collaborative aspects required during the development cycle. The CE teamwork in the center of the wheel ensures that both local or zonal iterative refinements and collaborative refinements take place during each concurrent track.

A SYNCHRONIZED WHEEL-SET FOR CE

All the above nineteen parts of CE put together creates a synchronized wheel-set for CE, as shown in Figure P1. The teamwork, with four cooperating components (technological teams, logical teams, virtual teams, and personnel teams), is in the middle ring. The 4Ms (models, metrics, measurements and methodology) form the center of this wheel. The center ring has four parts to it: *Information Modeling; Product Realization Taxonomy; Measures of Merit and IPD deployment methodology.* The 4Ms are shown in the center because it provides the methodology for guiding the product realization process. The two inner rings, which are same for both wheels, makes the wheels a synchronized set. The teams in the middle ring are the driving force of the methodology (4Ms listed in the center) and controller of the technologies (listed as sectors on the outer ring). *The emphasis of a team-centered wheel for CE is a departure from a conventional function-centered approach.* Outer rings of each wheel contain the remaining parts of an *integrated product and process organization*—PPO (volume I) and *integrated product development*—IPD (Volume II), respectively. The idea of this middle ring is to provide a team-centered 7Cs (*Collaboration, Commitment, Communications, Compromise, Consensus, Continuous Improvement*, and *Coordination*) interplay across layers of enabling technologies and methodologies. Everything is geared towards cutting and compressing the time needed to design, analyze, and manufacture marketable products. Along the way, costs are also reduced, product quality is improved and customer satisfaction is enhanced due to the synchronized process. There is, however, a finite window in which the benefits of time compression and cost cutting are available. As more manufacturers reduce lead time, what once represented a competitive advantage can become a weakening source. Fortunately, the CE wheel provides a continuum (dynamic) base through which new paradigms (process, tools, technology and 7Ts) can be launched to remain globally competitive for a long haul.

Before we take a closer look at the different parts of this wheel as different chapters of this book, it is important to note that all the parts of the wheel-set are not of the same kind. They emphasize different aspects of CE. The four major aspects are (see Figure P2):

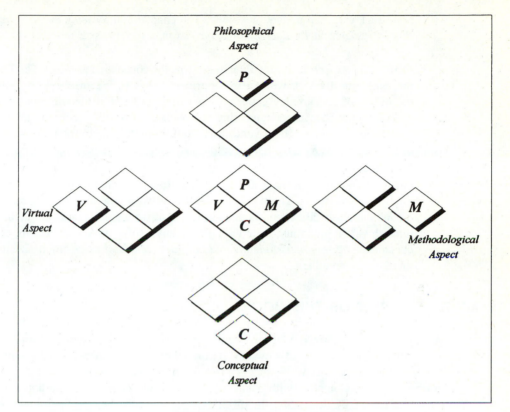

Philosophical Aspect

Virtual Aspect

Methodological Aspect

Conceptual Aspect

FIGURE P2   Four Aspects of CE

- Philosophical aspect
- Methodological aspect
- Conceptual aspect
- Virtual aspect

- *Philosophical Aspect:* Personnel CE team governs the philosophical aspects of CE. Philosophical aspect deals with the boundaries of the responsibility and the authority, culture, empowerment. It also includes team's make-up, program organization, supplier rationalization, management styles or philosophies, change management, workplace organization and visual control, physical proximity (collocation), management and reporting structure, etc. The chapters on *Cooperative Teamwork and Life-cycle Management* emphasize more of this aspect than others.

- *Methodological Aspect: This aspect* of CE is governed by technology team. Methodological aspect deals with system thinking, approaches to system complexity, system integration, transformation model of the manufacturing system. It also deals with CE enterprise system taxonomy, integrated product and process development, transformation system for product realization, pull system for product realization, track and loop methodology, etc. The chapters on *Systems Engineering, The Whole System and Product Realization Taxonomy* emphasize more of this aspect than others.

- *Conceptual Aspect:* Logical CE team governs the conceptual aspect of CE. Conceptual aspect mostly deals with the major principles of CE, concurrency and simultaneity, modes of concurrency, modes of cooperation. It also deals with understanding and managing change, reengineering approaches, work flow mapping, information flow charting, process improvement methodology, etc. The chapters on *CE Definitions and Process Re-engineering* emphasize more of this aspect than others.

- *Virtual Aspect: This aspect* of CE is governed by a virtual CE team. Virtual aspect mostly deals with capturing life-cycle intent, information modeling, electronic capture of CE invariants. These CE invariants deal with product model class, process model class, specification model class, cognitive model class, communication through virtual proximity, agile virtual company, artifact intent definitions, etc. The chapters on *Information Modeling and Life-cycle Mechanization* emphasize more of this aspect than others.

MAJOR FEATURES OF THIS BOOK

Whether you are a firm CE believer, or this is your first introduction to CE, this two volume (book) set provides a full view of CE from all of the above aspects and perspectives. The management perspective, which is a part of philosophical aspect, relates to organization and culture. Complete with a historical review and context, the author articulates these CE aspects by illustrating the differences between the best methodologies (or the best taxonomies) and what are being practiced in industries today.

Some examples of topics included in this volume are:

- What is required to control one's own process—identifying and satisfying the needs and expectations of consumers better than the competitions and doing so profitably faster than any competitor..

- You will understand why QFD is not enough for IPD.

- How to consider deployment of competing values simultaneously.

- You will discover why TQM is not enough to gain competitive edge in the global marketplace.

- Why is it not enough to deploy *"Quality"* into the product and expect the outcome to be a world-class?

- How to incorporate *"Voice of the Customers"* into all necessary tracks of the product development cycle.

- Why individual assurances of DFX specifications (one at a time) do not capture the most important aspect of Concurrent Engineering—the system perspective.

- How to build a product that optimizes a number of value objectives intrinsically, not just on the basis of *Quality*.

- A set of twenty-five metrics and measures for concurrent engineering.

- Three-layer structure for a CE logical framework to provide a flexible application development environment.

Integrated Product Development (volume II) deals with methodology—applying the process taxonomy for CE. Methodology (development and deployment) is necessary to adequately classify, integrate and automate core functions of a complex enterprise in a PD^3 process. The innermost core of this deployment methodology is its foundation, which has four supporting **M** elements: **m**odels, **m**ethods, **m**etrics and **m**easures as mentioned earlier. The Table P1 summarizes the major features of this second volume.

TABLE P1   Major Features of Volume II of the *CE Fundamentals* Book

Features of Volume II	How do these features benefit readers?	What chapters or sections of the book contain these features or examples of them?
a This is the first CE book that emphasized all aspects of Total Values Management (TVM) such as X-ability, cost, tools and technology, responsiveness and organization issues. What is required is a total control of one's process—identifying and satisfying the needs and expectations of consumers better than the competitions and doing so profitably faster than any competitor.	It is not enough to deploy *Quality* into the product and expect the outcome to be a world-class. The competitors are always finding better and faster ways of doing things. Catching up in quality only makes a company at par with its competitors in terms of inheriting some of their product quality characteristics but relatively speaking it gets you there a few years later.	Chapters 1 and 3 (see Figures 1.4, 3.1, and 3.5)
b In this volume, author has expanded the original definition of QFD, discussed in Volume I, to include parallel deployments. This provides a method to consider the deployment of competing values simultaneously. This volume calls this approach as *Concurrent Function Deployment (CFD)*.	The intent of CFD is to incorporate *"Voice of the Customers"* into all nine phases of the product development cycle, and finally into continuous improvement, support and delivery (see Figure 4.2, volume I) phases.	Chapter 1 (Section 1.4, see Figures 1.1 through 1.4)
c This is the first time this CE book points out that the deployment of many artifact functions (values) can proceed in parallel with what we know today as quality function deployment (QFD) or *quality FD*. CFD enforces the notion of concurrency and deploys simultaneously a number of competing artifact values, not just the *"Quality as found in QFD."*	CFD breaks the multi-year QFD ordeal by allowing work-groups to work concurrently on a number of conflicting values and compare their notes at common check points. CFD is a simple and powerful tool that leads to long range thinking and better communication across several value functions. Examples are: X-ability (performance), tools and technology, cost, responsiveness and infrastructure.	Chapter 1 (see Figures 1.4 through 1.9, Section 1.5)

(continued)

Features of Volume II	How do these features benefit readers?	What chapters or sections of the book contain these features or examples of them?
d No book has yet been published encompassing *concurrent function deployment, CE metrics and measures, total value management, product development methodology, frameworks and architectures, capturing life-cycle intent, decision support systems, intelligent information system, life-cycle mechanization, deployment methodology* and integration issues all described within a unified *IPD (integrated product development)* theme.	It allows the readers to consider a wider view meaning "integrating over the enterprise" while implementing CE. This eliminates the common problem of blindly automating tasks—meaning repeating the same mistakes but doing it more often and more quickly.	Chapters 1 through 10.
e This book, for the first time, identifies twenty-five CE metrics and measures. Metrics and measures are categorized into four groups: simulations and analysis, product feasibility and quality assessment, design for X-ability assessment, and process quality assessment. They are arranged in four file drawers of a file cabinet.	Individual assurances of DFX specifications (one at a time) do not capture the most important aspect of Concurrent Engineering—the system perspective or the trade-off across the different DFX principles. Product development teams (PDTs) can draw upon these metrics and measures to influence an enterprise PD^3 process.	Chapter 2 (see Figures 2.6 through 2.8).
f For the first time, this book proposes Total Value Management (TVM) as a concept to replace Total Quality Management (TQM). The six major recognized objectives of TVM are: Quality (function-wise), X-ability (performance-wise), Cost (profit-wise), Tools and Technology (innovation-wise), Responsiveness (time-wise) and Infrastructure (business-wise).	It allows the PDT groups to build a product that optimizes these six value objectives intrinsically, not just on the basis of *Quality*. How effectively, efficiently, and quickly the work-groups are able to succeed in this endeavor depends upon many factors that need to be considered. TVM is meant to provide a winning path to increase global market share and profitability.	Chapter 3 (Section 3.1, see Figures 3.5 through 3.6). *Quality in the aforementioned sense* plays only a minor role in fostering a total optimized product from a *world-class* perspective.
g The book introduces for the first time a concurrent process of quality engineering (QE) — wherein Quality begins with concurrent product and process design running in parallel with an off-line quality control. Inspection oriented QC methods are shown replaced by on-line quality control (QC) or quality process control (QPC) methods.	The design-oriented QC methods, shown as being part of the product design step, provide an important defect prevention mechanism. Quality circles or work-groups can establish a QE methodology following this concurrent approach.	Chapter 3 (many of these methods are shown in Figures 3.3 and 3.4)
h This book for the first time introduces *Process invariants* as	The invariants provide a common ground for the work-groups to	Chapter 4 (see Figures 4.2 and 4.3). The basic structure of model

Features of Volume II	How do these features benefit readers?	What chapters or sections of the book contain these features or examples of them?
key contributors of an IPD realization process that are constant or stationary (always present) in the process dimension of IPD. The *process invariants* are vertical cross sections of the IPD realization process. *Model invariants* are horizontal cross sections of IPD realization process.	represent enterprise or business-driven, product-driven, and process-driven works, activities, features, functions and decisions. The process and model invariants are linked by taxonomic relationships.	invariants and their interactions are shown in Figure 7.11 of volume I.
i The book for the first time views the IPD methodology as consisting of eight parts called IPD building blocks. The first four blocks provide a conceptual framework for understanding the IPD challenges and opportunities. The last four parts provide the building blocks for an analytical framework for decision making and improvements.	The purpose of this IPD methodology is to improve the performance characteristics of the product or process relative to customer needs and expectations. It builds the theory of knowledge through systematic revision and extension of the paradigms introduced in previous Chapters.	Chapter 4 (see Figures 4.7 through 4.9)
j The book introduces a three-layer structure for a CE logical framework to provide a flexible application development environment. The lowest layer is the computing platform. The second-layer—intelligent interface—provides the primary programming interface to application developers. The top layer consists of end-user applications communicating among themselves (horizontally) and to the intelligent interface (vertically).	When work-groups integrates the computing platforms with intelligent interface over the applicable standards, this results in a long life of the end-user applications developed on the top layer. The architecture shields end-user applications from possible downstream changes.	Figure 5.24 shows a logical view of this CE sub-architecture, which forms the basis for the flexible CE environment described in this book. Chapter 5 (see Figure 5.24)
k Benefits of life-cycle capture stem from a few basic CE principles. The book describes the three life-cycle capture languages on which life-cycle capture is founded. Languages are means of capturing the knowledge for the design and development of a product. These language-based systems use the intent-driven techniques to generically capture product life-cycle values. Such developments are dynamic in nature when it comes down to managing changes.	Models are the results of such knowledge capture. They are suited for altering a part geometry, say using variable dimensions, or capturing its engineering design intent. The primary goal of a knowledge-capture formalism is to provide a means of defining ontology. An ontology is a set of basic attributes and relations comprising the vocabulary of the product realization domain as well as rules for combining the attributes and relations.	In the present form, most C4 (CAD/CAM/CAE/CIM) systems are mainly suitable for analyzing a problem or for capturing an explicit, static geometric representation of an existing part. Chapter 6 (see section 6.3)

(continued)

TABLE P1 (continued)

Features of Volume II	How do these features benefit readers?	What chapters or sections of the book contain these features or examples of them?
l The types of decisions that engineers make today to solve design problems are bounded by a spectrum with cognitive aspect at one end and progressive aspect at the other end. The book for the first time describes two types of cognitive models and seven types of progressive models.	The work group can use these aspects to choose possible design models during decision making. Progressive models can be used to calculate, analyze or to evaluate design alternatives, or to come up with a new or revised product.	Chapter 7 (see Figures 7.2 and 7.3)
m The book for the first time describes how CIM plus CE equals IIS. Today, CIM systems are merely being applied to integration and processing (storage and automation) of data, communication, and processes (common systems and standards).	Intelligent handling of information through computer techniques can yield a better CIM system since it can monitor and correct problems. IIS reduces the need for frequent manual intervention. CE brings forth three missing links of CIM.	Chapter 8 (see Figure 8.2).
n The book for the first time describes the 8 enabling elements of Intelligent Information System (IIS) applicable to product development.	The effective implementation of product development process control strategies can be facilitated by a systematic collection and monitoring of relevant enabling elements of IIS.	Chapter 8 (see section 8.1, Figure 8.3)
o The book for the first time describes thirteen barriers that inhibit work groups regain full potential of manufacturing competitiveness.	The key to the successes of IIS is understanding the obstacles and barriers to unifying CE with existing CIM processes and identifying new opportunities for improvement.	Chapter 8 (see section 8.2, Figures 8.4 through 8.7)
p This book for the first time describes a network of 12 modules, which form the infrastructure for life-cycle mechanization process. Five modules belongs to CIM; four relates to automation; and three deals in CE topics.	The criteria of mechanization are global in nature (such as 7Ts, 4Ms, and 3Ps) with the overall company goal of making maximum profits and great product.	Chapter 9 (Figure 9.4)
q The book explains that the concurrent movement of 1990s is not just a "bunch" of concurrent programs. It is the realization that certain fundamental ideals need to be enforced during an IPD deployment. These ideals can have a profound impact on the long-term success of a business or for	A common implementation mistake committed by a concurrent work-group is to confuse a CE program with a CE Ideal. CE programs are the vehicles for implementing the ideals in an organization.	Chapter 10 (see Table 10.1).

TABLE P1   (continued)

Features of Volume II	How do these features benefit readers?	What chapters or sections of the book contain these features or examples of them?
ensuring manufacturing competitiveness.		
r The book offers a set of ten implementation guidelines for product redesign and development through its life-cycle functions. This "Ten Commandments" serves to guide the product and process iterative aspects of IPD rather than just the work-group collaborative aspects of a PD^3 cycle.	Deployment consists of a number of activity-plans arranged in increasing order of enrichment. The activity-plans overlap, and provide a structured approach to organizing product ideas and measures for concurrently performing the associated tasks.	Chapter 10 (IPD deployment is a multi-plan methodology as shown in Figure 10.1)

BEST PRACTICES

Sixty-six senior mangers from 33 progressive companies were surveyed in a NSF study to validate the importance of 56 "best practices" (see Table P3) for both new BS mechanical engineering (ME) graduates and for experienced MEs. The results indicated that [ASME/NSF, 1996]

- 53 of the identified 56 identified "best practices" *are in use* in more than two-thirds of the companies surveyed.
- "Concurrent Engineering" practice received the *highest number* of votes for *all the three* questions in the "Knowledge of PRP" category. The three questions that were asked are listed in Table P2.

TABLE P2   Product Realization Process Survey Results (66 Industry Respondents from 33 Industries)[1]

Question Number	What was the question posed?	Respondents, who answered Concurrent Engineering (CE) as their answer.	Ranking based on what respondents Judged (compared to answers in "Knowledge of PRP" best practice category)
#1	Are the following (56) PRP "Best Practices" currently used in your business unit?	88%	Highest "YES" answers
#2	How important is it for experienced mechanical engineers (5+ years) to have a working knowledge of the following (56) best practices?	91%	Highest number judged CE—very important
#3	How important is it for entry level mechanical engineers (new BS Graduate) to have a working knowledge of the following (56) Best Practices?	74%	Highest number judged CE—very important

[1] ASME/NSF, 1996, *Integrating the Product Realization Process (PRP) into the Undergraduate Curriculum*, New York: ASME Council of Education, NSF Grant # 9354772, New York.

TABLE P3 Ranking of Best Practices for New BS Graduates and Experienced ME's by 66 Industrial Respondents

Serial Number from ASME/NSF PRP Report	Elements of the PRP "Best Practice" identified by ASME/NSF [1996] study[1]	Section/Chapter/ Volume where material is covered or described	Serial Number from ASME/NSF PRP Report	Elements of the PRP "Best Practice" identified by ASME/NSF [1996] study[1]	Section/Chapter/ Volume where material is covered or described
1	Teams/Teamwork	5.1–5.7/I; 4.2/I; 4.5/I	29	Design for Service/Repair	2.6.6/II; 2.6.10–11/II; 7.9/II
2	Communication	1.5/I; 6.8.1/I; 7.5/I; 9.0/I; 1.2.2.1/I; 4.3.5/I	30	Product Testing	—
3	Design for Manufacture	1.8.1.2/II; 2.6.1-2/II;	31	Process Improvements Tools	3.7/I; 8.7.1/I; 3.7/II; 3.5.1/II; 7.1/II; 7.9/II
4	CAD systems	1.2.2.3/I; 2.8/I; 7.1.3.2/I	32	Tools for "Customer Centered" design	2.8.2/I; 2.8.3/I; 3.7/II; 7.3/II;
5	Professional Ethics	5.0/I	33	Information Processing	1.8.1.3/II; 7.3/I; 8.5/II; 9.4/II; 5.4/II
6	Creative Thinking	6.1/II; 7.1/II; 7.9/I	34	Leadership	5.1.1/I; 5.1.2.4/I; 5.2.4/I
7	Design for Performance	1.6/I; 3.6/I; 7.8/I;	35	Statistical Process Control	3.1/II;
8	Design for Reliability	2.5.6/II; 3.8/II	36	Test Equipment	—
9	Design for Safety	2.4.4/II; 2.7.4/I	37	Industrial Design	6.2–6.5/II; 9.5–6/II
10	Concurrent Engineering	4.1–4.6/I; 6.7/II	38	Design for Commonality-Platform	2.6.7/II; 2.6.4/II;

No.	Topic	Reference
11	Sketching/Drawing	7.1.3/I; 8.1/I
12	Design for Cost	1.4/I; 2.7/I; 2.6.5/II; 1.8.1.4/II; 7.5/II
13	Application of Statistics	3.7.2/II/ 2.7.2/II
14	Reliability	2.4.4/II; 2.5.6/II;
15	Geometric Tolerancing	2.4.3/II; 3.7.3/II
16	Value Engineering	3.6.1/II;
17	Design Reviews	2.5.2/I; 9.6.1/I;
18	Manufacturing Processes	1.5/I; 2.5.3/II; 6.8/I;
19	Systems Perspective	6.2–6.6/I; 4.6/II;
20	Design for Assembly	2.6.3/II
21	Design of Experiments	2.7.2–3/II
22	Project Management Tools	3.8–3.9/I; 6.7/I; 3.8.5/II
23	Design for Environment	2.2/I;
24	Solid Modeling/Rapid Prototyping Systems	7.3.2/I; 9.5.4/I;
25	Design for Ergonomics (Human Factors)	1.8.1.6/II; 9.1.2/II
26	Finite Element Analysis	7.1.3/I; 1.2.2.3/I; 4.3.3.4/I; 2.4/II
27	Physical Testing	—
28	Total Quality Management	3.8.4/I 3.1/II; 2.5.1/II; 2.7.1/II
39	Computer Integrated Manufacturing (CIM)	8.1–8.9/II; 9.12/II; 5.1/II
40	Design Standards	5.8/II; 5.4.5/II
41	Mechatronics	4.8/II
42	Testing Standards	2.4.1/II;
43	Electro-mechanical Packaging	—
44	Conflict Management	9.12.2–3/II
45	Robotics and Automated Assembly	2.5.5/II; 2.6.3/II;
46	Knowledge of the Product Realization Process	9.1–9.7/I; 4.3–4.7/II;
47	Design for Dis-assembly	2.6.9/II
48	Budgeting	
49	Project Risk Analysis	2.2.1/I; 2.4.5/II; 9.11/II
50	Competitive Analysis	3.5–3.6/I; 9.8.3/II
51	Process Standards	4.3.4.2/I; 5.8/II
52	Manufacturing Flow/Workcell Layout	
53	Bench Marking	3.1/I; 3.7/I; 9.6.2/I
54	Corporate Vision and Product Fit	3.7/I; 8.4.4/I;
55	Materials Planning—Inventory	1.9.3/II; 2.4.2/II; 2.5.4/II;
56	Business Functions (Mktg., Legal, etc.)	2.7/I; 5.1.2.4/I

¹ ASME/NSF, 1996, Integrating the Product Realization Process (PRP) into the Undergraduate Curriculum, New York: ASME Council of Education, NSF Grant # 9354772, New York.

The two volumes together contains 50 of those 56 best practices that were initially proposed for the new BS graduates and experienced ME's [ASME/NSF, 1996]. The primary sections or chapters, where those best practices are discussed in this book, are listed in Table P3.

TEST PROBLEMS

At the end of each chapter, test problems are included. The instructor may choose a set of problems (ten or less) that he or she has covered in the class for that week from each chapter. Most test problems are based on the materials covered in the chapter itself. Some are based on materials covered in the earlier chapters thus stretching the student's grasp and understanding of the subject matters covered so far. Only a few test problems require stretching the students' imagination beyond what is discussed in this book. A rich reference section is provided for professors to reinforce the materials beyond what is discussed therein. The generous use of self-explanatory illustrations and bullets makes this book an easy and pleasant reading for everyone. Illustrations provide a quick visual grasp of the materials without the use of long and wordy sentences and paragraphs.

Biren Prasad
Electronic Data Systems
General Motors Account
P.O. Box 250254,
W. Bloomfield, MI 48325, USA
Email: <bprasad@cmsa.gmr.com>

ACKNOWLEDGMENTS

Over the last several years, having associated with Concurrent Engineering: Research and Applications (CERA) Journal as a founding editor and having attended/organized numerous conferences dealing on this subject, I have steadily built up a massive collection of precious knowledge on concurrent engineering (CE). Many of the ideas set forth in this book are formulated based on the rigorous analysis of what has been reported in those journals and conferences, of what we found worked well in practice, and from our research of what we observed was essential and relevant for those that failed to be successful. In most cases, the materials in the book are mostly built on trying these ideas on problems facing the automotive, electronic, aerospace, and software industries (working with Ford Motor Company, General Motors, Electronic Data Systems, NASA, and other Delphi Automotive System customers). Many CE concepts contained herein, therefore, are reported for the first time. The others are extensions to the ideas—derived from various CE books, journal articles, and my research papers presented at various meetings—but never published. Relevant references are contained at the end of each chapter. Many thanks to those who supplied reprints of their articles and thesis included therein. The author wishes to acknowledge the contributors of the CERA Journal and the members of its two editorial boards with whom the author corresponded on numerous occasions, which helped solidify many of the concepts reported in this book.

A very important aspect of almost any technical publication is the exposition of key definitions of fundamental terms, and this book has a lot of them. In those areas, I can only take credit for bringing them together and packaging them in what is, I hope, a convenient format. To this end, many thanks are due to my professional colleagues in a number of fields relevant to CE, TQM, quality circles, QFD, knowledge-based engineering (KBE), and product design and development.

The author wishes to acknowledge the assistance of General Motors, Electronic Data Systems, and Delphi Automotive Systems for providing the environments and opportunities, and assistance of our Automated Concurrent Engineering (ACE) colleagues and customers, with whom I worked, particularly Mr. Pat Race. In addition, I thank the people who reviewed this manuscript including Ross P. Corbett and several others. Thanks are also due to my colleagues, who spent time in reviewing the Volume I and providing their candid comments, which are published in the front few pages of this volume. Thanks also to many of my close teaching associates, students and friends for the valuable guidance they gave me in making this book more useful to our readers and students. A special thanks is extended to my spouse Pushpa and to my lovely daughters—Rosalie, Gunjan, and Palak—for their patience and understanding throughout these never-ending years while I was busy writing.

Biren Prasad
Electronic Data Systems/General Motors Account
P.O. Box 250254,
W. Bloomfield, MI 48325, USA
Email: bprasad@cmsa.gmr.com

CHAPTER 1

CONCURRENT FUNCTION DEPLOYMENT

1.0 INTRODUCTION

While manufacturing philosophies have changed drastically during the eighties, the pace of such transitions from concept to practice has been very slow. Despite painful restructuring, reorganization, and even process re-engineering efforts, both the European and U.S. automotive industries have failed to attain parity in product cost, productivity, or throughput with Japanese producers and transplant operations. Earlier published work showed assurances that the competitive gaps could be closed using Quality Function Deployment (QFD) or similar programs. This had motivated abandonment of many traditional function values in favor of employee empowerment and autonomous multi-functional teamwork. Many such combinations have been tried with QFD, along with Pugh's concept [Clausing, 1994], voice of the customer (VOC), and product development team (PDT). They are discussed in volume I of *CE Fundamentals*. Though each QFD combination provided new opportunities and contributions toward cost and productivity improvements, such programs have encountered difficulties in making a parent company globally competitive. Furthermore, the gains that would seem obvious and feasible through the exploitation of QFD and its combination (in a quantifiable competitive sense) have not always been fully realized.

The application of QFD is a fairly old (over two decades) idea [Hauser and Clausing, 1988]. Historically, the concept of QFD was introduced by Japanese [e.g., Mizuno and Akao, 1978; and Aswad, 1989] in 1967. It did not emerge as a viable methodology until 1972 when it was applied at the Kobe shipyards of Mitsubishi Heavy Industries [e.g., Hales, Lyman and Norman, 1990; Taguchi, 1987; and ASQC, 1992] in Japan. American Supplier Institute (ASI) and GOAL/QPC (Growth Opportunity Alliance of

Lawrence, Massachusetts/Quality Productivity Center) [e.g., Akao, 1990; and King, 1987] have done a great job in publicizing it in the United States. QFD was originally designed to take the *voice of the customer* (called customer objectives) and translate them into a set of design parameters that can be deployed vertically top-down through a serial four-phase process [Sullivan, 1988]. The four phases, known as ASI's four-phase process, are product planning, parts deployment, process planning, and production planning. The overall objective of QFD, which was *quality* deployment when introduced in 1967, today is still the product's quality. Emphasis on quality was also the reason why it was named *Quality* Function Deployment by the Japanese producers [Crosby, 1979; Deming, 1986; Taguchi and Clausing, 1990]. Recently Don Clausing and others have introduced some structural changes in the way the QFD information is arranged. The new arrangement is commonly called the extended House of Quality [Hales, Lyman and Norman, 1990; Taguchi and Clausing, 1990]. However, the original emphasis has not changed at all.

1.1 COMPONENTS OF QFD

This extended house of quality (HOQ) consists of eight fundamental areas, all of which are not essential. Figure 2.23 of Volume I identifies the names of each area, and the door example in section 2.8.2.3 (see Figure 2.24) gives a glimpse of its full potential. In the following section, we visit each room of the extended HOQ and examine its essential features.

1.1.1 HOQ List-Vectors

Figure 1.1 identifies all rooms in the extended HOQ by its list-vectors and matrices. The four list-vectors—*WHATs, HOWs, HOW-MUCHes*, and *WHYs*—are described in the following sections.

1.1.1.1 *WHATs*: Customer Requirements (CRs)

The customer defines what constitutes *WHATs* in a QFD/HOQ. In simple terms, this is a list of customers' wants. In most consumer goods manufacturing companies, Voice of the Customer (VOC) is considered the market requirement. Customers are initially listened to and a list of customer needs and expectations are created. This list is called *WHATs* or customer requirements (CRs). Some typical examples of *WHATs* might be "pleasing to the eyes," "looks well built," "provides good visibility," or "opens and closes easily." *WHATs* normally define a set of end-user requirements about what a consumer wants or likes to see in a future product.

Dr. Noriaki Kano developed an expanded concept of quality. The Kano model of Quality or Features, as it is most frequently called, defines three types of *WHATs*: Basic, Performance, and Excitement. The Kano model relates customer satisfaction for each *WHAT* to its degree of achievement. An improved form of the Kano Model is shown in Figure 1.2. The degree of achievement is plotted against customer satisfaction. The two extremes of achievements are *fully achieved* and *not yet started*. The two extremes of

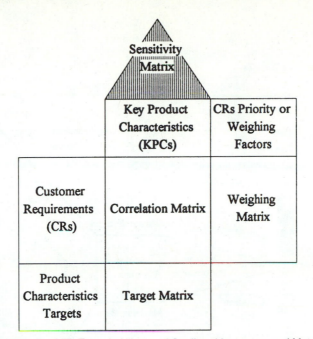

FIGURE 1.1 QFD Extended House of Quality—List-vectors and Matrices

customer satisfaction are *very satisfied,* and *very dissatisfied.* This divides the domain spanned by the achievement and satisfaction axes into four quadrants, which in itself are divided into three regions. The spoken performance called performance *WHATs* constitute the middle region and are shown in Figure 1.2 by a flashlight- or a speaker-shaped boundary. The upper region of the flashlight model describes the areas for excitement *WHATs*. The lower region in the flashlight model shows the domain of unspoken or basic *WHATs*.

- **Basic WHATs:** These are the basic set of *WHATs*—the core quality feature or function set that a customer normally expects in a product. This is schematically shown by a region below the speaker-shaped area in Figure 1.2. It is assumed that customers do not have to ask for these quality features or functions, hence called *unspoken*. These features are typically hidden or implied functions of the products. Such features exist as a natural part or normal function of the product in some form, unless the product fails to work. The basic set of *WHATs* hardly ever increases customer satisfaction, but when left unattended, it can adversely affect customer satisfaction. One example of this basic quality feature or function is a car engine starting and not stalling. A customer would not necessarily be happy just because the car started or it did not stall, but surely enough, he or she will be frustrated and dissatisfied if it does not start or if it stalls frequently. Another example of a basic quality feature/function is that a customer would normally assume that the oil pump works and would not even think about it in a marketing survey unless, of course, it gives trouble.

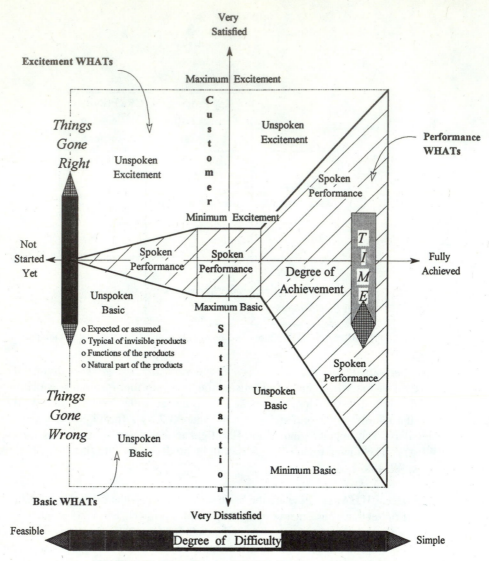

FIGURE 1.2   Three Types of *WHATs* (Customer Requirements)

- **Performance WHATs:** They are usually determined through market research and are one-dimensional. These are often called *spoken WHATs*. A set of performance quality features describes how well a product measures up to the customer's wants. This is represented in the middle region of Figure 1.2 by a flashlight or a speaker-shaped area. Customer satisfaction increases with the degree of achievement. Examples of *performance WHATs* would be fuel economy, quietness, and a comfortable ride.

- *Excitement WHATs:* These are *unspoken WHATs*. They include new innovations and thoughtful engineering, providing pleasant surprises to the customer. These are shown in Figure 1.2 by a region above the speaker-shaped area. They increase customer satisfaction, even though the customer would not necessarily be dissatisfied without them. The 3M Corporation calls these *WHATs customer delights*. Examples of these are split sun visors, cargo nets in the trunk to hold down plastic grocery bags, holes drilled for fixing toddler seats, and so on.

Over a period of time, a CR can shift character on the Flashlight model. This is shown in Figure 1.2 by a TIME arrow pointing down. Over time, a CR that was an excitement character at one point (a part of the upper half region) can take the form of a basic quality (a part of the lower half region) when products are matured. An example of this is power steering. This was an excitement CR in the 1960's, a performance CR in the 1970's, and now it is almost a basic CR. If a quality feature/function of a new product is contained in the upper half region, things are said to be going right for the product. The companies are expected to make a debut in the marketplace with such product introductions. On the other hand, if some quality features/functions fall below the speaker-shaped region, products are considered to be noncompetitive in the marketplace. The normal cycle of change is from a set of excitement quality to performance quality and then to a set of basic quality. The corresponding *WHATs* can further be categorized into primary (must have), secondary (may be), and tertiary (like-to-have) categories.

1.1.1.2   *HOWs*: Quality Characteristics Items

Manufacturers define what constitutes *HOWs* in a QFD/HOQ. This is represented by a list-vector in the Quality House marked as *HOWs* (see Figure 1.3). In simple terms, *HOWs* are a set of quality characteristics through which a set of *WHATs* can be realized. *HOWs* thus represent an array of design variables or alternate solutions, which may or may not be independent. Each of the *HOWs* provides a solution or alternative for attacking one or more *WHATs* (or CRs). Manufacturers do not know what magnitude of each of these *HOWs,* when considered as a unit, will lead to the realization of as many *WHATs* as possible. *HOWs* provide an operational definition for the market quality characteristics. Using this list, a company can measure and control quality in order to ensure *WHATs'* satisfaction. The *HOWs* are the methods or techniques to translate the *Voice of the Customer* into design evaluation criteria. Typical entries on the *HOWs* list-vector are parameters for which means of measurements or a measurable target value can be established. For example, a customer need for a *good ride* (a *WHAT*) is achieved through dampening, shock isolation, anti-roll, or stability requirements (the four *HOWs*). The *HOWs* determine the set of alternate quality features to satisfy the stated customers' needs and expectations (*WHATs*). For this reason, *HOWs* are also called Quality Characteristics (QCs). A typical *HOW* might be a length, a width, a height, a thickness, a usable surface area, a volume, a set of material characteristics or mass properties, and so on.

For every *WHAT* in the CRs list, there are usually at least one or more *HOWs* to describe possible means of achieving customer satisfaction. A best of the class product contains *HOWs* that satisfy a set of *WHATs* in some prioritized manner (see Figure 1.3). This is the chosen way in QFD of defining a relative priority for meeting the *WHATs* objectives. If a product solution (a *HOW*) exists today, a vector of such *HOWs* can be looked upon as strategic proportions in which customer requirements (*WHATs*) are satisfied. A *HOW* is the way to assess feasibility of the product in the marketplace. A *HOW* list helps to define the target solution in relation to the *WHATs* list.

1.1.1.3 *HOW-MUCHes*: Bounds on Quality Characteristics

This is a list-vector and normally identifies the bounds on the feasibility of *HOWs*. These entries are in the list-vector called *HOW-MUCHes* and represent the target values for each quality characteristic (see Figure 1.3). In other words, for each *HOW* on the list-vector, there is a corresponding value for a *HOW-MUCH* entry. The idea is to quantify the solution parameters into achievable ranges or specification table, thereby creating a criterion for assessing success. This information is often obtained through market evaluation and research. *HOW-MUCHes* capture the extremes—the permissible target values, positive or negative—depending upon the *HOWs* sentence construction or statements. A typical *HOW-MUCH* measures the importance of *HOWs*, a performance of Product X, or a set of target

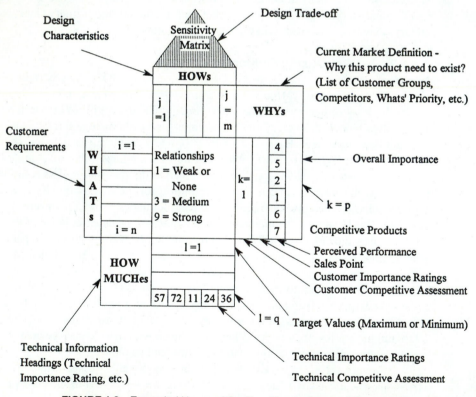

FIGURE 1.3   Expanded House of Quality—Terminology and Conventions

values. In an optimization formulation discussed in section 1.7, a row of *HOW-MUCHes* is used to collect upper and lower bounds for the attributes in the *HOWs* list-vector.

1.1.1.4 *WHYs*: Weighting Factors on *WHATs*

Similar to *WHATs* and *HOWs*, a set of *WHYs* is also a list-vector that describes the relative importance of current competitive products, referred to as world-class or best of the class products. Once these important values are multiplied with the corresponding set of *WHATs* and then summed, they can provide a single pseudo measurement index for overall customer satisfaction. In terms of optimization this can represent a weighted sum of *objectives*. An example of *WHYs* is a list-vector of relative importance with respect to customer wants for a *world-class product* of a major competitor. If the product is targeted to multiple customer groups, such as U.S., Asian, European, Japanese, and the like, this list must include these customer groups and their relative wants. *WHYs* are names of competitors, competitive products, market segments, or other items that describe the current market conditions. *WHYs* are also factors for weighing decisions that a future product must take into account. This usually translates into specifying weighting factors for *WHATs*. Setting priorities means specifying what is significant in the list of *WHATs* and what is not. A typical *WHY* might be a list-vector of overall importance, a vector list of importance to the world purchaser, or a set of world-class achievable performance of a product X.

1.1.2   HOQ Relational Matrices

The four relational matrices are described in this section. HOQ Relational Matrices employ either numbers or symbols depending upon the purpose of QFD and the context in which QFD is being used (see Figure 1.3). Two possible rationales have been traditionally proposed depending on whether a relational matrix is used for calculations or for visual aid.

- *Quantitative Reasoning:* Numbers are used for specifying magnitudes of HOQ matrices. This facilitates comparing magnitudes of computed list-vectors through mathematical means.
- *Qualitative Reasoning:* Symbols are used to represent list-vectors or matrices. This provides a better visual communication. Three symbols are often used to indicate the relationship between the entries of *WHATs* and the *HOWs*. A solid circle (●) implies a strong relationship, an open circle (○) a medium relationship, and a triangle (△) a weak or small relationship.

This process of evaluating expressions in QFD gives a concurrent team member a basic method of comparing the strengths and weaknesses, importance of column-vectors (*WHATs*, *WHYs*) or row-vectors (*HOWs*, *HOW-MUCHes*), and measuring interactions between them. The notations used here follow the convention adopted by the employees at the Kobe shipyards who incorporated the local horse racing symbols. By convention, each symbol in the relationship matrix receives a value. Table 1.1 shows a convention that is typically followed in defining QFD relational matrices.

TABLE 1.1   Standard Relationship Conventions (Weight and Symbols)

Matrix		Quantitative	Qualitative
WHATs versus *HOWs*	*Grade*	*Weight*	*Symbols*
	Strong relationship	9	Double or Solid Circle (●)
	Moderate relationship	3	Open Circle (○)
	Weak relationship	1	Open Triangle (△)
	None	0	Blank
HOWs versus *HOWs*	*Grade*	*Weight*	*Symbols*
	Strong Positive relationship	9	Double or Solid Circle (●)
	Medium Positive relationship	3	Solid Triangle (▲)
	Positive relationship	1	+
	None	0	Blank
	Negative relationship	−1	−
	Medium Negative relationship	−3	Open Triangle (△)
	Strong Negative relationship	−9	Open Circle (○)

1.1.2.1   *WHATs* versus *HOWs*: Correlation Matrix Relationship between Market Requirements & Quality Characteristics (QCs)

To get a relationship between market requirements and quality characteristics, a matrix is created by placing the *HOWs* list along the column of a matrix and the *WHATs* list along its rows (see Figure 1.3). The rectangular area between the rows and the columns then depicts the relationships between the set of *WHATs* and the *HOWs*. The matrix thus developed is called a relationship matrix. It correlates what customers want in a product and how an enterprise can achieve those objectives. The matrix—*WHATs* versus *HOWs*—is a core relational matrix of QFD. Relationships within this matrix are usually defined using a four level procedure: strong, medium, weak or none (see Table 1.1). An example is shown in Figure 1.3. This matrix may be densely populated (more than one row or column affected). This results from the fact that some of the quality solutions may affect more than one market requirement. For example, what a customer wants in good ride and good handling (*WHATs)* are both affected by quality characteristics like dampening, anti-roll, or stability requirements (*HOWs)*. The more densely populated and spread in ranks the correlation matrix is, the more valuable the information (relationship) is likely to be. A diagonal correlation matrix means there is none or very little interaction between the rows and the columns. *HOWs* could also include some design evaluation criteria such as how do we know we can satisfy our customers, how can we test, how can we control the quality, and so on.

1.1.2.2   *WHATs* versus *WHYs*: Matrix of Influence Coefficients

This is a relationship matrix that can be used to prioritize the *WHATs* based upon criteria for competitiveness. Usually, a list-vector in the matrix (e.g., a column) consists of one or more of the following items (see Figure 1.3).

- Marketing Information Ratings that identify the relative importance of each of the *WHATs*.
- Ratings that show how important the different customer groups perceive each of the *WHATs*. This is often referred to as Customer Importance Ratings (CIRs).
- Ratings that show how well a competitive product is perceived to meet each of the *WHATs*.
- Ratings that show where your product ranks or is perceived relative to the competition.
- Factors that a company would like to consider to be a world-class quality producer in defining its (a product) specification set.

The above criteria provide a set of possible options for identifying the stated importance ratings and factoring in how your product is perceived relative to competitions. These list-vectors help establish a realistic target that a company would like to shoot for in future product offerings. Most importantly, the above criteria can be used to determine a weighted average of *WHATs* as a single performance index. The latter can serve as a quantitative measure of an *overall customer satisfaction* for the product that the company is planning to launch.

1.1.2.3   *HOWs* versus *HOW-MUCHes*: A Feasibility Matrix of bounds on *HOWs*

This is a feasibility matrix that lets a team decide how much of each *HOW* you can possibly vary to meet customer wants. Typically, the data in this matrix (a row) consists of one or more of the following (see Figure 1.3). In this case, a row of matrix *HOW-MUCHes* of *HOWs* may contain:

- what an organization perceives its product ranks relative to its competitor's product (technical competitive assessment).
- ratings that identify the relative importance of each of the *HOWs*.
- how a competitive product performs relative to each of the chosen *HOWs* (benchmark data).
- estimates or calculations of realistic upper limits that a chosen *HOW* can take.
- estimates or calculations of a realistic lower limit that a chosen *HOW* can take.
- estimates of service repair cost data, direction of improvements, legal, safety, and other control items.
- computed values of Technical Importance Rating (TIRs). This is a weighted sum of Quality Characteristics (QCs) computed with respect to Customer Importance Ratings (CIRs).

Most commonly, a PDT team, through a row of feasibility matrix, establishes a set of realistic target values, both upper and lower bounds, for each *HOW*. Product values or target values identify engineering tolerances and specification limits on QCs. These can also be used to pinpoint whether or not a current technology is sufficient and where new manufacturing research would be necessary. Varying the appropriate combinations of *HOWs* in such a way that a set of TIRs lies within these bounds ensures that the product in question meets customer requirements (CRs).

1.1.2.4 *HOWs* versus *HOWs*: Sensitivity Matrix Relationship Between Quality Characteristics

This relationship is described by means of a sensitivity matrix that forms the roof of the house of quality (see Figure 1.3). The purpose of the roof is to identify the qualitative correlation between the characteristic items (*HOWs*). This is a very important feature of the quality house since, at times, the possible solutions could be redundant and may not add much value to customer wants. Other times it may be at cross-purposes (in disagreement) with each other. The sensitivity matrix helps identify the situations in such occurrences. If two *HOWs* help each other meet their target values (*HOW-MUCHes*), they are rated as *positive* or *strong positive*. If meeting one *HOW* target value makes it harder or impossible to meet another *HOW* target value (*HOW-MUCH*), those two *HOWs* are rated as *negative* or *strongly negative* (see Table 1.1). One case in point is when *0-60 MPH time* and *fuel economy* are two quality items. Efforts to decrease the *0-60 MPH time* would have an adverse effect on the *fuel economy* item. In this case, the two *HOWs* have a negative correlation.

      In actuality, correlation between QCs (solution parameters) could be positive or negative and in varying degrees: strong, medium, or none. When one *HOW* adversely affects another *HOW*, a qualitative negative correlation results. On the other hand, if it favorably supports a second *HOW*, a positive correlation results. For example, *fuel economy* and *gross weight* are considered as having a positive correlation because reducing gross weight will increase fuel economy keeping all other remaining parameters constant. Standard QFD practice uses the weights (9 for strong, 3 for medium, and 0 for none) as shown in Table 1.1. Symbols can be used to visually portray the different types of correlation.

      After the HOQ relationship matrices are developed, care is taken in reviewing its constructs. Blank rows or columns call for closer scrutiny. A blank row implies a potential unsatisfied customer and emphasizes the need to develop one or more QC items (*HOWs*) for that particular market requirement (*WHATs*). A blank column implies that the corresponding QC item does not directly relate to or impact any of the market requirements. It may in turn suggest an outright removal of a QC or a reallocation of resources or might pin-point a new set of customer wants that have not been identified.

1.2   LIMITATIONS IN DEPLOYING QFD

Early on, when Japanese producers became successful in bringing cars to market in record time, many automotive world leaders mistakenly assumed that their success was solely because of technical tools. This explains the initial flurry of activities (with QFD, SPC, Taguchi, Pugh, Kaizon, etc.) that American industries went through during the 1980s. As many American automotive industries failed on this front, manufacturers began to unearth the cause of their failures. It did not take very long to realize their apparent short-sightedness.

They discovered that many of the barriers to global competitiveness were rooted in their assumptions, that is, basing their PD^3 decisions on *quality* while ignoring other important aspects such as cost, design for X-ability, tools and technology, and infrastructure that have not been deployed simultaneously.

QFD does not specifically address the cost, tools and technology, responsiveness (time-to-market), and organizational aspects in the same vein as it addresses the *quality* aspect. While some consider the product design process as being independent from technology, design for X-ability, cost, and responsiveness, the reality is that these are tied together by a common set of product and process requirements. The design process only provides a product design from the perspectives of performance (i.e., *quality*). The product design performance requirements drive the product selection process, including system, subsystems, components, parts and material selection, and influence the selection of the fabrication method, process and production. Others have argued that while performing Quality FD, designers could choose to include requirements that belong to considerations other than quality in the original customers' list of HOQ [Dika and Begley, 1991; Carey, 1992; Kroll, 1992]. Accomplishing this through a conventional deployment process is not simple. Working on the multiple lists of requirements as part of a single function deployment (under Quality) is a much tougher problem.

> *First,* it would be a complex undertaking considering just the size of the resulting relational matrices.
>
> *Second,* deploying them serially would be a long, drawn-out process.
>
> *Third,* cascading the requirements all together as we did in the case of Quality functions would be so large that it would be difficult to handle.
>
> *Fourth,* there is no way of insuring that the design obtained through this combinatorial Quality FD process would not result in a sub-optimized design, that is, a product particularly designed for characteristics related to quality.

What is required in optimizing an artifact is designing with respect to all important functions that characterize a world-class product today. Normally in actual practice, information for these measurements is independently obtained and design often proceeds in parallel. Paralleling allows the combinatorial problems to be addressed in sizable chunks that, in turn, can be handled by a number of specialized work groups comfortably. Parallel deployment of values would allow concurrent teams to work independently, thus reducing the PD^3 cycle-time.

Major pitfalls of Akao's QFD approach are:

- *Conventional Deployment is mainly Quality focused:* One of the pitfalls of conventional deployment is that it is based on a *single* measurement, which has mostly been quality. Today manufacturing sectors are more fiercely competitive and global than ever. Consumers are more demanding, competition is more global, fierce, and ruthless; and technology is advancing and changing rapidly. The quality based philosophy inherent in Akao's *quality* FD style introduced during the early 1970s does not account for the time factor inherent in today's complex PD^3 process. The competitors are always finding better and faster ways of doing things. Catching up in quality is not enough. It only makes a company at par with its competitors in terms of inheriting some of their product quality characteristics but relatively speaking getting there a few years later. What is required is a total control of one's process—identifying and satisfying the needs and expectations of consumers better than the competitors and doing so profitably faster than any competitor.

• *Conventional Deployment is a Phased Process:* The conventional deployment process prescribes a set of structured cross-functional planning and communication matrices for building quality as specified by customers into a product. Such a methodology is described by Sullivan in 1988 and is based on the most popular four-phased deployment due to Macabe, a Japanese Reliability Engineer in 1970 [Aswad, 1989]. This is often represented in a cascade time-bound process where characteristics of a prior phase feed as requirements for a subsequent phase. The serial nature of deployment tends to make the QFD process sequential. If each phase of deployment is a *multi-part* process, the elapsed time can be significantly large. This elongates the total time this QFD would take for an artifact realization process. However, it is not essential that each phase of QFD be a hands-off process with no overlap between the consecutive phases.

• *Conventional Deployment is one-dimensional:* The roles of the organization and engineers are changing today, as are the methods of doing business. Competition has driven organization to consider concepts such as time compression (fast-to-market), concurrent engineering, design for X-ability, and tools and technology (such as Taguchi, Value Engineering) while designing and developing an artifact. Quality FD addresses major aspects of *quality* with reference to the functions a product has to perform but this is one of the many functions that need to be deployed during product development. With conventional deployment, it is difficult to address all aspects of total values management (TVM) such as X-ability, cost, tools and technology, responsiveness, and organization issues. It is not enough to deploy *Quality* into the product and expect the outcome to be a world-class. TVM efforts are vital in maintaining a competitive edge in today's world marketplace. The question is *how* to deploy all the aspects of this TVM.

Conventional deployment cannot account for the increasing complexities of our product and the conflicting requirements that need to be addressed. Hence, the best efforts of the concurrent team simply do not result in products that optimally meet customer requirements. It is not because the teams are not able to work closely, but the deployment vehicle is not robust enough to accommodate multiple functions deployment simultaneously. Conventional deployment lacks vigor while implementing simultaneously various conflicting value characteristics such as cost, responsiveness, quality, and so on. In the absence of any better deployment vehicle, the team repeats the conventional deployment process for each value one at a time. This elongates the PD^3 cycle-time into a multi-year ordeal.

1.3   CONCURRENT PRODUCT DEVELOPMENT

The first step in creating a great product is an understanding of what, exactly, makes a product great. Kim Clark defines a great product as one that meets all pertinent characteristics that are required to ensure product integrity. This is discussed further in Chapter 3, volume 2. Generally, development of a new artifact does include considerations for several life cycle values that are pertinent to meeting the customers' requirements. Many of these values are independent, that is, there is very little interaction between them. Through the course of investigations and study, the author has found that the deployment of many artifact functions (values) can proceed in parallel with what we know today as *quality FD*. Examples are X-ability (performance), tools and technology, cost, responsiveness, and infrastructure. Generally, these functions or values are independently specified or estimated. The results of experience can be used to specify the requirements and expectations for each of the values in parallel without having to wait until a deployment of *quality* FD is complete.

1.4   CONCURRENT FUNCTION DEPLOYMENT

In this chapter, we have expanded the original definition of QFD, discussed in volume I, to include parallel deployments. This provides a method to consider the deployment of many competing values simultaneously. We have called this approach Concurrent Function Deployment (CFD). The intent of CFD is to incorporate *Voice of the Customers* into all nine phases of the product development cycle, through mission definition, concept definition, engineering and analysis, product design, prototyping, production engineering and planning, production operations and control, manufacturing, and finally into continuous improvement, support and delivery (see Figure 4.2, volume I). In other words, CFD is customer driven PD^3 methodology.

CFD is a concurrent engineering methodology that enforces the notion of concurrency and deploys simultaneously a number of competing artifact values, not just the *Quality as found in QFD*. The artifact value deployment is through all its life-cycle phases. If a specification chart is being developed for the product, the taxonomy for requirements and constraints (RCs) must reflect all value considerations. RCs thus include customer requirements (CRs), VOCs and all other types of *WHATs* that one may encounter. There are many value characteristics (VCs) for artifact, such as quality, X-ability, tools and technology, costs, responsiveness, infrastructure, and so on. The RCs and VCs for an artifact can be plotted as shown in Figure 1.4. Such taxonomy will ensure that all important aspects for product and process design have been identified and included. The focus of CFD is on systematically capturing product information, such as market competitive analysis and customer satisfaction ratings, analyzing these ratings to improve product functionality (e.g., an X-ability element) and then adding an array of values that are important to the customers and to the company. CFD thus ensures concurrent product development. CFD breaks the multi-year QFD ordeal by allowing work groups to work concurrently on a number of conflicting values and compare their notes at common check points. CFD is a simple and powerful tool that leads to long range thinking and better communication across several value functions.

This Chapter presents a methodology for concurrently deploying a line of value objectives for successive product refinements leading to a world-class category. Also, since each TVM's life cycle value meets only a partial set of artifact specifications, the characteristics of the chosen TVM values will dictate the life-cycle needs to manufacture or fabricate the product.

1.4.1   CFD Architecture

CFD uses a three-axis approach for orderly deployment of functions/features (see Figure 1.4) spanning in three dimensions: horizontal (x-axis), axial (y-axis), and vertical (z-axis). Artifact values (AVs) are deployed along the x-axis, Value Characteristics (VCs) associated with each class of artifact values are deployed along the y-axis, and Requirements and Constraints (RCs) are deployed along the z-axis (see Figure 1.4). The components of axial and horizontal dimensions are arranged in a matrix and deployed concurrently, while vertical dimension is staggered in tiers. Value characteristics (VCs) for each value class are identified so that specifications developed, using this methodology, will yield an optimum product configuration the first time and every time the CFD is used. The CFD methodology drastically reduces dependence on trial and error methods, such as *prototype fabrication* or testing. The methodology is independent of the types of manufacturing processes and products to be designed.

Let us denote the following:

X_i represents an ith track artifact values (AVs) for horizontal deployment,

Y_{ij} represents a jth level VCs for axial deployment,

and Z_{ijk} represents a kth tier RCs for vertical deployment.

The following is the process used for concurrent function deployment.

1.4.1.1   Step 1: Horizontal Deployment

The CFD process starts with a horizontal deployment of an artifact value, X_i. The team chooses a set of artifact values (along the x-axis) that need to be deployed. Deployment is concurrent, meaning deployment of corresponding VCs can proceed in parallel.

X_i represents an ith track of an artifact value. The following artifact values are commonly found relevant during product development.

A typical X_i for a class of 6-value set ($i = 1, \ldots, 6$) may look like:

$$X_1 = \text{Quality (functionality)} \tag{1.1}$$

$$X_2 = \text{X-ability (performance)} \tag{1.2}$$

$$X_3 = \text{Tools \& Technology} \tag{1.3}$$

$$X_4 = \text{Cost} \tag{1.4}$$

$$X_5 = \text{Responsiveness (time-to-market)} \tag{1.5}$$

$$X_6 = \text{Infrastructure} \tag{1.6}$$

1.4.1.2 Step 2: Axial Deployment

The second step is to identify a set of axial (y-axis) value characteristics (VCs), Y_{ij}, for axial deployment corresponding to each X_i. This process is concurrent, meaning the VC functions corresponding to an artifact value can be deployed simultaneously.

$$Y_{ij} \text{ for } 1 \le i \le I \text{ and } 1 \le j \le J \tag{1.7}$$

where Y_{ij} is a matrix, j takes the value from 1 through J, and J is the maximum number of VCs level selected for an ith value track. A typical Y_{ij} for a matrix of size ($I = 6$ and $J = 5$) is shown in Figure 1.4 for illustration.

1.4.1.3 Step 3: Vertical Deployment

The third step is the vertical deployment of Y_{ij} in relation to RCs for the tier k. X_i and Y_{ij} are the AV and VC functions that were identified in Steps 1 and 2, respectively. There are three tiers to CFD deployment, tier $k = 1$ through tier $k = 3$. A tier structure means a line of vertical (z-axis) RCs deployment proceeds before the next tier of RCs deployment begins. This means there is an overlap between tiers. It does not require reaching an end of one deployment before starting another (not phased in as in QFD). From the above definitions,

$$Z_{ijk} \text{ for } 1 \le i \le I \, ; \, 1 \le j \le J \text{ and } 1 \le k \le K \tag{1.8}$$

where Z_{ijk} represents a kth tier for vertical deployment, k assumes the value 1 to 3 corresponding to tiers 1, 2, and 3, respectively.

A typical Z_{ijk} for a 3 tier RCs ($k = 1, \ldots, 3$) may look like as:

$$Z_{ij1} \equiv \text{Product Planning (Tier 1)} \tag{1.9}$$

$$Z_{ij2} \equiv \text{Process Planning (Tier 2)} \tag{1.10}$$

$$Z_{ij3} \equiv \text{Production Planning (Tier 3)} \tag{1.11}$$

Steps 1 through 3 form a trio. Deployment through a particular tier (1, 2 or 3) completes a CFD pass. CFD is complete if a series of trio (horizontal-axial-vertical) deployment is carried out for all passes and for all value tracks, X_i.

1.4.2 Trio Deployment Technique

As discussed, the three-step CFD architecture utilizes a trio (horizontal-axial-vertical, $\ldots, \ldots$) deployment technique (see Figure 1.4) to arrive at the end of the first pass. This results in a product design validated with a manufacturing process concept. During step 3

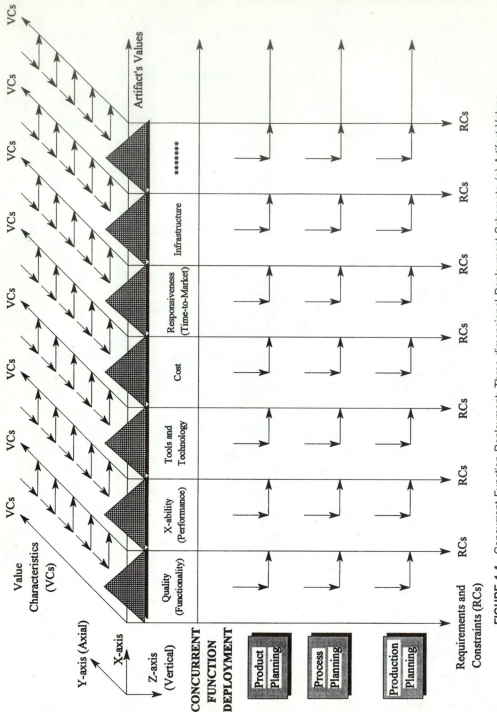

FIGURE 1.4 Concurrent Function Deployment: Three-dimensional Deployment Schemata (a) Artifact Values along x-axis, (b) Value Characteristics (VCs) along y-axis and, (c) Requirements and constraints (RCs) along z-axis.

above, each tier completes a pass for a CFD. The first pass is horizontal-axial-vertical deployment for tier 1. The CFD trio is repeated for tiers 2 and 3.

$$\text{First Pass} \equiv \text{(horizontal-axial-vertical) for tier 1} \tag{1.12}$$

$$\text{Second Pass} \equiv \text{(horizontal-axial-vertical) for tier 2} \tag{1.13}$$

$$\text{Third Pass} \equiv \text{(horizontal-axial-vertical) for tier 3} \tag{1.14}$$

However this process of trio deployment is concurrent. There are overlaps between vertical (z-axis), axial (y-axis) and horizontal (x-axis) passage from timing perspectives. The CFD methodology interweaves the three-axis deployment with several other concurrent engineering techniques, such as TQM, goal oriented management, integrated product development, cross-functional teams, and so on. It is a concept of three-dimensional (concurrent trio structure) deployment. This quickly allows many of the downstream steps (*WHATs* and *HOWs*) of the PD^3 process to be brought in earlier and satisfied at the first available opportunity during a CFD pass of deployment (see Figure 1.5). Other *WHATs* and *HOWs* are further addressed in greater details in subsequent passes. The process leads to a selection of the best design and process (*HOWs*) for the overall product specifications (*WHATs*). CFD's *WHYs* and *HOW-MUCHes* metrics support this selection with sound analytical rationale and targets for quality (functionality), cost (profitability), X-ability (performance), tools and technology (innovation), responsiveness (time-to-market, flexibility, etc.), and infrastructure goals performed simultaneously.

1.5   CFD METHODOLOGY

Concurrent function deployment (CFD) is a methodology that allows designers and manufacturing engineers to communicate early and work in parallel during various stages of a PD^3 process. One critical new tool to facilitate this early communication is *house of values*, which is a concept similar to the *house of quality* that was introduced in Akao's QFD. However, the term *values* is not used here to mean just *quality*. It ranges from *quality* as it was in QFD to other values, such as X-ability, tools and technology, cost, responsiveness, infrastructure, and other similar type functions. The concept gives rise to a line of concurrent houses; namely, House of Quality, House of X-ability, House of Tools and Technology, House of Cost, and so on. House of Quality, thus, becomes a degenerate or a special case of this series—*House of Values*—template.

1.5.1   Three-dimensional House of Values (HOV)

The basic tool of CFD is the *relational matrix* concept. Matrices are schemata to generically define and directionally relate multiple lists of identifiers, often referred to as line- or list-vectors. The basic matrix of CFD is the *house of values*, so named to resemble *house of quality* that forms one of the many objectives of CFD. The relational matrix in CFD translates the corresponding requirements and constrains (RCs) into value characteristics (VCs). Figure 1.6 is a schematic view of a *house of values (HOV)* template. This

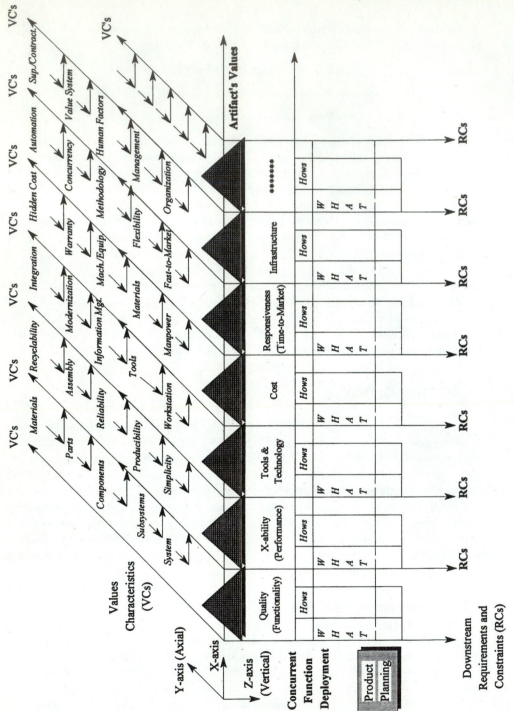

FIGURE 1.5 Concurrent Function Deployment: X-axis—*WHATs* and *HOWs*

18

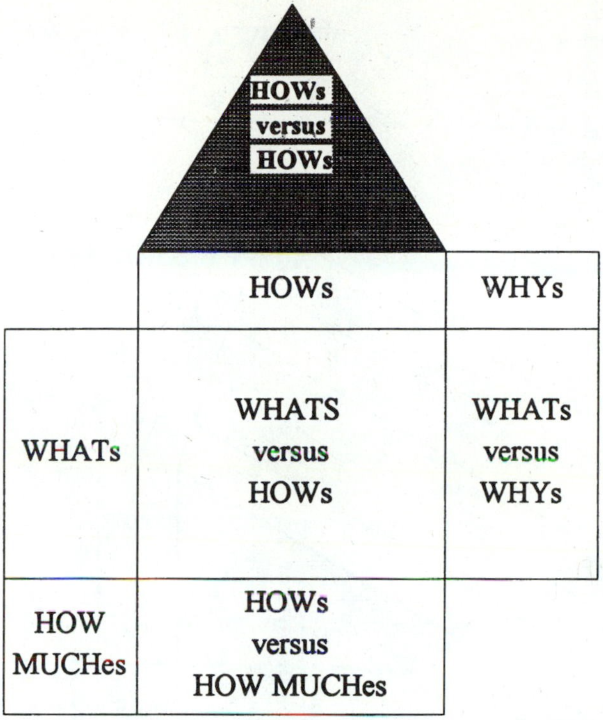

FIGURE 1.6 Typical House of Values (HOV) Template

template (HOV) has 8 rooms. Four of the rooms form the basic perimeters of the house. These are two row-rooms, *WHATs* and *HOW-MUCHes*, and two column-rooms, *HOWs* and *WHYs*. Concurrent HOV also encompasses relationships between these four list-vectors resulting in four *relational matrices*. These are:

> *HOWs* versus *HOWs*.
> *WHATs* versus *HOWs*.
> *HOWs* versus *HOW-MUCHes*.
> *WHATs* versus *WHYs*.

The relationship between CFD components are shown in Figure 1.7. The three-dimensional matrix takes the form of three roofs and three relational matrices as shown in Figure 1.7. It has three list vectors: artifact values (AVs) list, value characteristics (VCs) list, and requirements and constraints (RCs) list. Eight elements of AVs, nine elements of VCs and three major elements of RCs vectors are shown for example purposes. These lists may contain any number of values as necessary. The three relational matrices are

> RCs versus VCs.
> RCs versus AVs.
> AVs versus VCs.

This completes the concurrent deployment of artifact values, AVs, along the three axes.

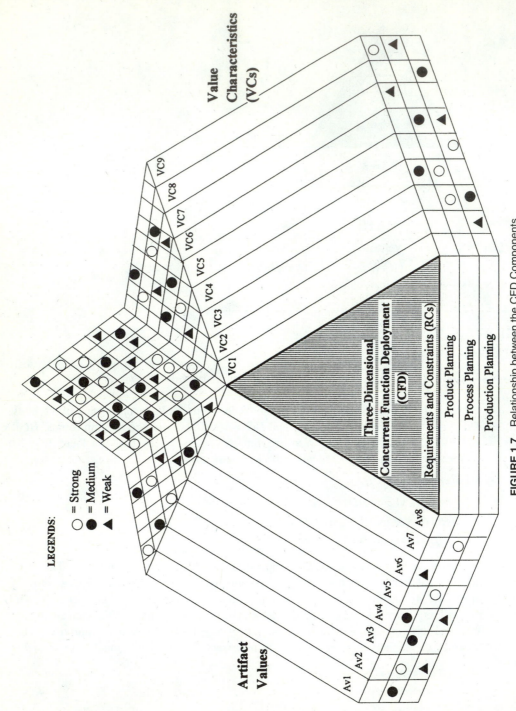

FIGURE 1.7 Relationship between the CFD Components

20

1.5.2   Life-cycle Deployment

Figure 1.8 compares the QFD and CFD approaches in great depth. QFD is actually a subset of CFD. In traditional deployment processes (as in QFD), quality is generally associated with manufacturing, and for which several quality measurement tools, such as SPC, QPC, CMM, and the like, are typically employed. For instance, activities, such as performance measurements, dimensional control, and others, are often used to check quality compliance during manufacturing. In reality, such efforts need not be limited only to manufacturing. Quality is not just a manufacturing problem alone. It is a cumulative outcome of decisions that are made during an entire product life-cycle. It is, therefore, important to affect all such decisions. What would be a more appropriate place to affect these decisions than during the individual processes where these decisions are made?

In the CFD process, quality begins with the quality of the introduced requirements and constraints (RCs). RCs in this context are not only those requirements which are specified by the customers, but also includes those that are introduced directly by the cooperating

Quality Function Deployment (QFD)	Concurrent Function Deployment (CFD)
QFD is a phased (or a serial) process	CFD is a concurrent process
QFD is an inside-out focussed	CFD is an outside-in holistically focussed
QFD works with the pieces of objects (car door, roof, hinges, etc.)	CFD works with the whole product rather than its pieces
QFD is mainly successful in solving pockets of problems	CFD is used in conjunction with company mission deployment principles
QFD is a problem solving process (e.g., a rusty car door, a leaking seal, etc.) or sometimes a redesign process	A systematic approach to handling all life-cycle values relevant to the product
QFD focuses on technical parameters that are not necessarily looped back to the whole product	Customer requirements are considered separately from QCs values
QFD provides a technical importance rating for quality characteristics (QCs)	CFD provides a value index - cummulative effectiveness rating
QFD deals with pieces of product or pieces of requirements	CFD optimizes the system with consistency of purpose as target goals
Because of its serial nature of processing, QFD is perceived to take a long time	Because of its concurrent processing, CFD is conceived to be faster than QFD

FIGURE 1.8   Comparison of QFD and CFD

CE teams [Prasad, Morenc and Rangan, 1993] (see Figure 2.22 of volume I). The burden of poor outcome of a design has been shifted from the work groups expertise in product manufacturing to the teams' choice (or selection) of RCs at each CFD transformation step. If appropriate methods can be employed in systematically classifying, deploying, and solving the transformed problems, the assurance of VCs' considerations becomes merely a scheduling and distribution job. Quality considerations are ensured by the proper selection of RCs and methods for solving the constrained problems. Satisfaction of RCs and VCs at each transformation state is what constitutes a product's *values deployment*. By following this methodology, the taxonomy of transformation leads to a *world-class valued* product, the quality characteristics of which are appropriately distributed across the various levels of transformation. At each level, differences between proposed RCs and the computed outputs provide a measure of the differences that exist among alternative trial designs. Dealing with quality at loop level is straightforward, since problem definitions (number of RCs, inputs, and the transformation matrix) are small and are at a manageable scale. Satisfaction of RCs during these early loop levels (e.g., feasibility or product synthesis loops) is easier when problem definition is more explicit in form than when the product is somewhat mature, say, after the product has crossed several decision boundaries.

The next section illustrates a degenerate case of CFD, that is, of deploying a quality FD through a CFD trio process. This concept is virtually equivalent to a QFD. However, the process applies to other values as well.

1.5.3 Quality FD

Products are often divided into logical hierarchical blocks depending upon their complexity levels. Different parallel teams can work in these different hierarchical groups. Work groups at each level can work concurrently. Some dependencies can exist between the levels. Establishing common quality standards for communications and definitions of VCs can allow parallel work groups to work concurrently. The most commonly employed quality characteristics, Y_{ij}, are [Prasad, 1993]:

where Y_{ij} for $i = 1$ (Quality FD) and $j = 1, \ldots, 5$,

1.1 Assembly

1.2 Sub-assemblies

1.3 Components

1.4 Parts

1.5 Materials, and so on

Figure 1.5 shows the above quality characteristics spanned along the axial (y-axis) dimension.

Figure 1.9 illustrates the CFD concept of vertically deploying (along the z-axis) quality RCs often embedded in the *voice of the customer*. The three tier deployment structure is shown in Figure 1.9 for quality FD. Tier 1 is for product planning, tier 2 is for process planning, and tier 3 is for production planning. The same three-dimensional trio process is repeated for each tier. For example, during product planning, customer require-

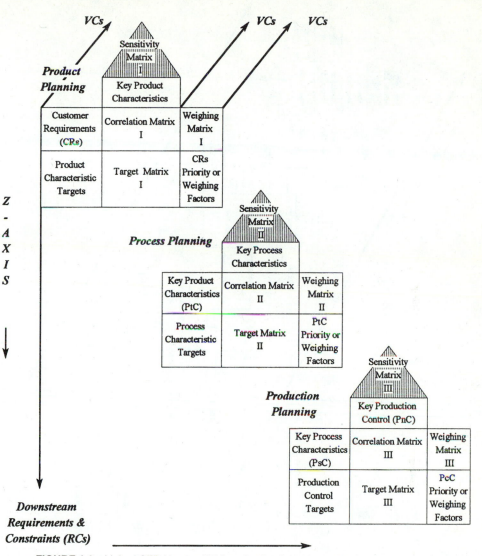

FIGURE 1.9   Linked CFD House of Values for Quality—Vertical Deployment (Z-axis)

ments (CRs) or *WHATs* are related to key quality characteristics for which a list of *WHYs* and a list of *HOW-MUCHes* are then identified. *HOWs* define the desired key product characteristics (PtCs) of a product to counter the *WHATs*. *WHYs* are the overall evaluation criteria used within the organization to define acceptability of the product. Targets for the PtCs (*HOW-MUCHes*) are established based upon competitive benchmarks and the customer's competitive assessment. Such deployment methodology is followed for tier 2 and the tier 3 trio sequences.

In Figure 1.10, the *quality* value for CFD tier 1 is further spanned axially (along y-axis) into its characteristic (VCs). This axial expansion corresponds to the five VCs for

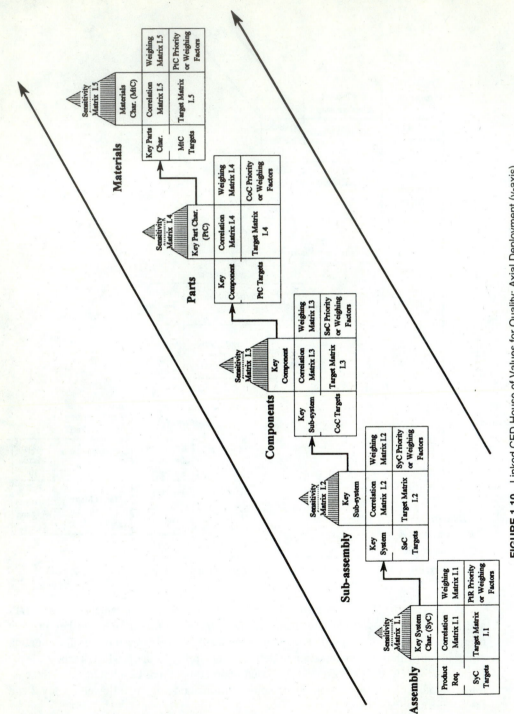

FIGURE 1.10 Linked CFD House of Values for Quality: Axial Deployment (y-axis)

quality that were listed in Figure 1.5. The axial expansion of the Product Planning tier (Y_{ij}, $j = 1, \ldots, 5$; levels I.1 through level I.5) uses the PtCs defined in tier 1 to evaluate alternatives and filter a design that meets most of the customers' demands. At the end of tier 1 (level I.5), a set of product PtCs is identified that represents best of the class (see Figure 1.9). Process planning (tier 2) deals with selection of process concept and identification of critical operation parameters, here called key process characteristics (PsCs) that can cause the product PtCs identified in tier 1 to be satisfied. Production planning (tier 3) identifies production PnCs (control requirements, maintenance requirements, mistake proofing, education and training issues, etc.) in line with the process (PsCs) identified in tier 2.

In the example described herein only a three-tier trio (horizontal- axial-vertical) structure for CFD is shown. This is the most common. However, such a CFD structure can have as many tiers as needed. In these diagrams, the filtering process is shown through a solid pipeline connecting the *characteristics* (*HOWs*) room to the *WHATs* room (see Figure 1.10). It ensures that VCs (namely PtCs, PsCs and PnCs) that are critical to meeting the product, process, and production objectives (RCs) are given proper and early attentions (during y-axis deployment). It also ensures that *HOWs* are further deployed into their root or key characteristic factors during the subsequent vertical tiers (z-axis) deployment.

1.6   APPLICATIONS OF CFD

The four key relational matrices in the house of quality discussed in section 1.2 are also common to HOV. These matrices are useful in drawing conclusions about the relative importance of *WHATs*, *WHYs*, *HOWs* and so on. There are some computer programs and software that allow you to enter these matrices interactively. They also provide a variety of sorting and matrix analysis algorithms such as weighted average, ranking, technical importance, normalized ratings, sum of *WHYs* or *HOWs* matrix column, weighting factors, graphics utilities (bar charts, line chart), and so on [QFD/Capture, 1990].

Before we show how CFD can be applied to solving a variety of problems, let us establish some nomenclatures for CFD list-vectors and relational matrices.

Let us denote the following:

$$\text{HH} \equiv \text{A sensitivity matrix defined by } HOWs \text{ versus } HOWs. \qquad (1.15)$$

$$\text{WH} \equiv \text{A correlation matrix defined by } WHATs \text{ versus } HOWs. \qquad (1.16)$$

$$\text{WW} \equiv \text{An influence matrix defined by } WHATs \text{ versus } WHYs. \qquad (1.17)$$

$$\text{HM} \equiv \text{A feasibility matrix defined by } HOWs \text{ versus } HOW\text{-}MUCHes. \qquad (1.18)$$

In the next section, some general usage of CFD charts are described to assess competitors' products.

1.6.1   Competitor Product Assessment

Competitor product assessment charts are used to assess two things: to rate the requirement of a competitive product and to rate the quality characteristics for the same competitor's product. There are two types of competitive assessment.

Customer Competitive Assessment (CCA) is developed from customer surveys. Customers are asked to rate the requirements (CRs) of a competitive product. They are asked to identify what they liked in a competitive product and what they did not like including their preferences of one requirement with respect to the other. In the *House of Value*, this is entered in a column of a *WHATs* versus *WHYs* relationship matrix. Customer competitive assessments rate the *WHATs* (perceived response).

$$CCA \equiv \{ww_k\} \quad \text{for } k = 1, \ldots, p \tag{1.19}$$

or
$$CCA_i \equiv ww_{ik}; \quad \text{for } i = 1, \ldots, n; \tag{1.20}$$

where $\{ww_k\}$ represents a kth column of the *WHATs* versus *WHYs* matrix.

Technical Competitive Assessment (TCA) is also developed from customer surveys. In a similar fashion, the quality characteristics *(HOWs)* for the same competitive product are rated here from a technical perspective. The customers are asked to rate the features they find interesting in a competitive product including the ones they did not like. TCA represents the customer's opinion of quality characteristics (that is, interesting features) found in a competitor product in a particular marketplace. In the *House of Quality* this is found in a *HOWs* versus *HOW-MUCHes* relationship matrix. Engineering assessments quantify the *HOWs* (engineered or measured outputs).

By definition,

$$TCA \equiv \{hm_l\} \quad \text{for } l = 1, \ldots, q \tag{1.21}$$

or
$$TCA_j \equiv hm_{lj}; \quad \text{for } j = 1, \ldots, m; \tag{1.22}$$

where $\{hm_l\}$ represents lth row of the *HOWs* versus *HOW-MUCHes* matrix.

These two competitive assessments are said to be in conflict when corresponding to a location *WHAT, HOW*—wh_{ij}, there exists a strong positive CCA *(WHAT, WHY)* value—ww_{ik}, at the same time there exists a strong negative ECA *HOWs* versus *HOW-MUCHes* value—hm_{lj}. In mathematical terms, a conflict occurs for an (i,j) combination if there exists a row l in matrix HM and a column k in matrix WW, such that

$$[ww_{ik}/hm_{lj}] \cong -1; \tag{1.23}$$

corresponding to a location (i,j).

Note, a conflict occurs only when the ratio is closer to a negative unity. If the ratios are positive no conflict occurs. For example, if the ratio is a positive number, it represents a strong unity and the existence of good supporting data. Any negative values other than -1 indicate existence of a minor discrepancy for that location (i,j) in the correlation matrix.

When the two assessment values (CCA and TCA) are in conflict, it is often a result of failing to understand the *voice of the customer*. In such a case, the *HOWs* list must be amended to reflect customer perception. This is most often resolved by letting team members get directly involved in the process of comparing the in-house and the competitive products.

1.6.2 Company Product Assessment

Competitive product and company product assessment charts are used in HOV to compare the requirements and the quality characteristics of a competitor product with a company product. The following are two types of ratings commonly used in CFD to rate the importance of requirements (CRs) and the importance of QCs of a company product.

Customer Importance Rating (CIR) is derived from the field or customer surveys. The customers (users of the company products) are asked to rate the requirements that customers perceive are important for the manufacturer to consider in a product. They are also asked to identify what they would consider important in a future product and what they would not. These ratings are for each *WHAT* based on overall evaluation of the products in the field. It is posted as a column of the *WHATs* versus *WHYs* matrix. If that column is a *k*th column of matrix WW,

then by definition,

$$\text{CIR} \equiv \{ww_k\} \quad \text{for } k = 1, \dots, p \tag{1.24}$$

or

$$\text{CIR}_i \equiv ww_{ik}, \quad \text{where } i = 1, \dots, n \tag{1.25}$$

Technical Importance Ratings (TIR) are the results of calculations from the CFD matrix defined earlier. It is not a direct result of customer surveys as in the TCA. The customers are only required to rate the requirements (*WHATs*) and prioritize the importance of each with respect to the rest. This is done in the above step called CIR. The information is used later in the calculation of TIR. The steps used in the calculation of TIR are as follows:

1. Assign a numerical value for each symbol used in the correlation matrix *WHATs* versus *HOWs*. The relationship conversion listed in Table 1.1 is most commonly employed.

2. Corresponding to each *WHAT*, multiply the *WHATs* versus *HOWs* equivalent numerical value of the correlation matrix by the *WHATs* versus *WHYs* CIR value.

3. Repeat the results of multiplication in step (2) for all *WHATs* and add the results in each *HOW* column.

4. Enter the results of step (3) into a *HOWs* versus *HOW-MUCHes* row. The row of computed numbers stored in a *HOWs* versus *HOW-MUCHes* matrix represents a TIR for each *HOW*. If that row is a *l*th row of HH matrix,

then, by definition,

$$\text{TIR} \equiv hm_{lj} \equiv [\{wh_j\}^T \times \{ww_k\}]$$

$$\text{TIR} \equiv hm_{lj} \equiv \sum_{i=1}^{n} [(wh_{ij}) \times (ww_{ik})] \tag{1.26}$$

where $i = 1, \dots, n$ and $j = 1, \dots, m$.

Ratings stored in a *l*th row of "*HOWs* versus *HOWs MUCHes*" matrix are a relative comparison of each *j*th element, provided the *k*th row contains the computed value of the equation 1.24 shown above. Equation 1.26 represents an overall measurement of the quality of the final product.

Figure 1.11 illustrates the sequence of steps involved in computing the **Technical Importance Rating (TIR).** A WW column in Figure 1.11 contains an example of the customer importance rating (CIRs) values obtained through surveys. A typical correlation matrix may have symbolic representation. If so, they are first converted into a quantitative value matrix using the conversions shown in Table 1.1. As shown, 9 is the equivalent of a solid circle symbol and 3 is an equivalent quantitative value for a weak (an open triangle) symbol. These quantitative values in a cell (i,j) are multiplied by the ith customer importance ratings (CIRs) (4 and 7 stored in the kth column of the WW matrix), resulting in an importance value for each (i,j) location in the matrix. The technical importance rating for each *HOW* (jth location) is then found by adding together the importance values in each jth column. As an example, in the first column of the matrix in Figure 1.11, the first relationship has a value of 36 ($= 4 \times 9$), and the only other relationship has a value of 21 ($= 7 \times 3$). The technical importance rating for this column (a *HOW*) is therefore the sum of these two values, 57 ($= 36 + 21$). Technical importance ratings are used to determine which of the *HOWs*

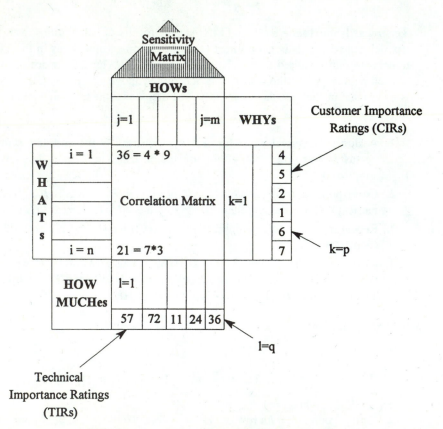

FIGURE 1.11   CFD Application to Determine Ratings

should receive the most resources and are particularly useful in trade-off decisions. The numbers 4 and 7 chosen in Figure 1.11 have no significance except to show the computational steps involved. The technical importance rating is finally stored as an element of the *HOWs* versus *HOW-MUCHes* matrix. Other elements might include service/repair cost data, technical difficulty, safety control items, newness, reliability, timings, cost, and so on (see Figure 1.3). It is impossible to fill-in CIRs and complete TIRs effectively without multidisciplinary teams. Additional tables can do the same with suppliers and end-users of the product.

A significant outcome of a successful CFD application is establishing a process of smooth communication between the work groups that let the 7Cs take place.

1.7 FORMULATION OF CFD AS AN OPTIMIZATION PROBLEM

CFD is a methodology for capturing the information necessary for making decisions on future product upgrades and/or analyzing competitors and their competing products. Though these applications of CFD are useful tools for the management and marketing personnel in decision making, CFD has not been applied to solving an optimization problem, although the basic methodology for CFD is quite general [Prasad, 1993]. In the following, the author has found a way to extend CFD methodology to solve a class of optimization problems. As stated before CFD captures four list-vectors: *WHATs, HOWs, HOW-MUCHes* and *WHYs*, and four relational matrices: *WHAT* versus *HOWs; WHAT* versus *WHYs; HOWs* versus *HOW-MUCHes;* and *HOWs* versus *HOWs*. A methodology has been developed in this section to capture the optimization formulation as a part of CFD process. It employs the same structure as HOV list-vectors and relational matrices to reformulate the CFD problem into a mathematical programming setting.

1.7.1 General Statement of the Mathematical Programming Problem

A general statement of the optimization problem can be posed as finding a vector of design variables $v \subset D$ that minimizes the objective functions, $F(v)$, while satisfying a set of constraint equations, $g_j(v)$. In mathematical notation,

$$F(v) \Rightarrow \text{Minimum or Maximum.} \tag{1.27}$$

$$\text{Subject to } g_j(v) \geq 0; \quad \text{for } j = 1, 2, \ldots, m \tag{1.28}$$

$$V_{\min} \leq v \leq V_{\max} \tag{1.29}$$

In most cases, $F(v)$ is termed the objective function and $g_j(v)$ the Constraints. $V_{\min}$ and $V_{\max}$ denote lower and upper bounds on design variables.

Note, the question we are posing is *how* can CFD be used in this setting. What elements of CFD represent design variables, constraints, and objective function? Notice that

rows of *WHATs* in CFD constitute customer wants or CRs. It is therefore similar to specifying a set of objective functions that a company must strive for, and pay attention to, in
order to please the customers. Let us introduce some additional notation.

$$WHATs \equiv \{f_i\} \quad \text{for } i = 1, \dots, n \tag{1.30}$$

Clearly, all *WHATs* are not independent. It is also safe to assume that one or more *WHATs*
can be satisfied individually by choosing the right *HOWs*. It will be very difficult, however, to satisfy all *WHATs* simultaneously by trial and error. Since *HOWs* are possible solution parameters that can be chosen to address *WHATs*, they are like perturbation variables of the original problem. In an optimization setting, vector *HOWs* of CFD will be
like *design variables*. This means

$$HOWs \equiv \{v_j\} \quad \text{for } j = 1, \dots, m. \tag{1.31}$$

If *HOWs* represent system variables, in order to achieve an optimal solution, we must
have these variables independent of each other. From CFD matrices we know that all
HOWs are not independent. The correlation matrix, the *roof* of CFD denotes the dependencies of one *HOW* on another. In an optimization formulation we can pose such relationships between *HOWs* as constraints of the problems. This situation can be handled
through what we normally call a *linking technique* in optimization. The correlation matrix
gives a linking matrix for the design variables. This matrix can be formulated as follows:

$$\{v_s\} = [HH] \{v_r\} \tag{1.32}$$

where *HH* is a square matrix of size (m, m),

or $$v_s = \{HH_{rs}\}^T \times v_r \tag{1.33}$$

where *HH* is a triangular square matrix whose diagonal terms are all zero. This is because
in the CFD correlation matrix for *HOWs* versus *HOWs*,

$$H_{rs} = 0 \ \forall \ r = s, \tag{1.34}$$

where r and s are the rows and columns of the *HOWs* matrix.

The above relation simply says that if we consider *HOWs* as a set of design vectors,
all of the terms in the *HOWs* vector are not independent of each other. Equation 1.32 or
1.33, thus, serves as a linear set of equality constraints to the corresponding optimization
formulation.

Let us define a term of HM matrix as hm_{lj}. Since, in equation 1.31, we denote a row
vector of *HOWs* as $\{v\}$, by definition of a *HOWs* versus *HOW-MUCHes* matrix, a value
stored in hm_{lj} would represent a limit on v; If we denote two vectors of the HM matrix
($l = L$ and U) representing permissible upper and lower bounds on *HOWs*, the following
statements can be made,

$$hm_{Lj} \leq v_j \leq hm_{Uj} \quad \text{for } j = 1, \dots, m \tag{1.35}$$

where hm_{Lj} denotes a lower bound on v_j and hm_{Uj} denotes an upper bound.

It is interesting to note that equation 1.35 represents an equation of bounds on the
system variable for an equivalent optimization formulation.

So far we have introduced the concept of design vectors, bounds on the design variables, and a series of objective functions defined by a *WHATs* vector. The only term left in completing an optimization formulation is a weighted objective function. Earlier we also noted that a list of *WHATs* represents a series of objective functions. We also noted from the matrix WW that a column of this matrix contains the weighting factors associated with CRs, the *WHATs*. If a *p*th column of WW matrix contains this factor, ww_{ip}, a cumulative or weighted function, $F(v)$, can be formed using TIR formulation (equation 1.26) as follows:

$$F_{pj}(v) \equiv [[wh_{ij}]^T \times \{ww_{ip}\}] \tag{1.36}$$

or in a matrix form,

$$F_{pj}(v) \equiv [\{wh\}_j^T \{ww\}_p]. \tag{1.37}$$

Note that in Equation 1.36 or 1.37 a weighted multi-objective function has been formed that is the weighted sum of all its components. The result can be stored as a *q*th row of an HM matrix.

That is,
$$hm_{qj} \equiv F_{pj}(v) \qquad \text{where } j = 1, \ldots, m. \tag{1.38}$$

In *WHATs* versus *WHYs,* sub-section 1.1.2.2, many ways of obtaining these weighting factors were described. Equation 1.38 represents one of many cumulative forms that are possible. Now, if F(v) is employed as an objective function the general statement of optimization previously described in section 1.7.1 can be transformed in CFD context as described in the following section.

1.7.2   CFD based Optimization

The original optimization problem as stated in section 1.7.1 can now be transformed.

Find a vector of design variables $v \in D$, where $\{v\} \equiv HOWs$, that minimizes the objective functions $F_{pj}(v)$,

where
$$F_{pj}(v) \equiv [[wh_{ij}]^T \times \{ww_i\}_p] \tag{1.39}$$

or
$$\{F(v)\}_p \equiv [WH]^T \{ww\}_p \tag{1.40}$$

where $i = 1, \ldots, n; j = 1, \ldots, m$; and p is a column index of a WW matrix, while satisfying the following constraint equations

$$\{v_s\} = [HH] \{v_r\} \tag{1.41}$$

where r and s range from 1 to m. The bounds on the design variables can be expressed as:

$$hm_{Lj} \leq v_j \leq hm_{Uj} \tag{1.42}$$

or
$$\{hm_L\} \leq \{v\} \leq \{hm_U\}. \tag{1.43}$$

In equations 1.39 to 1.43, the vectors $\{wh\}$ and $\{ww\}$ belong to CFD/HOV relational matrices, WH and WW respectively. Lower and upper bound vectors $\{hm_L\}$ and $\{hm_U\}$ are the rows of the CFD/HOV relational matrix, HM.

1.8   HORIZONTAL DEPLOYMENT

In sections 1.5.3, a basic case of CFD for quality FD was described. Similar to a transformation of a Quality FD from a qualitative approach to rigorously applied quantitative
methodologies, each of the TVM life-cycle values can also be transformed following this
approach. Figure 1.5 shows the 6 artifact values (AVs) along a horizontal x-axis and five
value characteristics (VCs) along a y-axis. The z-axis shows the vertical deployment
process of each of these (x-y) combination groups through its three tiers: the product planning, process planning, and production planning tiers. The vertical arrows (z-axis) show
the sequence of steps during the *tier deployment* and the horizontal arrows (x-, and y-axis)
show a process of *concurrent deployment*. A typical list of *WHATs* for *Quality FD* values
is shown in the next section. In the following sections, *WHATs* have been identified along
with a set of VCs. This is useful if a separate CFD matrix is required for each artifact
value. If there are not many items for VCs, some of the items can be combined.

   One of the goals of CFD is to develop products flexible enough so that many life
cycle TVM perspectives may be run in parallel without violating or contradicting any one
of the previously assigned *WHATs*. The optimum result would be derived from deploying
a great number of TVM values simultaneously, some of which—new materials, cost estimates, process technology—may not always be available.

1.8.1   First Level Deployment Metrics
of Artifact TVM Values

In the following sections, the first tier Function Deployment (FD) metrics are defined for
each TVM value. For the initial tier, a generic set of *WHATs* (requirements) and *HOWs*
(characteristics) are identified. Besides the two list-vectors, future concurrent FD metrics
must include quantifiable measures that can be used directly in performing values' trade-
off analysis. Only with the set of quantifiable measures, would the product performance
designer, cost estimator, X-ability checker, responsiveness keeper, and infrastructure
manager be able to adequately evaluate an artifact's conformance to TVM requirements.

1.8.1.1   Quality Deployment
Quality is the first artifact value that needs to be deployed. Products designed without
built-in quality during design and quality assessments after production-release suffer from
a number of shortcomings. Quality problems translate into higher manufacturing costs,
particularly when some defects get out to the customers. The two list vectors, *WHATs* and
HOWs for a HOQ matrix, are listed in Table 1.2.

1.8.1.2   X-ability Deployment
Most companies set goals on a limited set of X-ability considerations. Product performance and quality are generally considered the basic needs. They are separately specified
and explicitly dealt with. Besides the major key ones, such as functionality, reliability and
maintainability, there are many other factors that need to be quantified. These are parts

TABLE 1.2   Examples of *WHATs* and *HOWs* in a Quality Deployment

Examples of *WHATs*	Examples of *HOWs*
Component packaging	Managing manufacturing precision or quality
Concept design	DFM guidelines
Design criteria	Computer Aided Design (CAD)
Detail design	Design axioms
Difficult-to-handle parts	Managing detailed dimensions, roundness, etc.
Experimental testing	Just-in-Time (JIT)
Material handling	Continuous improvements
Material properties	Ensure robustness
Customer satisfaction	Process description rules
Nonstandard routing	Specify minimal eccentricity
Pre-production	Formal design reviews
Process planning	Informal design reviews
Product shape	Minimum parts assessments
Production	Material and process selection
Prototype	Assembly evaluation method
On-time shipment	Managing tolerances
Validation	Quality Control (QC)
Warehousing	Good surface finish, texture, etc.

count simplicity, parts joining efficiency, mating parts assemblability, sheet metal parts formability, plastic parts producibility, and so on. Tables 1.3 and 1.4 outline a list of such WHATs and HOWs belonging to this X-ability class. The problem is really in deciding what is important about an aspect of design and expressing the corresponding value X-ability (*HOWs*) attributes in a way that addresses the customer objectives.

1.8.1.3   Tools and Technology Deployment

In Section 2.8 of volume I a number of life-cycle tools were outlined. They were categorized as computer tools and technology tools. (See equations 2.86 and 2.87 of volume I). The impact of a tool deployment on the resulting organization is not the same. Table 1.5 outlines a list of *WHATs* and *HOWs* belonging to this tools and technology class.

1.8.1.4   Cost Deployment

In Chapter 2, volume 2, of this book, Design for X-ability (DFX) is described. The techniques employed during DFX methodologies, are either handbook type procedures, or a collection of rules and guidelines (heuristics) of what might be considered the best design practice. The aim of such DFX techniques is usually to meet the provision of X-ability guidance one at a time.

In concurrent engineering, the problem of collectively evaluating a design is more fundamental than provisions for judging the designs one at a time. DFA, like many another Design for X-ability, represents only a single consideration for obtaining a better design. In DFA, when two separate components are integrated into one, the latter is considered better because it ensures easier assembly than the old ones. Often such recommendations lack quantification. The designer is usually not told how the new design is better

TABLE 1.3   Examples of *WHATs* in an X-ability Deployment

Examples of WHATs	**Brief Descriptions**
Simplicity	Simplicity is perhaps the most significant attribute of a successful design. Simplicity signifies many things to many persons. It may mean fewer parts, parts that perform multiple functions, parts that are easier to assemble (manually or by robots), or parts that can be made by simple processes. Most engineers and designers are intrigued by the complexity of parts. Very few really know how to design for simplicity.
Reliability	Manual product assembly may suffer from dimensional deficiency (e.g., lack of adequate fits and clearances), variability (e.g., precision machining) and non-uniformity in assembly (e.g., lack of adequate accuracy and consistency), etc. They ought to be minimized. Impact of equipment reliability on process needs ought to be minimized. Designers must consider manual options in all facets of production as stand-by alternatives in the case of castrophic failures.
Maintainability	Normal wear and tear, unexpected malfunctions, and alike necessitate maintenance. To increase product reliability, maintenance considerations must include a checklist for troubleshooting, such as how to bring the fault equipment quickly on-line, and normal inspection as standard DFM procedures. Total preventive maintenance is usually considered a part of design for maintainability.
Serviceability	Serviceability includes items such as: Repair parts Testability Inspectability
Parts Consolidation	Besides many other benefits, parts consolidation reduces the need for using different materials for various parts of a product. It also increases the value of a single product to the collector of recycled parts. The receiver of the recyclable parts will not have to decipher each part to recover materials. • *Material Selection Strategy*: With high volume products, broader use of the same material and with low volume products, market value of the higher end material (such as resins), make disassembly and recycling more attractive. • *Ease of assembly*: Parts consolidation or incorporation of multiple parts into a single part reduces the number of materials to be processed in the assembly and recycling process. It also focuses on eliminating unnecessary assembly operations since these parts need not be assembled. • *Ease of Disassembly*: Parts after consolidation should not be made so complex that the process of disassembly becomes difficult and time-consuming. If that occurs consolidation is counter-productive.
Assembly/Disassembly	Fastening methods (employed during assembly) dictates the level of difficulties during disassembly. There are two ways to disassemble a part, by reverse assembly or by brute force. The following are various methods used: • Adhesives • Heat Staking • Induction Welding • Inserts • Screws • Snap-fit Latches • Thermal Methods
Producibility	Material selection strategy and its appropriateness impact significantly the cost and part performance and also influence the selection of manufacturing processes. • Process classification is dictated by the approach taken for product manufacturing such as net shape or near net shape machining. Near net shapes require secondary operations of finer finish and tolerances.

Examples of WHATs	Brief Descriptions
	• Manufacturing process selection should be made after detailed evaluation of all economic and competing life-cycle factors. • Design geometry features establish the shape of the part. • Production quantity such as minimum production volumes is a function of the required level of product quality, such as tolerances and surface finish.
Product Usability	Product usability emphasizes making the work group members the focal point of the design, stressing ease of use including skill requirements for learning how to operate. Products should be designed to fulfill work group needs.
Availability	The work groups should have access to design histories, old CAD files, design database, technical memory and library of parametric parts, etc.
Supportability	This includes both the personnel support and the virtual support extended by the network of computers (both hardware and software).
Installability	The virtual library of software programs should be installed on all hardware platforms, firmware, and operating systems, etc. and work seamlessly as if it was developed for that platform.
Upgradeability	The new version of all electronic computer programs should be upwards compatible. The old data files or objects should be able to work well with new programs or new objects without problems.
Disposability	If materials are health hazards or inflammable, work groups should find ways to devise a waste or fire management system for their safe disposal.
Recyclability	When large parts use the same materials, or when less disassembly is required to recover the recycled materials (such as plastics), the recycler is guaranteed a better opportunity to recover investments through recycling.

TABLE 1.4   Examples of HOWs for X-ability Deployment

Examples of HOWs	Brief Descriptions
Assembly and System Considerations	This relates to design for assembly or system design using consistency of purpose.
Design Criteria	Design criteria are considered a set of constraints to be satisfied.
Engineering Change Order/Note (ECO/ECN)	Design in its life cycle are subject to a number of ECOs and ECNs.
Features and Tolerance	Certain features and tolerance results in better design or better representation for the design capture.
Parts Classification	Classification of parts helps in concurrency and downstream operations such as GT-based process planning, etc.
Quality Function Deployment (QFD)	QFD stream sorts the relevant requirements from those that are redundant or unnecessary.
Quality Inspection and Tooling	Quality inspection and tooling is a combination of error-monitoring, measurements, and feedback.
Risk Management	This is described in section 2.2.1 of volume I.
Standardization	This is discussed in Chapter 5, section 5.8 of volume II.
Systems Engineering	This is discussed in Chapter 6 of volume I.
Value Engineering	This is discussed in section 3.6.2 of volume II.

TABLE 1.5 Examples of *WHATs* in Tools and Technology Deployment

Examples of *WHATs*	Brief Descriptions	Examples of *HOWs*
Workstation	This includes bringing in MIPS power to reduce the computational burden and wait time. Design and operator training should account for variations in population, skill levels, and other human stamina and comfort needs.	Workstation—includes distributed computing, work-group computing, collaborative decision making, electronic design notebooks, open architecture, virtual collocation, etc.
Information Management	Information Management consists of two parts: information modeling and intelligent information management. Information modeling is discussed in Chapter 7 of volume I. Intelligent information management is discussed in Chapter 8 of this volume.	Information Management—product data management, data exchange standards (PDES, IGES, STEP), data handling utilities, requirements manager, data protection and security, etc.
Team-based Requirements	Team-based requirements is discussed in Chapter 5 of volume I.	Team-based requirements include things like share information, collocate with networks, improve inter-team communications.
Product Information Management	Product information management is discussed in Chapter 5, section 5.4 of volume II.	Product information management specifies capturing corporate history, interface database, project coordination, maintain data integrity and expert monitoring of data, collaborative decision making, coordinating teamwork, etc.
Modernization	Modernization—readiness assessment, decision tractability, constraint management, bring-in multiple decision views, real time display, real-time machining and NC, e.g., stereolithography prototyping.	Modernization—wrapper system, concurrent engineering, Taguchi method, design tools and CAD, group technology, rapid prototyping, benchmarking, design of experiments, life cycle values (design for assembly and manufacturability), etc.
Integration	Integration includes capturing the product/process, capture representation of team members' viewpoints, product and process requirements, integrating frameworks, tools and services.	Integrated PD^3, enterprise engineering, computer-based engineering, CAD/CAM, early manufacturing involvement, design for X-ability, integrated tool sets, virtual collocations, cooperative product development, etc.
Failure Modes, Effects and Criticality Analysis	FMECA (often referred to as FMEA is a systematic procedure for evaluating the reliability of a design. In FMECA, (a) every possible failure mode is identified along with (b) causes of each failure mode (often called "failure mechanisms") (c) the effects of each failure mode on the operation of the system and (d) the criticality of each failure mode.	Examples include failure analysis, failure modes analysis, fault-tree analysis, DFMEA, fatigue analysis, reliability, statistical based methods, etc.
Design Tool-kit	Design tools kits are a number of integrated tools brought to capture the design or life-cycle intent of the product or its realization process.	Examples include, CAD, CAM, CAE, CAPP, CIM, CAT, etc.
Time-to-Market Tool-kit	Tools that provide different degrees of concurrency, teamwork, facilitate re-engineering strategies, synthesis and optimization.	Example includes design team, cooperative teamwork, supplier participation, management involvement, rapid prototyping, benchmarking, Taguchi quality engineering, quality function deployment, requirements analysis, forward engineering, design reviews, etc.
DFX Tool-kit	Tools that enable design for X-ability considerations.	Examples include: design for assembly, disassembly, life-cycle, maintainability, quality, recyclability, reliability, serviceability, variability, variety, etc.

in terms of cost. In the previous DFA example the integral design may be complex and costly to manufacture. The latter may still be better if the saving in assembly cost is higher than the increase in manufacturing cost. For such analysis it is enough to consider the cost difference or so called *increment in cost* for introducing a change in state.

Cost, therefore, represents a common denominator for evaluating and comparing designs and an important criterion for measuring the relative goodness of a product. In this regard TVM value functions, such as what is described earlier as *Design for X-ability,* are analogous to performing local optimization while ignoring the global picture or the global trade-off. The underlying reason for employing relative cost as value function is twofold.

1. Cost provides a common denominator in converting dissimilar life cycle values into common quantifiable terms, allowing the comparison of designs with varying attributes.
2. Cost difference is an important design metric by itself.

Prasad [1993] proposes a causal relationship between design decisions and costs through a QFD representation and computational scheme. In this context we have developed a similar scheme called *Cost Function Deployment* (CoFD). This is a unified data structure that can be applied at various levels of designs, for example, design of assembly, sub-assembly, components, parts and materials, to encompass different cost aspects. Along this line a CoFD is characterized here as an implicit function of four vector parameters: function, form, context and cost. In this equation, cost is a change variable associated with realizing the given task, following a given mode or path, for example, machining, threading, milling, and so on, to accomplish a change in status from the current state (machine rate, physical properties of design, etc.), to the changed state (materials removed, machine time, etc.).

CoFD is a function of

$$\text{CoFD} \Leftrightarrow \text{CoFD (task, path, initial state, change in state, incurred cost)} \qquad (1.44)$$

where the first four arguments are the inputs and the fifth is an output. Obviously the form is intentionally quite general to capture various types of CoFD situations. Once individual deployment costs are identified, total deployment cost can be computed as:

$$\text{Total Deployment Cost} = [\text{CoFD (quality)} + \text{CoFD (X-ability)} \\ + \text{CoFD (tools \& Technology)} + \text{CoFD (infrastructure)}]. \qquad (1.45)$$

Examples of *WHATs* in costs deployment are shown in Table 1.6 and examples of *HOWs* in costs deployment are shown in Table 1.7.

1.8.1.5   Responsiveness Deployment

In today's global marketplace, time is becoming a major competitive force. Quality used to take this place but not anymore. Today's customer sees quality and everything else as given (taken for granted). Every automotive company wants to get its new car models on the dealer's showroom floors in record time. Airlines spend billions of dollars in maintenance facilities and repairs to cut the non-flying downtime. Overnight delivery carriers are handling more and more packages with faster service than ever before. I believe even

TABLE 1.6 Examples of *WHATs* in Cost Deployment

Examples of *WHATs*	Brief Descriptions
Labor-force or Manpower	• People or manpower (work-force) • Standard Direct Labor—involves wages of the people who do tasks in a manufacturing process. There are two parts: necessary labor, and labor waste, if the work is not value-added to the function being performed. • Indirect Labor—refers to the wages of people who do tasks not directly related to the manufacturing process for a specific product. For example, the costs of janitorial services, forklift operators, general maintenance people, machine-repairs people, general inspectors, etc.
Materials Cost	• Standard Direct Material—includes costs to purchase material or parts that will ultimately go into manufacturing the product. There are two parts to standard direct material cost: necessary material and material waste (or scrap) that was incurred during the process. Examples of necessary material are cost of purchased raw materials, fasteners, etc. • Indirect Materials—includes the cost to purchase materials that will not ultimately be part of a manufacturable product. These may include items such as cutting and grinding fluids.
Manufacturing Cost	Manufacturing cost is given by the following equation: $$\text{Manufacturing Cost} = \text{Parts Cost} + \text{Assembly Cost} + \text{Tooling Cost.}$$ • Parts cost—includes material finishing, design dimensions and tolerances, tooling type, process volume, standardization of packaging, etc. • Assembly cost—depends upon time for assembly, parts handling, levels of assembly, volume, time for assembly design, labor rates and standardization, etc. • Tooling cost—depends upon process, volume, tooling materials, dimensions and tolerances, standardization, etc.
Machines and Equipment Cost	• Facilities Costs—such as real estate, taxes and insurance—should be less because less equipment and facilities will be needed. • Storage Costs—Implementation of JIT reduces the amount of space for storage of inventory. • Inventory Costs—Fewer parts and lower scrap and rework allow less total inventory to be carried.
Warranty and Maintenance Costs	• Warranty Costs—Fewer parts should mean fewer failures in service, improving customer satisfaction and lowering warranty costs. • Repair Costs—Simpler and fewer parts result in less repair parts needed and lower costs. • Maintenance Costs—Cost of maintenance should be less with simpler equipment and reduced facilities. • Inspection Costs—Better quality facilitates reduced inspection effort and lower costs.
Hidden Costs	• General and Administrative Costs—includes all costs associated with the operation of the plant that are not directly related to the product, labor or materials. Such costs may be incurred due to machine precision, multi-step, and environmental impact. This often causes increased burden for inspection, test, rework or scrap. • Standard Overhead Expenses—The two costs, direct labor and direct material, combined are known as prime cost. There are two types of overheads: necessary overhead and overhead waste that is not directly related to the specific product. • Variability Costs—are caused by many different kinds of variations, such as product variances, labor variances, material variances, etc.

TABLE 1.7   Examples of *HOWs* in Cost Deployment

Examples of *HOWs*	Brief Descriptions
Maintain Schedule	This allows managing time most effectively.
Reduce Inventory	This saves inventory costs.
Minimize Product Variations	This reduces the number of parts and improves reliability.
Control Batch Size	This allows the parts to be tailored to customer expectations and desires, similar to customization.
Lifetime buys	This means long term contractual arrangement or profit-sharing partnerships.
Cost Estimates	Estimating the life-cycle costs—planning, design, manufacturing to disposal.
Cost Models Variable Dimensions Modeling (VDM)	VDM allows capturing the information models into a generative form of geometry.
Monitoring and Measuring at the Source	This means institutionalizing error-proofing techniques, monitoring variation at the source rather than hunting for the cause of the problems.
Promotion & Marketing	The results should be communicated to all CE parties concerned.
Just-in-Time Scheduling	Scheduling the manufacturing processes in such a way, that work is not hindered and mission is accomplished JIT, without parts in the shop for long.

food chains, such as pizza parlors are competing for home delivery on the basis of time rather than taste. The duration for responsiveness and time-to-market is declining every time a new product is introduced. The following Table 1.8 lists some *WHATs* examples of responsiveness and Table 1.9 gives key *HOWs* examples of responsiveness, respectively.

TABLE 1.8   Examples of *WHATs* in Responsiveness Deployment

WHATs	Descriptions
Fast-To-Market	• Set-up Time—Simpler and fewer parts result in less total time spent in setting up machines.
	• Throughput Time—Concepts like JIT implementation as well as TQM and QC should decrease lead time for processing an order.
	• Production Rate—JIT tactics should also increase the rate of production, lower unit costs, and increase units available for sale.
	• Capacity—The amount of capacity that is available should increase with JIT tactics.
Environmentally Responsible	• Secondary Operations—such as paints and insulating foams that may not yield environmentally responsible designs.
	• Recyclable Materials—The choice of materials should be such that it is economical to recycle. Such materials originate from post-industrial materials, scrap materials that didn't make good parts, and post-consumer materials, such as soda cans.
Flexibility	• Flexibility—Process change reorganization, phased development, overcoming cultural barriers (NIH syndrome, File transfer barriers, evaluate early product or process concepts, etc.).
Methodology	• Methodology—Virtual team approach, virtual collocation, CE standards, benchmarking, goal oriented pert management, early manufacturing involvement, design for assembly, design for manufacturability, design for simplicity, design for X.
Automation	• Automation—Readiness assessment, capturing design intent, knowledge-based engineering, parametric programming, language-based system, design decision support, capturing product and process requirements, decision traceability, etc.

TABLE 1.9 Examples of *WHYs* in Responsiveness Deployment

Examples of *HOWs*	Brief Descriptions
CAD/CAM/CIM	Computer aided design/computer aided manufacturing and Computer integrated manufacturing.
CAE/CAX (X = analysis, simulation, engineering, FEA, prototyping, etc.)	Computer aided engineering/computer aided everything X.
Collaborative Engineering & Teamwork	Recent developments in collaborative techniques such as E-mail messages flow to executives and to co-workers without hesitation even across the organizational boundaries.
Continuous Process Improvements (CPI)	The continuous improvement—"support and delivery" is an ongoing coordination track, which runs for the full life-cycle. This track provides besides normal project management functions, sequencing, cooperation and central support to other tracks.
Design for X (X = Assembly, Disassembly, etc.)	By bringing in techniques such as DFX, an integral part of CE, the manufacturing team can bring a number of fundamental cost-cutting tips to the attention of the design team during the early part of product development.
Process Improvements	Process improvement is a pervasive set of renovation activities that form the life blood of the company's regenerating profit potential.
Flexible Manufacturing System (FMS)	Flexible manufacturing operations promote the quick changeover of lines or entire plants to accommodate rapidly changing market demands. Technology elements in FMS include reusable processes and machines, versatile controls to empower operators, advanced sensors to let operators know in real time what is going on, and advanced control algorithms to give the company a sustainable, proprietary competitive edge.
Group Technology (GT)	In group technology, features common to many parts of a product are identified and classified in groups. This practice not only saves in process planning time for machining but a number of parts can be machined in a batch mode.
Groupware	Today multi-media equipment, such as groupware, have accelerated the penetration of electronic reviews and electronic forms of communication.
Just in Time (JIT)	No in-process inventory is required, materials are purchased just-in-time and finished products are delivered on schedule and on cost.
Knowledge-Based Engineering (KBE)	KBE is a process of implementing knowledge-based systems in which domain-specific knowledge about a part or a process is stored along with other attributes (geometry, form features, etc.).
Project Management	CE requires a new approach to project management. Each team must work closely with other teams to identify and develop techniques that are more cost-effective, innovative, and simple to use.
Parametric Engineering	Parametric engineering is referred to as a method for determining a geometry (e.g., size and orientation of geometrical elements), whereas relationships are defined using a cyclic graph and related geometric (line or arc) constraints governing the system.
Matrix Methods (QFD, Pugh's Method, etc.)	Matrix methods give a concurrent team member a basic method of comparing the strengths and weaknesses, importance of column-vectors (*WHATs*, *WHYs*) or row-vectors (*HOWs*, *HOW-MUCHes*) and measuring interactions between them.
Rapid Prototyping	Rapid prototyping helps consistently develop and subsequently redesign and re-engineer a finished manufacturable product.
Quality Programs and TQM	Quality programs should not only be directed towards minimizing defects in production, but also augmenting the capability of the products to monitor and correct their own operations.
Work group Computing (WC)	WC implies a transparent access of data/system resources to the linked applications or work groups, independent of their locations, installations, processor hardware, operating systems, and programming languages.

TABLE 1.10   Examples of *WHATs* in Infrastructure Deployment

Examples of *WHATs*	Brief Descriptions
Value System	Value System—Business practices, how the team will work, disciplines, culture, etc.
Organization/management	Organization/management—The product team, skill of the product team, matrix organization, goal oriented management practices, strategic business units, profit sharing, empowerment, etc.
Human Factor Issues	Human factor issues deal with perspectives such as product usability, product safety and liability, quality control and inspection, product assembly, maintenance, impact of equipment reliability, and workstation design and operator's training.
Product Safety and Liability	Product Safety and Liability—in terms of its life cycle use such as user/operator training, assessing the risk of injury or fatality, etc., must be considered by the designer.
Quality Control & Inspection	Quality Control and Inspection—is often a manual process. The Inspectors' efficiency and variability (such as Inspectors' visual activity, age and experiences, lighting, visual aids, social pressure, length of no-break-time) should be monitored, controlled and supplemented by modern tools such as on line feedback quality control, etc.
Supplier and Contractors	Suppliers and contractors bring specialized knowledge to the work-groups that can have significant impact on the design of the product or the allied processing systems.
Education and Training	Education and training provides work groups with a basic understanding of the available infrastructure, how to effectively manage the project, including training in the use of hardware and software.

1.8.1.6   Infrastructure Deployment

Flexible infrastructure is the wave of the future. In short, it is not necessary to reinvent the wheel every time a challenge arises. Creating new solutions from existing systems is a more cost-effective and responsive approach. Flexible assembly operations promote the quick changeover of lines or entire plants to accommodate rapidly changing market demands. Infrastructure includes rationalizing hardware/software, CAD/CAM tools, and their deployment for improved communication and efficiency. A cooperative, seamless information and systems environment is a major component of concurrent engineering infrastructure. Table 1.10 gives examples of WHATs in an infrastructure deployment. Table 1.11 gives key HOWs examples.

1.9   CFD TIER-BASED VERTICAL DEPLOYMENT

CFD is accomplished by multi-disciplinary teams employing a series of matrix-based techniques to propagate critical customer wants throughout the product life-cycle. A three-tier vertical (z-axis) cycle of CFD: product planning, process planning and production planning is proposed. This is shown in Figure 1.12. This is different from the American Supplier Institute's four phases that consider Parts' Deployment an additional phase besides the above three. In CFD we do not need to worry about parts deployment since it is now handled concurrently with the Quality FD. The vertical deployment uses a tier-based filtering technique, tier-based because process planning can start even before

TABLE 1.11   Examples of *HOWs* in Infrastructure Deployment

HOWs	**Brief Descriptions**
Project Phases	This involves defining the life-cycle phases of the project, such as definition, design, develop, testing and implementation.

Set the Stage
  Develop Mission
  Finalize Issues
  Develop Strategy
  Long Term Objectives
  Integrate Objectives
  Financial Projections
	Define the rest of the organization elements, including, work-groups, mission, goals and objectives, strategies, and projects including cost projections.

Process Analysis	Conduct a process analysis. This may consist of both an as-is analysis and a to-be.

Develop Detailed Plans
  Manufacturing Plan
  Marketing Plan
  Financial Plan
  Product Development Plan
	A WBS is really a series of interrelated work tasks initially set in motion by the planning track. New tasks are added or created by the subsequent tracks when put into motion.

Supplier Certification
  Supplier Criteria
  Initial Contact
  Plant Visit
  Implementation
  Supplier Appreciation
	Identify the selection process for suppliers and partners including certifications.

Outsourcing
  Outside Tooling
  Vendor
  Purchasing
	This consists of outsourcing of jobs, such as outside tooling, vendor subcontracting and purchasing.

Process Verification	Institutionalize a process for periodical verification and continuous upgrades.
Concurrent Scheduling	Facilitating the transfer of work information among work groups is an essential organizational task of any company. Concurrent resource scheduling involves scheduling the distributed activities so that they can be performed in parallel.
Manufacturing or Assembly	Mobilizing the internal resources so as to get the products manufactured on time and budget.

Strategic Tools and Technologies	Examples of strategic tools and technologies in an infrastructure include:
	  Fourth Generation Languages (4GL)
	  Imaging Systems
	  Enterprise Data access and communications
	  Product and Process Data Exchange
	  Manufacturing Data Collection and Control
	  Engineering Workstations (EWS)
	  Technical Publications
	  Distributed Computing
	  Machines and facilities

product planning is completely finished. Similarly, production planning can overlap with process planning as shown in Figure 1.12. The three tiers are:

Tier I: Product Planning

Tier II: Process Planning

Tier III: Production Planning.

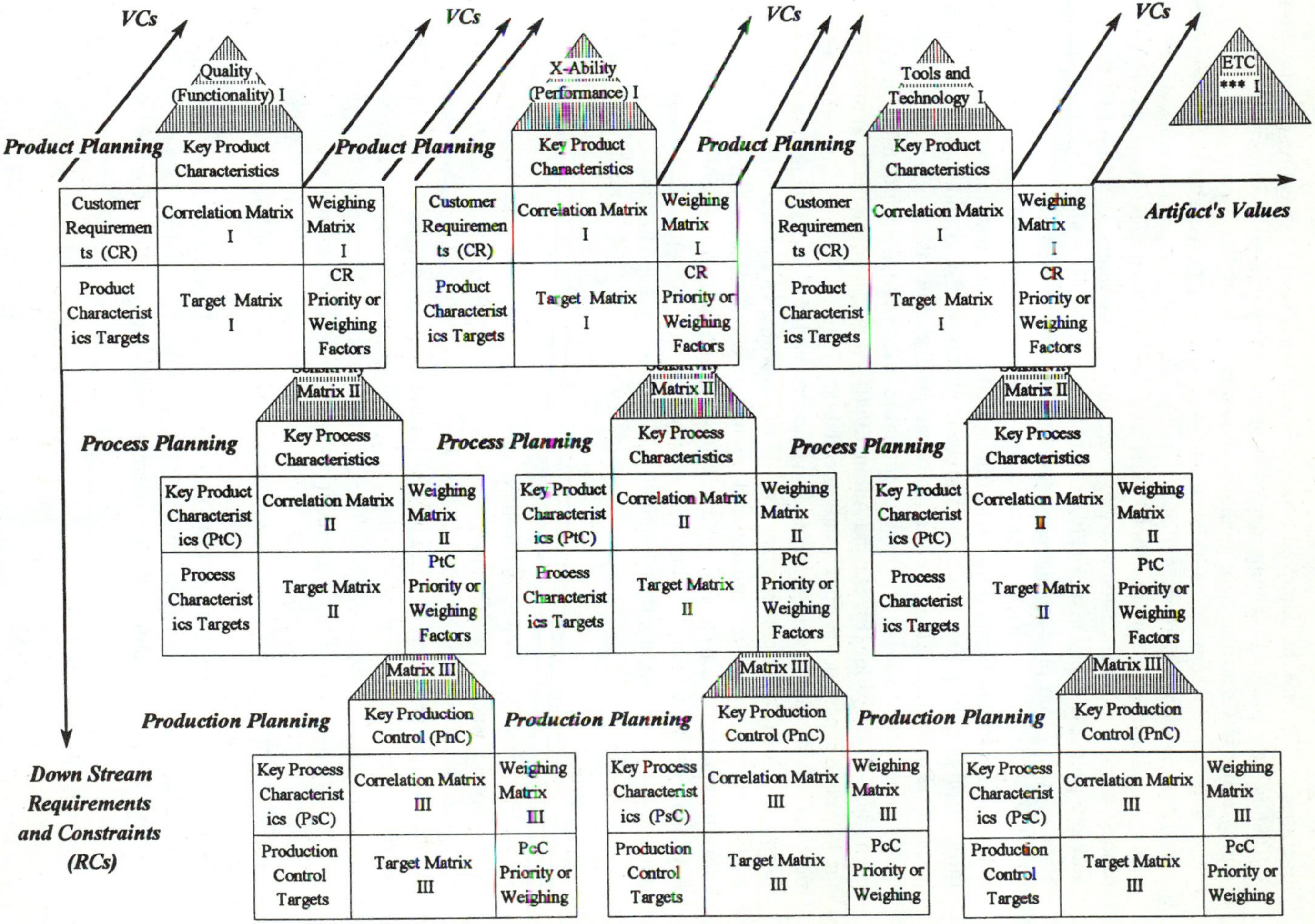

FIGURE 1.12 Linked CFD House of Values: Vertical Deployment (Z-axis)

Each tier can, in turn, be deployed concurrently along a vertical direction (z-axis). Figure 1.10 shows the product planning deployment of quality along the vertical (z-axis). The deployment of RCs attributes of quality is shown in Figure 1.13

This enables teams of multidisciplinary work-groups to focus their planning operations (product planning, process planning and production planning) in concert with each other's needs and expectations. This is shown in Figure 1.14 through an ellipse chart. The process ensures that the customer wants (*voice of the customers*, CRs) are first translated into key product characteristics (PtCs). Initial PtCs serve as product specifications for a planning tier. From product planning, key product characteristics are identified and selected. The chosen key product characteristics are then transferred into the process planning tier as shown in Figure 1.14. Selection, specification, and transfer processes continue through product planning (PtCs) into process planning (PsCs) and through production planning (PnCs) tiers. Along the way PsCs can go directly to the production machine operator or service employee and back again, cross referenced at each translation step with each trade-off decisions [Freeze and Aaron, 1990]. At the end of production planning tier, the key control parameters (PnCs) are selected (see Figure 1.14). The three-tier CFD vertical cycle is repeated as many times as needed providing a traceable path of CFD documentation. Such documentation can provide teams with a vast amount of *interface* knowledge between key product, process and production characteristics. On its basis, one can make trade-off decisions and remove inconsistencies arising from assumptions regarding key characteristic parameters that were selected initially. Key product, process, and production characteristics are, thus, continuously enriched and updated in the cyclic process.

1.9.1   Filtering

As the name implies, filtering is a method used in CFD to narrow down the list of available *HOWs* options (including VCs and AVs). In the beginning tiers, all sorts of *HOWs* (VCs and AVs) that can possibly affect any *WHATs* (RCs) are usually included. A predetermined set of objectives or criteria is initially used as a filtering technique for CFD. This concept is shown schematically in Figure 1.15 through a *funnel* pipeline. Some use criteria, such as *important, new, difficult* and *high risks* [ASQC, 1992] that can be employed for this purpose. The characteristics can also be filtered based upon cost, time, features or most economical use of resources. In a recent paper, a filtering technique based upon a *weighted multi-criterion optimization* formulation was proposed [Prasad, 1993]. Whatever the objectives may be, the end effects of all these filtering and concurrent deployments are to reduce (or to narrow down) the initial list of *HOWs* into a critical set. This critical set of *HOWs* then becomes a new set of *requirements* for the subsequent iteration of the CFD process. This is a recursive process of obtaining the key characteristics (PtCs, PsCs, PnCs) in each tier.

1.9.2   Flow of CFD Process

The CFD is an analytical planning tool for an orderly translation of what the customer wants (CRs) initially into what can be provided through a PD^3 process. The CFD process flows parallel to the PD^3 cycle. It provides a traceable path beginning with customer

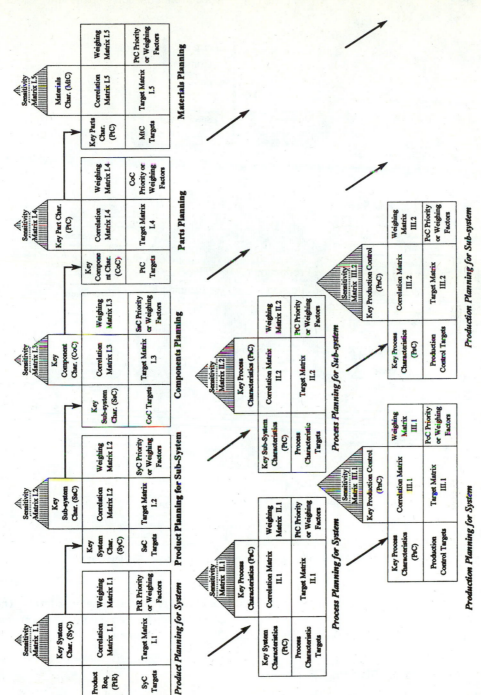

FIGURE 1.13 Linked House of Quality—Deployment of RCs Attributes for Quality along the z-axis

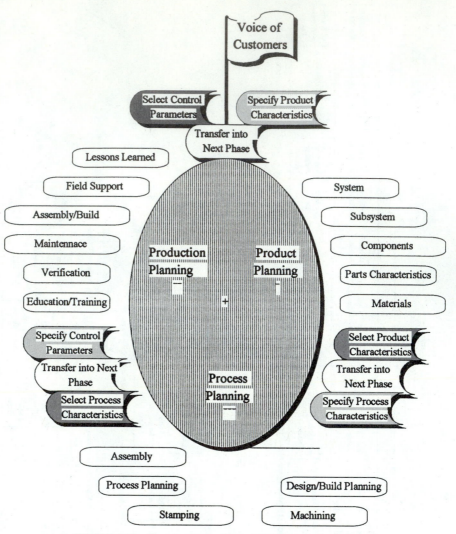

FIGURE 1.14   Planning Operations of CFD (z-axis Deployment)

requirements through each tier of product, process, and production planning levels. The basic approach CFD uses is conceptually similar to the step by step product development approach used in the CE process discussed in Chapter 9 of volume I.

1.9.3   Typical Planning Parameters

A typical set of generic information that is required during vertical deployment (three tiers of a CFD) is defined in Table 1.12.

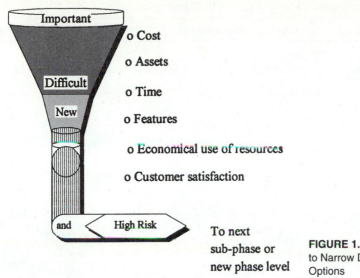

- o Cost
- o Assets
- o Time
- o Features
- o Economical use of resources
- o Customer satisfaction

To next
sub-phase or
new phase level

FIGURE 1.15   Filtering Method in CFD to Narrow Down the List of Available Options

1.10   IMPLEMENTATION ISSUES

Accurate creation of the CFD house of value (HOV) vectors and matrices is not a simple task. The data has to be consistent and should represent real world situations. Implementation requires putting together a team of experts familiar with the products, processes and the competition. Inputs are required from various sources including management council, engineering steering committee, quality control directors, marketing product managers, manufacturing steering committee and strategic business units. It takes a long time to come up with substantiated data that everyone is comfortable with. Each must understand the mission of the product introduction, its goals, strategy, and key success factors. Assumptions and ground rules should be laid down to guide each team member into the same goals and directives.

In this chapter, a framework for deployment is described. The CFD methodology exploits the independence of units that manifest itself in Strategic Business Units, TQM, and Total Enterprise Management concepts that are now emerging. It considers parallel deployment, as contrasted to the phased deployment, for example, American Supplier Institute's QFD concept [Sullivan, 1988], in meeting life cycle values. It is not based on using a single measurement, such as *quality* as in QFD. The present approach is more versatile than Akao's GOAL/QPC concept. In the present setting, ASI's QFD emerges as a specialcase of concurrent function deployment. It enables the Planners and Strategic Decision Makers to deal early with tradeoffs among the crucial factors of artifact values. Six concurrent values, namely functionality (Quality), performance (X-ability), tools and technology (innovation), cost, responsiveness, and infrastructure (delivery) are considered simultaneously rather than serially. Three-dimensional Value Characteristics Matrices (VCM) are employed to ensure that both the company and the customers' goals are optimally met. The key artifact values are

TABLE 1.12   Planning Information Required during Vertical Deployment Tiers

Examples of Product Planning	Examples of Process Planning	Examples of Production Planning
Voice of the Customer: gathering requirements, interview, survey, etc.	Product characteristics: determination of go/no go for design	Process characteristics
Marketing: needs of market, customers, collection of technical data, information	Process audit improvements	Production audit improvements
Planning: product strategy, long term technology development, sales plan, etc.	Network of assurances	JIT; daily production control
Research and Development: develop ideas and off-the-shelf technologies.	Part Classification: group technology	Purchasing/outsourcing
Develop: development target, critical control, technical development plan	Plant Designation: the person or the work group in charge of production	Manufacturing and Assembly
Design: making of part and audit of design	Processing/equipment	Mass production
Analysis: Failure mode and effects analysis (FEMA), fault tree analysis (FTA), etc.	Pre-manufacturing: QA charts, mistake proofing	QC process charts, parts inspections, shipping inspections
Engineering: Engineering the product functions	Process Analysis	Sales
Prototyping: prototype design, prototype order, etc.	Process Design: Process Deployment, control plan, NC, Machining, grinding, pre-production prototype	Claims Handling
Testing: prototype testing, evaluation, etc.	Pre-production evaluation performance test, process capability study	Maintenance and service, delivery and support
Design standards: methods and 3Ps, features, and parameters	Process standards: methods & procedures, process continuous improvements	Production standards: methods & procedures, production continuous improvements

deployed in parallel tracks, stressing their importance and making it less likely to have them ignored by omission.

CFD is a customer driven process for concurrent product design and development. CFD is a system of identifying and prioritizing customers' needs obtained from every available source. By using CFD, engineers can methodically analyze the details of design and process improvement to meet those needs. CFD is a highly versatile engineering tool translating the *Voice of the Customer* through all phases of the product life cycle. In this chapter we have shown that the concept of CFD can be used to develop a world-class product in the least possible duration (elapsed time). The key feature of this approach is that it is based on generally accepted QFD practices, conventions, and nomenclatures. It uses the same set of list-vectors and relationship matrices that were previously defined.

The concept, however, is much more useful and versatile than conventional deployment as in QFD. The trio techniques allow the users to optimize concurrently all sorts of product characteristics under a unified set of *WHATs* that may belong to a number of tracks, such as X-ability, cost, responsiveness, tools and technology, and infrastructure, among others.

REFERENCES

AKAO, Y.A. 1990. *Quality Function Deployment—Integrating Customer Requirements into Product Design*. Cambridge, MA: Productivity Press Inc. Also in *Quality Deployment,* a series of articles edited by Yoji Akao, Japanese Standards Association, translated by Glen Mazur. Methuen, MA: GOAL/QPC.

ASQC (American Society for Quality Control). 1992. Automotive Division, American Supplier Institute (ASI), GOAL/QPC, 1992. *Transactions from the fourth symposium on Quality Function Deployment (QFD)*, June 15–16, Novi, MI.

ASWAD, A. 1989. "Quality Function Deployment: A Tool or a Philosophy." SAE Paper No. 890163. Society of Automotive Engineers, International Congress and Exposition, February 27–March 3, 1989.

CAREY, W.R. 1992. *Tools for Today's Engineer—Strategy for Achieving Engineering Excellence*: Section 1, Quality Function Deployment, SP-913, SAE Paper No. 920040. Proceedings of the SAE International Congress and Exposition, February 24–28, Detroit, MI.

CLAUSING, D. 1994. *Total Quality Development: A Step-by-Step Guide to World-Class Concurrent Engineering*. New York: ASME Press.

CROSBY, P.B. 1979. *Quality is Free: The Art of Making Quality Certain*. New York: McGraw Hill.

DEMING, W.E. 1986. *Out of Crisis*, 2d ed. Cambridge MA: MIT Center for Advanced Engineering Study.

DIKA, R.J. and R.L. BEGLEY. 1991. "Concept Development Through Teamwork—Working for Quality, Cost, Weight and Investment." SAE Paper No. 910212, pp. 1–12. Proceedings of the SAE International Congress and Exposition, February 25–March 1, Detroit, MI: SAE.

FREEZE, D.E. and H.B. AARON. 1990. "Customer Requirements Planning Process CRPII (beyond QFD)." SME Paper No. MS90-03, Mid-America '90 Manufacturing Conference, April 30–May 3, 1990, Detroit, MI: SME.

HALES, R., D. LYMAN and R. NORMAN. 1990. "Quality Function Deployment and the Expanded House of Quality." Technical Report, pp. 1–12. OH: International TechneGroup Inc.

HAUSER, J.R., and D. CLAUSING. 1988. "The House of Quality." *Harvard Business Review*, (May–June) pp. 63–73. Volume 66, No. 3.

KING, B. 1987. *Better Designs in Half the Time—Implementing QFD Quality Function Deployment in America*. Methnen, MA: GOAL/QPC.

KROLL, E. 1992. "Towards Using Cost Estimates to Guide Concurrent Design Processes." PED-Vol. 59, pp. 281–293. *Concurrent Engineering*, ASME, edited by Dutta, Woo, Chandrashekhar, Bailey, and Allen, Proceedings of the Winter Annual Meeting of ASME, November 8–13, 1992, Anaheim, CA: ASME Press.

MIZUNO, and Y. AKAO. 1978. (ed.): Quality Function Deployment, *JUSE* (published in Japanese).

PRASAD, B. 1993. "Product Planning Optimization using Quality Function Deployment." Chapter 5 In *AI in Optimal Design and Manufacturing*, edited by Z. Dong, and series Editor Mo. Jamshidi, Englewood, NJ: Prentice Hall, 117–152.

PRASAD, B., R.S. MORENC, and R.M. RANGAN. 1993. "Information Management for Concurrent Engineering: Research Issues." *Concurrent Engineering: Research & Applications*, Volume I, No. 1 (March 1993).

QFD/CAPTURE. 1990. *Users Manual*. International TechneGroup Inc., Version 2.2.

SULLIVAN, L.P. 1988. "Quality Function Deployment." *Quality Progress*, Volume 21, No. 6 (June).

TAGUCHI, G., and D. CLAUSING. 1990. "Robust Quality." *Harvard Business Review*, Volume 68, No. 1, pp. 65–75, (Jan.–Feb.).

TAGUCHI, S. 1987. "Taguchi Methods and QFD." *Hows and Whys for Management*, American Supplier Institute, Dearborn, MI: A.S.I. Press.

TEST PROBLEMS—CONCURRENT FUNCTION DEPLOYMENT

1.1. In QFD, there are 4 phases that deploy Voice of the Customer (VOC) to get to an improved product. What are the components of QFD? Explain each of the four QFD phases and give examples.

1.2. Besides QFD, Taguchi and Pugh concept selection, can you think of other engineering tools that can help in concurrent engineering? What prevents a team from not using them during an early product introduction stage?

1.3. What makes QFD powerful? How can one use Value Engineering with QFD? Show a flow-chart of the two working together.

1.4. When is QFD not useful? Many of the *HOWs* (so called quality characteristics) in QFD are not all independent. How is this condition handled in QFD?

1.5. When customers want to buy a product, what qualities (values) of a product influence them to purchase it? What is voice of the customer (VOC)? When do they differ from so called customer requirements?

1.6. How can the Kano Model be used to prioritize a set of customer requirements (CRs)? How does a CR shift character? When does that happen?

1.7. What are the rooms of HOQ? Why are Technical Importance Ratings (TIRs) listed under a *HOW-MUCH* list-vector?

1.8. What is the significance of weighting factors in computing TIRs? How can manufacturers use TIRs to prioritize the quality characteristics of a product yet to be launched?

1.9. What are the limitations of deploying a QFD? What is required in optimizing an artifact to be recognized as the best in every class?

1.10. What are the pitfalls of a conventional QFD? Why is *quality* deployment not enough to lead a company to be a world-class product manufacturer or a service provider?

1.11. What makes a great product? Can deployments of many value functions proceed in parallel? What will be their impact on the time-to-market aspect?

1.12. What is CFD? How does it differ from QFD? How does CFD support long-range thinking and better communication across several value functions, work groups, etc.?

1.13. Describe a CFD architecture. How does it differ fom a QFD architecture? What are the three main dimensions of a CFD deployment?

1.14. Why is CFD concept with regard to *quality FD* equivalent to a QFD concept? Explain a trio deployment methodology for CFD and explain how it works.

1.15. CFD results into a three-dimensional house of values (HOV). Describe its major rooms. Where is HOV different from a *house of quality* (HOQ)?

1.16. What benefits does CFD present during quality deployment that are not evident during standard QFD process?

1.17. Show an example of a linked CFD House of Values for quality during vertical deployment (z-axis, as shown in Figure 1.9)

1.18. Show an example of a linked CFD House of Values for quality during axial deployment (y-axis, as shown in Figure 1.10).

1.19. How would you deploy cost functions for design and development of a 100-pin stapler?

1.20. How would you deploy X-ability functions for design and development of a multi-color ball point pen?

1.21. How would you deploy tools and technology functions for a cooperative CE work groups in a CE-based organization

1.22. How would you deploy infrastructure functions for the cooperative CE work groups in a CE-based organization?

1.23. How would CFD look for a service organization? What service values would you consider for a CAD/CAM or CAE service organization?

1.24. Describe the significance of *filtering* in QFD? What other dimensions would filtering take when applied to CFD?

1.25. How would you use CFD in formulating an optimization-based design problem?

1.26. How can CFD be used for company product assessment? Outline two rating schemes for this assessment?

1.27. List five major advantages of CFD that can be used for the life-cycle management of a product. What should manufacturers do to become a world-class product manufacturer?

1.28. How can CFD be used for competitive product assessment? Outline two rating schemes for this assessment?

1.29. What are the implementation issues of CFD?

1.30. If CFD uses the same set of QFD practices, conventions, and nomenclatures, why is it a better concept compared to QFD?

CHAPTER 2

CE METRICS AND MEASURES

2.0 INTRODUCTION

In Concurrent Engineering, although the activities run in parallel, the enrichment of information occurs in a traditional fashion, from product conceptualization to design, to production, and to delivery, in a single thread of continuity (Figure 2.1). Such an information enrichment continuum can be viewed as a collection of overlapped steps made out of many related cycles, including product and process loops (1-T, 2-T, or 3-T). The loops actually run in parallel, as discussed in Chapters 8 and 9 of Volume I. Individual processes, activities, and steps in such realization loops can be measured, managed, and improved. If a team member cannot measure what he or she is talking about and express it in quantitative or qualitative terms, the team knows nothing about it. One cannot impact what one cannot measure. However, if team members can measure it and would be able to express it in numbers or in sets, they can improve it. Measurements are not new to design engineering. Traditionally, to ascertain confidence at an early stage, designers are accustomed to physical aids such as hardware prototypes, wood/clay models, conceptual models, model making, mock-up, and so on. These physical aids measure the compliance with respect to the stated specifications. Furthermore, in traditional systems, designers have used documentation (engineering drawings, sketches, prints) to manage the traditional process, and are quite familiar with the process of design review to improve its functionality. If design changes were necessary, the annotated design was returned to the drawing board and the process of measurement and improvement was repeated. There were rules of thumb that the designers had, over time, become accustomed to using while adding or selecting a design option or a feature. Today, drawing has been, or is, in the process of being replaced by a 3-D CAD system to manage the product design, development, and delivery (PD^3)

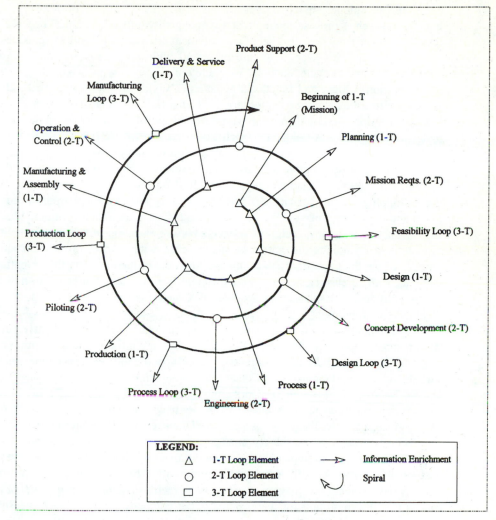

FIGURE 2.1   Information Enrichment Spiral

process. Most ad hoc metrics, known then as a multitude of good design practices, are formalized today as design for X-ability (DFX, such as design for manufacturability, assemblability, maintainability). X-ability is a generic reference to a life-cycle measure or concern (e.g., ease, economy, flexibility, efficiency, effectiveness) for a product. In computer software products, the term X-ability refers to things like usability, portability, scalability, interoperability, stability, and so on.

Selecting and satisfying a set of ad hoc practices and DFX methodology prevents a company from reinventing the wheel. It makes sure that the proposed design contains what is considered a better set of 3Ps (policies, practices, and procedures) in each domain.

Most DFX techniques are based on heuristics (e.g., rules of thumb), or some type of scoring method such as process coding, cost drivers, quality indexes, technical importance ratings, and so on. They provide a set of measurement criteria for judging the goodness of an artifact's design. However, if we are presented with a *virtual system* with a shallow knowledge of its life-cycle functions, we could only begin to speculate on the resulting artifact, its morphology, and functional behaviors [Nevins and Whitney, 1989]. Given all DFX specifications, each of the technology teams (DFX specialist) in a particular field (e.g., a DFX domain) could come up with different speculative responses based upon their understanding and their fields of specialization. Each response may represent an acceptable design in its own right, meeting all of the corresponding basic DFX specifications. A synthesis of these speculative responses may yield some clues to a complete *real* artifact system. This resembles a story of seven blind-folded persons, who were asked to examine the morphology of an elephant. Six were allowed to touch and feel a portion of an elephant and the seventh person, the whole elephant. Based upon their own individual experiences and observations, each person gave a correct narration of their findings, but individually, their findings did not make much sense with respect to the elephant as an animal. Given the opportunity to examine all aspects of the elephant, the seventh person came up with an answer that was quite different from the rest, but closer to how an elephant really looks. A measure on a single DFX metric is like examining a portion of the elephant's morphology.

CE requires that all life-cycle DFX issues be considered simultaneously during the design stage. Inevitably, this process generates design conflicts among the multiple life-cycle (DFX) issues. Individual assurances of DFX specifications (one at a time) do not capture the most important aspect of Concurrent Engineering—the system perspective or the trade-off across the different DFX principles [Shina, 1991]. While satisfying these DFX principles in such manner, only those that are not in conflict are occasionally met. Concurrent engineering views the design as a part of PD^3 and evaluates the artifact as a system that has a wider impact than just sub-optimizing each sub-system within its own domain.

Consequently, an approach to adequately measure these conflicts and a resolution methodology for making high-level design-trade-off between the issues is required. For product development this may mean establishing metrics and measuring scores of product values that are important for the customers, the company, or both. Such measures can focus internally on internal customers, or externally on external customer requirements. In addition to metrics and measures, one can establish corresponding goals or targets for attaining these product values. Monitoring and tracking progress against established targets results in identification of product realization gaps. The gaps, in this context, are used here to signify the difference between the actual and the desired product features. Gaps can also result from the existence of conflicts or conflicting specifications. There can be many measurement gaps during product realization when two or more of the processes are carried out in parallel, or when their decomposed tasks are overlapped (see Figure 2.2). Metrics are measures that indicate (in a relative or an absolute sense) when results make sense in terms of assessments and evaluations. Concurrent Engineering needs both a series of measurement criteria that are distinct to each process and a set of metrics to check and validate the outcome when two or more of the processes are overlapped or re-

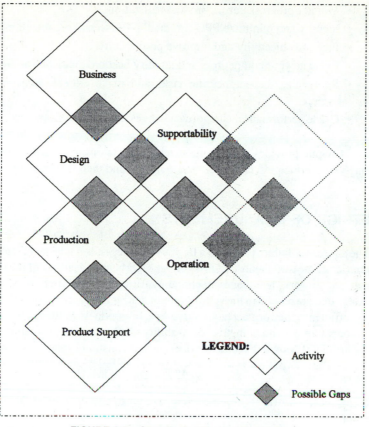

FIGURE 2.2   Gaps During Product Realization

quired to be executed in parallel [Shina, 1991]. Since product realization transformation involves concurrent processes that occur across multiple disciplines and organizations, appropriate metrics and the methods of qualifying them vary considerably. Choice of the appropriate metrics depends on the availability of data, its incompleteness, degree of overlap, ambiguity, and so on. Metrics change with time as new data or taxonomy pictures emerge. In addition to their use in the product realization situations, effective metrics do the following:

- Identify process bottlenecks and eliminate root causes of defects.
- Serve as a management tool for assessing and evaluating performance, and efficiency.
- Help teams understand engineering processes better.
- Determine when and where to apply the 7Ts (talents, tasks, teamwork, techniques, technology, time, and tools).

- Monitor progress during product realization.
- Identify and minimize PPO (product, process, and organization) complexity.
- Increase objectivity and improve productivity.
- Evaluate your competitors and identify best product features and practices.
- Reorganize the engineering tasks and make critical decisions earlier in the life cycle.
- Grade performance, categorize changes, and move toward trade-off or optimization.

This Chapter provides insight into mechanisms to ensure effective trade-off across different system principles, customer requirements, and their inclusion in the PD^3 process.

2.1 METRICS OF MEASUREMENTS

The success or failure of CE, to a large extent, depends on the team's ability to define useful metrics of measurements (MOMs). Most MOMs include many of the so called 7Ts (talents, tasks, teamwork, techniques, technology, time, and tools) characteristics. They measure things that are related to the state of completeness of specifications, transformation feasibility, efficiency, performance, effectiveness, or goodness (a fitness function) of outputs. If Σ denotes a union-sum of metrics of X-ability measurements, its magnitude will equal 1 when the artifact is complete (content-wise) and the constraint space is empty.

$$\text{If } \Sigma \equiv \cup \, [\{\text{MOMs}\}] \tag{2.1}$$

then $\Sigma = 1$, if artifact is complete (content-wise) and the corresponding constraint space is empty.

In equation 2.1, it is assumed that each MOM is a normalized set.

The scope of the transformation covering product and process completeness includes the following ranges.

- **Range of Specification:** As the product and process design reach maturity, specifications to the realization model, or the information contained in the artifact's definitions, will be less and less abstract. At the beginning of the realization (stage 1), the design exists in pure specification form and the artifact has no content (see Figure 8.9, volume I).

$$\text{Stage 1:} \quad \{S_1\} \Rightarrow \{\bullet\} \text{ a Full State} \tag{2.2}$$

$$\{O_1\} \Rightarrow \{\varnothing\} \text{ a Null or an Empty Content} \tag{2.3}$$

where S denotes the specification set and O indicates the outputs. As various tasks within this taxonomy are performed, designs (meaning product and/or process designs) begin to take shape. The set of specifications changes over time. At the end of the nth stage (when estimated time is up) all specifications have been implemented.

$$\text{Stage } n: \quad \{S_{n+1}\} \Rightarrow \{\varnothing\} \text{ or Null or an Empty State} \tag{2.4}$$

$$\{O_n\} \Rightarrow \{\bullet\} \text{ a Full Content} \tag{2.5}$$

At the conclusion of product realization, the set of specifications becomes null. All the specifications (inputs, requirements, and constraints) transform into an output form that is an artifact (see Figure 2.3). At that point, the product and process designs are at the *full content* and specifications are at the *null state*. The CE metrics and measures block in Figure 2.3 is what provides the feedback to the realization model.

- ***Range of Assessments:*** The product realization process is not complete if certain types of product and process design assessments are not carried out or their results are not satisfactory [McNeill, Bridenstine, Hirleman and Davis, 1989]. An assessment may exist in qualitative (as design guidelines) or in quantitative manner. Quantitative measures provide a degree of objectivity in the range of assessments. This may also include the existence of certain types of information that are essential for manufacturing, customer satisfaction, or company profitability. The range of assessments for mechanical components may include:

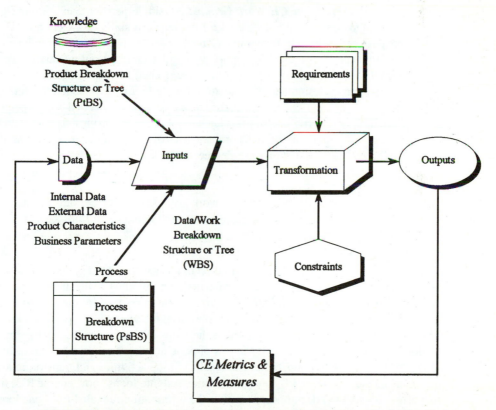

FIGURE 2.3   Showing CE Metrics and Measures as a Feedback to a Realization Model

- Performance (basic geometry design, functionality, performance design, component's design).
- System Assembly (assembly modeling, DFA/DFM assembly design [Boothroyd and Dewhurst, 1988b]).
- Manufacturing Precision or Quality (detailed dimensions, roundness, eccentricity, surface finish, texture, quality control (QC), material and process selection, and tolerances).
- Robustness (insensitivity to manufacturing, material, and operational variations).
- Ownership Quality (ergonomics, reliability, diagnosability, testability, and serviceability).
- Product Retirement (disassembly, reuse, recycling).
- Logistics (purchasing, inventory, international use, environmental standpoint, lead time, cost-drivers) and so on.

$$\text{Range of Assessments} = U \text{ [Performance, System Assembly,} \\ \text{Manufacturing Precision or Quality, Robustness, Ownership Quality,} \quad (2.6) \\ \text{Product Retirement, Logistics, etc.]}$$

- ***Range of Appropriate CE Work Groups:*** No single person, team, or work group is expected to comprehend or possess all the required skills of a CE trade. The products designed today are too complex. In order for the product output to be complete, the right expertise is required to provide inputs so that the outputs are balanced. For example, experts in design work groups are needed for configuration design activities. Experts in design and manufacturing work groups are needed for performance type activities, and experts in quality work groups are needed for inspection and prevention type activities. Appropriateness of CE work groups for each level of abstraction is important for the completeness of the product and process designs. Qualitative measures provide a degree of subjectivity in coalescing the range of assessments.

$$\text{Range of Appropriate CE work groups} = U \text{ [Technology Teams,} \\ \text{Virtual Teams, Logical Teams, and Personnel Teams]} \quad (2.7)$$

In other words, there is a need for a unified set of CE metrics and a system level methodology to aid in the process of product realization. The product realization taxonomy, discussed in Chapter 9, volume I, listed some of the measures (see Figure 9.5) to use during the realization process, but their actual descriptions were omitted. It is assumed that the appropriate metrics and measures are used in the loops of the product realization process as feedbacks as shown in Figure 2.3. Metrics provide answers to a broad range of questions related to the formulation, operation and results of loop activities and transformation. Such questions include: how is a PD^3 process decomposed into loops? how are the RCs (requirements and constraints) decomposed? how is the information funneled from one loop to the other? how are RCs satisfied within a loop itself (inter-loop realization)? how are some

RCs satisfied across the loops (intra-loop realization)? A metric within a loop is aimed at describing one or more of the following steps:

- Identify the initial state of the design.
- Measure the initial state of the RCs in terms of efficiency, effectiveness, etc.
- Control the process, i.e., establish timings of the various iterative steps applied to the initial design state.
- Measure the state of the modified design.
- Compare the performance of the various processes with respect to the chosen norms.
- Minimize effects of the external uncontrolled processes.

Each step may deal with one or more elements of 5W1H (i.e., what, when, why, who, where, and how). Such metrics must comprise several life-cycle perspectives, each representing a supplement or an add-on to this methodology. Each must contribute to the overall effectiveness of the methodology.

2.2   ESTABLISHING LIFE-CYCLE MEASURES

At the heart of any good PD^3 process, there lies the CE focus on satisfying the interests of both the customers and the company. The customer focus shows up in measures, such as market research targets, performance, field or warrantee measures that a company imposes in response to what customers desire in a product. The company focus shows up in measures, such as built-in prevention measures by design, on-line process measures, inspection measures, diagnostic measures. This assesses the company's ability to manufacture a quality product in record time and cost. Such measures are called life-cycle measures. Life-cycle measures generally fall into 7 categories (Figure 2.4).

- *Market Research Targets:* These determine the extent to which customer satisfaction prevails in product development. This is commonly listed in the *WHATs* column of the QFD matrices. Examples of market research targets are strategic planning, product plans, organizational goals, meeting goals, objectives, and so on.
- *Built-in Prevention Measures (by Design):* These are measures that are factored-in when the parts were initially conceived to prevent any future mishaps. Examples of built-in measures are error-proofing, design for consistency, design for insensitivity to parameter variations, and design for reliability, and so on.
- *On-line Process Measures:* These are metrics that determine the cause of a process malfunction, such as deterioration of product or process area quality, machine failures, and so on. Metrics are internally focused.
- *Diagnostic Measures:* These are metrics that ascertain why a product or process is failing to perform as expected. Diagnostic measures determine which features of the structure part, or of the design prototype, are the causes of failures and are introduc-

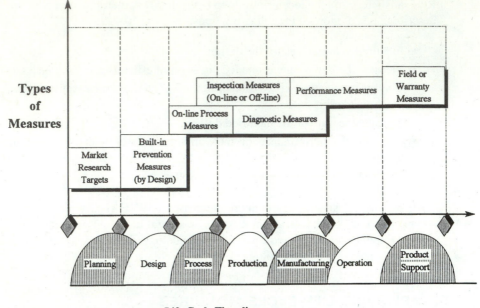

FIGURE 2.4   Common Life-cycle Measures

ing out of norm behavior. In the product area, diagnostic measures might include test results, MTBF (mean time between failures) analysis, FMEA (failure mode and effect analysis), reliability checks, quality indices, and so on.

• *Inspection Measures (on-line or off-line):* Inspection measures are less desirable because they commonly deal with fixing a problem. They do not eliminate the cause of the problems or detect and eliminate the source of the problems. Because of these reasons, inspection measures are sparingly used in aggressive corporations.

• *Performance Measures:* Performance measures are high level metrics that assess the overall performance of product, process, the team, or the enterprise. Performance measures are generally associated with product performance in the field, or in customer use of the products compared to their competitors. These measures are customer focused and are externally based. Examples include productivity, responsiveness, cost, time-to-market, quality content, and so on. Eight performance indicators were mentioned in section 1.6 (Figure 1.11). An indicator represents a combined outcome of doing two major efforts in a company: *doing things right* and *doing the right things. Doing things right* is measured by the corresponding efficiency of doing 7Ts, 3Ps, 4Ms, or 7Cs. *Doing the right things* is measured by the corresponding effectiveness of doing 7Ts, 3Ps, 4Ms, or 7Cs. The desired result is the product of the two lists. The items in each of the two list categories and the desired result list are outlined in Table 2.1. The order of items in the lists is immaterial.

TABLE 2.1   Measuring A Performance Indicator

Desired (Examples) Result	=	Doing Things Right	×	Doing the Right Things
Fewer unscheduled changes		Measured by the corresponding		Measured by the corresponding
More overall productivity		efficiency of doing the		effectiveness of doing the
Less time-to-market		following:		following:
Less cost-to-quality		Integrated Product Development		Total Value Management
More profitability		(IPD)		Concurrent Function Deployment
Less inventory		Integrated Product and Process		(CFD)
Better quality		Organization (PPO)		QFD, TQM, C4, etc.
Great product		7Ts		7Ts
Increased safety		3Ps		3Ps
Increased stability		4Ms		4Ms
Increased flexibility		7Cs		7Cs
Increased market growth		*		*
More customer satisfaction		*		*
		*		*

- *Field or Warranty Measures:* These are metrics that assess the field use of the product in terms of its maintenance, upkeep, and warrantee costs. Most measures are customer focused. Examples include customer-found faults, maintenance costs, customer satisfaction index, and so on.

Some of the above measures are required for an organization to become lean, while others are to become agile. Metrics for leanness do not imply agility. They are simply a necessary condition. Organization needs a lot of lean capabilities to become agile. Both lean and agile are measurable, but leanness, in particular, is more observable. You can visualize a just-in-time (JIT) system by looking at work-in-progress (WIP) inventory, floor space, or cycle time. Agility is not directly observable (in real time) because it represents flexibility or the ability to change. An analogy could be distance/velocity. Leanness may be analogous to distance. When someone traverses a distance, its path can be observed. Speed or velocity cannot be observed, but is the rate of change of distance. Agility may be considered the change in rate in moving from one lean state (a distance) to another. This gives the sense of direction. Agility provides a measure of *dynamics*, how fast the change can take place (that is distance traversed) and *which direction to traverse*. Because agility cannot be easily observed (compared to leanness), it is a difficult concept to measure and for the CE management to grasp. Leanness, on the other hand, is easy to understand because it deals with eliminating wastes, and can be measured and observed using some lean metrics.

2.3   VALUE CHARACTERISTIC METRICS (VCM)

The first step in CE is to develop predictors (metrics for the object systems) and the supporting analysis for assessing product and process behaviors. Types of analysis, tools or concepts required to fill the product realization gaps are contained in Figure 2.5. They are

Requirements	Identify	Analyze	Plan	Evaluate	Performance-to-plan
Business	Competitive Assessment, VOC	Market Research	Cost-to-quality	QFD	Organizational Performance
Design	Production Definition Problem Decomposition Multi-disciplinary System Specifications TQM	Robust Design	Concept Generation Cost-to-design	Product Planning Concept Development DFMA Interchangeability	Statistics
Supportability	Specification History	2nd Level Analysis Parametric Optimization	Process Planning Cost-to-supportability FMEA/DFMA FTA	Taguchi's Method Design of Experiments, RMS	Product's Agility Modular Design Reliability
Production	Tolerances	Variational Analysis (VSA) Simulation	Error-proofing Cost-to-error-proofing Mistake-proofing	FMECA	Statistics
Operation	Gathering Data "As-is" Data Flow		Corrective Actions RCM, SPC "To-be" Data Flow	LORA	SPC SQC QC
Decision Support	Economic Analysis	Trade-off Analysis	Cost/Benefit Verification	Cost/Benefit Tracing Tools or Concepts	Cost Benefits Monitoring

FIGURE 2.5   Types of Analysis Tools or Concepts Required

categorized according to the level of analysis details required: identify, analyze, plan, evaluate, and performance-to-plan. Many of them are off-the-shelf tools, which a company can buy and integrate. Some are product specific, others are process specific. The required analysis tools are categorized in accordance with the needs and purpose—where during the PD^3 process such tools are used and the purpose of using them. The six needs identified during a PD^3 process are: business, design, supportability, production, operation, and decision support [Gladman, 1969]. The purposes of using the tools have been categorized in accordance with the types of actions taken—to identify, to analyze, to plan, to evaluate, to validate, or check performance-to-plan. Four types of CE metrics and measures are contained in Figure 2.6. They are arranged in four file drawers of a file cabinet. Product development teams (PDTs) can draw upon these metrics to influence the PD^3 process. Metrics and measures are categorized into four groups. For example, design for X-ability assessment metrics, such as design for manufacturability (DFM) [Boothroyd, 1988] or design for assembly (DFA) [Boothroyd and Dewhurst, 1988b], design for flexibility can be effective in reducing the number of parts or processes. Metrics for modular design for sub-assembly, design for interchangeability, and design for flexibility can be effective for reducing cost. Product feasibility and quality assessment matrices are used to furnish voice of customers into products,

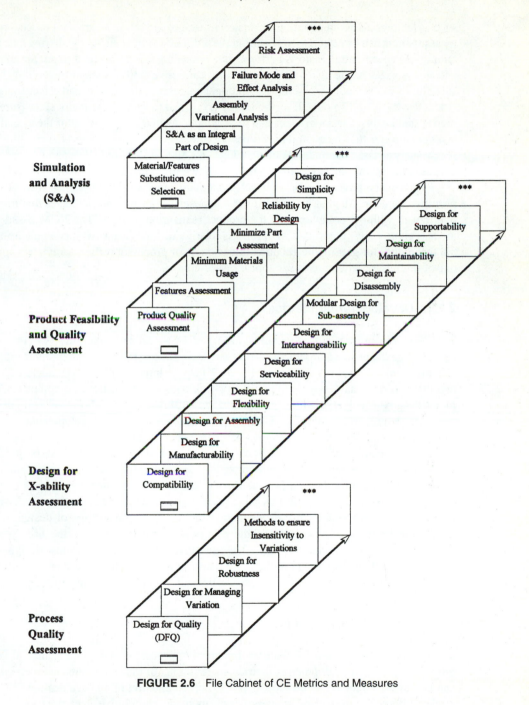

FIGURE 2.6   File Cabinet of CE Metrics and Measures

such as features assessment, minimum materials usage, and so on. Metrics for process quality assessment can be effective for ensuring the product's agility, such as gathering data pertaining to a specification history, performance, precision, tolerances, and so on. Simulation
and analysis (S&A) are also MOMs for driving corrective action, such as material features
substitutions or selections, assembly variational analysis, failure mode and effect analysis,
risk assessment, and so on. CE methodology is defined to keep these metrics, measures of
merits, and analyses tasks in focus and to provide a desired output. Most of these analysis
tasks or concepts that are quantitative types are contained in the file cabinet (see Figure 2.6).
A few analysis tasks or concepts not included in the file cabinet could be a part of a general-
purpose conceptual library (similar to what is shown in Figure 2.5).

A primary advantage of value characteristic matrices VCM, rather than ad hoc
primitive modeling, is that it formalizes and exposes errors and inefficiencies that may be
overlooked with the complexity of the product realization process. The VCM based PD^3
process continually monitors CAD progress relative to specifications. It contains influen-
tial techniques for getting the attention of designers or processors when parameters appear
out-of-bounds, or when processes appear out-of-control.

2.3.1 Developmental Metrics

The development of VCM—(contained in Figure 2.6)—depends upon the 3Ps, 4Ms, 7Ts
and 7Cs prevalent in an enterprise. However, their successful use requires their integration into computer-based models (see Figure 7.12 of volume I). The two (models + VCM)
together can serve as measures of merits in checking a variety of DFX compliance. Compliance can be checked for robust design, design optimization, collaborative work, design
for manufacture, and design for assembly, to name a few. These developmental metrics
can be used for risk reduction, allowing new product concepts to be investigated earlier in
their design cycle by all members of an integrated PDT. The clear advantage of developing MOMs, such as metrics, computer models or simulations, is that changes or improvements to the total product and process design can be made earlier, when costs of such
changes are less. X-ability measures can be extracted and captured as a part of such simulation models. This enables use of simulation models at an early stage of design, rather
than being forced to use them to handle problems discovered later in the life cycle or
when the design is relatively set. The question is how the various designs for
X-abilities can be incorporated in the PD^3 process at an early life-cycle stage without im-
pairing important features, functions, ease, efficiency, flexibility, or feasibility.

2.3.2 Implementation Metrics

An implementation of new CE features, functions or processes in an organization are evolutionary. Like the developmental metrics, there is a set of measurements and metrics that
can be devised to measure the success of CE implementation in an organization. This includes criteria for measuring success based upon the cumulative lead time reductions.
One of the discrepancies obviously is in the scale of measurement that is changing with

time. The percentage of the efforts involved in various stages of the product life-cycle may not be a good criterion for comparison either. This is because in CE, the stages are often restructured, re-engineered, and/or redesigned. Hence, there remains very little, if any, similarity between a redesigned and an old process and their breakdowns. This is discussed in volume I (Chapter 3, Process Re-engineering). The implementation matrices are contained in Chapter 3, Total Value Management.

2.4   SIMULATIONS AND ANALYSES

Complex product design teams increasingly use simulations and/or analyses as an integral part of their PD^3 process. A simulation, or analysis, is becoming more and more common in all walks of product development, including testing. Instead of the physical testing of parts, test cases are simulated via derivative simulation to avoid the cost of re-testing parts in a similar situation. Predictive simulation is used to predict parts' behavior more quickly than conducting a series of tests or actual experiments. Enterprise modeling and simulation are common techniques to monitor operations within an enterprise and to identify conditions that could produce defects. Simulation has also been found useful for comparing alternative designs, organizational structures, facility layouts, and cell configurations. Traditionally, an analysis or simulation tackles a multi-physics problem with a single physics approximation. The physical domain of a problem is first segregated and then simplified. Many models, such as fluids, solids, thermal, structural, are obtained for each physical phenomenon. These models are then used independently in studying the effects of an isolated physical presence on product components. With the advancement of computer power and interactive graphics, it is now becoming possible to consider interactions between these models. It is also possible to predict how a product will actually perform under a multi-physics (closer to a real world) situation.

The ability to simulate multiple interacting physics makes it possible to analyze a new range of problem types not previously envisioned. Such problems belong to three main groups: interactive, coupled, and concurrent.

- *Interactive Problems:* Such problems develop between two or more parts when their corresponding geometric regions interact with one another at their boundaries through the transfer of physical effects, such as motion or force. Earthquake or wind effects on civil structures, such as buildings, bridges, and so on, the aeroelastic behavior of aircraft wings, and nuclear reactors are representative examples of interactive problems.

- *Coupled Problems:* They arise when a product behavior is independently linked to two or more physical phenomena and a single set of equations (models) is not enough to predict its outcome. For instance, the aerodynamic shape of a wing is determined simultaneously by the wing's structural characteristics and the fluid flow characteristics around the wings. The shape of the wing initially determines the flow, but the flow creates pressure on the wings that affect its shape.

- *Concurrent Problems:* Such problems arise when behavior is characterized by multiple but independently acting physical phenomena. Each subset of a problem can be solved independently and the results are brought back to satisfy a common set of objectives. For instance, in the case of an automobile, the teams can simultaneously study engine efficiency, interior room, and body style, if these can be carried out as independent activities.

Problems = $\exists$ [Interactive Problems, Coupled Problems, Concurrent Problems] (2.8)

2.4.1   Analysis or Simulation as an Integral Part of Design

In Chapter 9, volume I, a loop methodology was presented as a part of a formal taxonomy for the product realization process. Each loop in the taxonomy included a set of analyses that provided some quantitative information. Analysis served as a predictor for the assessment of goodness of a product or a process behavior at different points along its realization process. There were a number of analysis procedures available to be used during the design and engineering loops of the PD^3 process. However, there were only a few methods available for the beginning loops, such as planning or feasibility loops of product development. The methods applicable during the production domain (process planning, production, and manufacturing loops) were even smaller. In many cases, the appropriate models of the products or design-to-analysis interfaces were missing or lacking. Figure 2.7 lists a pallet of activities required to be accomplished for the product management cycle. These represent product related metrics and measures. Many of these models may have an inherent interdependency and thus require common interface models to be mutually shared. The typical measures during a product management sub-domain are design functionality, design assessment, reliability, producibility, cost assessment, risk assessment, compatibility, and other design for X-ability considerations. Likewise, the typical activities in a process management sub-domain are: process planning, manufacturability, testability, assembly assessment, investment/tooling/cost assessment, serviceability, maintainability, recyclability assessment, and other design for X-ability considerations. The process sub-domain activities are shown in Figure 2.8. They represent process related metrics and measures. Determination of optimum system design, based on both product and process sub-domains, requires a series of interconnected model abstractions. The series of abstractions guide the design from one sub-domain to the other. As CE workgroups move from one sub-domain to the other, new parameters are added and old parameter values are updated. The *corrector* portion of each loop uses the analysis information (from the predictor) to optimize the resulting system (see Figure 9.21).

2.4.2   Materials/Features Substitution or Selection

Materials and features are often the common elements enumerating parts in a product realization process. They appear as key breakdown elements in a class of physical decomposition (see Figures 6.3 and 6.4 in Chapter 6 of volume I). Today, engineers often rely

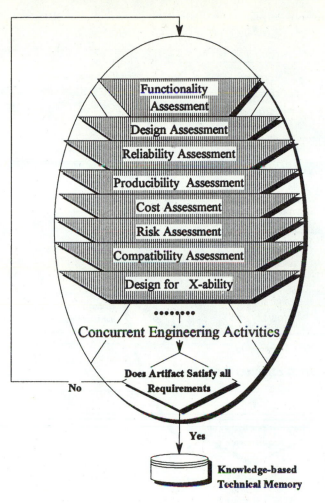

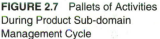

FIGURE 2.7 Pallets of Activities During Product Sub-domain Management Cycle

on comfort zone, or familiar materials, or features. Yet, when design requirements exceed the limits of such materials or exceed the limits on material properties or functions of such features, teams must consider alternative material or feature substitution. With direct on-line access to material or feature information (such as material database or a features' library) the teams could optimize material or feature selections to design parts that are lighter, stronger, and lower in cost. Assuming the impact of such substitutions can be analyzed or simulated, the teams could easily do the following:

- Make an optimum choice of features for a desired product functionality, for example, snap-fitted rather than riveted components.
- Make an optimum choice of materials for the processes at hand, for example, cold-headed and finish-machined rather than machined from bar stock components.
- Conserve materials for each process and minimize material waste.

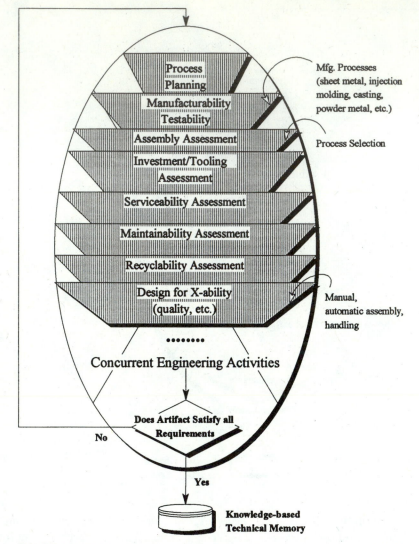

FIGURE 2.8   Pallets of Activities During Process Sub-domain Management Cycle

2.4.3 Assembly Variational Analysis

It is a myth that if a part is manufactured within a prescribed specification (tolerance limits) it will turn out to be a good assembly at the end. Though it is a good practice to specify a tolerance range for the parts, tolerance stack problems obviously will occur as parts are assembled into components, components into sub-systems, and subsystems into a system. In some cases, stacked tolerances may cancel out, while, in others, they may be built-up (cumulative). The greater the degree of complexity in the assembled system, the greater is the potential that the parts will be difficult to assemble. The resulting assembled system may fail to meet system specifications, even though each individual part is *good*, meaning each part falls within an acceptable tolerance range.

In most design cases, Variation Simulation Analysis (VSA) deals with variations of part dimensions. Besides part variations, the VSA method employs simulation of assembly process variations to determine if design intent can be met within a range of prescribed tolerances. VSA can identify the cause of variations and obtain ranking. This technique can also help teams choose the most cost-effective changes to the product or the process dimensions to reduce total variability.

2.4.4   Failure Mode and Effect Analysis (FMEA)

Failure Modes and Effects Analysis (FMEA), also known as Design Failure Mode Analysis (DFMA) or Fault Tree Analysis (FTA), is a common technique for the analyses of design functions, potential failure modes, and effects of failure. Together they are referred to as Failure Modes, Effect and Criticality Analysis (FMECA). FMEA constitutes a cornerstone for reliability, maintainability, and supportability (RMS) analysis efforts. FMECA is commonly used to challenge new design alternatives, to expose their weaknesses, and to find potential design problems. The basic intentions of FMECA are to *design-out* the failure modes whenever and wherever feasible and as early as possible and *institutionalize* corrective measures (such as error-proofing) when failure modes cannot be removed. The maintenance requirement is accomplished through Reliability Centered Maintenance (RCM) and Level of Repair Analysis (LORA). FMEA is ordinarily considered a reliability tool. Broadly, it can be used to identify what could go wrong with the design. The mode of failure is first ranked from the most to the least impact on part functions and then addressed one by one during an FMEA redesign in that order. Corrective steps can be established analytically to reduce failures while the design is still conceptual. The basic method applies to both products and processes.

2.4.5   Risk Assessment

The risks and rewards associated with various aspects of product development are shown in Figure 2.13 of volume I. Detecting failures before a part is physically produced or before parts are put in field use (such as before assembly) can be very difficult. Assessment of risks involves assessing the difficulty level in detecting a failure and its impact on design. Impact of failure can be severe if it involves safety considerations or involves human injury. Failure in a part can have a domino effect in triggering malfunction in other parts. Integration of probabilistic reliability assessment methods with input data requirements, process controls, life-cycle cost, confidence, and robustness can provide some measure of the level of risks. Refer to section 2.2.1 of volume I for a detailed discussion of risk management. Let us denote:

I = the index for measuring an impact of a failure,

D = the difficulty level in detecting a failure,

P = the probability of failure of the part itself.

Further, we can associate a value (in the range of 0 to 1) with D, I and P.

$D = 0.1$ (or $I = 0.1$) indicates an easy to detect failure (or a minor overall impact of a failure).

$D = 1$ (or $I = 1$) indicates a very difficult to detect failure (or a major overall impact of failure).

P is usually rated in between 0 and 1 as well.

Then, a critical or risk priority index (RPI) can be expressed as:

$$RPI = P \times I \times D \tag{2.9}$$

2.5   PRODUCT FEASIBILITY AND QUALITY ASSESSMENT

There are many product feasibility and quality assessment tools (such as quality function deployment (QFD)) [Hauser and Clausing, 1988; Prasad, 1995], Pugh's Selection Method [Pugh, 1991], Taguchi's Method [Taguchi, 1986] that are useful for planning the design of products or services. Product planning is based on the understanding of the voice of the customer, the competitors, and the product itself to determine a set of desired product characteristics. QFD provides a set of matrix-based techniques to quantify the organizational capabilities and identify quality characteristics (QCs) that would meet customer expectations and needs [Hauser and Clausing, 1988]. The concept selection matrix initially proposed by Pugh is another matrix-based approach to quantify and measure product QCs. Like QFD, Pugh's selection method is based on a list of product and customer requirements. They are discussed in section 2.8.3 of volume I. The steps in Pugh's methodology are shown in Figure 2.25 of Chapter 2, volume I.

2.5.1   Product Quality Assessment

Quality assessment is a measure of product and process conformity to requirements. Merely acquiring a CAD tool or blindly designing everything on a CAD system does not assure quality. Quality must be built into the product through compliance with requirement specifications at the beginning stages of design rather than through inspection of design during later stages [Hays, 1992]. The aims of quality improvement programs is to look for ways to make a product better, more reliable, and durable.

Prevention means keeping quality problems from happening. Prevention is a process of finding, proofing, and tracking possible error situations that may adversely affect quality or customer satisfaction, and could result in waste. There are three ways to approach prevention.

- The *first* is to concentrate on the design of the product itself (whether it is a hard product or a service).
- The *second* is to work on the production process [Gladman, 1969].

- The *third* is to concentrate on error proofing (EP), also referred to as fool-proofing or mistake proofing. Error proofing employs techniques or devices for early detection and prevention. Some major typical steps in error-proofing are
  - Identify and classify each operation or process for error proofing needs and techniques.
  - Use monitoring techniques or devices that detect errors during work in progress rather than when the product or process is complete.
  - A priori design parts that cannot be incorrectly manufactured or installed.
  - If an error is detected, change the process, document it, and make sure (meaning implanting EP methods) that this error will not occur again.
  - Find and eliminate error situations, such as minimizing waste (Japanese term—Muda), reducing overburden (Muri), controlling unevenness (Mura), and reducing possible variation. If error situations can be prevented, defects that cause problems can be eliminated. Preventing problems from occurring has any positive effects, including an increase in quality level and a decrease in operating costs.

There are several commonly used computer tools and methods that can help teams focus on prevention.

- Total quality management (TQM)
- QFD
- Failure mode and effect analysis (FMEA)
- Inventory Control (Just-in-Time)
- Ordering method for production (Kanban system ordering)
- Quality control methods and inspection (Andon system)
- Delivery method for production (Pull system) is one of the better known tools and methods.
- Other frequently used methods are
  - Preventive maintenance.
  - Continuous improvements (or Kaizen concept).
  - Failure prevention analysis, pioneered by Kepner Tregoe.
  - Product mix or variety programs.
  - Balance of options (Katashiki system).
  - Balance of work flow (Heijunka system).
  - Fool proofing for production (or pokayoka Jidoka concept).

2.5.2   Features Assessment

The teams should be careful in selecting features. They should avoid employing form features that are unnecessary and/or expensive to manufacture. By features, we mean both design features and manufacturing features. Examples of features that should be avoided include: specifications of surfaces smoother than necessary, wide variations in wall thickness of an injection molded component, or an internal aperture too close to the bend line of a sheet metal component, and so on [Nevins and Whitney, 1989].

2.5.3 Manufacturing Process Selection/Substitution

The National Research Council [1994] has categorized manufacturing processes into five distinct process families based on manufacturing trait changes:

- *Mass-change processes:* This involves removing or adding material by mechanical, electrical or chemical means. Examples include traditional processes, such as machining, grinding, plating, and nontraditional processes, such as electrodischarge and electrochemical machining.
- *Phase-change processes:* This involves producing a solid part from materials that were originally in the liquid or vapor phase. Typical examples include casting of metals, composites by infiltration, and injection molding of polymers.
- *Structure-change processes:* This involves altering the microstructure of a work-piece, either through its bulk or in a localized area such as its surface. Heat treatment and surface hardening are some typical examples.
- *Deformation processes:* This involves altering the shape of the solid work-piece without changing its mass or composition. Typical examples include metal working processes such as rolling, forging, deep-drawing, ironing, and so on.
- *Consolidation processes:* This involves combining materials such as particles, filaments, or solid sections to form a solid part or a component. Typical examples include powder metallurgy, ceramic molding, polymer-matrix composites, welding, brazing.

Teams make the selection of the right process based on the results of the following steps:

- An assessment of the design and form features of the part and its complexity, volume of production, etc. This is discussed in section 2.5.2.
- Assessment of process requirements, such as level of acceptable surface finish quality, volume, etc.
- Identification of the candidate manufacturing processes on the basis of changes to materials and volumes and their feasibility in manufacturing the part. This is discussed earlier in this section.
- Cost of tooling, tools utilization, machine availability, dies, molds, etc.
- Computation of alternative cost for manufacturing the part if some manufacturing processes were replaced by others.

2.5.4 Minimize Material Usage

The desired configuration for a part is obtained by one of the following actions:

- Removing material, for example, cutting (drilling, lathe cutting, laser cutting), turning, machining, grinding, ECM, EDM, LBM, etching, boring, and so on.

- Redistributing material, for example, forging, stamping, rolling, spinning, drawing, extrusion, power forming, sheet-forming, metal working, and other classical bulk-forming, and so on.
- Adding material, for example, injection molding, molding, casting (sand casting, die-casting, investment casting), plating, infiltration, and so on.
- Retaining materials, for example, filament winding, braiding, lay-up, die position, and so on.
- Uniting dissimilar materials, for example, joining, assembly, welding, laser-welding, and so on.

Figure 2.9 shows a classification of manufacturing processes based on action similarities. The manufacturing processes have been classified into the above five categories. The three columns in Figure 2.9 list the results of applying the processes on three quantities, material change, volume change, and material wasted or extra effort required. The amount of material wasted, extra work performed, or energy spent is typically quite different for different processes. In machining, waste results from chip generation and in forging it comes from flash removal. In stamping, waste results from materials in binders that have to be trimmed, and in casting, waste is due to materials to be removed in the runners.

The material usage efficiency is defined as

$$\eta_{material} = \frac{\text{Volume of the Finished Part}}{\text{Intake Volume of Raw Material}} \times 100. \qquad (2.10)$$

If this index is close to 100, material utilization is high and most are consumed in making the finished part. The difference between 100 and the material usage efficiency, $\eta_{material}$, indicates the percentage of materials that are wasted in the process (material and volume transformations).

2.5.5   Minimum Part Assessment

One of the most important concepts embodied in the design for assembly method [Boothroyd and Dewhurst, 1988, Boothroyd, Dewhurst, and Knight, 1994] is the idea of a theoretical minimum number of parts. It is not difficult to imagine how this information can be used to improve a design. Knowing this minimum number, designers could begin searching for ways to reduce or eliminate excess parts. The steps shown in Figure 2.10 elaborate the procedures involved in determining the minimum number of parts. The following are some key steps that can be followed during minimum part assessment.

- *Identify the product fixture, minimum number of parts, materials:* The first step is to look at the initial design, identify the product fixture, number of parts, materials, and so on. Determine the theoretical minimum number of parts for the chosen initial design, product fixture, materials, and so on. In many situations, the product concept or technological approach determine how low this number can be.
- *Identify reasons for requiring separate parts:* Determine from the previous study the reasons for requiring separate parts. The number of parts in a design does drive cost, and design part simplification leads to improved reliability. If the parts can be com-

	Actions Types of Manufacturing Processes	Results of Applying the Processes		
		Effect on material change	Effect on volume change	Material wasted or Extra work performed
P R O C E S S C L A S S I F I C A T I O N S	Cutting (Drilling, Lathe-cutting), Turning, Milling, Machining, Grinding, ECM, EDM, LBM, Etching, Boring, Laser Cutting, etc.	Removing materials	Reduction in volume or shapes (local)	Chip generation
	Forging, Stamping, Rolling, Spinning, Drawing, Extrusion, Power Forming, Classical Metal Forming, Bulk-forming, Sheet-forming, etc.	Redistributing materials-	Volume or surface deforming	Flash removal (forging), materials trimmed in binders
	Injection Molding, Molding, Casting (Sand Casting, Die-casting, Investment Casting), Plating, Infiltration, etc.	Adding materials	Volume filling solidification	Materials wasted in runners
	Filament Winding, Braiding, Lay-up, Deposition, etc.	Retaining (constant) materials	Volume build-up shape-change	Extra work performed
	Joining, Assembly, Welding, Laser-welding	Uniting dissimilar materials	Parts positioning and orientation	Extra work performed

FIGURE 2.9   Classification of Manufacturing Processes Based on Action Similarities (Materials and Volume Transformations)

bined, the product is redesigned and reanalyzed and then they are checked to verify whether or not they are easy to manufacture and assemble. If both conditions are met, the product is then analyzed for production volume and assembly operations.

- *Design parts for multiple use:* The following are the four frequently asked questions during the assembly operation of a product.
  - Does a part move freely with respect to all other parts? It is easier to combine parts if a part moves bodily with respect to parts already assembled.
  - Does a part have to be of a different material from an adjacent part? If not, perhaps the two parts can be combined.
  - Why is the part needed? or why do parts have to be separate from all other remaining parts?
  - What special function does a part provide that cannot be performed by modifying the remaining parts?

- *Choose alternate product concepts:* Recompute the theoretical number of parts. Keep on modifying the design or changing the theoretical approach and recomputing the theoretical minimum until a favorable aspect of an alternative concept is reached for which the theoretical minimum number of parts is as low as possible. Choose the outcome as a new test case or a candidate design.
- *Analyze the design for other technological considerations:* Question the design to provide a desired range of product model variations. Check all parts for functions and seek innovative ways to assemble the remaining parts. Seek ways to modify the design to minimize part variations.
  - Eliminate parts which do not add functional values to the product (redundant parts).
  - Combine those parts that do not need to be separate for theoretical or business reasons.
  - Reduce the number of extra parts beyond the possible theoretical minimum level.
- *General assembly, de-assembly sequence order:* This consists of analyzing for manual assembly, automatic assembly, and robotics assembly [Boothroyd and Dewhurst, 1983]. Types of assembly operations also affect the cost. Determination should be made early in the design process whether the product under consideration could be assembled using methods based on general purpose assembly robots, special purpose assembly equipment, such as an indexing or free-transfer, rotary, or non-synchronous device, or by manual assembly, in that order (see Figure 2.10). Each of these methods has its appropriate range of conditions for economic benefits, depending on the number of parts to be assembled, the production volume, and so on.
- *Select appropriate assembly:* The outcome of the previous step is to select appropriate assembly. If the test or candidate design is acceptable for chosen assembly then the design is accepted, otherwise the candidate design is reevaluated. The lowest assembly cost can be achieved by designing the product so that it can be economically assembled by the most appropriate assembly method [Boothroyd and Dewhurst, 1988a].

2.5.6   Reliability by Design

Reliability is a measure of the following:

- How well a product has been designed to perform its intended function, and
- How durable is (or will be) the product in field or service use.

Design for reliability means maximizing the probability of the product to perform as expected over a given period of time, while functioning under a given set of operating conditions. Performing as expected means adhering to a given set of product performance characteristics. If P(system) denotes the probability of survival of the product, then Reliability by design means

$$\text{Reliability-by-design} \equiv \text{Maximize } [P\text{ (system)}]. \tag{2.11}$$

If a product is decomposed into its constituents: sub-systems (Su_i), components (Co_i), parts (P_i), and so on, in such a way that the following two conditions hold true:

1. The failure of a constituent is independent of the failures of the other constituents, and
2. The failure of any constituent in the product will cause the product to fail, then

the probability of the product survival, P(system), can be calculated by the equation

$$P \text{ (system)} = P \text{ (sub-systems)} \times P \text{ (components)} \times P \text{ (parts)} \times P \text{ (materials, etc.)} \quad (2.12)$$

where $P(y)$ is the probability that everything will go right for y, where y stands for sub-systems, components, parts, materials, and so on.

Since each of the probabilities of survival is a fraction (a number between 0 and 1), the probability of survival for the system will actually be less than the probability of survival for the individual constituent sets (sub-systems, components or parts, etc.). That is,

$$P \text{ (system)} < P\text{(sub-systems)} < P\text{(components)} < P\text{(parts)} < P \text{ (materials)} \ldots \ldots \quad (2.13)$$

Let us assume that for a decomposed set of sub-systems, components, parts, and so on, the following is true.

If failure of an element in the kth set of constituents will cause another element of the kth set constituent to fail, the probability of survival of the corresponding set can be written as:

$$P(\text{sub-system}_k) = P(Su_{k1}) \times P(Su_{k2}) \times P(Su_{k3}) \times \ldots \times P(Su_{ki}) \times P(SS_{k\,i+1}) \times \ldots \ldots$$

$$P(\text{components}_k) = P(Co_{k1}) \times P(Co_{k2}) \times P(Co_{k3}) \times \ldots \times P(Co_{ki}) \times P(Co_{k\,i+1}) \times \ldots \ldots$$

$$P(\text{parts}_k) = P(Pa_{k1}) \times P(Pa_{k2}) \times P(Pa_{k3}) \times \ldots \times P(Pa_{ki}) \times P(Pa_{k\,i+1}) \times \ldots \ldots \quad (2.14)$$

$$P(\text{materials}_k \ldots) = P(Ma_{k1}) \times P(Ma_{k2}) \times P(Ma_{k3}) \times \ldots \times P(Ma_{ki}) \times P(Ma_{k\,i+1}) \times \ldots \ldots$$

where P denotes the probabilities of survival and

Su_{ki}: represents the ith element of the kth sub-system,
Co_{ki}: denotes the ith element of the kth component,
Pa_{ki}: denotes the ith element of the kth part, and
Ma_{ki}: represents the ith element of the kth material.

This is the worst case scenario. This occurs because each element of the constituents is linked in a series. The probabilities of survival can be improved in each case if some critical sub-systems, components, or parts can be set up in parallel with other elements. This way failure of some non-critical elements of a constituent will not cause failure of the complete constituent set. In other words, out of all those critical parts, if at least one part does not fail, the system will not fail. This is because the part that didn't fail represents a parallel back-up system to keep the constituent running. The individual probability of such a critical element is:

$$\text{Probability of Survival for a Critical Element} = 1 - P \text{ (element)}. \qquad (2.15a)$$

The reliability of the whole system can still be computed on the basis of the above equations, if the occurrences of $[P$ (elements)$]$ in the above equations (2.14) can be replaced by $[1 - P$ (element)$]$, wherever such critical element is used in parallel. By using n elements in parallel, reliability of the components can now be based on

$$\text{Probability of failures for a set of } n \text{ elements} = \{1 - [1 - P \text{ (element)}]^n\} \quad (2.15b)$$

if all probabilities are the same.

Designing parallel elements of a constituent to increase reliability becomes critical when risks to human lives or human safety is involved. The formulas, equations 2.15a and 2.15b, are useful to start a preliminary reliability estimation that must be refined as more and more information becomes available about the interdependence of constituents or the elements within a constituent. In the absence of actual data, each of the individual probabilities can be assumed to follow a pattern similar to an exponential distribution.

$$\text{that is,} \quad P = e^{-t\lambda} \qquad (2.16)$$

where P is the period of failure-free operation, t is the time specified for the failure-free operation and λ is the failure rate.

Reliability engineers are generally integral participants of the concurrent PD^3 process. They carry out a series of trade studies in order to come up with an optimum solution that is close to the critical requirements. Reliability is ensured by design and not by chance. For example, reliability engineering methods—failure analysis, environmental stress screening (ESS) used during design and development, or development testing—are performed as part of reliability-by-design. Two major elements of the traditional practice of mechanical system reliability are probabilistic design analysis and reliability database for mechanical parts.

2.6   "DESIGN FOR X-ABILITY" ASSESSMENT

A great deal is known about some key elements of DFX such as design for manufacturability and assembly (DFMA). DFX generally relates to the early design phase, feasibility, and product designs. There is also considerable research on DFMA and more is being done on its use in the manufacturing process, such as injection molding, stamping, casting and forging [Boothroyd and Dewhurst, 1988]. Information is usually sketchy about designing for other DFX issues, such as testability, serviceability, recyclability, design for green, design for logistics, and so on. Design for green (DFG) means considering life-cycle usage of the product from an environmental standpoint while the product is being designed. Design for logistics (DFL) means influencing the design teams to use various performance indicators, such as cost-drivers, lead-times, and so on, while the product is being developed.

In CE, measuring overall performance is not easy. Measuring performance is not like introducing a change (for example, a perceived improvement) and then measuring its effects on overall performance. The following are the reasons for such difficulties:

- *The process is time-bound:* Many of the overall effects cannot be measured until after a significant passage of time with the accumulation of a prolonged history of performance.
- *Simultaneous occurrences of change:* Most companies, while making changes in the design portion of the PD^3 process, also simultaneously make changes in production operations, including in-process development and control, and product support operations. It is difficult to separately correlate benefits with such changes.
- *Individual contributions are obscure:* Since all CE parties (work-groups) are committed at all times, and each measure influences the company's overall performance improvement, it is difficult to predict changes in overall results due to a single measure.
- *Most measures cannot be easily quantified:* If measures can somehow be quantified, then they can be used during an early PD^3 process.

The development of CE metrics is not an easy task. They, however, become important tasks if downstream requirements are to be considered upstream. The cost of assembling a product is related to both the underlying design complexity of the product and the assembly method used for its production. According to the National Research Council (NRC), "development of more accurate, quality and cost models, especially at the preliminary or conceptual design stages, is essential to support effective concurrent design" [National Research Council, 1991]. Design for manufacture and assembly (DFMA) is a popular DFX technique that goes way beyond articulation of the best practice axioms for design work groups. Figure 2.10 shows the steps involved in using DFMA. Following the steps of Figure 2.10 can result in a design that is both cost-effective for the chosen assembly method and, at the same time, is in compliance with most DFMA principles.

In practice, DFX techniques are used to encourage designers and manufacturing engineers to work together, thus fulfilling the premises of Concurrent Engineering. The following are some additional DFX techniques used for product design.

2.6.1   Design for Compatibility

In design for compatibility (DFC), the teams look at every set of specifications that may be independently imposed on the product from its life-cycle standpoint. For instance, a set of specifications may have come from the design side in the form of its features, functions, complexity, materials, and so on, and another set of specifications may have come from the process side in terms of manufacturing operations, machine functions, availability, and so on. It is important that these sets of specifications are mutually compatible both from the product and process standpoints [Milacic, 1989]. Figure 2.11 shows examples of design incompatibility with respect to manufacturability of parts. Design for compatibility provides, early in the design process, a set of checks on consistency of specifications and a set of methods to evaluate alternatives if those checks are not compatible. Some key factors that measure DFC are:

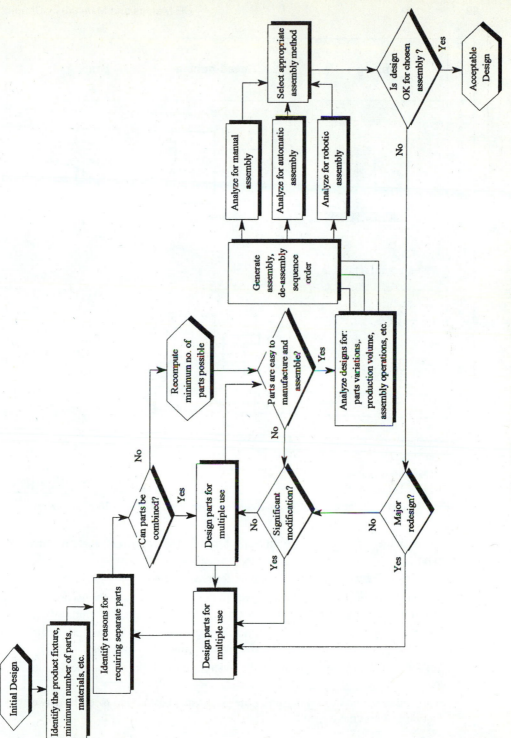

FIGURE 2.10 Design for Manufacture and Assembly

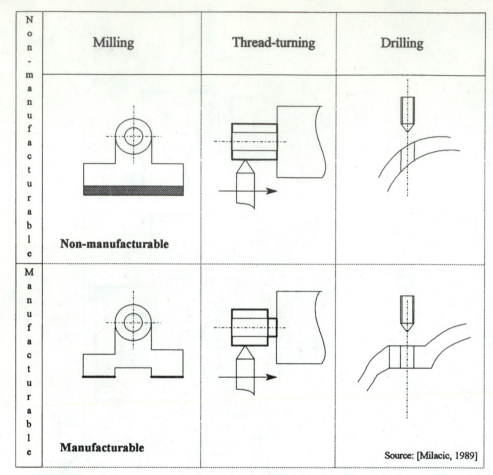

FIGURE 2.11   Examples of Design Incompatibility with respect to Parts Manufacturability

- Compatibility of part shapes with respect to the chosen process, form features, and so on.
- Interference between the parts and their desired functions.
- Processability of materials with respect to the chosen manufacturing process (e.g., stamping draw or stretch; forging, machining, cutting, rolling, molding, casting, etc.).
- Compatibility of the features (form features, geometry, undercut, draft angle, etc.) with respect to the chosen process, such as machining and molding.
- Ability of the process to provide the desired surface finish, hardness, tolerance, color, look, etc.
- Selection of the process to minimize the processing cost and to meet the chosen production and volume targets.

2.6.2 Design for Manufacturability

In order to improve design and to reduce the time for assembly, it is important to know how much time and effort an existing design of the product takes during assembly. One way to determine this is to perform a manual design for assembly (DFA) set-up and record the time each operation takes. Figure 2.12 shows a set of nine operations in a manual DFA. Operations are typical of a manual DFA analysis. The details may differ but the nine steps shown in Figure 2.12 are generally required. Knowing this information, it is easier to redesign the parts and eliminate or automate those operations that manually are

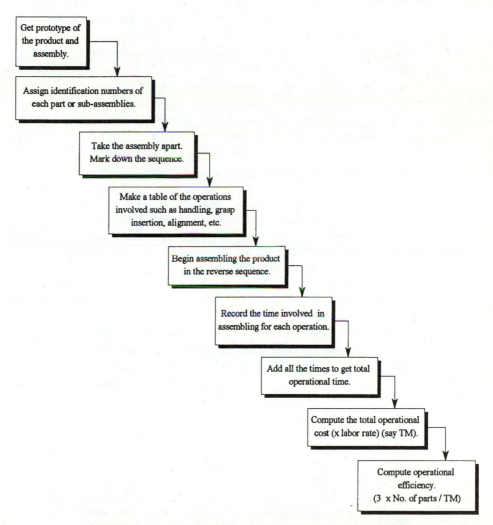

FIGURE 2.12 Analysis of Steps in Manual "Design for Assembly"

either very time consuming or labor intensive. Some of the operations are not related to
either design deficiency or lack of automation, but are often carried out due to poor assembly and manufacturing practices, for instance, the design incompatibility shown in
Figure 2.11 [Milacic, 1989]. Others are included due to bad organizational 3Ps (policy,
practices and procedures). In this assessment category, the following DFM guidelines
are considered the most useful.

- *Never fight with gravity:* It is easy to work with lighter components and operations,
  such as up and down movements to snap-fit in vertical slot openings, screws, and so
  on. Putting a screw on the underside or sideways in a part amounts to fighting
  against gravity. Such actions may give rise to accessibility problems as well.
- *Work with fewer parts:* Every increase in part number means an additional increase
  in design and manufacturing cost. Non-existent components cost nothing to qualify,
  purchase, assemble, and test. Then why employ a large number of parts unnecessar-
  ily if the work can be performed with only a few?
- *Design parts for ease of fabrication:* Teams should design parts so that (a) tolerances are compatible with the assembly method employed; and (b) fabrication costs
  are compatible with targeted product costs. This eliminates the possibility of part re-
  jections or tolerance failures during assembly.
- *Reduce purchased parts:* Purchased parts do eliminate the development cost of designing parts and manufacturing them in house. However, whether it is advantageous from the overall assembly point of view is the issue to consider. Replacing a
  few parts of an assembly by a set of purchased parts may not be the right decision if
  there are major interactions between the in-house designed parts and the parts being
  out-sourced. On the other hand, if the purchased part is a replacement for an in-
  house independent designed part and the purchased part is performing similarly or
  is better than the in-house part, replacement may be desirable. Whether a part is designed in-house or not, every part needs documentation, control, and almost essentially an inventory. Reducing purchased parts not only reduces the chances of failures but also reduces all direct and associated indirect costs.
- *Drive the design toward more functionality per part:* The goal is not to contend
  with fewer parts but to have those parts perform more and more functions. For example, avoid separate fasteners if an existing part can incorporate this function. The
  idea is to use fewer parts while providing all necessary design functions. Cost is
  minimized since product design is often a primary driver of its manufacturing costs.
  There is, however, a danger of going overboard with this. Combining a highly functional part with another function may in turn boost up its serviceability cost. It can
  irritate customers since servicing would be more costly than it ought to be. Though
  a new multi-functional part may save money on the factory floor, it may also require additional research and development. This may elongate design development
  time, and may lengthen the time-to-market window as well.

2.6.3   Design for Assembly

DFA is an analysis technique that focuses on the relationships between mating parts. DFA provides an early quantitative method for determining the assembly cost of the design, whether the part is new or carry-over (see Figure 2.12). There are three objectives of DFA.

1. *Minimize Part Counts:* The most important objective is to minimize part counts either through elimination of some parts or their concatenation into a few multi-functional unit. Minimum counts mean fewer parts manufactured and fewer parts that can fail. This is related to designing parts to be multi-functional or designing parts for multiple uses. The parts count efficiency is defined as

$$\eta = \frac{\text{Theoretical minimum number of Parts}}{\text{Actual Number of parts}} \times 100. \qquad (2.17)$$

   Clearly, when the number of parts is equal to the theoretical minimum, the parts count efficiency is at 100%. It may be noted that minimizing the number of parts does not necessarily imply a more manufacturable design.

2. *Minimize Interfaces:* The second objective is to minimize interfaces between the remaining parts. Physical breakdown of products (PtBS) and processes (PsBS) should yield self-contained modules and assembly lines. Fewer interfaces between mating parts means fewer assembly or de-assembly issues to worry about. More part interfaces means more chances for the product failures.

3. *Ease of Assembly:* The third objective is to design the remaining parts and their interfaces in such a way that they can be assembled easily [Boothroyd and Dewhurst, 1994].

During the analysis of product assembly, the following DFA guidelines can be used.

- *Employ automatic inserters:* Teams should specify parts that can be automatically sequenced and inserted using DIP inserters, a variable center device (VCD), axial part inserters, or Selective Compliance Assembly Robot Arm (SCARA) insertion robots. If none of the above is possible, minimize setups and reorientation.

- *Employ "pre-oriented" parts:* Parts that cannot be supplied *pre-oriented* in reels, tubes or matrix arrays for easy insertion should be avoided. The key to successful automatic insertion is *pre-oriented* parts.

- *Minimize sudden and frequent changes in assembly directions:* Following a uni-directional assembly, such as design for top-down assembly, is the most desirable.

- *Maximize process compliance:* This consists of designing with standard parts, standard processes, ease of assembly, and so on. Use processes which are easy to install and maintain, for example, snap-fits instead of fasteners. Sometimes a fastener is okay if parts are not internally manufactured. However, if snap-fits are designed on an existing part by combining the two features functionally, the integral part can boost the tooling costs very high. If fasteners have to be used, standardize fasteners by types and sizes. Use low-cost irreversible fasteners only where skilled service people work.

- *Maximize accessibility:* Drive the design for easy replacement and accessibility. Designers should provide adequate clearances for disassembly or for accessing the part for future repair or replacement.
- *Minimize handling:* There are two aspects to minimize handling: design of parts for ease of feeding (insertion) and design of parts so that they are easy to grasp, manipulate, or orient. Both of these aspects apply whether the part is a manual assembly or a machine assembly. Boothroyd, Dewhurst and Knight [1994] have discussed several means to orient and transfer parts automatically using a variety of vibratory and non-vibratory feeders to the assembly workstations. Other ways to minimize handling are to limit size, minimize weight, eliminate, or simplify adjustments, and so on.
- *Avoid flexible components:* Flexible components are generally difficult to handle.

In other situations these guidelines offer a check on what makes a design better from DFA perspective.

2.6.3.1 Major DFMA Usage and Benefits
There are three primary usages of DFMA.

1. Analysis of a product assembly while the product is still in the design stage.
2. Redesigning a product already in production or for which a prototype is already available.
3. Refurbishing a product that requires face lifts. In this case, a time and motion study is conducted, and time and efforts of assembling and de-assembling are carefully noted. Motions corresponding to parts handling, for example, feeding and orienting, attachment or insertion, are monitored. DFA cost drivers are often found for various motions in data tables through software [Boothroyd and Dewhurst, 1992]. Such cost drivers can be used to predict assembly costs. If the product is a carry-over part, changing the design of the product simply to reduce assembly costs should be reviewed in perspective. The assembly savings should be weighted against the increase in investment cost in new tooling. If, however, products are made in large quantities so that automation might be feasible, changing the design may make more sense.

The following are the major DFMA benefits.

1. The University of Rhode Island [Boothroyd, Dewhurst and Knight, 1994], Hitachi, and GE have all developed quantitative methods for enhancing DFMA capability for product designers. Using these methods, both assembly costs and component manufacturing costs can be predicted with varying degrees of sophistication at the earliest points in product design.

2. There have been some spectacular applications of DFMA techniques to product design. When IBM developed its Proprinter to the Epson MX80, it designed a printer with 79% fewer parts and 83% fewer assembly operations. The new design had only 32 parts including four printed circuit connectors. Most parts were snapped together during final assembly, including two motors and a power supply. IBM transferred assembly operations from Charlotte, North Carolina, to direct labor lines at its typewriter facility, where doing the job manually took only 3 minutes [Wilson and Greaves, 1990].

3. Applying DFMA techniques, electronic firms like Philips have reduced the number of parts by 75% in their compact disc players. Philips have simplified VCRs by a 55% parts reduction, tuners by 45%, and amplifiers by 45% [Kumpe and Bolwijn, 1988]. Swatch has reduced the number of components in its watches from 150 to 51. Less dramatically, though more significantly perhaps, automotive companies have reduced the number of parts in its average car from 30,000 to about 20,000. Similar percentage reductions have been reported in the design of microwave ovens, personal computers, office copiers, and air compressors [Wilson and Greaves, 1990].

2.6.4 Design for Product Variety

Variety means packaging together a number of functions/features in a single part. As functions per part increase, each part becomes more complex but the total number of parts in a product decreases. Product variety has the effect of decreasing the volume related costs, while increasing the complexity related costs. It costs more to manufacture a part that has many functions and features built into it. This is shown in Figure 2.13. Three curves are shown.

1. *Curve a* shows the variation of complexity related cost with respect to the number of functions/features that can be packaged in a part. *Curve a* is shown by solid line.

2. *Curve b* shows the volume related cost. *Curve b* is shown by a chain line. This cost depends on how many parts can be packaged in a typical product. If a large number of functions/features can be packaged in a single part, fewer parts are required to manufacture a product. The corresponding volume-related cost will be small. However, if the number of functions per part is small, a large number of parts will be required to provide an equivalent product functionality. The volume related cost will be high.

3. *Curve c* shows the variety related cost. A typical *curve c* is shown by a series of dotted lines. A series of vertical straight lines are drawn in Figure 2.13 connecting the points along the *curves a* and *b*. A point along this vertical line shows the range of product variety at a particular combination of product complexity and volume. These lines on the right represent the variation of costs when functions/features per part are high and the lines to the left when functions/features per part are low. The dotted *curve c* shows a mean variation of the variety cost as complexity

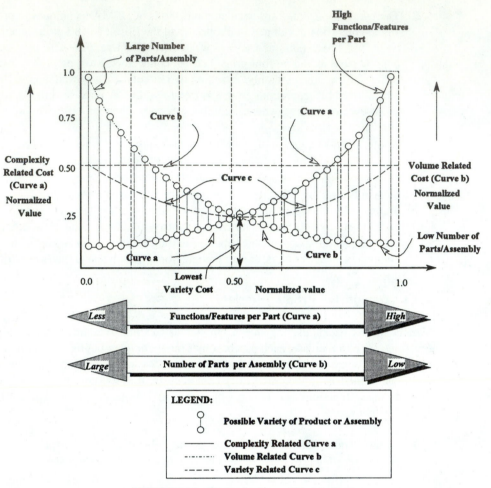

FIGURE 2.13   Design for Lowest Variety Cost

(functions/features per part) is increased. Clearly the mean variety cost is lowest when functions/features per part are either not too high nor too low.

Design for variety (DFV) is associated with minimizing the cost of providing variety options. There are three main factors that affect the cost of providing variety [Ishii, Juengel and Eubanks, 1995].

- *Number of options in a product variety:* The cost of manufacturing is proportional to the number of product variations. The fewer are the variations the lesser is the cost for its manufacturing.

- *How far away the product is from its finish (job no. 1) stage, when a variety program is implemented:* If a product variety option occurs closer to the end of the manufacturing process (say job no. 1), it will have less impact on upsetting any of the upstream processes. However, if variation occurs in the early phase of manufacturing, such a variety option may amount to performing a greater number of subsequent operations, increasing the complexity and cost of all related downstream processes.

- *How "painful" it is to change from one variety set up to another:* An example would be changing a die or a paint color. If the change requires activating a number of things, such as a different plant layout, an additional production line, and a different supplier, it would be more time consuming. The more parts or components involved, the more costly it will be to manufacture the product for a large number of variety options.

If a parameter α_1 is associated with number-of-options, a parameter α_2 is associated with time measured from the finish stage (job 1), and a parameter α_3 is associated with change-over-efforts, a rough measure of cost of variety index can be expressed as:

$$C_v = \prod_{i=1}^{i=3} (\alpha_i) \qquad (2.18)$$

where α_i denotes a mapped parameter for an ith cost factor. In the case of cost of variety, the above factors were also considered relevant by Ishii, Juengel and Eubanks [1995]. The cost factors, number-of-options, time measured from finished stage, and change over efforts, are parameters corresponding to a current value of an ith cost factor. The current value of an ith factor is governed by the following parametric equation.

Current value for an ith cost factor = (Minimum-value for the ith cost factor)
$(1 - \alpha_i) +$ (Maximum-value for the ith cost factor) (α_i) (2.19)

The current value of a cost factor is mapped into these parameters, α_i, in such a way that if α_i equals zero, the current value is the minimum value of the cost factor, and if α_i equals 1, it reaches its maximum value. The parameters, α_i, thus, take a value between 0 and 1

$$\text{where } 0 \leq \alpha_i \leq 1.0 \text{ ; } i = 1, 3 \qquad (2.20)$$

The term α_i indicates the factors in equations 2.18 and 2.19. C_v is smaller if either number-of-options are large, or stage-in-manufacturing is early, or the efforts required to change over to a new set-up takes more time. The actual cost of variety for producing a part with specifications α_1, α_2, and α_3, can be computed as follows:

Cost of variety = Minimum Cost of manufacturing an assembly $\times (1 - C_v)$
+ Maximum Cost of Manufacturing an assembly $\times C_v$ (2.21)

where C_v is given by equation (2.18).

2.6.5 Design for Flexibility

Design for Flexibility (DEF) means providing a made-to-order part as demanded by a royal customer or a part that customers like in the shortest possible time. This means creating a flexible production system that can handle a variety of diverse products in small lot sizes. A production system's flexibility is often measured by a set of variety reduction program (VRP) indices. DFF indices include: part index, process index, production process index, and control points index.

- **Part index (P_{ti})** is the product of the number of parts times the number of part types that go into a particular product.

$$\text{Parts index } (P_{ti}) = \text{Number of parts} \times \text{Number of Part types} \qquad (2.22)$$

- **Process index (P_{ci})** is the product of the number of processes times the number of process types that go into processing a particular product.

$$\text{Process index } (P_{ci}) = \text{Number of processes} \times \text{Number of Process types} \qquad (2.23)$$

- **Production process index (P_{pi}):** The production process index indicates how many lines and how many processes use a particular part or a product. It is the product of the number of lines times the number of processes at each production site.

$$\text{Production Process Index } (P_{pi}) = \text{Number of Lines} \times \text{Number of Processes} \\ \times \text{Number of Sites} \qquad (2.24)$$

- **Control points index (C_{pi})** is based on the number of controls or changes in the processing, assembly, or production process. Examples include flow of drawing, flow of materials, flow of parts, and so on.

$$\text{Control Points Index} = \text{Function of (number of drawings, number of material} \\ \text{change, number of parts change, number of process change, etc.)} \qquad (2.25)$$

DFF divides the production costs (C_{pv}) into variety cost (C_v), function cost (C_f), assembly cost (C_a), and control cost (C_c) and the indices. Variety cost originates in part types and process types. The variety costs are comprised of parts variety costs (C_{tv}) and process variety costs (C_{cv}). They vary depending upon the changes in production output for each product and production process variety. Function cost is the cost of the parts and materials required to meet the necessary specification for functions or features that are associated with a product or process. Assembly cost is the cost of processing (machining, welding, stamping, etc.) and/or assembly of parts. Control costs occur due to control tasks associated with the volume of the parts and processes including material handling, ordering costs, item control cost, quality control cost, and so on.

Using the above indices and the unit cost associated with each activity, one can get an estimate of the total cost as follows:

$$\text{Production Cost } (C_{pv}) \propto [(P_{ti} \times C_{tvi}) + (P_{ci} \times C_{cvi}) + (P_{ti} \times C_{fi}) \\ + (P_{pi} \times C_{ai}) + (C_{pi} \times C_{ci})]. \qquad (2.26)$$

2.6.6   Design for Serviceability

While design for reliability and assembly has received wide attention, timely dissemination and utilization of information related to design for serviceability (DFS) has not kept the same pace. The three issues (reliability, assembly and serviceability) conflict in many ways. Consequently, there is a need to understand its (DFS) ties with other X-ability issues and establish appropriate trade-offs between them [Makino, Barkan, Reynolds and Pfaff, 1989]. During the early design phase, designing for serviceability often means changing product features that may in turn affect reliability, change safety, and increase operating expenses. DFS deals with issues such as standardization of components, component accessibility, failure rate, fault diagnostics, component identification, safety, and so on. The goal of DFS is to enhance the customer's perceived value of the product, making it attractive for anyone to acquire, own, and operate.

2.6.7 Design for Interchangeability

Interchangeability indicates a set of common or standard functions, features, or parts suitable for use by multiple product lines or within a product family itself. Following are some key design for interchangeability (DFI) guidelines.

- *Employ interchangeable parts:* Parts could be designed in such a way that they may not be identical in looks, esthetics, and so on, but could still be interchangeable. Parts that are not identical can be interchanged if the boundary constraints between the parts and other parts of the system remain intact. For example, there are a large variety of electrical plug outlets available in hardware stores for household use, but they all fit into a 110 Volt wall-socket present in every house. In the computer world, there is a large line of PC products (PS1 lines or PS2 lines) that are IBM compatible, but they all share the same DOS diskettes that were initially designed to run on IBM PCs. In the early 1980s, PCs were based on Intel 8088/8086 and 80286 processors. They made the integrated circuit architectures open and, thus, became the model for a new category of clones. Now there are a variety of IBM compatible machines (PCs) available that are either Intel 80386-based, 80486-based, Pentium-based, or P6-based. They all have different integrated circuits inside, but data and peripherals are quite interchangeable. The clone computer product market has increased manifold and is still growing. In the electrical and mechanical world, products are being designed for interchangeability with peripheral parts requiring only minor adjustments or configuration settings. Such adjustments and settings also usually follow some defacto or industry standards and most are factory set. Factory setting of options are less susceptible to field service problems at installation time and seldom go undetected.

- *Commonality:* Commonality provides a basis for the interchangeability of parts across several components. Using common parts across several product lines drives down the cost of the product and increases volume. It also limits any future maintenance or repair costs. An idea related to DFI is part mixes and product options (Katashiki system). Similar to the DFI of parts, the interchangeability of options across multiple product lines can provide tremendous savings in product costs and repairs. Closely associated with DFI idea is the notion of designing parts for the international market (DFIM). DFIM seeks a constituent of a product that can be quickly adapted to a particular country, a particular market, or a multi-lingual users' group with the least amount of rework. Using common processes across several product lines can also drive down the cost of manufacturing. This notion of common process across parts also enables the work-group members to be interchanged. A team member could be trained for an equivalent job or a part in a very short time if the only thing changing is the part, not its underlying process.

- *Use standard parts and option packages:* Using non-standard parts creates a tidal wave of costly transactions that can cripple cost advantage in a hurry. Working within the constraints of a company best practices, such as coming up with a list of standard parts, can result in much smaller cost and inventory. Standard parts and option packages require less of everything including documentation and control.

- *Standardize suppliers' parts:* Standardize the parts purchased from outside suppliers. This way the cost of manufacturing and tooling is distributed to many suppliers from whom the company has purchased parts. The indirect costs are also shared across multiple customers. The company pays only a fair share and the *risks* of failure are also minimized. It allows companies to leverage expertise, coordinate production of parts with vendors' capabilities, and to minimize in-house inventories.

2.6.8   Modular Design for Sub-assembly

Modular design for sub-assembly (MDFS) consists of identifying sub-assemblies, designing components and parts with fabrication and assembly process in mind, and designing an assembly sequence to minimize waste of wait and part movement. Modularity is obtained through a concatenation of parts into components and components into sub-assemblies. The idea is to design the product as a series of semi-independent sub-assembly modules that can be built-up off-line, perhaps parallel with other modules, either in a plant or by an outside supplier, and delivered to the main assembly line just-in-time. This has the effect of reducing the number of individual assembly operations on the main production line. Each sub-assembly, by design, (using DFA principles) may have many detailed components or discrete parts that can easily be integrated into a sub-assembly module. Modularity has the effect of reducing both the number of parts and their complexity when each sub-assembly module is designed. The modular assembly approach, thus, offers potential for reducing the number of assembly operations that must be scheduled and handled by an in-house assembly plant.

2.6.9   Design for Disassembly

Design for disassembly (DFD) is an important first step of any recycling process. Recycling is becoming one of the most important fields that any CE work group has to deal with during a PD^3 process. The number of consumer products to be recycled during its life cycle are increasing dramatically every year. Pressure for recycling is also rising due to the rising cost of waste disposal, increased demands of the consumers for environmentally benign products, clean corporate image, pro-public image, and stricter environmentally conscious regulatory requirements. The three main goals of DFD are to:

1. Design for ease of disassembly.
2. Reduce the cost of disassembly by simplifying the recycling process.
3. Improve the humane working environment in disassembly factories by using a highly flexible and mechanized process.

2.6.10   Design for Supportability

Like human factors, design for supportability (DFSu) has only recently become an important CE metric for product realization. It has a major impact on the customer satisfaction and cost of ownership. In the broader sense, design for supportability deals with such issues as standards across the various product lines and operator work practices (or standardized works). More specifically, DFSu deals with Logistics Support Analysis (LSA). It examines relationships and interfaces of parts and their impact on other LSA disciplines, such as vendors, collaborative partners, purchasing agents, and customers. Design for supportability also provides a foundation for Reliability Centered Maintenance (RCM), such as preventive maintenance philosophies and programs across different product lines.

2.6.11   Design for Maintainability

Design for maintainability is a concept similar to design for reliability or design for supportability since they all form an integral part of the so called RMS (reliability, maintainability, and supportability) system. Sound engineering techniques (a set of 3Ps, policies, practices and procedures), applied early in the design of the system, can result in significant reduction in eventual cost of supporting the RMS system. Maintainability may range from development of a maintenance logic, a maintenance concept, to a set of qualitative and quantitative maintainability requirements.

2.7   PROCESS QUALITY ASSESSMENT

It is generally not possible to manufacture a set of parts constantly identical, that is, each part having exactly the same nominal dimensions. This inability to produce identical parts is due to an inherent variation in the manufacturing process caused by noise or other environmental factors. The major sources of a product variation are contained in Figure 2.14. They are grouped into three main categories, system variations, constituent variations, and external variations. Common causes of variation in the constituents of a product or a manufacturing process can cause difficulties in assembling a final product. One solution is to tighten the manufacturing process tolerances. However, this may result in unnecessary costs for inspection, increased scrap rates, increased cost of machining, and so on. Common causes cannot be fixed completely by keeping the system or the process intact. For instance, tolerance build-up in an assembly is a function of its constituent variations, such as tolerance stack-up. This can be expressed in functional equation form as:

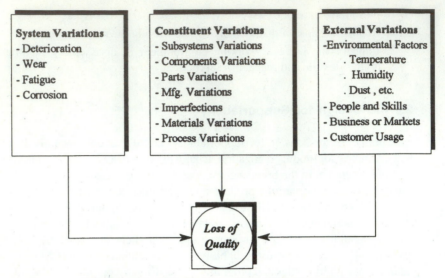

FIGURE 2.14   Major Sources of Variations (Noise Factors)

$$Y = f\ (p_1, p_2, p_3, \ldots, p_n) \tag{2.27}$$

where Y is the output and $p_1, p_2, p_3, \ldots, p_n$, are a set of constituent parameters or variables. Some of these variables in a product, when mass produced, will deviate from their nominal values. There are two kinds of variables that a work group often encounters during a PD^3 process.

1. ***Control variables, (p_i^c):*** Control variables (e.g., $\{p_i^c\}$) are those variables whose magnitude (e.g., nominal values) or sources of variation can be controlled or easily perturbed.

2. ***Noise variables, (p_i^n):*** Noise variables (e.g., $\{p_i^n\}$) are those variables whose magnitude (e.g., nominal values) or sources of variation cannot be controlled through normal means (e.g., by redesign) and require someone to pay a hefty penalty.

that is, $\{p_i\} \cup [\{p_i^c\}, \{p_i^n\}]$ (2.28)

The analysis of this functional equation is called the *robustness or tolerance analysis.* Hence, robustness analysis is the study of the effects of individual constituent variation, $p_1, p_2, p_3, \ldots, p_n$, on the total variation of the complete product assembly.

Figure 2.15 shows a mechanism for identifying a variable type, $\{p_i^c\}$ or $\{p_i^n\}$, depending upon whether its magnitude or source of variation can be controlled or not. As shown in Figure 2.15, a variable is definitely a control type if its magnitude can be changed and its source of variation is known or can be effectively managed. It is definitely a noise type if neither can be accomplished. This is shown by the bottom right region of the rectangle. In the two remaining cases, a variable can be a noise type or a con-

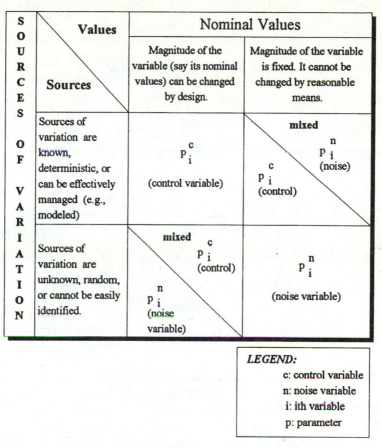

S O U R C E S O F V A R I A T I O N	Values / Sources	Nominal Values	
		Magnitude of the variable (say its nominal values) can be changed by design.	Magnitude of the variable is fixed. It cannot be changed by reasonable means.
Sources of variation are known, deterministic, or can be effectively managed (e.g., modeled)		p_i^c (control variable)	mixed p_i^n (noise) p_i^c (control)
Sources of variation are unknown, random, or cannot be easily identified.		mixed p_i^c (control) p_i^n (noise variable)	p_i^n (noise variable)

LEGEND:
c: control variable
n: noise variable
i: ith variable
p: parameter

FIGURE 2.15   Identifying the Types of Variations

trol type depending on whether or not a work group can change its magnitude, identify its source or control it effectively. This is shown by a partitioned rectangle in Figure 2.15. Variables such as environmental factors, temperature, humidity, and light, for instance, are noises, since they are difficult to control by design.

2.7.1 Design for Quality

Design for quality (DFQ) programs are commonly directed toward achieving 100% prevention and 0% inspection goals. DFQ means building quality control into the product using error-proofing techniques or statistical means, such as six sigma, when the product is first conceived and designed. Achieving six sigma means catching defects 99.9997 percent of the time (see Figure 2.16). DFQ also means minimizing the teams' dependence on some of the quality control or inspection techniques, such as the Andon method. DFQ as-

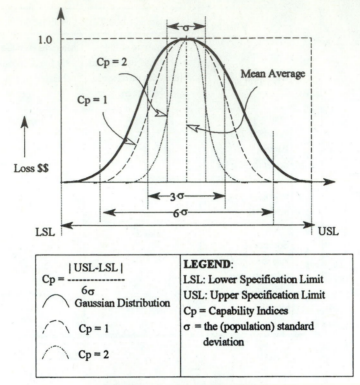

FIGURE 2.16   What is Six Sigma?

sures that suitable quality standards are reached in meeting performance, reliability, main-tainability, durability, operability, safety, economy of manufacture, and operation targets. Product quality is governed by the teams' choice of 7Ts (talent of the work force, tasks, teamwork, techniques, technology, time, and tools). The corresponding options in each category are quite large. The DFQ options in the "techniques" category, for instance, may include QFD, DFMA, FMEA, Taguchi, SPC, Six Sigma, and so on [Green and Reder, 1993]. The following items are some typical DFQ objectives.

- The tolerance range can be increased by applying DFQ techniques to increase the design latitude. Anticipate possible quality problems during early phases of product design and prevent them from occurring by implementing error-proofing techniques that could assure correct outcomes.
- The manufacturing variability can be decreased by applying DFQ to the manuafc-turing processes. Anticipate possible quality problems during early phases of process design and take preventive steps to assure correct assembly and minimize variation to assure repeated performance.
- Apply lessons learned, past experiences, and the team review process to detect qual-ity problems early and consciously (by design) prevent them from recurring.

The basic Six Sigma metric is the capability index for bilateral limits (e.g., nominal is the best). The process capability index (C_p) for DFQ is defined as the ratio of the difference between USL and LSL, and 6σ (see Figure 2.16).

that is,
$$C_p \equiv \frac{(USL - LSL)}{6\sigma \text{ (total range from "}-3\sigma\text{" to "}+3\ \sigma\text{")}} \qquad (2.29)$$

where USL is upper specification limit, LSL is lower specification limit, C_p = Capability Index.

The numerator in equation 2.29 is the customer functional limit tolearance range for a design parameter chosen in a product or a process. The denominator is the measure of the manufacturing variability of the chosen parameter. Three sigma quality is achieved when $C_p = 1$. However, it is inadequate for most products. Six Sigma quality is achieved when $C_p = 2$. A process capability index of $C_p \geq 2$ implies designs and processes that are typical of Japanese manufacturing. Figure 2.16 also shows two Gaussian distribution curves in relationship to the tolerance limits when $C_p = 1$ and when $C_p = 2$.

2.7.2   Design for Manufacturing Precision Quality

When exploring the correlation of manufacturing precision quality and cost, there is a prevalent misconception that precision machining quality automatically entails high cost, meaning if the manufacturing work groups try to increase the given part precision (e.g., machining quality, bore roundness, surface finish, texture, etc.), the manufacturing expenses go up with it. This is not necessarily true if alternate manufacturing processes are considered as part of this equation. Figure 2.17 shows a series of curves for a number of alternate processes. Typically, a given manufacturing process (e.g., machining or milling) follows a hyperbolic curve of interdependence between performance and precision quality. An increase in productivity is usually at the expense of precision quality. In manufacturing machining, performance can be measured in terms of feed rates or stock removal. The variable for precision in machining could be bore roundness or surface finish. Most manufacturing groups establish a set of comfort zones in which the parts are to be produced. This zone is very much tied to a particular manufacturing process or a machine capability. This means in the above example, when for instance, a reamer has to finish a bore with a finer finish (precision), the respective feed rate has to be increased. This means the dependencies of precision and performance are only valid within this comfort zone. In order to improve both performance and precision, without sacrificing one for the other, either the process configuration has to be changed or machine capability has to be improved. This might entail phasing in new manufacturing processes or a changeover to an advanced cutting tool system. This signifies the importance of considering alternate manufacturing processes when considering the trade-off between diverging criteria, such as performance, precision quality, robustness, cost, and so on. Tooling system suppliers must be part of the CE work groups to direct their attentions to the individual aspects of manufacturing. Economical aspects of manufacturing, for instance machining, involves simplifying and optimizing:

- Workpiece configuration.
- Cutting materials.

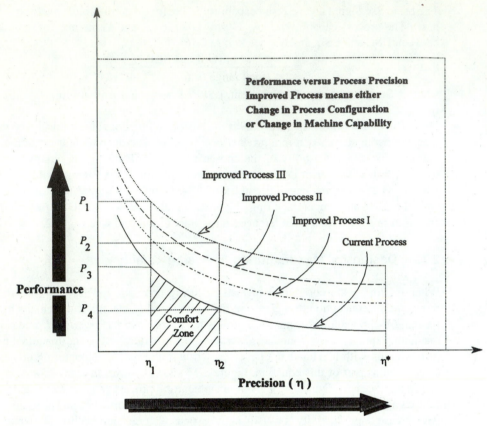

FIGURE 2.17   Design for Manufacturing Precision Quality

- Machining parameters.
- Realistic finish requirements.
- Tool setting and verification.
- Coolant management.
- Tooling systems.

The goal is to obtain products at their best performance (e.g., least cost), while securing the best manufacturing precision quality in fit, form, function, and service.

2.7.3 Design for Managing Variation

Design for managing variation is based on a principle that variation is an inherent quality problem and occurs in most processes. The objective is to achieve uniformity around targets so that the desired outputs (product or process) are insensitive to minor variations in the design parameters.

There are five steps to managing variations, shown in Figure 2.18.

- *Involve the teams:* This is the first step and it is the foundation of most variation reduction programs (VRPs) on which all other remaining steps are built. It deals with getting teams in the concerned process to become organized in dealing with variation.

- *Identify type and sources of variation:* The next step is to study the theory of variation and understand reasons for common and special causes. Identify types and sources of variations in the concerned process. This may include deploying process measurement tools to detect or to measure the amount, cause, and source of variations.

- *Control sensitivity:* The third step is to evaluate results and institutionalize process control techniques to reduce or eliminate undesirable effects of variation and incorporate desirable effects.

- *Formulate strategy:* The fourth step is to analyze the results of control sensitivity and formulate a modification strategy for variations that cannot be easily handled through control means, such as tightening tolerances, and so on. The most cited example of such a modification strategy is a change from *conformance to specifications* to a philosophy of *uniformity around targets*. This may mean reduction of variations (narrower control limits) or the movement of the mean to a higher or to a

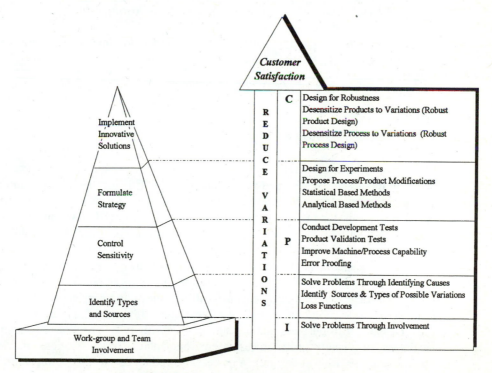

FIGURE 2.18   Steps in Managing Variations

lower level. Reanalyze the product design and process plan, improve it according to the strategy adopted, and reiterate if necessary. Significant modification of the process may be necessary to reduce multiple causes of variation at the same time or to minimize the sensitivity of their variations.

• *Implement innovative solutions:* The last step is to introduce a set of innovative solutions (e.g., robustness techniques) to reduce the variation beyond the usual control or modification levels. During this step, innovative solutions are sought to reduce sensitivity to process variations from multiple sources, causes, and uncertainties. Many other popular measurement tools, like Taguchi's method [Taguchi and Clausing, 1990], axiomatic approach, QFD, Pugh's concept [Pugh, 1991], and parameter design [Taguchi, 1986], can be employed for this purpose.

The types of tools and techniques required at each step depend upon the severity of the problem. Figure 2.18 also lists the popular types of tools and techniques (a partial list) that could be used to control variation and improve overall customer satisfaction. The higher are the steps, the more sophisticated tools, techniques, and knowledge of the process are required to manage variations. Involvement of people and teamwork, of course, is a must at each step. Measures of variability determine the extent of data spread. If data is normally distributed, standard deviation can be used to estimate the percentage of scores that fall within a specified range. The larger is the standard deviation, s, the larger is the spread of data. In most normal distribution cases the following is true:

• 68% of the scores fall within a range whose limits are: (mean − 1s) and (mean + 1s)
• 95% of scores fall within a range whose limits are: (mean − 2s) and (mean + 2s).
• If the standard deviation, s, is small, a higher percentage of data falls closely around the mean.

2.7.3   Design for Robustness

Implementing design for reliability (DFR) during a design phase requires that tolerance limits on the design be specified as large as possible. This increases the domain of operability of the designed parts in various functional modes. However, during a manufacturing phase, design for manufacturability and assemblability (DFMA) implementation requires that tight tolerance limits be imposed to control or reduce the variability of the manufactured parts. The conflicting requirements of the design specifications and the manufacturing specifications on parts are pictorially shown in Figure 2.19. Design for robustness is based on a third set of criteria. It does not relate to either maximizing or minimizing the tolerances, but to desensitizing the uncontrolling parameters so that their effects on tolerance variations are nullified (see Table 2.2).

Design for Robustness is a measure of insensitivity to variations in design due to factors such as manufacturing processes, environmental factors, operations, and deterioration through use. Robust design is a process of product design that is tolerant of uncontrolled variation, that is, it consistently conforms to the original design intent in spite of external and internal noises that may be present. Such product designs are classified as

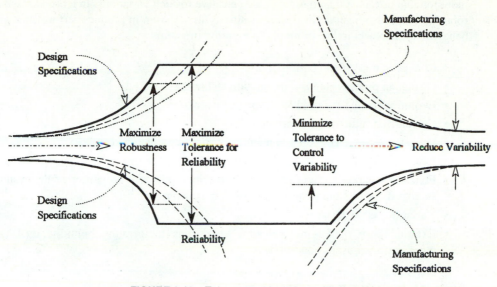

FIGURE 2.19   Tolerance—Maximize or Minimize?

TABLE 2.2   Examples of Maximize, Minimize, and Desensitize

Minimize	Maximize	Desensitize
Lead Times	Fast Delivery	Variations
Time-to-market (Responsiveness)	Standardization	
	CALS, PDES, EXPRESS, CE	
Design changes during downstream operations	Design simplicity	Design Variations,
	Design intent capture	Tolerances
Number of Revisions	Life-cycle capture	
Unscheduled Changes		
Use of Critical Processes	Process reliability and predicatbility	Process Variations
	Manuafcturing Strategies and standards (DFA, DFM, DFX, DFQ, etc.)	Process Variations
Unit Costs, Cost-to-quality	Profitability (ROI)	Material Properties
Material Removal	Performance	Variations
	Economics	Cost Variations
Obsolete Technology	Advanced technology	Variation in Common
	Use of CAD/CAM, CAX, EDI, etc.	systems
	Use of KBE, AI, Fuzzy Logic	
Defects Per Million	Quality	Variation in Quality
Controversy	Cooperation	Human Factors
	Supplier invovement, Teamwork	
	Management involvement	
Inventory	Customer Satisfaction	
	Productivity	
Wastes and Reworks	Throughput	Variation In throughput
	Agility	

more reliable or robust than those that are sensitive to such variations. In a broader sense, robustness has also been defined as insensitivity to variation in product performance and behavior with respect to one or more of the following items.

- Change in market and customer needs.
- Change in process plan and production technology.
- Evolution path of the product developed over time.
- Current and future model variation.
- Piece to piece variation due to manufacturing, imperfection, etc.
- Future change in manufacturing conditions.
- Functional variation due to changes in the inner inherent properties of the system, such as deterioration, wear, tear, fatigue, corrosion, etc.
- Variation between sub-units.
- External noise, such as environmental conditions (temperature, humidity, dust) and other customer usage conditions.

2.7.4 Methods to Ensure Robustness

The above definitions clearly indicate that opportunity for quality enhancement exists through proper assessment of robustness of a product concept in the early stages of design. There are two methods of assessing robustness.

- *Analytical based methods:* In this approach it is assumed that it is possible to obtain an analytical model of the design in terms of its variables (problem parameters) in explicit or implicit forms (see Figure 2.18). CAD and FEA are examples of an implicit model of a design. Some mathematical models, such as a set of equations, a computer program, or a procedure, relate an output response to a set of control or noise variables. They are examples of an explicit model of a design. In both cases, appropriate mechanisms exist to determine the effects of parameter variations without performing a series of costly experiments.
- *Statistical based methods:* When an analytical model of the design cannot be obtained, the other recourse often used by designers is to carry out a series of statistical experiments (see Figure 2.18). A subset of observations is randomly selected from a larger set of observations. The larger set of observations is called population and the smaller set is called a sample. Since the population can be extremely large, work groups often examine the sample (e.g., by computing mean, variance, and standard deviation of the sample) to make inferences about the mean, variance, and standard deviation of the large population. Calculation of correlation coefficients is one way of determining relationships between variables and their importance. Another approach involves a multiple regression analysis. Other statistically based methods include concurrent use of techniques such as probabilistic analysis, statistical decision theory, and design of experiments during product or process designs [Taguchi, 1993].

2.7.4.1 Analytical Based Methods

Computation of Output Variation   Equations 2.27 and 2.28 represent an explicit form of representations relating a model output, Y, to a vector of constituent variables, $\underline{p}$. Y could also be in any implicit form, as it appears for deflection, stress, or stiffness results in the FEA analysis model.

If $\underline{p}_{i0}$ is a vector of nominal values for the vector $\underline{p}$, then a new value of an output response Y at any (p_i) can be expressed using the first order Taylor series expansion.

that is;

$$Y(\underline{p}_i) = f(\underline{p}_{i0}) + [\{\partial Y/\partial p_i\}\,(p_i - \underline{p}_{i0})] + \cdots \atop \big|_{\underline{p}_i = \underline{p}_{i0}} \tag{2.30}$$

or,

$$Y(p_i) = f(p_{10}, p_{20}, p_{30}) + \cdots + [\{\partial Y/\partial p_1\}\,|\,(p_1 - p_{10})] + \cdots$$
$$+ [\{\partial Y/\partial p_i\}\,|\,(p_i - p_{i0})] + \cdots$$
$$+ [\{\partial Y/\partial p_n\}\,|\,(p_n - p_{n0})] \tag{2.31}$$

Sensitivities when the design is at the nominal $\equiv \{\partial Y/\partial p_i\}\ \big|_{\underline{p}_i = \underline{p}_{i0}}$
for $I = 1, 2, 3, \ldots, n$
$\qquad\qquad\qquad\qquad\qquad\qquad\qquad\qquad\qquad\qquad\qquad\qquad\quad$ (2.32)

Note, the first order derivatives (see equation 2.32) are the sensitivities of Y with respect to p_i when the design is at the nominal. The variation of the output response Y is related to the variations in p_i. Some interesting conclusions can be drawn about sensitivities depending upon when the derivative function is evaluated, meaning whether the with-respect-to variable p_i is a control $(p_i \to p_c)$, or a noise variable $(p_i \to p_n)$, see Figure 2.20.

Note, p_i, is an ith variation, which may be a control type or a noise type. That is,

$$p_i \in (p_i^{\,c}, p_i^{\,n}) \tag{2.33}$$

or
$$p_{i0} \in (p_{i0}^{\,c}, p_{i0}^{\,n}). \tag{2.34}$$

Equation 2.30 can also be rewritten as

$$Y(\underline{p}_i) = f(\underline{p}_{i0}) + [\{\partial Y/\partial p_i^{\,c}\}\,(\underline{p}_i^{\,c} - \underline{p}_{i0}^{\,c})] \atop \big|\ \underline{p}_{ii} = \underline{p}_{ii0}^{\,c}$$
$$+ [\{\partial Y/\partial p_i^{\,n}\}\,(\underline{p}_i^{\,n} - \underline{p}_{i0}^{\,n})] + \cdots \atop \big|\ \underline{p}_i = \underline{p}_{i0}^{\,n} \tag{2.35}$$

In order to minimize the variation in output response, we must choose the variable, p_{i0}, in such a way that the second and third terms of equation 2.35 tends to zero, that is,

$$Y(\underline{p}_i) \Rightarrow f(\underline{p}_{i0}). \tag{2.36}$$

To nullify the contributions from the second and third terms in equation 2.35, a work group must choose a set of nominal values for control variable p_{i0} such that one of the following (a) or (b), or both, are true.

Existence with respect to	Taylor's Coefficients	
	Evaluated at $p_i \longrightarrow p_i^c$	Evaluated at $p_i \longrightarrow p_i^n$
When control variable is changed or varied (p_i^c)	$\left.\dfrac{dY}{dp_i}\right\|_{p_i = p_{io}^c}$ exists $(p_i^c - p_{io}^c)$ is finite and nonzero	$\left.\dfrac{dY}{dp_i}\right\|_{p_i = p_{io}^n}$ is unknown $(p_i^n - p_{io}^n)$ can be computed
If we could perturb the noise variable (p_i^n)	$\left.\dfrac{dY}{dp_i}\right\|_{p_i = p_{io}^c}$ is unknown $(p_i^c - p_{io}^c)$ is unknown	$\left.\dfrac{dY}{dp_i}\right\|_{p_i = p_{io}^n}$ is unknown $(p_i^n - p_{io}^n)$ is known

LEGEND:

c: control variable
n: noise variable
o: original point
p: parameter
Y: response function

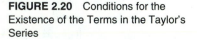

FIGURE 2.20 Conditions for the Existence of the Terms in the Taylor's Series

(a) Choose a set of control variables in such a way that the sensitivity of the response with respect to a control variable is negligible or zero, that is,

$$\partial Y/\partial p \Rightarrow 0.0 \; \forall \; (p_i = p_i^c) \atop |\underline{p} = p_{i0} \tag{2.37}$$

where p_i is given by equation 2.33.

(b) The variation in the noise variable as a result of variation in the control variable is negligible, that is,

$$(p_i^n - p_{i0}^n) \Rightarrow 0.0 \; \forall \; p_i \in p_i^n. \tag{2.38}$$

The challenge is to find a set of values for the control variables p_i where both of the above conditions hold true. In that case a work group achieves a robust design, and,

$$Y(p_i) \cong f\,(p_{i0}) \tag{2.39}$$

The mechanism to ensure robust design or to provide robustness is to minimize the deviation in the original output response subject to changes in the control variables. This translates into finding a set of nominal values (p_{i0}^c) for the control variable so that a large number of the terms in condition (a) or (b) hold true. Condition (b), in optimization terms, is equivalent to "targeting" or setting the bounds on variations. Together they form a robust design optimization problem.

If the noise variables cannot be checked to nullify the contributions from the second and third terms in equation 2.35, work groups must choose a set of nominal values for control variables (p_{i0}) such that the following (c) or (d), or both, are true.

(c)   We choose a set of control variables in such a way that the variation of control variables as a result of noise variable is negligible, that is,

$$(p_i^c - p_{i0}^c) \Rightarrow 0.0 \ \forall \ p_i \in p_i^c. \tag{2.40}$$

(d)   For those set of p_i for which the noise variable cannot be checked, the other option is to change the nominal values of the control variables that interact with the noise variable to reduce sensitivity:

that is; $$\partial Y/\partial p_i \mid \Rightarrow 0.0 \ \forall \ p_i \in p_i^n \tag{2.41}$$
$$\mid p_i = p_{i0}.$$

In either case, we search for the nominal values of the control variables that yield a flatter spot on the sensitivity curve in one domain and a less sensitive design in another domain.

This concept is very different from tightening the tolerances. However, if none of the control variables interact with a particular noise variable in providing results similar to (a) through (d), the only recourse available is to tighten the tolerances corresponding to noise variables.

Sensitivity Computations   In explicit analytical forms of representation, such as FEA, a finite difference method can be employed, using forward difference:

$$\partial Y/\partial p = \frac{[Y(p_{i0} + \Delta p_{i0}) - Y(p_{i0})]}{\Delta p_{i0}}. \tag{2.42}$$

The advantages of analytical based methods are manifold. They do not require hardware for testing or evaluation and thus can be conducted quite early in the design stage and even integrated into a mainstream design process.

2.7.4.2   Statistical-based Method

The measure of output variability of a function, Y, in equation 2.30 is called a *variance*. There are many approaches used to measure this variability. There are corrective techniques that first measure the effects of such variation and then correct them. These measures can be on-line or off-line depending upon the purpose and its stage of application during a PD^3 process.

Three major ways to statistically characterize data are the arithmetic average (or mean), the variance, and the standard deviation. The mean indicates the central tendency of data, while the latter two describe the spread of the data. The formula for the arithmetic average is

$$\overline{p} = [\sum_{i=0}^{i=n} \{p_i\} / n].$$ (2.43)

The $\sum$ notation (called sigma) indicates summation of variables, and n equals the number of elements in the data set or sample. The equation for the mean can also be expressed as

$$\overline{p} = \frac{[p_1 + p_2 + p_3 + \ldots + p_i + \ldots + p_n]}{n}.$$ (2.44)

For the present equation, sigma indicates that all the observations going into the calculation of the mean will be added together.

The formula for the variance, σ, is

$$\sigma = s^2 \equiv \frac{[\sum (p_i - \overline{p})^2]}{(n-1)}.$$ (2.45)

The numerator is the sum of the squared deviations of each observation from its mean. As the observations are widely spread out from the mean, the numerator increases. Therefore, a large variance would indicate that observations are widely spread out (and could be improved), and a small variance indicates that the observations are tightly packed around the mean. The denominator stands for the degrees of freedom. In general, the degree of freedom for a particular experiment is rendered by the number of observations in the data set minus the number of estimated parameters (m) used in the equation. In most cases, the total number of observations used in the calculation is the sample size (n) [Hays, 1992].

That is,

$$\text{Degrees of freedom} = (n - m).$$ (2.46)

Degrees of freedom ($n - m$) are used to make inferences about the population variance from which the sample is drawn. The number of estimated parameters used in the calculation is often one ($m = 1$). The sample standard deviation, s, which is also a measure of variability, is the square root of the sample variance.

$$s = \sqrt{\sigma}$$

Control Charts: The control chart is the visual display of overall service or quality (see Figure 2.21). The vertical axis of the control chart represents the measurement of the overall quality characteristic of a particular quality dimension. The horizontal axis represents either the sample number or the time. The control chart includes:

1. A center line representing the average value of the quality dimensions over the entire range of samples.
2. An upper control limit (UCL) and
3. A lower control limit (LCL).

The latter of the two runs parallel with the center line.

With the notations defined earlier, the general formula for the control charts are

$$\text{UCL} \equiv \overline{p} + k\,s$$ (2.47)

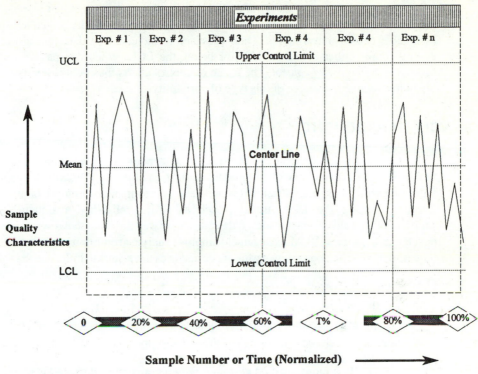

FIGURE 2.21   Example of a Control Chart

$$\text{Center Line} = \overline{p} \tag{2.48}$$

$$\text{LCL} \equiv \overline{p} - k\,s \tag{2.49}$$

where k represents a constant used to determine the distance of UCL and LCL from the center line, and s is the standard deviation. It is standard that k equals 3.0. Taguchi's method uses signal-to-noise ratio as the measure of robustness. The signal-to-noise ratio is the ratio of the mean value of performance p (obtained by experiments or by simulation) to the associated standard deviation s of the performance values [Taguchi, 1986].

$$\text{Signal-to-noise ratio} = [\overline{p} \,/\, s] \tag{2.50}$$

Taguchi's method directs CE teams to select a set of design variables (called control factors) that as nearly as possible maximize the signal-to-noise ratio.

Control charts can monitor the manufacturing processes that generate the data in the chart. Each data in the control chart indicates how the process is running at a given time. Over time, because of the natural variability in the process (called common causes of variations), we expect that not all data points will fall right on the center line. Under normal circumstances, we expect the data points to lie somewhere between the UCL and LCL lines. When this pattern occurs, the process is said to be in statistical control or *under control*.

There are other sources of variability that are due to the events that represent potential problems in the process. The sources of this extreme variability are called special or assignable causes of variations. Their presence is indicated by data falling outside the control limits, either above the UCL or below the LCL. When this pattern occurs, the process is called *out of control*. The usage of control charts thus helps identify and possibly helps remove the source of this type of variability.

2.8 VCM MANAGEMENT

Value Characteristics Matrices (VCM) management encompasses integrating all metric activities under a unified view of the product realization objectives. VCM management provides a measure of estimating fielded (system level) performances through examining the interactions among the DFX matrices such as RMS (reliability, maintainability, and supportability) value characteristics. It describes how well they fit into the system design and development process. VCM management employs analytical and simulation modeling techniques to predict fielded system performance and uses the results to influence the following:

1. DFX matrices selection.
2. Interface definitions.
3. Hardware or software configuration design.
4. Support system design during the life-cycle support process.

This gives rise to a more satisfied customer with a realized product that is more likely to function free of fault, and a product easier to restore to full term and functionality at a lower production cost and scheduled time. One way to achieve effective integration of metrics and measures into the product development process is to perform product mock-ups or demonstrations. Demonstration techniques are needed at all levels of an enterprise, beginning with the strategic enterprise level and the product and process realization cycles, to the factory floor where real-time performance measurement is done [Dika and Begley, 1991]. The demonstration provides a method of proving that a planned system is working and a method of establishing what corrective actions may be needed. Program tracking and management is a common technique for assessing performance, risk, and control. Multi-media demonstration and presentation of the integration methodologies for mock-ups and system integration analysis (SIA) are useful techniques to validate the system integration and supportability concepts. Demonstration or mock-up often involves creating a working electronic prototype or a video mockup of computer-aided electronic pictures for product, process, fixtures, layout configuration, assembly, reassembly, and art-to-part scenarios.

REFERENCES

BOOTHROYD, G. 1988. "Making it Simple: Design for Assembly," *Mechanical Engineering*. Volume 110, No. 2, pp. 28–31.

BOOTHROYD, G., and P. DEWHURST. 1983. "Design for Assembly: Manual Assembly," *Machine Design*. (November 10) pp. 94–98 and (December 8) pp. 140–145.

BOOTHROYD, G., and P. DEWHURST. 1988a. "Production Design for Manufacture and Assembly," *Manufacturing Engineering* (April 1988) pp. 42–46.

BOOTHROYD, G., and P. DEWHURST. 1988b. *Design for Assembly: Handbook.* Amherst, MA: University of Massachusetts.

BOOTHROYD, G., P. DEWHURST, and W. KNIGHT. 1994. *Product Design for Manufacture and Assembly.* New York, NY: Marcel Dekker, Inc.

DIKA, R.J., and R.L. BEGLEY. 1991. "Concept Development Through Teamwork—Working for Quality, Cost, Weight and Investment." SAE Paper No. 910212, pp. 1–12, Proceedings of the SAE International Congress and Exposition, SAE, February 25–March 1, 1991, Detroit, MI.

GLADMAN, C.A. 1969. "Design for Production," *Annals of the CIRP.* Volume 17, pp. 5–12.

GREEN, N., and M. REDER. 1993. "DFM boosts U.S. Car Quality," *Machine Design.* Volume 65, No. 9 (May 14), pp. 61–64.

HAUSER, J.R., and D. CLAUSING. 1988. "The House of Quality," *Harvard Business Review.* Volume 66, No. 3 (May–June) pp. 63–73.

HAYES, B.E. 1992. *Measuring Customer Satisfaction—Development and Use of Questionnaires.* Milwaukee, WI: ASQC Quality Press.

ISHII, K., C. JUENGEL, and C.F. EUBANKS. 1995. "Design for Product Variety: Key to Product Line Structuring." *Proceedings of the 1995 ASME Design Engineering Technical Conferences,* Volume 2, DE–Volume 83, pp. 499–506, 9th International Conference on Design Theory and Methodology, Edited by A.C. Ward, New York: ASME.

KUMPE, T., and P.T. BOLWIJN. 1988. "Manufacturing: the New Case for Vertical Integration," *Harvard Business Review.* (March–April 1988) pp. 75–81.

MAKINO A., P. BARKAN, L. REYNOLDS, and E. PFAFF. 1989. "A Design-for-Serviceability Expert System," *Concurrent Product and Process Design.* DE-Volume 21, PED-Volume 36, Proceedings of the ASME Winter Annual Meeting, San Francisco, California (Dec. 10–15) pp. 213–218.

MCNEILL, B.W., D.R. BRIDENSTINE, E.D. HIRLEMAN, and F. DAVIS. 1989. "Design Process Test Bed," *Concurrent Product and Process Design.* DE-Volume 21, PED-Volume 36, pp. 117–120, Proceedings of the Winter Annual Meeting of the ASME, San Francisco, California (Dec. 10–15) edited by Chao and Lu, ASME.

MILACIC, V.R. 1989. "Fundamental Principles for Design and Produciability," *Concurrent Product and Process Design.* DE-Volume 21, PED-Volume 36, pp. 9–20, Proceedings of the Winter Annual Meeting of the ASME, San Francisco, California (Dec. 10–15) edited by Chao and Lu, ASME.

National Research Council (NRC). 1991. *Improving Engineering Design: Designing for Competitive Advantage.* National Academy Press, Washington D.C.

National Research Council (NRC). 1994. *Unit Manufacturing Process: Issues and Opportunities in Research.* National Academy Press, Washington D.C.

NEVINS, J.L., and D.E. WHITNEY. 1989. *Concurrent Design of Products and Processes.* New York: McGraw Hill.

PRASAD, B. 1995. "JIT Quality Metrics for Strategic Planning and Implementation." *International Journal of Operations and Production Management.* Special Issue on Modeling and Analysis of Just-in-Time Manufacturing Systems, Guest Edited by S.K. Goyal and A. Gunasekaran, Volume 15, No. 9, pp. 116–142.

PUGH, S. 1991. *Total Design-Integrating Methods for Successful Product Engineering.* Addison-Wesley Publishing Co. Inc.

SHINA, S.G. 1991. *Concurrent Engineering and Design for Manufacture of Electronics Products*. Van Nostrand Reinhold.

TAGUCHI, G. 1986. *Introduction to Quality Engineering*. Tokyo, Japan: Asian Productivity Organization.

TAGUCHI, G. 1993. "Taguchi on Robust Technology Development—Bringing Quality Engineering Upstream," Translated by S.-C. Tsai, ASME Press Series on International Advances in Design Productivity, New York: ASME Press, 1993.

TAGUCHI, G. and D. CLAUSING, 1990, "Robust Design", *Harvard Business Review*, Volume 68, No. 1 (January–February), pp. 65–75.

WILSON, P.M., and J.G. GREAVES. 1990. "Forward Engineering—A Strategic Link between Design and Profit," *Mechatronic Systems Engineering*. Volume I, No. 1, pp. 53–64.

TEST PROBLEMS: CE METRICS AND MEASURES

2.1. Why does one need measures of merits (MOMs)? What kinds of measurements can CE work groups use? Why is it important to use MOMs during a PD^3 stage?

2.2. List the activities carried out during a product-subdomain management cycle. What are some of the X-ability requirements desired during this cycle?

2.3. List the activities carried out during a process-subdomain management cycle. What are some of the X-ability requirements desired during this cycle?

2.4. Explain differences between design for manufacturability (DFM) and design for assembly (DFA). Which one would be performed first in the design process, if they cannot be performed concurrently and why?

2.5. What are the world-class CE practices to strive for and how do you quantitatively benchmark metrics relevant to most common challenges faced during a PD^3 life-cycle?

2.6. How can the power of metrics be harnessed to monitor product and process variations (or changes) and subsequent implementation of error-proofing means? How can such metrics be used to prevent quality problems from occurring (or discovered through inspection)?

2.7. What are the general principles of DFA? What are the steps to minimize the number of parts for an assembly? Chart the steps followed during minimum parts assessment.

2.8. Why does a product work group design for interchangeability? How would you go about designing a family of mechanical pens, having different size leads and softness? What would you need to do to interchange some components for a ball point pen? Describe areas where interchangeability of parts would result in significant cost savings.

2.9. Describe a design for experiment method for product and process optimization? Discuss the drawbacks to collecting data during production test and/or field applications.

2.10. The success of CE can depend upon the teams' ability to define MOMs. Discuss the useful metrics of measures and how to aid the process of product realization? How can metrics be used to measure the effects of CE implementation (before and after) on the organization?

2.11. What is Design for Quality? How can other design for X-ability considerations be used to support design for Quality? How can a work-group measure manufacturability?

2.12. Design for Robustness is a concept that has been in existence for a long time. Explain why it is so important and repeatedly cited? What is achieved by using Taguchi's efficient design technique?

2.13. What are some of the potential advantages and disadvantages of using simulations as an integral part of a design process? How can material usage be minimized? What are the benefits of using stochastic simulation as opposed to determining performance analytically? Give examples of simulation models used for each problem type: interacting, coupled or concurrent.

2.14. Obtain an inexpensive or broken artifact, such as an automatic pencil sharpener, inexpensive camera, or child's toy, and disassemble it. What are the benefits achieved after going through a product disassembly? Discuss some of its necessary trade-offs. Create drawings of your disassembly sequence.

2.15. From question 2.14, think about assembly being done by a robot. What parts would you combine or change and why? How would your changes affect the fabrication complexity and cost? Create drawings of your redesign and include DFX justification for each design change.

2.16. What are the general principles of DFX? Using the design of a kitchen blender and mixer, apply several of the DFX methods to the product design and development and explain how the PD^3 process for the kitchen blender would benefit from DFX?

2.17. What are the issues involved in Design for Serviceability? List those issues you need in your redesign for a manual pencil sharpener and specify how you satisfy them?

2.18. Why are compatibility requirements important for product realization?

2.19. What are the five steps to managing variations? Explain how such steps would help in the design of an automobile engine that must run on unleaded gasoline in the summer and, to cut winter pollution, a mixture of ethanol and gasoline in the winter. List some sources of variations that exist on the production floor.

2.20. Why is it important to have measurable matrices in CE? What are the key measurements used to determine the success of a CE program? How would you change the process for a part studied in exercise 2.14? What are the primary advantages of VCM matrices?

2.21. Measurement gaps can occur when different activities are overlapped. Using CE metrics and measures, how can these measurement gaps be resolved? What tools can you use to change the design and manufacturing approach from a reactive mode to proactive mode?

2.22. How do design and manufacturing teams view tolerances differently? Does robustness mean maximizing or minimizing tolerances? Explain your answers.

2.23. Figure 2.2 lists a number of gaps during product realization. What can a company do to keep these gaps from occurring?

2.24. How do CE metrics and measures provide feedback to a realization model? How important is this feedback? What are some of the feedback examples? How many of these are inspection type measures?

2.25. How can the modular design concept control a set of production variety and, at the same time, allow a number of product variety? What roles do simplicity and modular design play in cutting costs and why?

2.26. Explain how DFM helps in reducing the manufacturing cost of a product? What are some examples of things to consider during DFM study? If you are creating a design of a single part, what should you be concerned with and why?

2.27. How can MOMs be used to verify that the system has constancy of purpose? List nine areas that should be concentrated on *up-front* to assist in assuring that a product meets some five objectives.

2.28. Do MOMs ensure concurrency? How do these measures differ from traditional business practices? Categorize some measures used in CE transformation of a part you know into preventive and inspection types.

2.29. Is FMEA a useful tool for product quality assessment? If so, explain what FMEA is, and give some examples of its usefulness. Draw a basic process diagram followed in FMEA.

2.30. Slip planes and design gaps are some brute means for controlling dimensional integrity of the finished part. What are some of the design means to managing variations?

2.31. Describe the guidelines for three X-ability considerations that you are familiar with. Create a table listing the terms commonly substituted for X in DFX.

2.32. What are the important components of RMS and how does each component contribute to RMS objectives?

2.33. Graphically layout a SE (Serial Engineering) activity plan and a corresponding concurrent engineering (CE) activity plan. Show a spreadsheet computation of time savings. Give a real life example where some or all principles of CE are applicable.

CHAPTER 3

TOTAL VALUE MANAGEMENT

3.0 INTRODUCTION

Besides declining profits, companies today are facing a variety of new challenges. Persistent among these are global competition, increasing labor costs, rising customer expectations, shorter product life cycles, and increasing government regulations. The old techniques of coping with *short-term fixes* in *reactionary mode* are not enough to sustain the continuing competitive pressure. Companies are slowly realizing the need to focus on precautionary measures, that is, to concentrate on problem preventions rather than fighting fires most of the time. There is a need to plan ahead, combine the available corporate talents on marketing, management consultants, design engineers and manufacturing staff to work closely together and somehow plan a product that has all relevant life-cycle values. By designing and manufacturing products that reflect the customers' desires and tastes, everybody wins. Customers see the benefits and are willing to purchase the products. Manufacturers bag more profits. Today, many companies are interested in improving their competitive position in the world marketplace. It is important for these companies to bring new product innovations and value-added services to the market in a timely fashion. This is because those companies that introduce new concepts at high quality levels often grab the largest share of the market. Timely product development benefits a company in many ways:

- By early introduction, the company gains the customers' confidence, who see their needs filled and buy the products. The company gains easy market share, giving itself a competitive advantage.
- Customers become familiar with the products and, therefore, develop loyalty and are less likely to switch.

111

- The company gets on the learning curve ahead of their competitors.
- The company is able to set the product price and reap its profits much longer.

In order to develop a product that customers like, companies must know the wants (*must-have*), needs (*like-to-have*), and desires (*wish-to-have*) of their customers—the end users of products or services and their business. Total Value Management (TVM) is a methodology designed to build values into an entire product realization process. This way, customers get what they are interested in, employees are happy and proud of their contributions, and the company makes a fair profit.

This is schematically shown in Figure 3.1. Two objectives are shown crossing each other at a 45 degree angle (1) company profits and (2) customer and/or employee satisfaction.

Two parallel lines are shown separating a company's current level from its competition's level. They cross in the middle creating four cross-points (A, B, C, and D). Each cross-point on the chart (see Figure 3.1) reflects the state of a company depending upon its choice of strategies.

- *Level A:* This is the bottom-most cross point. This level reflects the most poorly performing company. At this level, there exists minimum customer and employee satisfaction and a marginal profit. This is often the case when companies are desperately trying to win back the lost customers and are not performing well (showing a poor return on investments). When a product does not measure up to the customer's expectations, it does not sell well. This often forces companies to introduce incentives to move their product lines faster. Some of these techniques are quite expensive, since it cuts into their profit margins. Companies that engage in those tactics reach a level B or level C on the chart.
- *Level B:* Some companies reach level B by adjusting price, increasing sales commission, advertising heavily, improving public relations, carrying extra inventory or by other similar incentives. Resorting to such fixes does improve customer satisfactions temporarily. However, it comes at the expense of the company's gross profits. There are other kinds of fixes that can also help a company move away from level A.
- *Level C:* At this juncture, any increase in profit margin happens temporarily and that too at the expense of customer satisfaction. This is shown by level C in Figure 3.1. Examples of such fixes are performing integration, improving service, cutting costs, value engineering, reducing waste, applying just-in-time, reducing headcount, or other similar means. In either situation (level B or C), the company is not able to achieve both—that is increase in customer/employee satisfaction and increase in profit margin.
- *Level D:* At level D, both the mismatches in customer/employee satisfaction and profits are removed. TVM is a technique that lets you achieve both these goals.

The major goals for TVM are to incorporate customer voice during conceptual design [Cagan and Agogino, 1987], upgrade enterprise infrastructure, and improve quality, functionality (X-ability), innovations (tools & technology), and responsiveness.

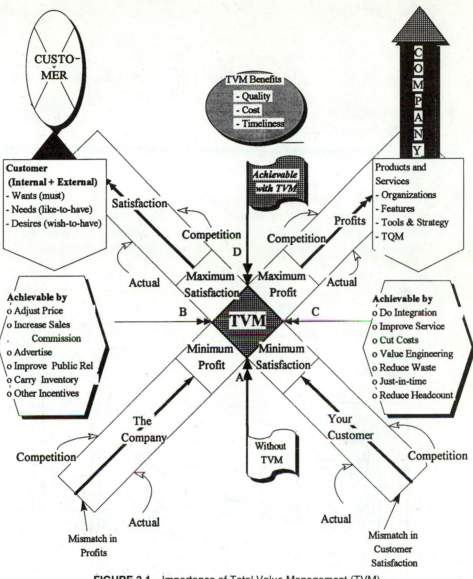

FIGURE 3.1   Importance of Total Value Management (TVM)

3.1   TOTAL QUALITY MANAGEMENT

Typical of a company going through a *total quality process* would be a realization that *quality* does not limit itself to *product quality*, but also to the reputation of the company as a quality provider (i.e., TQM). Fundamental to TQM are the ideas that everyone in the organization has a customer, internal or external. That improvement comes from understanding and improving business processes and applying a set of governing quality principles (see Figure 3.2).

- Process of continuous improvement.
- Appropriate 3Ps (policies, practices and procedures) deployment at all process levels: culture, organization, teamwork, facility, etc.
- Employee involvement.
- Management leadership.
- Tools and technology deployment; Andon, Kanban, Pokayokal Jidoka, etc.
- Measurements of quality deployment.

TQM is about institutionalizing a process of continuous improvement for individual employees, work groups, and organizations. What differentiates TQM from other management techniques is its emphasis on *concurrent process re-engineering*—establishing a change process for improvement. The major tools used in this process are:

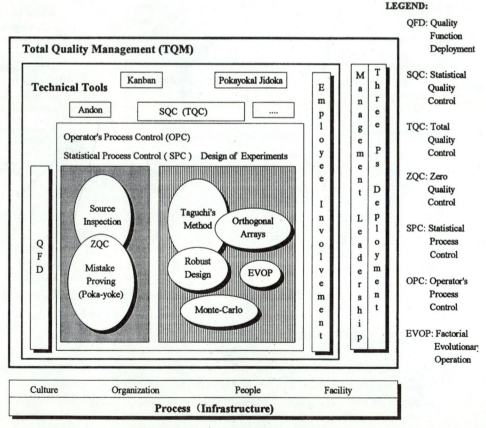

FIGURE 3.2   Quality Management Principles

Quality function deployment (QFD) [Clausing and Hauser, 1988].

Operator's process Control (OPC), such as zero quality control (ZQC), statistical process control (SPC), design of experiments (DOE) and Factorial Evolutionary Operation (EVOP).

Statistical Quality Control (SQC) [also referred to as Total Quality Control (TQC)].

Fool-proofing process for production (or Pokayokal Jidoka concept).

Andon system.

There are several books that discuss each of these major tools in great detail [Ishikawa and Lu, 1985; Feigenbaum, 1990; and Juran and Gryna, 1993]. The definitions of OPC are shown in Figure 3.2 through a Venn-type diagram. Preliminary details of the tools are contained in Chapter 7 of volume II. Another aspect of TQM is employee involvement, that is, quality has to be led by every employee in the company, including suppliers and customers (see also sections 4.2.3 and 5.3, volume I). Each team member needs to know what to do, and how to do it. Each member must be empowered to make suggestions, be able to measure performance, to input or receive feedback, and to have the right tools to do his and her jobs. Management leadership is discussed in Chapter 5 (volume I) under section 5.5, "Management Styles or Philosophies." Types of CE organization are discussed in section 5.4. The measurement of quality is described next.

3.1.1   Quality Measurements

A desired process for a quality engineering operation is depicted in Figure 3.3. Quality measures are shown to be present in some form in every major step of the product design, development, and delivery (PD^3) process, such as product design, process design, manufacturing planning, and production. Statistical quality control (QC) methods are shown applied both before and after a product is manufactured. The methods applied before a design is completed are called design-oriented QC methods (e.g., fool proofing, the Pokayokal Jidoka concept [Ishikawa and Lu, 1985]). Those applied after a product is manufactured are called inspection-oriented QC methods (e.g., the Andon concept, Quality Circles, etc.) shown in Figure 3.3. The *built-in quality by design* methods are shown to be applied during product design and process design steps. Examples of such built-in measures include the Pugh method, quality function deployment (QFD) [ASI, 1989], design for manufacturability and assembly (DFMA), failure mode and effects analysis (FMEA), design for quality (DFQ) techniques, and Taguchi methods [Carey, 1992]. The design-oriented QC methods, shown as being part of the manufacturing planning step, provide an important defect prevention mechanism. They are not commonly found in most traditional systems. At the management level, Deming's QC philosophies are employed to provide a set of common and consistent goals and to ensure an organization with consistency of purpose.

Let us assume

pt_o is the output of the product design and

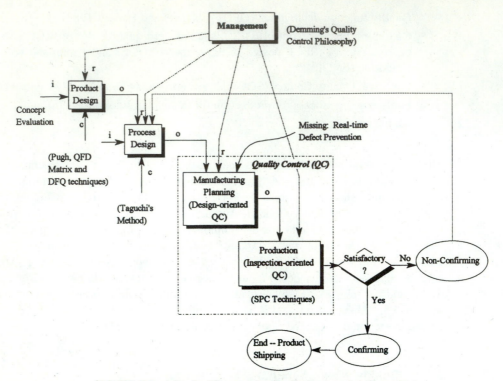

FIGURE 3.3   A Desired Process for Quality Engineering Operation

pt_i, pt_r, and pt_c are the inputs, requirements, and constraints, respectively, for the product activity block.

As shown in Figure 3.3, outputs are functions of inputs, requirements, and the constraints. In equation form, this means:

$$pt_o \equiv f_{product}\ (pt_i, pt_r, pt_c) \tag{3.1}$$

where $f_{product}$ indicates a product function.

In a similar manner, let us denote ps_o as the output of the process design and

ps_i, ps_r and ps_c are the inputs, requirements, and constraints, respectively for the process activity block.

Following Figure 3.3, outputs for process design can similarly be expressed in the following equation form:

$$ps_o \equiv f_{process}\ (ps_i, ps_r, ps_c) \tag{3.2}$$

where $f_{process}$ is a process function.

A concurrent process of quality engineering (QE) with many of these methods is shown in Figure 3.4. Quality begins with concurrent product and process design running

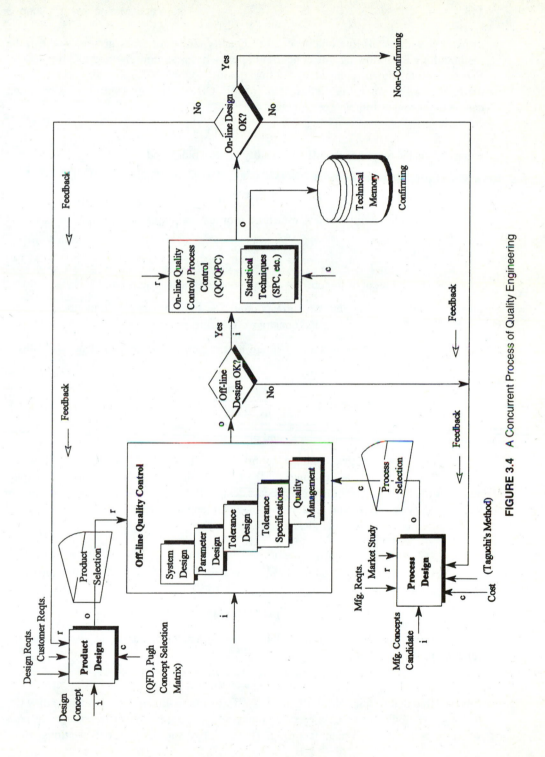

FIGURE 3.4 A Concurrent Process of Quality Engineering

in parallel with an off-line quality control. Quality circles or work groups establish the methodology for following this concurrent approach. Inspection oriented QC methods in Figure 3.3 are shown replaced by on-line quality control (QC)/process control (QPC) methods. On-line control is the last stage of QE. Off-line quality control transformation can be expressed as follows (see Figure 3.4):

$$\text{off}_o \equiv f_{\text{off-line}}(\text{off}_i, \text{off}_r, \text{off}_c) \tag{3.3}$$

where off_o is the output of the off-line quality control block and

$$\text{off}_i, \text{off}_r \text{ and off}_c \text{ are the corresponding (off-line) inputs, requirements,}$$
and constraints respectively.

Similarly, on-line quality control transformation can be expressed as follows (see Figure 3.4):

$$\text{on}_o \equiv f_{\text{on-line}}(\text{on}_i, \text{on}_r, \text{on}_c) \tag{3.4}$$

where on_o is the output of the on-line quality control/process control and

$$\text{on}_i, \text{on}_r \text{ and on}_c \text{ are the corresponding (on-line) inputs, requirements,}$$
and constraints respectively.

From Figure 3.4, it is clear that all the inputs and outputs are not independent. The following relationship exists:

$$pt_o = \text{off}_r; \tag{3.5}$$

$$ps_o = \text{off}_c; \tag{3.6}$$

$$\text{and off}_o = \text{on}_i. \tag{3.7}$$

Furthermore, there are two feedback loops. The on-line outputs are distributed at two places. The distribution is governed by the types of the on-line measurement's outputs and their relationship to product and process design. If the measurement discrepancies can be handled by perturbing the product parameters, then that portion of the output is passed as a requirement to the product design system:

$$pt_r \in \{\text{on}_o\} \tag{3.8}$$

and a portion of the output which is more process related is fed as a constraint to the process design system:

$$ps_c \in \{\text{on}_o\}. \tag{3.9}$$

As shown by equations 3.1 through 3.4, the measurement of quality (Y) is, therefore, a function of the following:

$$Y = F(f_{\text{product}}, f_{\text{process}}, f_{\text{off-line}}, f_{\text{on-line}}) \tag{3.10}$$

It is possible that in spite of all off-line QC efforts, the on-line quality outcomes may differ. A good product or process by design may not guarantee a defect-free library of parts when processed through manufacturing. There are two sources of such variations, variation added at an individual process step and variation transmitted from the previous steps. One way to ensure a fail-safe process is to institutionalize a check and balance system. Due to the highly complex and dynamic nature of manufacturing processes in the past, many companies had taken an easy route. Inspection-oriented QC methods were used to serve this purpose. QC methods were primarily used to detect the *out of control* conditions of a process after products were manufactured. A Failure Reporting, Analysis, Corrective Action System (FRACAS) was often deployed to collect data during a production test and/or field application. This approach has a number of drawbacks.

- The inspection-based QC methods rely on inspecting parts. A large sample of experiments may be needed to detect a real process shift or variation.
- When a process is out of control, a fix can only be found after the fact. This is non-preventive—it keeps on producing defective products. Some of the recourses available are to look for causes, adjust, rework, or scrap the defective products. All these consume time, money, and resources.
- A large number of defective products may be produced before corrective steps can be taken.
- There could be a long delay in the communication of information from the traditional inspection division to the manufacturing division that has the authority to fix the problems.
- It is often difficult for the manufacturing division to find the cause of the problem, devise a reasonable solution, or fix the initial problem. They may require additional data, which needs to be processed by the inspection department, causing further delay.

Many manufacturing companies, therefore, do not employ inspection-based QC methods during manufacturing. Some of the main reasons for inspection-based QCs unpopularity are:

- On-line inspection-based process is subject to influence by many variables and disturbances.
- Process data are often correlated. However, the common post-processing QC tools, such as process or control charts, assume that measured data are independent.
- Statistically it is more difficult to deal with inspection-based data than taking select samples or using filters.
- Adequate capabilities for measuring on-line variation are not easily available or may require laborious fine tuning.

The above difficulties do not constitute a good basis for not using an on-line feedback-based QC method. Most off-line methods (e.g., design of experiments) are also based on the assumption of independence of sampled data. One of the main features of an off-line method is that they can be applied quite early in the process and the feedback can be automatic. Furthermore, the on-line feedback-based QC features eliminate many of the problems associated with inspection-based QC methods. Off-line QC methods assure many of the same benefits as the on-line feedback-based QC methods, but have the following additional features.

- The product and process can proceed in parallel. There are two parts to this concurrent process: product design and process design. They refer to the two sub-domains: product management and production management cycles and loops that were described in Chapter 9 (Figure 9.13 of volume I). The off-line QC methods are shown outside of product and process design loops for clarity reasons. During this quality engineering (QE) process, the design engineering work group obtains information related to product quality specifications such as reliability, safety, performance, variability, and cost goals. Likewise, the manufacturing engineering work group obtains information related to material, process, environmental conditions, cost, and so on. The product selection from the product sub-domain cycle and process selection from the process sub-domain cycle is then fed into an off-line quality control system as shown in Figure 3.4.

- The concurrent QE process connects the off-line quality control methods (e.g., QFD, Taguchi, etc.) with the SPC and feedback controls to build an on-line feedback quality control (QC) or process control (QPC). Verification is required to determine whether the upstream resource capacity is capable of flowing product or process information at the rate needed to support the next loop of transformation. If discrepancies are found, the choice is to either change resource capacity or throughput rate of upstream loops, or change the input specifications.

- The QE process includes decision blocks for continuous improvement (Kaizen concept) instead of stopping abruptly at an unacceptable quality.

There are two decision blocks in this revised quality engineering process (see Figure 3.4). The first decision block is to ascertain that the design following the off-line QC methods are satisfactory. If not, then depending upon the situation, the control moves to one of the two sub-domains, product or process sub-domains. If the results of the off-line checks are satisfactory, it then moves to on-line QC/QPC checks. Note that this is a new block replacing the old inspection based QC method block. If the results are satisfactory, the manufacturing process is stored in the technical memory and the product proceeds to shipment. If the results are not satisfactory, the discrepancies are fed back to the product or process cycles, as the case may be, for possible fix. Next, extensive interactions between the two sub-domains and the off-line QC methods take place for producing a satisfactory design that meets all (product and process) quality specifications. This is followed by an on-line QC/QPC block with a feedback mechanism as shown in Figure 3.4. The tools required to support on-line quality control (QC) and process control (QPC) are described in Figure 3.2.

3.1.2 Why TQM Is Not Enough

One of the most common mistakes that manufacturers make when considering and implementing new modernization programs is a failure to consider total enterprise or a complete set of objectives as work groups are developing their TQM solution. Work groups want to rapidly introduce new TQM features, new technology, and new QC fixes into a specific area of a product line, or to quickly re-engineer a particular process that has been the cause of an unacceptable quality. The tendency is to seek a TQM solution which fits one's own unique environment or to seek improvement for a specific objective, such as quality. Usually, little or no consideration is given to what impact such a solution would have on other functional areas of the company or how that solution will affect other company objectives. A proposed solution might temporarily quench the thrust for a modernization program or even get an initial problem cured completely, but this may also adversely affect other areas or other goals of the company. The fact is—a continued dependence on a set of conventional three Ps, conventional four Ms, and conventional seven Ts is likely to yield conventional results (see Figure 3.5). The net result may be an overall reduction of the enterprise's efficiency or it may affect the net profit margin. However, if the dependence is continued with a right or modern set of 3Ps, a right and modern set of 4Ms, and a right and modern set of 7Ts, this is likely to yield right results (Figure 3.5)—meaning great products. Great products all share a common set of properties (built into a product) that account for its greatness. Steven C. Wheelright and Kim B. Clark [1992] of the Harvard Business School call it product integrity. Integrity is what

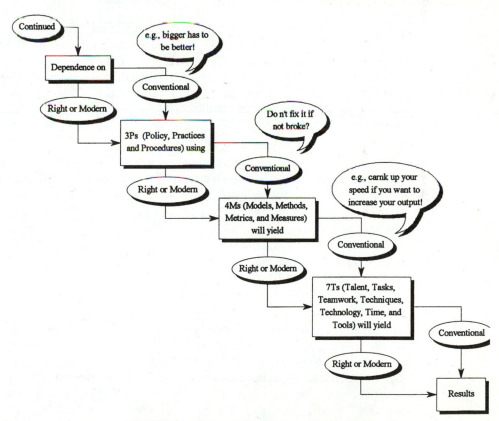

FIGURE 3.5   Why Is Value Management Important?

causes users to exclaim about the greatness of a product in words like—"They got it right!", "This is the best I have ever seen!", "This is cute!", and so on.

From studying Japanese product and process designs [Ishikawa and Lu, 1985], particularly Dr. Taguchi [Cullen, 1987], QFD [ASI, 1989], and their recent emphasis on quality recognition [Nemoto and Lu, 1987], and from reading reports of the American management visits to Japan [Ohno, 1988; Ziaja, 1990], one can draw the following conclusions: There are five striking differences between the most and the least successful companies. They can be found on both side of the hemispheres—in Japan, United Kingdom, United States, and so on.

The most successful companies are those that manufacture products with functional integrity and performance excellence. To manufacture such great products a company must concentrate on the following integrity aspects in pursuing a system solution.

- *Voice of the Customers (VOCs) at every Stage of Product Development:* Pay strict attention to and focus all efforts on the voice of the customers throughout every stage of the product development [Freeze and Aaron, 1990], from product design, definition to delivery (PD^3) cycle. The customers in this case are not limited to those who buy the product but include internal and external customers (See Figure 2.22, volume I)

- *Strong Fit:* There is a strong relationship between the needs of the product users and the features of the product itself. A product should fully satisfy those stated needs [Wheelright and Clark, 1992]. In fact, great products go beyond the simple satisfaction of stated needs. They often satisfy needs that users never thought of until they started using the product, thus, often surprising and delighting the users.

- *Multi-objective Optimization:* Work groups optimize their products for a number of considerations simultaneously: cost, weight, responsiveness, timeliness, X-ability, and various other objectives, not just the quality.

- *Parameters Desensitization:* Work groups desensitize the products subject to a wide range of variations (i.e., variation in parts, process variation, tolerances, manufacturing variations, computational approximations including human factors variation as well).

- *Function as a Coherent Unit:* Work groups take appropriate steps or precautions so that all parts and constituents of the product, including those physically outside the product, for example, interfaces, plug-ins, work together seamlessly. The product functions as a unit. Examples include integrating work force (both horizontally as well as vertically) or organizing teams cross-functionally.

It is usually more cost effective to have work groups consider all of the above integrity aspects when pursuing a solution even though only one aspect may be targeted initially. Thus, when the time comes to modernize through other integrity aspects, one feels assured that previously introduced technologies or new processes will not adversely affect any other area, or will not violate any other previously assigned objective, or will not cause any adverse situation. Considering one particular aspect, or one functional area of a company, or one particular objective for improvement often leads to a sub-optimized design of a product.

3.2   TOTAL VALUE MANAGEMENT

Total Value Management (TVM) is proposed here as a concept to replace Total Quality Management (TQM). The six major recognized objectives of TVM are: quality (function-wise), X-ability (performance-wise), cost (profit-wise), tools and technology (innovation-wise), responsiveness (time-wise), and infrastructure (business-wise). These objectives are examples of what characterize a *world-class* product. Clearly, *quality,* in the afore-mentioned sense, plays a minor role in fostering a total optimized product from a world-class perspective. Building a product that optimizes these six value objectives intrinsi-cally, not just on the basis of *quality,* ought to be the dream of any progressive company. How effectively, efficiently, and quickly the work groups are able to succeed in this en-deavor depends upon many such factors that need to be considered. TVM is meant to pro-vide a winning path towards a global market share and profitability.

3.2.1 Basis of Value Management

Value management requires a commitment to incorporate integrity characteristics at all lev-els of human-computer interactions, that is, with respect to the product, process, and the or-ganization (PPO). Besides pursuing on the above six integrity characteristics, products with strong integrity (in terms of functions and performance excellence) are usually designed and developed around a clear and a core set of Ms (i.e., 4Ms, methods, models, metrics, and measures) as discussed in volume I. Normally, these Ms are defined early in the life cycle of the product and followed religiously throughout a PD^3 process. Many books are written on this topic. The book *Out of the Crisis* by Edward Deming presents a forceful statement on the corporate philosophy needed to make meaningful quality improvements [Deming, 1986]. The emphasis is placed on five fundamental areas (see Figure 3.6):

- *Teams:* Work groups are the primary source of value improvement (e.g., quality or productivity), limited only by their corporate knowledge, expertise, and talent—what they do on the job and how they do it.
- *Value System:* Every work group must understand what the ultimate corporate vi-sion is, what value system work groups are expected to operate under, the corporate philosophy, and so on. For example, a work-group goal should be to minimize the total cost over the life cycle of the product, not just to save on the purchased parts.
- *Customer:* How to determine what customers would like and what to build in a product that customers would buy.
- *Organization:* Most problems arise from the inflexible process (e.g., 3Ps), not from the people or the work groups who are executing it. Management can fix this by changing the process or through empowerment. Management must support and pro-vide resources, show time commitment, break down barriers if needed, and create a set of consistent goals/objectives with a clear constancy-of-purpose vision in mind.
- *Methods:* Determine how people develop their skills, their knowledge (e.g., analyti-cal or statistical knowledge), and how the pride work groups take in their contribu-

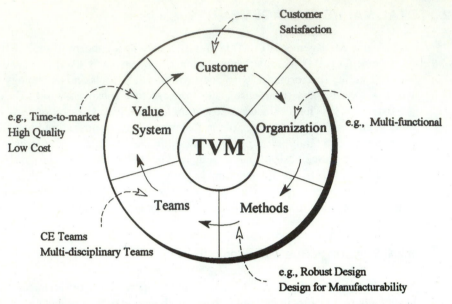

FIGURE 3.6   Basis of Value Management (based on Dr. Deming's Philosophy)

tions (education and training). Are the work groups' decisions formed on a sound engineering or rational basis? Methods also include setting up the right processes up-front, for instance, establishing a multi-disciplinary review team during analysis and design stage makes it more likely to catch defects soon after they are made.

Total Quality Control by Armand Feigenbaum [1991] addresses a complex nature of quality across an organization. It covers a wide variety of topics, including the basics of statistical process control, how to set up systems to monitor quality, and how to design better quality products. It also discusses the economics of improved quality and how to get large savings and profits from a modest investment [Feigenbaum, 1990].

The next section focuses on a total value management (TVM) methodology and a set of deployment steps necessary to yield a *world-class* product design.

3.3 METHODOLOGY FOR TVM

One of the requirements in developing a world-class product is to keep the *voice of the customers* in focus. World-class product means the developer is providing the best class of values, low product variation, and at the same time, maintaining the lowest possible variable cost, component weight and manufacturing investment. To accomplish this without a methodology or a systematic process would be a formidable task. Work groups need a methodology to

- Deploy the voice of the customer throughout each step of the PD^3 process.
- Help make right decisions as work groups go on developing the product.
- Reflect the voice of the customer in every piece of the product or service that a work group provides.

TVM is such a methodology. TVM is a six step process.

1. ***Empower a cross-functional team:*** In section 3.1, it was pointed out that quality methods and computer tools cannot stand alone, other value methods must be considered simultaneously. Tools and methods by themselves cannot transform a company into a world-class product producer or a world-class service provider. Many who attempted have learned through the school of hard knocks that unless members of cross-functional teams are working with each other, excellence in service and product cannot be attained to its fullest extent. At the present time, it is not possible to automate all decisions or capture all knowledge about the products (in technical memory or knowledge-based systems (KBS)). Chapters 6 and 7 of volume II discuss KBS and life-cycle intent capture in more detail. Employee or work group involvement is a part of teamwork. The best of the successful companies have found that work groups working in cross-functional teams focused more on the benefits to the whole organization than their individual departments or groups. To succeed, an organization needs both a cross-functional team and a team-oriented constancy-of-purpose management style.

2. ***Transform what customers say they want into a design build:*** This is the key to building a *world-class* product at competitive prices. This step can further be broken down into four sub-steps.

    - Listen carefully to the voice of the customers (VOCs), what the customers tell us, and make those VOCs *expressible* in their language.
    - Translate those VOCs into technical specifications that are *actionable* by all different work groups involved in the PD^3 process.
    - Make those technical specifications *understandable* by work-group members in every related nook and cranny of the multi-functional organization.
    - Find ways to make sure technical specifications are *executable* into the product. For example, during the process of translation, planning work-group must ensure that in an attempt to satisfy customer requirements other work groups have not inadvertently created any secondary affects or lost a portion of the customer's original intent.

    The incorporation of customer wants and needs into successful products and services is the essential activity of successful companies. Along the way, employees and stockholders require that the company stay in business and make a fair profit (see Figure 8.8, Chapter 8, volume I).

3. ***Prioritize activities within the product development process:*** In view of what makes the most sense to the customer, employees and the business. The activities

that are prioritized should be based on principles of work classifications or distributions that are not ambiguous. Work group managers involved in product development know that it is no longer enough to improve product quality and responsiveness to customers, or to take steps to eliminate waste, rework, and to reduce labor costs. They must also continuously improve their methods, processes, and tools to drive the costs down based on discriminating facts, not just hear-say.

4. *Deploy parallel tracks of value characteristics:* Deploy parallel tracks of quality, X-ability, productivity, tools and technology, infrastructure, and so on. The goal is to bring technologically superior quality products to the market at competitive prices before anyone else does (see Figure 8.8, Chapter 8, volume I). Many organizations believe that value objectives like quality, cost, X-ability, and responsiveness are conflicting issues, which is true. However the two presumptions that (1) they cannot be traded against each other, and (2) they cannot be achieved simultaneously, are not true at all. The degree to which a work group can overcome these two presumptions depends on which stage of product development a team is in. The natural conflicts among these objectives demand that they all be considered in a systematic and scientific manner as early in the process as possible. The chances that an organization can achieve a best overall product value early in the life cycle is more likely than during a later (e.g., during the detailed design) stage. For example, Japanese manufacturers have shown many times that they can produce cars of far better quality than their United States counterparts.

5. *Desensitize parameters to variations:* Apply design, manufacturing, and production principles that emphasize reduction of variations around target values. The goal is to insulate parts from all types of variations so that each part is consistently robust. This is accomplished by desensitizing the product to variation in parts, manufacturing variations, and customer uses; and desensitizing the manufacturing processes to variation in equipment operators, and materials. The latter can be accomplished by finding out the ranges of variation (upper and lower limits) in parameter values for which the response function is insensitive. Insensitivity indicates that variations in functional values are very small or negligible. Figure 3.7 shows a plot of several response functions when a concerned parameter value is changed. There are portions on each curve that are marked as "insensitive range". Clearly one can then choose a set of respective parameters for which the response falls within this range assuring that the corresponding functional value will not change very much even if one cannot hold the parameter strictly at its nominal value. What work groups want to design are products that satisfy the customer needs, perform well under a wide range of usage conditions and that are not difficult to manufacture and/or assemble properly [Freeze and Aaron, 1990].

6. *Function as a coherent unit:* The product realization process is the foundation to building a TVM system the first time and every time thereafter. Because the realization process is iterative, work groups are not reinventing the wheel with each new product line program or with each model release of the product. The realization process allows the teams to draw upon an existing technical memory including historical data and digital models, to apply what makes sense and redesign what does not,

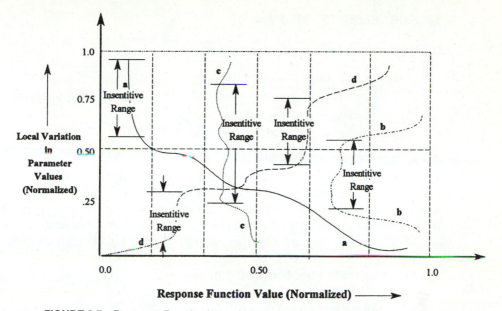

FIGURE 3.7   Response Function Insensitivity to Local Variation in Selected Parameters

thus repeating past successes, not past failures. Toyota achieved success with its fa-
mous vehicle development process called Kozokeikaku (or K-4 in short) because the
employees were culturally trained to review structures of a similar line of recent Toyota models whenever they had to come up with a new design, and identify the features
that could be applied [Martin, Sawyer, and Sorge, 1995]. The Toyota development
process in conjunction with the trained work force in the K-4 culture, in reality, provided "a culturally prescribed landscape of looking at the vehicle engineering relationships and how the whole structure snaps together." The K-4 process worked well
because the employees were trained in the Toyota vehicle development process
(VDP) and had a full understanding of vehicle histories. The product realization
process does the same thing but it does it more systematically and consistently. The
burden on the part of the K-4 employees to follow a prescribed procedure is shifted
to the product realization process (taxonomy) itself. The taxonomy accounts for
ensuring that all needed checks are performed and forces the teams to draw upon technical memory (knowledge-bases) to support the infrequent decision making process.

CFD is a deployment methodology to deploy several value objectives concurrently. TVM
is a management methodology to incorporate or infuse total values into the entire system.
This means concurrently considering up-front all relevant values and ensuring world-class
performance in each value class. There are other popular tools or methods that can be part
of this deployment methodology, namely, QFD, SPC, value engineering, FMEA, Taguchi
methods, parameter design, design for experiments, and so on [Carey, 1992]. They are
discussed in section 3.6 of this chapter.

3.4 MAJOR ELEMENTS OF TVM

Total value management is a concatenation of

1. Concurrent Function Deployment (CFD)
2. Seven Ts (talents, tasks, teamwork, techniques, technology, time, and tools)
3. Four Ms (methods, models, metrics and measures)
4. Three Ps (policies, practices and procedures)

It is crucial to have all four of the above elements in place to successfully manage and renovate a company. These four aspects of TVM can be thought of as four legs of a stool. The Seven Ts is referred to as the *employee involvement leg*. Four Ms is referred to as the *management leadership leg*, and the three Ps is referred to as the *business process leg*. Without any one of these legs, the stool tips over. Thus, all four elements are required in a successful company. They are discussed individually in the following sections.

3.4.1 Concurrent Function Deployment

In Chapter 1 of this book, the author described a concurrent function deployment (CFD) architecture. This methodology explained key aspects for preparing an environment for such an architecture, including types of work groups, monitoring, refining, and measuring the CFD process; and extending the effectiveness to the company's principal trade partners. The relationship of QFD to CFD was discussed with a particular emphasis on continuous process improvement (CPI).

Although concurrent function deployment focuses on product and process values (e.g., cost, inventory and investment reduction), these may not have to come at the expense of the business process (both management and organization). Improvements in the product, process, or organization (PPO) content that deliver a limited set of values (such as quality, cost, etc.) cannot be expected to provide improvements in all other business aspects, such as organization 3Ps, culture, human factors, and so on. Work group or team members must view other members of the work group or team as internal customers. There is a definite need to manage the business process. The requirements of the internal customers must be met, at the same time, they must work together to continually rethink or re-engineer the business process with which they are accustomed. TVM provides the needed management process infrastructure, the missing link of this PPO value-chain continuum.

3.4.2 Seven Ts and Employee Involvement

The first challenge is to energize the work force so that they buy into the concept of value improvement (not just the quality improvement) in every aspect of their actions. The second challenge is to organize the work groups so that employee efforts are aligned with the company goals. The way to meet these two challenges is cooperative teamwork. Teamwork consists of four elements: virtual teams, technology teams, personnel teams and logical teams (see Chapter 5 of volume I). One part of teamwork is familiarizing the work groups with the proper use of the technical tools, the other part is employee involvement. With all sorts of empowered tools, it will not do much good if employees are not motivated. The third challenge is to bring the right kind of talent to the right kind of tasks. We can reorganize the tasks in the best possible way but it will do little good if the right talents are not put to work on them. Furthermore employees, no matter how motivated they are, may not be able to function well until the right techniques are in place and are supported by the right set of technologies. Such core enablers of CFD are arranged in this book as chapters. Figure 3.8 shows the relationship between these core enablers of CE and the influencing agents.

3.4.3   Four Ms and Management Leadership

The four Ms (Methods, Models, Metrics and Measures) play a crucial role in defining TVM goals. For example, if management is actively involved, then possible value thrusts that have been developed so far cannot be preempted. It cannot be overshadowed by other

FIGURE 3.8   Relationship between CE Enablers and the Influencing Agents

programs of the month with which marketing often spends most of its time. Management leadership plays a key role in policy deployment. Management must provide clarity in leadership through mission statements, standards, constancy of purpose, incentives, and adjustments to the rewards system. Management leadership includes three elements.

1. **CFD Deployment:** It involves deployment of a target value set, which addresses the voice of the customer. The Kano Model, requirements prioritization process, and their integration into product, process, and production strategic planning are parts of the CFD deployment.

2. **Management Style:** A major part of management is encouraging employee involvement in the entire PD^3 process. When management teams subscribe to the right management style they create a climate that supports employee involvement and aims at employee discretionary effort.

3. **Culture:** Culture is a deep-rooted behavioral thing. The general belief is that one cannot simply change a company's culture by focusing just on culture per se. It may be a formidable task to change the company culture, but it is not at all difficult to change the processes that created it. Set-based method provides one such mechanism to easily change the process. Taxonomy is a part of set-based methods. Taxonomy shifts the pressure from the employees who are culturally venerable to a loop-based product realization process that the work groups may find culturally acceptable. As highlighted in this Chapter, part of this change is altering management style, another part is providing leadership by establishing a direction, and enforcing standards, set-based methods, policy deployment goals, constancy of purpose, and a mutually supportive action plan.

3.4.4 Three Ps (Policies, Practices and Procedures)

The concept of TVM does not limit itself to the values of the product (e.g., quality, design for X-ability (DFX), cost, etc.), but also includes the values of the 3Ps—the business processes that created the product such as infrastructure, organization, internal and external customer satisfaction, work-group cooperation, culture, etc. CFD operates best within a TVM environment since TVM fosters a conducive culture that provides harmony between customers, work groups (employees), and the business. TVM is based on the philosophy that the best way to expand sales and to increase profit potential is to provide customer satisfaction through superb products and services.

3.5 TVM IN THE PRODUCT DEVELOPMENT PROCESS

TVM methodology offers a systematic way of developing a product from its inception to finish. TVM techniques can be applied at several steps during the PD^3 process, such as concept development [Begley, 1990], engineering, piloting, manufacturing, and product support. Each product development stage ties together with the corresponding deployment tools (e.g., QFD, CFD, etc.) or management (SPC, OPC, etc.) planning tools. This results in the selection of the best application of design, process, and production capabilities that is possible at each step. TVM supports this selection with sound numerical targets for quality, cost, weight, investment and process capability at each point in the PD^3 process. TVM is based on the principles of concurrent engineering and employs cross-functional teams. It is a method for incrementally developing a product from art to part. At each step, TVM simultaneously considers many of the parallel states of product development. As work groups move from step to step, the governing attributes are refined, while TVM deploys the corresponding state's value functions leading the evolutionary design to be the best. The three dimensional staged deployments discussed in Chapter 1 for CFD are a snapshot of how a best practiced method provides a PDT (product development team) with a set of well thought-out alternatives in support of the overall TVM program. The best design should not only stand up to the benchmarks of each step, but also be optimized for all its value functions.

As work groups become more experienced in utilizing TVM, it will naturally take less time to complete each step. Since many of the key concerns or conflicts will be identified earlier, it is believed that unexpected problems during a latter part of the PD^3 process will not occur. Thus, teams would notice a shortening of the overall PD^3 cycle [Dika and Begley, 1991].

3.5.1 TVM Objectives

Typical TVM objectives, moving from general to specifics are

> CFD Deployment.
> Value Engineering/Analysis.
> Minimize Variations.
> Eliminate eight typical wastes and rework.

3.5.1.1   CFD Deployment

To deploy CFD certain assumptions are made and certain conditions are satisfied. Most problems are system-related and not people-related, hence multi-disciplinary teams are often recommended. Traditional processes are altered to accommodate CFD's heavy emphasis on concurrency and the voice of the customer. More time is spent upstream to develop the correlation matrices. Tools, such as statistical process control, design of experiment and other product and process improvement concepts, are encouraged during staged vertical deployment. This is thoroughly discussed in Chapter 1, volume II.

3.5.1.2   Value Analysis/Engineering

The term *Value Analysis* applies to a disciplined, step-by-step thinking with specific approaches for mind setting, problem setting, and problem solving [Fowler, 1990]. It was developed to determine whether or not an artifact/system performs the way it was supposed to perform. When applied to product design, it is often used to identify unnecessary cost in an existing design. Value analysis, like systems analysis, is a method of identifying, analyzing, and predicting the functional worth-to-cost ratio of having an activity in a process [Miles, 1982]. It evaluates whether or not an activity adds value to the work function or service that is being performed. From both customers' and the company's perspectives, value is defined as:

$$\text{Value} = \frac{\Sigma\,(\text{Functions, features or Activities})}{\Sigma\,(\text{Costs})} \qquad (3.11)$$

Giving the customer more value means increasing the number of customer-desired functions, features, and so on, while reducing the cost of providing them. If the product has n functions or features, and F_i represents an ith function or feature that is provided at a cost of C_i, then the above equation can be expanded as:

$$\text{Value} = \frac{F_1 + F_2 + \ldots + F_n}{C_1 + C_2 + \ldots + C_n} \qquad (3.12)$$

System analysis is a process of analyzing an engineering system for its value content to the customers and the share holders, and *system engineering* is a process of improving or maximizing the total value content or its impact. The process flow chart and the associated process description sheets are used as a means of capturing the manufacturing process, just as an engineering schematic and bill-of-materials capture the design process. The main idea is to study the functional worth of each activity in a process and to analyze whether an activity is adding any value to a product system or not. The activity that adds value is considered useful and what does not add value to the product is considered waste. The repeated application of this analysis can lead to improving the process worth. Therefore, value engineering can be considered a method to identify and eliminate waste. There are three types of values used in value analysis: customer perceived value, process value, and company-perceived value.

- *Customer-perceived value:* This value is considered a major driver of increasing sales or market share. Customer perceived value can be increased by providing more of the following: ease of use (functions properly), options (types of features or characteristics), esthetics (style, color, convenience, and look), performance (e.g., less frequent servicing or low repair history), and salvage value (exchange price or trade-in).

- *Process value:* This is a minimum set of value-adding tasks or activities that are needed to transform an input into a customer-usable output (perform type activities). Often such process values are design dependent and cannot be achieved without spending valuable resources (time, money, or expertise). Examples include accuracy, speed, consistency, simplicity, and suitability.

- *Company-perceived value:* Besides process values there are some company perceived values that may not be relevant to the current product, but are essential for the long-term survival of the company and its competitiveness. Some examples of company-perceived value are: reusability for other products, modularity, commonality, exchangeability, agility, and marketability (warranty, field support, etc.).

In customer-perceived value, the concurrent teams focus on two groups of customers, internal customers and external customers, and asks *what* would *delight* them. Delight means being best at what matters most to customers. The different types of customers are listed in Figure 2.22 of volume I. Every step that a work group takes during a product design affects what external customers get or think they get. Each design decision taken by the initiator (requester) or the supplier of functions or services indirectly affects what external customers eventually get. Such decisions can serve as a source of potential value or a point of competitive differentiation. Value engineering (VE) is a method of analyzing a process, identifying the value attributes that are associated with it, and eliminating hidden waste.

- Value engineering adds value to the functions that concurrent teams are seeking to be implanted in the product, or
- With VE the services work groups are performing will enable the product to have features that are attractive to both internal and the external customers.

Accuracy, cycle or turnaround time, consistency, timeliness, and conciseness are some of the quality attributes that are often valued by internal customers. Errors, rework, delays, high costs, and low quality services are some example measures of poor attributes for the external customers. Appropriate choice of strategies and design of process are often used to maximize the benefits and overcome the obvious negatives. The types of strategies that one can use are given in section 3.7 of volume I. The various types of tools that are used for life-cycle management are contained in section 2.8 of volume I. Customers' value criteria are often reflected in a QFD requirement matrix. The customer wants provide the basis for the value criteria that, in turn, influence the process specifications. In product development, targeting the requirements of internal and external customers, organizing an integrated PD^3, selecting corporate strategies, adding value and creating the winning edge are the primary focus. For the improvement process, Goldratt and Cox [1986] emphasize the importance of seeking out goals. Dr. Shingo [1989] divides the pursuit of goals into three parts (see Figure 3.9).

1. *Focus:* Focusing is the idea of uncovering goals that are deeper than what is immediately obvious on the surface.
2. *Identifying multiple goals:* Here the idea is not to block the various design possibilities (alternatives) before they are fully analyzed with respect to the stated goals.
3. *Pursuing goals systematically:* This refers to the idea of looking at current goals from broader perspectives that work groups or teams often tend to overlook. This could lead to spectacular improvements. An example cited by Shingo is the removal of burrs. When a company noticed burrs occurring in a machining operation, they put their efforts into how burrs could be removed quickly from the machined surfaces. This, however, did not eliminate the burrs problem. Since burrs were generated where tools leave the materials, burrs kept on occurring. The company then began tracking

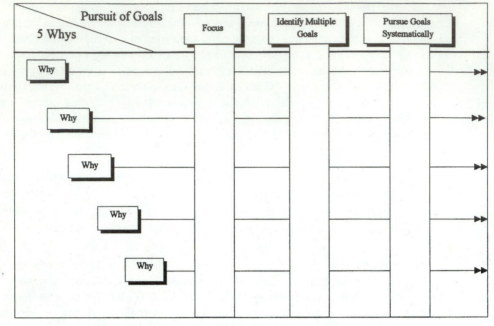

FIGURE 3.9   Scientific Thinking Mechanism for Pursuit of Goals

the root cause. They asked 5 whys questions. Once they found the root cause, efforts for quick removal of the burrs were rapidly replaced by efforts to prevent the burrs from occurring in the first place.

3.5.1.3   Minimize Variations

Figure 3.10 shows the statistical methods for ensuring stable processes. It depicts a flow chart for managing variations and for identifying sources of variations. It also lists the primary statistical methods applicable in stable and unstable processes. The major sources of product variation are contained in Figure 2.14, volume II. They are grouped into three categories: system variations, constituent variations, and external variations. Setting aside the product, variation is found almost everywhere, between work groups, in teams, in persons, in outputs, in inputs, in service, and so on. It is a fact of life. Common causes of product variations are the capability or limitation of the machine, variations of incoming raw materials, overall gage accuracy, lack of operator training, environmental fluctuations, and variations in parts [Deming, 1986].

3.5.1.4   Minimize Wastes

In the automotive industry as in other industries, such as the aerospace industry, waste can take the form of unnecessary product weight, which not only drives up the product cost but erodes fuel economy and performance [Ohno, 1988]. Waste can also take the form of needless investment in machinery, process equipment, and automation when simpler or existing equipment could be enhanced to meet the higher quality or performance requirements.

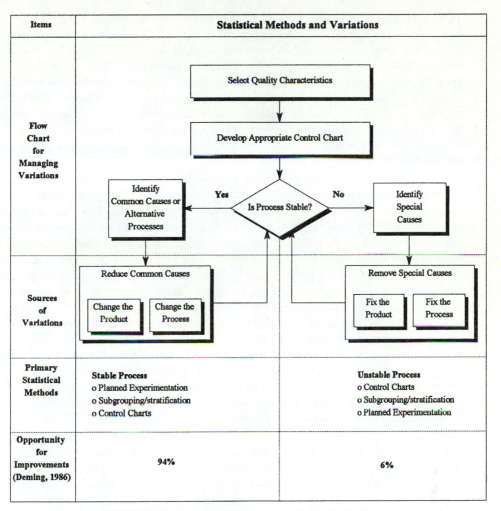

Items	Statistical Methods and Variations	
Flow Chart for Managing Variations	Select Quality Characteristics → Develop Appropriate Control Chart → Is Process Stable? Yes → Identify Common Causes or Alternative Processes; No → Identify Special Causes	
Sources of Variations	Reduce Common Causes: Change the Product / Change the Process	Remove Special Causes: Fix the Product / Fix the Process
Primary Statistical Methods	**Stable Process** o Planned Experimentation o Subgrouping/stratification o Control Charts	**Unstable Process** o Control Charts o Subgrouping/stratification o Planned Experimentation
Opportunity for Improvements (Deming, 1986)	94%	6%

FIGURE 3.10   Statistical Methods for Ensuring Stable Process

3.6   TVM MEASURES OF MERITS

The biggest challenge in applying a TVM to a PD³ process is defining the measures of merits (MOMs) for the realization of the total system as opposed to MOMs that are based on individual life-cycle concerns. As described earlier, the foundation of concurrent engineering is teaming and trading. Teaming and trading can be viewed as being positive since a PD³ function is no longer being performed in isolation from the other functions or teams. They can be considered negative in that the cost or time it takes to perform a new concurrent activity compared to an old one in a particular phase may increase. However, teaming has the effect of making things much easier and more cost effective elsewhere (in other life-cycle phases).

This is called the shadow effect. An example of this is in the design activity itself where the initial effort required (on the front end of the PD^3 process) to improve design definition quality may consume more time and cost than required during a traditional process. However, this extra effort may have some positive effects in reducing labor, material, and time elsewhere (in production, support, procurement, etc.). If the criteria for success were based on the design lead-time alone, some CE managers may argue that Concurrent Engineering does not really work because design activity now costs more or takes longer to complete than the way it was traditionally performed. This is because the far reaching effects of these early changes are normally not as severely felt on initial stages as on the downstream processing functions.

One way to handle this is to consider the cumulative effect of the incremental changes in costs over the entire product life-cycle as true measures. It does not matter whether an individual activity (like a design activity) costs more or less or takes more time to perform using Concurrent Engineering than before. It is the overall cost or time that matters [Margolias and O'Connell, 1990]. When concurrent engineering principles are imposed on the work groups, teams, or an entire organization in an enterprise, a great many things can change. The change usually brings pain with it, pain in learning new ways of doing things, pain in accepting new and expanded responsibilities, and pain in having to fix things when new processes and techniques do not work exactly as planned. There is often a desire to improve individual components or to assess their impact on downstream operations. Work groups must continue to learn from these exercises and push forward to expanding new frontiers. In TVM, the issues are not communication or teamwork, it is the creation of the total value content. For this reason, measures of merits (MOMs) of product elements, MOMs of process elements, MOMs of work group and business process performance must be devised to permit a more rapid deployment of the most effective entities in operations, to affect a maximum return in secondary (artillery) performance, as well as to yield the maximum cumulative return in total system content. In order to affect the old entities in product, process, or organization (PPO), it is important to determine the differences between the old entities and the new ones. Any new modification to an old practice is based on two considerations: (a) what effects the modification brings to a PD^3 process on its own, and (b) how it affects the rest of the downstream operations. The metrics of measurement provide a justification for each individual change, such as people, data, process, timing, or knowledge. The changed PD^3 process is not developed in a vacuum, it is often based on incremental gains or continuous improvements over the past situation. These gains or improvements may relate to benefits other than time or cost. All sorts of measures are, therefore, necessary to determine its true impact on the total product.

3.6.1 Design for Total Value

Design for value (DFV) is a powerful technique that allows concurrent teams and work groups to determine systematically the total value of the product over its lifetime in conjunction with appropriate analysis tools. Failure mode and effect analysis (FMEA) is used to identify and prioritize potential problems and design failure mode analysis (DFMA) is used to find alternate solutions. In conjunction with these tools, the value analysis/value engineering concept (VA/VE) is employed to improve the total value of a product or a service. Tools, such as activity-based costing, can help evaluate strategic planning, process redesign, or business process re-engineering (BPR) needs. Improving the total value has two meanings.

1. *Retaining the current value at a cost lower than before:* This applies to basic benefits the customers receive directly from improvements. For example, improving value with respect to functionality means developing goods or services that perform the required (or basic) functions at a lower cost, time, or manpower.

2. *Enhancing the current value at a minimum additional cost:* This applies to values that are not associated with basic functions of the product or the system that produces it. For example, in the case of maintainability and reliability, the customer does not see a direct benefit except when it comes to frequency of product maintenance and repair.

3.6.1.1 Steps in DFV

Design for value is based on activity charting. Activity charting is an important method for building quality into a PD^3 process. This is discussed in section 3.4 of volume I. It focuses on the total CE process and its interfaces, rather than its individual components. The steps involved in a DFV process are:

- *Get all the facts:* A work group or team gets information about specialty products, materials, processes, and vendors by talking to a number of customers, company partners and industry experts. It helps define the functions work groups are performing or seeking.

- *Draw a process flow chart:* A process flow chart represents the work flow laid down in minute detail identifying what is planned and how it will be done. A structured format is used to emphasize the impact of sources of variation on the process. The emphasis is on the procedure, not on the employee or the agent who is running the procedure.

- *Candidate process model or value tree:* The result of this flow charting is the development of some work or task specifics called *candidate* process models.

- *Draw value graph - process description sheets:* The work groups should now be able to break down the process flow into smaller tasks in the form of actions and process description sheets to accurately describe the associated manufacturing method.

- *Identify flow types:* This identifies the connections (sequential, parallel, alternative, join, and loop) between any two activities, and times for the completion of tasks (see Figure 3.14 of volume I).

- *Identify indirect expense and control parameters:* The next step is to determine indirect expenses and control parameters associated with those tasks. For example QC, SPC scheme, and so on, associated with just-in-time (JIT) manufacturing.

- *Identify process parameters:* During this step, process assumptions are identified, required machines and teams are outlined, and critical process characteristics (PtCs) are confirmed.

- *Identify new investment:* New machinery or equipment cost is also identified and estimated, and noted against the appropriate entry in the bill-of-materials.

There are many benefits that can be derived from DFV. Examples of benefits experienced by the NCR engineering and manufacturing team of Cambridge, Ohio, are contained in an NCSU Videotape [SME, 1989]. In the production of their 2060 terminal aimed for the hospitality industry, the multi-functional team focused on DFV before production. The motto "do it right the first time, and every time thereafter" paid off. Parts were reduced from 117 to 16. Vendors were reduced by 80%. Communications among players improved and the morale of the engineering team soared to an all time high.

3.6.2 Measuring Total Value

There are many ways to quantify or measure a value associated with an activity. It is, however, very difficult to define a single overall measure that is appropriate for all life-cycle considerations. In the evolving, highly competitive global marketplace, recurring customer satisfaction is essential for long term survival. Competitive products, whether they are consumer goods or for the defense industry, largely depend upon satisfying customer expectations. Customer satisfaction is achieved not through a single act, but through a coordinated array of actions, each contributing a useful and interesting dimension toward an artifact's overall performance. For example, the off-line and on-line methods of quality are a supplement to, but not a substitute for, sound engineering and manufacturing practices. Other contributors to customer satisfaction are attributed to efficiency gain and a reduction in the total resource requirements for the life-cycle support of the product. There is a difference between what is important to the customer and what is considered important by the customer for a life-cycle support. For example, cost cutting may not be an important aspect to the customer vis-à-vis worth spending time on, but the end-cost of the product is. There are five measures of savings associated with value engineering (see Figure 3.11). The following lists in each case what factors are considered important by the customers.

1. *Quality:* Quality to the customer means improved fits and clearances, no defective parts, shiny paint, a reduced number of parts, improved quality comfort and superb performance. What product manufacturers do to create a design and build quality parts has very little significance.

2. *Functional Worth:* Functional worth is a measure for defining the product's worth to the company and/or the customer. It measures the consumer utility in terms of functional-to-cost worth. Very often, functional worth is defined as the value per unit cost of the product. The savings due to functional worth are affected by work-in-progress (WIP) inventory, machine utilization, floor space, superior product design, finished goods inventory, materials overhead, and so on (see Figure 3.11). Benchmarking is a method for assigning values to the 3Ps (practices, procedures, and policies) and processes associated with developing a product. One of the purposes of benchmarking is to increase a product's functional worth. Benchmarking is also used to
   • measure the subject's part performance against that of the best-in-class companies.
   • determine the best-in-class features or functions.
   • achieve the best-in-class performance levels.

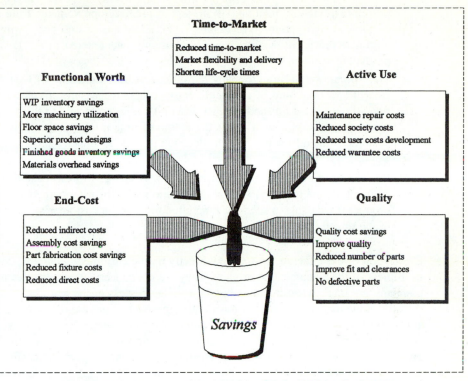

Time-to-Market

Reduced time-to-market
Market flexibility and delivery
Shorten life-cycle times

Functional Worth

WIP inventory savings
More machinery utilization
Floor space savings
Superior product designs
Finished goods inventory savings
Materials overhead savings

Active Use

Maintenance repair costs
Reduced society costs
Reduced user costs development
Reduced warantee costs

End-Cost

Reduced indirect costs
Assembly cost savings
Part fabrication cost savings
Reduced fixture costs
Reduced direct costs

Quality

Quality cost savings
Improve quality
Reduced number of parts
Improve fit and clearances
No defective parts

Savings

FIGURE 3.11   Measures of Savings Associated with Value Engineering

Many use information obtained from benchmark studies for setting their own company targets. Benchmark studies are also useful for strategic planning (to be used in QFD), determining product or process implementation plans, and performing value analysis/engineering.

3. *End Cost:* This defines a set of cost measures based on the end product's competitiveness. Two common cost measures are (a) how much the product costs to deliver as compared to its predicted (and sometimes contracted) cost and (b) how this cost compares to what the customer judges its fair market value to be.

Accurate cost estimation is neither essential nor feasible during early design stages. For example, for early design improvement purposes, it may be enough to know which of the two alternatives leads to lower cost of production than to know their actual absolute costs. It is, therefore, quite helpful to develop relative measures based on preliminary design descriptions that can predict the associated degree of X-ability. End-cost is affected by direct and indirect costs, assembly cost, part fabrication cost, fixture cost, and so on (see Figure 3.11). The design options (such as design configurations, material properties, manufacturing processes) also affect costs. It is unnecessary to spend a lot of time and effort to obtain an accurate cost estimate for each design option in order to suggest a design change. It is more appropriate to identify relative cost drivers to predict improvements from among the possible design options (see Figure 3.11). Early use of end cost estimations can eliminate unwanted design changes commonly encountered in the later stages of product realization.

4. *Time-to-market:* There are many definitions of time-to-market (TTM). Some consider TTM a measure of competitiveness, others a measure of customer satisfaction—how close a finished product or a unit comes compared to the customer's realistic desires. Actually, TTM is the length of time it takes to put a product in the customers' hand from the time the decision is made to launch a product.

5. *Active Use:* Active use of a product implies one of the following two situations: (a) what portion of the purchase price that is charged to the customer relates to maintaining the product in working condition and, (b) what percentage of time the product is available in such working condition for the customer's use as a function of the time it is kept in his or her possession.

Clearly, all of these factors are focused directly on the customers' end cost, delivery, and usefulness of the manufactured product. They are not concerned with the details of how a company got there. Measurements involving effectiveness of the teaming concepts or of the cross-functional department interactions on product values are not evident. Such measurements are usually in the form of the number of engineering change orders, mean time between failures, remaining time for ramp-up to part production, and so on. It does not make any difference to the customers whether engineering releases the design on time or not. The intermediate PD^3 process does not produce and capture happy customers. What most customers are interested in is getting the best valued product at the lowest price that anybody can offer. The best valued product ensures a continuation of the company's current share in the marketplace. The other most common measure that is important to a company is through worth-to-actual time ratio.

Worth-to-actual Time Ratio: This is defined as the ratio of the net worth time to the amount of time actually spent on an activity to generate an output. The net worth time is the time spent only on adding values to the activities with respect to the five measures of savings. The actual time spent on an activity will always be more than its worth, since it includes some non-value added time. The non-valued portion will be the measure of wasted time. A corollary to this is the wasted-to-actual time ratio. This is defined as the amount of time wasted on an activity to the amount of time actually spent to generate an output.

$$\text{Worth-to-actual time ratio} = \frac{\text{Net worth time}}{\text{Actual time}}.$$

$$\text{Wasted-to-actual time ratio} = \frac{\text{Non-valued added time}}{\text{Actual time}}.$$

The two are related as:

$$\text{Activity wasted-to-actual time ratio} + \text{Activity worth-to-actual time ratio} = 1. \quad (3.13)$$

Depending upon the activity and whether time spent on this activity is value added for the customer, company, or the process, there will always be a worth-to-actual time ratio for an activity. Temporarily, a wasted-to-actual time ratio for an activity as large as 1.0 can be acceptable if that activity adds value to most of the other dependent activities. The cost of the product can be computed using the following rules:

$$\text{A process actual time} = Tw / \sum [\text{activities for worth-to-actual time ratios}] \qquad (3.14)$$

$$\text{A department actual time} = Tw / \sum [\text{processes for worth-to-actual time ratios}] \qquad (3.15)$$

$$\text{Where } Tw = \sum_{I=1}^{n} [\text{Activities for net worth time}], \text{ and}$$

where n is the number of activities in a process or department as the case may be.

$$\text{Cost of the Product} = \sum_{i=1}^{m} \{[(\text{department actual time})_i \times (\text{Cost factors})_i]\}. \qquad (3.16)$$

where m is the number of departments or units.
The cost factors represent the cost per unit time spent in performing an activity.

The appropriateness and profitability will depend upon the following questions.

- How much is the user willing to pay for the product?
- What will it cost to produce what the customer wants?

The ideal profitability situation will be when, in a product, most activities worth-to-actual time ratio is closer to 1.0, and, at the same time, most activities are strong contributors to all three types of values, customer-perceived, process, and company-perceived.

Break-even Points: The cost of design and development of a product changes with time. At the beginning when a company does not have any competing product or would like to have some targeted features in an announced product, the cost of design and development is often too high. In Figure 3.12, three curves are shown.

A development expenditure curve: It shows the variation in unit cost of developing a product over its life-cycle time. This is shown in Figure 3.12 by a solid line.

Let us denote C as the unit cost of producing the product, then C represents the functional value of the expenditure curve. At the product launch time C is zero, since there has been no expenditure incurred as of yet. The development costs start building up the moment the product is launched. It peaks when the product is ready to be delivered to the customer. After that point, unit cost of delivery steadily decreases. The unit cost reaches a minimum when the maintenance and upgrade are at its minimum. This point is shown in Figure 3.12 by Co.

$$Co \text{ as the minimum cost when } \partial C/\partial t = 0 \qquad (3.17)$$

The development expenditure curve deals with costs on a per unit scale, while the cash flow curve deals with revenues from sales and the total cost for the delivery of the product to the customer.

A cash flow index curve: This curve depicts the net profits, total revenue minus the total product cost over its life cycle. Cash flow index is a measure of profitability. Cash flow index is defined as the net income.

$$\text{Cash Flow Index} = \frac{(\text{Total Revenues} - \text{Total Cost of delivery})}{C_o} \qquad (3.18)$$

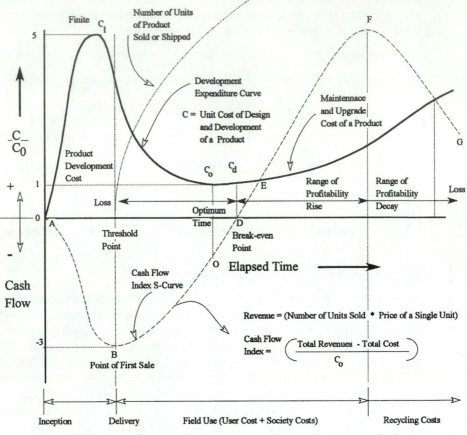

FIGURE 3.12   Range of Profitability during Product Development Cycle

$$\text{Total Revenue} = \text{Number of units sold} \times \text{Price of a single unit} \qquad (3.19)$$

The total cost of product delivery includes cost of sales, marketing, and administration.

The cash flow curve is shown in Figure 3.12 by a dotted line. C_o is merely used as a scaling factor.

A curve showing the number of units of product sold or shipped: This is shown in Figure 3.12 by a chain line. The first sale starts after the production phase is completed and when the product is delivered to the customer. Before that point, this curve is undefined.

Some interesting points along these curves are explained in the following:

- *Point A—R&D begins:* During the research & development stage, since there is no product or sales, there is no income. This stage is represented by point A on the cash flow curve in Figure 3.12. The cash flow curve starts at the origin and decreases representing the development expenditure of making the product. The expenditure curve shows the unit cost of reshaping the product for the market. During the inception to the delivery time duration, there is no sale up to this point, the negative cash flow curve in a large part is the result of the expenditure curve. The two curves are not exactly opposite since there may be some intangible gain not directly associated with the product. Hopefully, soon after this delivery point, a time will come when the first unit of the finished product will be sold.

- *Point B—First Sale:* This represents a threshold point—a point of first sale. This is the minimum point on the cash flow curve. With the first sale, the company begins to earn back its investment. This also marks a point on the development expenditure curve, the cost at which the company has sold its first product. This point corresponds to the minimum time required to design and develop the product to a point of sale. As time lapses, technology changes and the cost of successfully designing a physical part also changes. This is shown by a concave shaped expenditure curve in Figure 3.12. The number of units sold steadily increases as shown by the parabolic nature of the chain curve (c).

- *Point O—Optimum Production:* This is the point when the number of units sold is climbing but the rate of expenditure has leveled off. Production is now at its lowest possible unit cost. The cash flow index of the company may still be negative but soon this is bound to change.

- *Point D—Break-even Point:* This is the point when the company has recouped its investments. The corresponding time along the abscissa represents the total lapsed time from the inception of the project launch to the beginning of positive cash flow. This is also referred to as *time to profitability*. The shorter this time is, the sooner the company can start making money and get a better return on its investment. Profitability depends on the number of units sold. The more units sold, the better is the profitability index. The challenge for a company is to decide how early to start the product development process so that the net loss is minimized. It will be even better if the product is the first of its kind in the marketplace. Optimum time is the time when cost is at its minimum and the product timing is at its best. The company would gain the maximum cost benefits by bringing the break-even point D closer to optimum time. This way, profitability can start earlier than the lowest possible expenditure cost and the product can sustain a longer sales life.

- *Range DEF—Profitable Period:* This is the time period during which the product is profitable. The fixed costs are now borne by fewer units sold. However the unit cost is beginning to rise. This is primarily due to the higher cost of maintenance and product upgrades. Other factors causing costs to rise are inflation, higher cost of advertisement, marketing and sales, new materials costs, technology, and reduced market shares due to increased competition. This is shown by the portion EF.

- *Point F—Maximum of cash flow index curve:* This is indicated by the cash flow index curve hitting the maximum. It is indicative of the time to phase out the current product line. Further maintenance of the product would cost more money than sales. As shown in Figure 3.12, after this time (point F), sales may taper off due to competitors either introducing an advanced product in the marketplace, or maintenance and warranty costs getting out of control.

Cash flow is the measure of profitability since it measures the net income, the difference between the revenues and the cost. This way, the product income is the area under the cash flow curve DEFG and the x-axis. Similarly, the development cost is the area under the cash flow curve ABOD. In order for the company to benefit from this line of products, the product must give a positive return on investment, that is the following must be true:

$$\text{Product Income} >>> \text{Development Cost} \tag{3.20}$$

or   Area under the Cash Flow curve DEFG portion >>> Area under the
Cash flow curve ABOD portion

or $$A_{\text{DEFG}} >> A_{\text{ABOD}}. \tag{3.21}$$

The net profit is the difference in areas between the two zones of the cash flow curve. The rate of investment (ROI) is a function of the following factors:

Shape of the cash flow curve.

Shape of the development expenditure curve.

Longevity of the product in the marketplace.

Price of a single unit of the product.

Volume (number of units) of products sold or shipped.

Launch timing or timing of product introduction to the marketplace.

If the two curves are what is shown in Figure 3.12, then the profit only marginally exceeds the cost of development. The delay in the start of the product shipment has the effect of moving the whole cash flow curve ABODG to the right while keeping the development expenditure constant. This can obscure the net profits from the sales of the product.

3.7 VALUE MANAGEMENT TOOLS

Value management tools are useful in obtaining a mapping, a relative ranking, or ordering of the candidate concept alternatives based on a set of common criteria. There are a number of ways a candidate concept can be mapped or ranked.

* Pugh concept.
* Design for experiments.
* FMEA.

Some tools are useful in ordering/ranking candidates based on the weighted sum of various attributes; some in ranking of measures based on the weighted linear sum of various attributes; and others are based on the conceptual representation, such as the Spider Chart, of scores in multiple attributes.

3.7.1   **Stewart Pugh Concept**

Referring to the example presented in section 2.8.3, Figure 2.25 of volume I, it would seem that on the basis of quality, the team would pick alternative D. On the basis of cost alone, it would pick alternative C. On the basis of weight, it would pick alternative B or C. On the basis of investment, it would pick the datum A. The team might also conclude that if it can find ways to utilize some of the features of alternative D to improve the quality of alternative C, and if they can keep the incremental investment to a minimum, that some variation of alternative design C would be attractive. Such considerations would require a multi valued analysis. Different alternatives may have identical components, but due to varying features, the cost, weight, investment, and reliability numbers may be different. Those sub-components or features that are different, modified, added, or deleted in the hybrid concepts (H_1, H_2, or H_3) will show up in the concept selection matrix as having different numerical values for the criteria ratings. In multi-valued analysis, these individual criteria numbers for each alternative are multiplied by a weighting factor and summed up to obtain a cumulative value index. Through this process, the team not only feels more confident in its plus and minus analysis, but also begins to quantify the amount of benefit or penalty associated with each alternative or each hybrid concept.

Dr. Pugh recommends that the alternatives considered should be represented by a simple sketch at the top of the appropriate column in the concept selection matrix [Pugh, 1991]. Each sketch is drawn to the same level of detail. The simple sketch or schematic would list the fundamental sub-components of the design and indicate the way they would work together to provide system function. From these simple sketches, it is easy to list the sub-component names or features for each alternative, and against each sub-component or feature catalog the weight, cost investment, and reliability estimates identified during benchmarking. If this data is not available for a given sub-component, it can be acquired or estimated based on a description of its characteristics or the knowledge of how it differs from a known sub-component.

3.7.2   **Design of Experiments (DOE)**

DOE is an off-line technique for improving product and process performance. It is a systematic method of simultaneously changing a number of factors following a predetermined experimental plan. DOE effectively evaluates many sources of variation simultaneously as opposed to the "changing one variable at a time" method of experimentation. A full factorial design examines the effects of all factors and their interactions (2-level and higher). Figure 3.13 shows an example illustrating this concept for a two variable experiment (say Material A, Material B, and rates X and Y). Based on the four experiments, it is clear that Material A with a rate of X will give an above average finish compared to a world-class finish. In the 1920s, R.A. Fisher of the United Kingdom developed the idea of analysis of variance (ANOVA) [Hartley, 1992]. It was developed as a method of testing a number of variables in a minimum number of tests. Later came the idea of orthogonal arrays which further reduced the required number of experiments. For example, Taguchi's

Condition No.	Sensitivity Analysis		Quality Characteristics (say finish)
	Material	Rate	
1	Material A	X	q1
2	Material B	Y	q2
3	Material C	X	q3
4	Material D	Y	q4

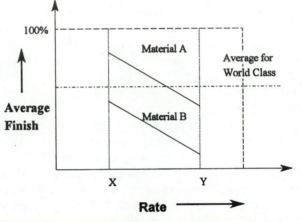

FIGURE 3.13   Two-Variable Example of a Design for Experiment

L9 orthogonal array requires only 9 tests against 27 for Fisher's Latin square technique, and 81 for a full factorial ($3 \times 3 \times 3 \times 3$).

3.7.3   Taguchi's Robust Design

Dr. Genichi Taguchi introduced a new approach using standardized orthogonal arrays and linear graphs to facilitate the use of DOE. His linear graphs have been described as a short-cut method to experimental design, which he then integrated into a robust design methodology. Taguchi's method is a popular off-line quality control technique for building quality by design. It starts with an orthogonal array. Figure 3.14 shows a list of typical orthogonal arrays. The number after L in Taguchi's orthogonal array indicates the number of tests required. For instance, an L9 requires 9 tests; L12 requires 12 tests, whereas the number of full factorial tests required for L9 is 81, and for L12, 2, 048. Clearly, there is a significant reduction in the number of tests required. This is shown in Figure 3.14. On-line quality control methods are described in Section 3.1.1. Taguchi views the robust design of a product or a process as a five-step program (see Figure 3.15).

Array	Number of Factors and Levels	Number of Trials in Full Factorial
L4	3 x 2 Levels	$2 \times 2 \times 2 = 8$
L8	7 x 2 Levels	$2^7 = 128$
L9	4 x 3 Levels	$3^4 = 81$
L12	11 x 2 Levels	$2^{11} = 2,048$
L25	6 x 5 Levels	$5^6 = 15,625$

FIGURE 3.14   Typical Taguchi Orthogonal Arrays

Robust Design Process

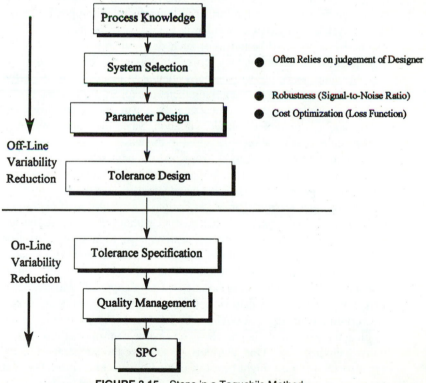

FIGURE 3.15   Steps in a Taguchi's Method

1. System Selection.
2. Parameter Design.
3. Tolerance Design.
4. Tolerance Specification.
5. Quality Management.

The first three parts introduced by Taguchi are collectively recognized as a cycle of never ending improvements for the purpose of designing a robust product/process design [Taguchi, 1986].

System Selection: System design is a phase when new concepts, new ideas, new methods, and so on, are generated to provide a set of new or improved product alternatives for further refinements. One way to remain competitive in the world economy is to research new ideas, insert new technology, and keep on introducing innovation into the products companies manufacture. This soon gives the industries recognition as leaders in utilizing new innovation and technology. The advantage in developing a new system is that the design can be protected by patents. However, this can be a short lived advantage, if the ideas can be refined or improved. A similar situation occurs if a competitor can fabricate the old idea in a more uniform, efficient, or cohesive manner. The technological advantage may then disappear quickly. This is the time when system design and parameter design can be found useful in further refining the maturing product concept. The very first step in system design is to consider all possible system alternatives that can perform the required functions. Although most such determinations are based on qualitative rather than quantitative reasoning, they form an important basis in selecting a new or a better system. In this process, the work groups attempt to achieve certain performance objectives from each candidate system applying their specialized skills and knowledge. For example, a system objective might be to design a TV power circuit to convert a 110 volt AC into a 12 volt DC plus/minus 1 volt DC variation. Since the transformer technology is widely known, no new innovation is required in this case. In most companies, this is where the research and development work stops and the next step is to apply a parameter design. This is the most important step in developing a stable and reliable product and is the very essence of Taguchi methods [Wilson and Greaves, 1990].

Parameter Design: Parameter design is generally used to further improve the uniformity (e.g., functionality) of the product without changing its basic elements. Certain parameters of the product or process design are set to make the performance less sensitive to causes of variations. Parameter design is useful in improving quality without controlling or eliminating causes of variation. Controlling or eliminating causes of variation may be expensive compared to the parameter design approach. Most experimental approaches look upon all factors as causes of variations. However, in many design situations, there are certain parameters that ought to be left alone. For example, a product is sensitive to ambient temperature variation. How can anyone control or eliminate temperature in a customer's environment? One alternative is to leave the temperature alone. Without a lot of expense, temperature might be a difficult parameter to control. The premise is that ambient temperature is a natural phenomenon. One can control temperature in a plant to eliminate the adverse effects on machine operations, but when the part is exposed to the atmosphere, temperature can get out of bounds. Parameter Design, so named by Taguchi, involves selecting parameters and target values for design variables that make the products and processes robust. Without the knowledge of Taguchi's experimental techniques, an average design team might well be intimidated by the monumental task of trials and simulations that they face in order to optimize a design. Formal education in experimental techniques is far more widespread in Japan than in the West. On the contrary, the use of tolerance in quality control instead of parameter design is much more widespread in the West than in Japan. Whether this results in an improved product quality or results in the production of reliable products for the west is questionable. Quality control does add to the cost and if the design could not be improved, it could leave customers dissatisfied. Figure 3.16 shows the results of parameter design from an analysis of the difference in performance of two Sony plants (one in Japan and one in the United States) making color television sets (CTVs) from the same design [Cullen, 1987]. At the Japanese plant, manufacturing teams produced CTVs where the number of sets with color density adjustments was distributed about the mean as illustrated. By design, only 0.3% of the sets were

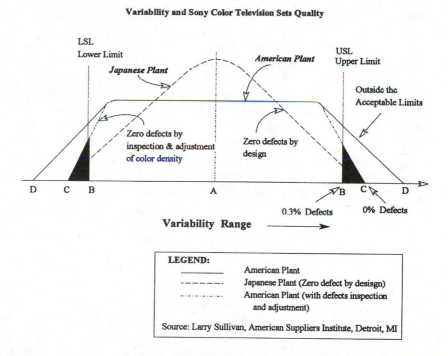

FIGURE 3.16   Reduction of Defects and Variability by Using Taguchi's Method and Its Application to a Sony Color Television Set

shipped with color density falling out of the range (identified customer satisfaction limits).

At the U.S. plant, the manufacturing team produced the same sets but with a much wider spread about the target value. Later all color television sets falling outside the mean were adjusted to fall inside. Even then, statistically, more U.S. sets were closer to customer dissatisfaction limits than falling within the satisfaction limits. The result was that the U.S. plant suffered the following [Wilson and Greaves, 1990].

- One percent (1%) cost premium per set.
- Higher number of customer complaints.
- Perceived reputation for poor quality.

Tolerance Design: One way to improve quality is by tightening a set of tolerances on product or process parameters. This reduces the performance variation. Tightening tolerances, however, may be expensive. It may require the use of expensive materials, new components, or use of expensive processes. The objective of tolerance design is to decide the trade-off between desired quality level and the cost of designing a new system. Consequently, this should be done only after a parameter design step is performed. A loss function is a tool used in parameter design. It describes the losses that a system suffers from choosing values for a set of adjustable parameters. In tolerance design, the loss is the excessive design and manufacturing cost and the parameter is the tolerance. For all practical purposes, the loss function output against a manufacturing tolerance takes the form of a parabola, with zero loss when manufacturing tolerance is at the optimum tolerance value. Departure from this optimum tolerance causes loss. Optimum tolerance may be unnecessary, since the same level of quality can be attained through a parameter design. By doing parameter design, the need for tightening tolerances may become unnecessary for the desired quality. In tolerance design, total manufacturing tolerance must fall within a specified customer tolerance limit, as shown in Figure 3.17. The factory (manufacturing) tolerance or statistical process control (SPC) limits are considerably smaller than the customer functional limits. Both customer and manufacturing tolerances follow the same pattern, a typical quadratic loss function. If y is the quality characteristics of a product and m the mean value for y; the quality loss function is given by:

$$L(y) = k(y - m)^2 \tag{3.22}$$

where k is a constant called the quality loss coefficient. From Figure 3.17, when $y = m$;

then $L(m) = 0$, that is, loss is zero. This is quite appropriate,
                          since m is the mean or average value for y.

The most important use of loss function is to help work groups change from the world of specifications (that is, to meet certain specifications) to a continual reduction of variations about the target value through improvement of processes [Deming, 1993]. This is also the basis of a parameter design. Work groups can move a short distance to the right or left of the point of this tangency, that is moving away from the optimum point, and suffer only imperceptible loss. But if teams move far out from this optimum point, there will be a substantial loss. Dr. Taguchi calls this a *loss to the society*. All loss functions are not usually symmetrical. Sometimes, it is steep on one side or the other. Two such loss functions are also shown in Figure 3.17:

- The flat loss function represents the cost of manufacturing with the aid of *gadget for automatic adjustment* meant to hold parts within a range of specifications. With *gadgets off*, the loss function tends to be a parabola and the loss is much smaller.

- The second loss function is meant to indicate the cost incurred from *no defect* satisfaction that is merely meant to hold a set of *zero defect* specifications. In this case loss function is straight up and down at both specification limits. This is a discontinuous curve with an abrupt rise at two extreme points and no loss in between. If we perform a test with two no-go gauges having the same upper and lower limits, the shape of this *no-go gauge* loss function would be similar to *zero defect* loss function. The same is true for the idea of decomposition of sum of squares in design of experiments and analysis of variance (ANOVA).

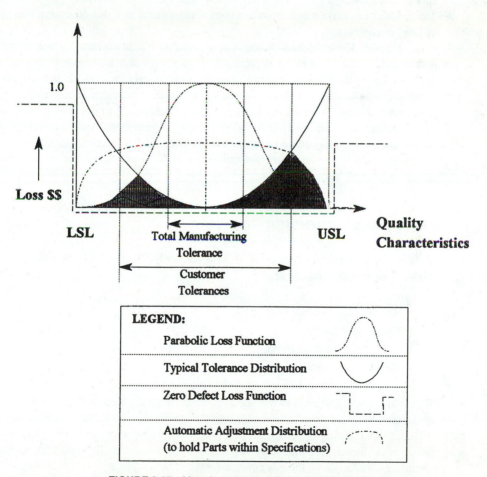

FIGURE 3.17   Manufacturing and Customer Tolerances

Taguchi's loss function can thus be used to relate customer specifications with design and manufacturing tolerances.

Tolerance Specifications: During tolerance design, the trade-off between cost and quality loss determines the upper limits on manufacturing tolerance specifications (dimensional and geometric). There are two types of manufacturing tolerances: local and global. Local tolerances are governed by local variations and are independently specified. Simple measures of local (geometric) tolerances are: min-and-max on diameter, length, distance, linearity of straight-side or cylinder-side, flatness, straightness, and so on. They are mainly governed by the machine process capabilities. Global tolerances are influenced by datum and the setup selection in process planning. Examples include dimensions between features, perpendicularity, parallelism, cylindricality of bore, concentricity of bore or cylinder-side, perpendicularity of cone-front, angularity of cone-side, run-out and so on. By appropriate selection of datum and setup in manufacturing, it is possible to improve global tolerances with the same setup for machines and tools. The tolerance specification goes hand in hand with tolerance design. Once the tolerance limits are finalized, the grades of materials and limits are commonly placed as notes on design blueprints and related documents.

Quality Management: When a new process, a new concept, a new material, or a method is installed and made operational, work groups must evaluate its performance on an ongoing basis. It is important to monitor continuously the statistical distribution of its critical dimensions. A feedback control system with gauges can be designed for measuring quality levels. The results can be monitored frequently to determine the best quality management levels at the least cost of high grade components and materials. Concurrent teams and employees can be an integral part of this quality management process. They all could share in the essential responsibility to identify and eliminate wastes—all kinds: Muda, Muri and Mura.

3.8 CONCURRENT PROCESS FOR TVM

A four step concurrent process for TVM is shown in Figure 3.18. The value management consists of four concurrent steps.

- Reduce the variable.
- Control the product.
- Control the process build.
- Control the operation.

Taguchi's five stage process may be applied during each step of this four-step concurrent process. The types of activities that one can perform during each step are contained in Figure 3.18. Having applied this process to a current state, a work group is likely to achieve an increased probability for success (e.g., repeatability and consistency) and a significant reduction in the margin of errors.

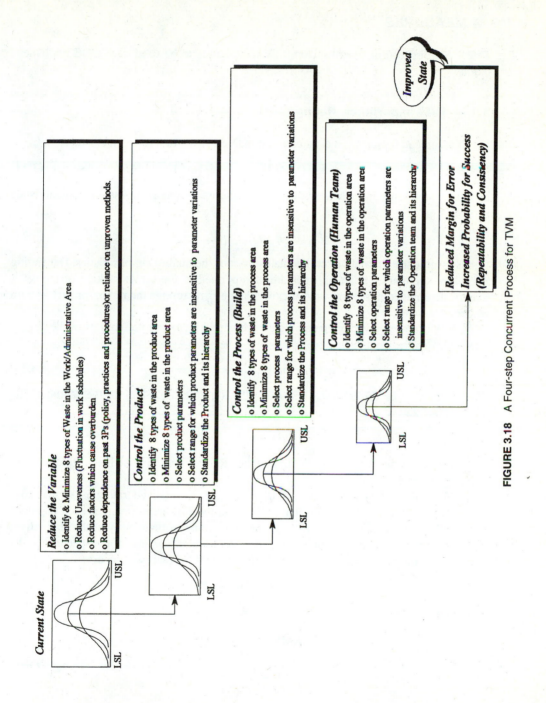

Current State

Reduce the Variable

o Identify & Minimize 8 types of Waste in the Work/Administrative Area
o Reduce Uneveness (Fluctuation in work schedules)
o Reduce factors which cause overburden
o Reduce dependance on past 3P's (policy, practices and procedures)or reliance on unproven methods.

Control the Product

o Identify 8 types of waste in the product area
o Minimize 8 types of waste in the product area
o Select product parameters
o Select range for which product parameters are insensitive to parameter variations
o Standardize the Product and its hierarchy

Control the Process (Build)

o Identify 8 types of waste in the process area
o Minimize 8 types of waste in the process area
o Select process parameters
o Select range for which process parameters are insensitive to parameter variations
o Standardize the Process and its hierarchy

Control the Operation (Human Team)

o Identify 8 types of waste in the operation area
o Minimize 8 types of waste in the operation area
o Select operation parameters
o Select range for which operation parameters are insensitive to parameter variations
o Standardize the Operation team and its hierarchy

Improved State

Reduced Margin for Error
Increased Probability for Success
(Repeatability and Consistency)

LSL   USL

FIGURE 3.18   A Four-step Concurrent Process for TVM

153

3.9   TVM MEASURES

The following section describes some TVM measures to be used during CE implementations.

3.9.1   TVM Importance Rating (TIR)

Let us assume a_k represents a candidate alternative in a set of alternatives named A.

$$A = \{a_1, a_2, a_3, \ldots a_i, \ldots, a_n\} \tag{3.23}$$

and the q_j represents an associated quality characteristic element in a set of characteristics named Q.

$$Q = \{q_1, q_2, q_3, \ldots q_j, \ldots, q_m\} \tag{3.24}$$

Let us assume the alternative set A contains most of the quality characteristics. Depending upon how the quality characteristics are incorporated in the set A, ranking of the set A may be determined in several ways. Let us assume r_{ij} represents the rating for each a_i with respect to an attribute q_j.

If W is a weight vector containing the weighting factors or preferences of the quality characteristic elements, where

$$W = \{w_1, w_2, w_3, \ldots w_j, \ldots, w_m\}, \tag{3.25}$$

then a basis can be formed for determining the importance ranking of the candidate designs as in 3.26.

$$(\text{TIR})_i \equiv [r_{ij}] \times \{w_j\}. \tag{3.26}$$

Note that the higher the TIR (TVM Importance Rating) value is, the better ith design alternative it represents.

Normalized TVM Importance Rating (NTIR): Normalized rating is similar to a weighted TVM importance rating. The numbers in the TIR are normalized such that none of the values exceed 1 (one). NTIR is a relative measure of the quality of the alternatives with two extremes: 0 (zero) representing the most unfavorable end of the spectrum and 1 representing the most favorable end of the spectrum. If NTIR is written as:

$$\text{NTIR}_j = \text{TIR}_j / \text{RMS} \tag{3.27}$$

where RMS denotes Root Mean Square Value.

$$\text{RMS} = [\sum_{j=1}^{m} \text{TIR}_j^2]^{1/2}$$
$$\text{TIR} = \{t_1, t_2, t_3, \ldots t_j, \ldots, t_m.\} \tag{3.28}$$

where
$$[0 \le t_j \le 1].$$

Instead of subjective ratings, quantitative rating r_{ij} can also be used.

There are several techniques of conceptually presenting the results. Section 7.1 of volume I describes several of these techniques. Amoeba Chart (also referred to as polygon graph or spider chart) is often used to represent these ratings in a compact form. This is shown in Figure 7.3, Chapter 7 of volume I. The spider chart graphically displays multivariate scores of various characteristics radially along the circumference of a unit circle.

3.9.2   Configuration Management Measures

Integral configuration control is a key to success in managing information back and forth between various life-cycle aspects of concurrent design. The enabling systems, loops, and processes described for product realization taxonomy, when appropriately applied, provide measures for configuration management as an integral part of their operation.

3.9.3   Resource Management Measures

The enabling taxonomy for product realization described in chapters 8 and 9 of volume I has built-in mechanisms to query-in on the work done, and to analyze schedule compliance without adding work or intruding on the activities of the work groups. The taxonomy must also provide measures to reveal the status of the work of the various other multi-disciplinary work groups throughout the organization.

3.9.4   Control Management

Another problem that needs to be managed in CE is to keep a balance between management control and teams' autonomy. Inadequate management control and too much work-group autonomy may result in designs that keep management out of focus. Too much autonomy may also result in designs that are inconsistent with the organizational or business goals. When the final output of a CE process is a long awaited big-bang or an all-at-once deliverable, it is difficult for management to understand status, schedule, and to provide adequate budget and support. Managers who cannot affect the outcome of the design process often get very nervous. A concurrent holistic design does not have to be this way. It is often desirable to add artificial releases at the intermediate points, (e.g., 20%, 40%, 60% or 80% of completion points), to reduce risk and, at the same time, maintain adequate control. This approach provides some outputs to show besides the usual abstract status reports to the management while being at a point of X% complete, creating some level of confidence. These artificial releases can add work to the teams but could steer control back into the holistic approach process. These releases minimize the risks while compensating for adequate control, and more importantly, open up a window for management involvement as a concurrent participant.

3.9.5   Program Tracking and Management

Today, program tracking and management functions are not based on attributing costs to the associated tasks responsible for completing the functions, but, instead, tasks are logged where they are actually incurred. This is essentially a very important point for a concurrent engineering operation. When one moves away from a traditional, functionally oriented operation to a CE driven process, the company must likewise shift management from trackng activities to managing an activity-based process. Activity-based costing (ABC) is an example of such a cost-tracking process.

3.9.6   Time-to-market Measures

CE often extends the time for completing some front-end tasks in a product or service life-span, but its effect on the overall time-to-market aspect is significantly reduced. In a taxonomy-based approach, if all of the 2-T and 3-T loops were performed during a product realization, the time-to-market (TM) from inception to finish can be computed following the time series analysis shown in Figure 3.19. The relative saving by applying a CE loop methodology for product realization can be measured or estimated as follows:

$$\text{TM}_{ce} = [\sum_{i=1}^{i=11} T_i] \tag{3.29}$$

where T_i represents the time duration in finishing an ith operation and
TM_{ce} represents time-to-market for CE process when all 2-T and 3-T loops are performed.

In actual implementation, one or more of the 2-T or 3-T loops will not be performed. Time-to-market will actually be smaller. When only a set of 2-T loops or 3-T loops are strictly followed, a lower bound on the TM_{ce} (time-to-market for CE process) results. This can be computed based on:

$$\text{TM}_{2\text{-}T} = T_1 + T_3 + T_5 + T_7 + T_9 + T_{11} \tag{3.30}$$

or
$$\text{TM}_{3\text{-}T} = T_2 + T_4 + T_6 + T_8 + T_{10} \tag{3.31}$$

where T_1 through T_{11} represent the individual time duration in finishing the corresponding concurrent operations.

$\text{TM}_{2\text{-}T}$ or $\text{TM}_{T\text{-}3}$ denotes, respectively, the time-to-market when a set of operations meant for either 2-T or 3-T loops are performed.

Please note that there are five concurrent 2-T loop operations and six concurrent 3-T loop operations. They are discussed in Chapter 9 of volume I. Thus,

$$\text{TM}_{ce} \geq \text{TM}_{2\text{-}T} \geq \text{TM}_{3\text{-}T} \tag{3.32}$$

If a serial engineering (SE) process has been used, the time-to-market can be computed using the serial analysis shown in Figure 3.20. This is shown in equation 3.33.

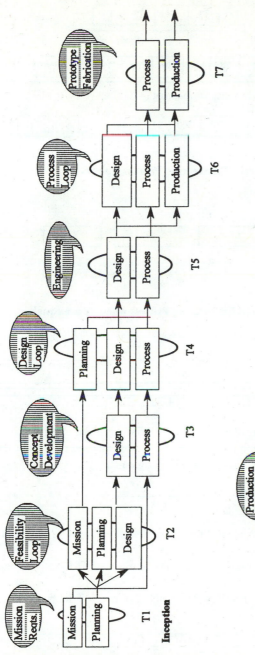
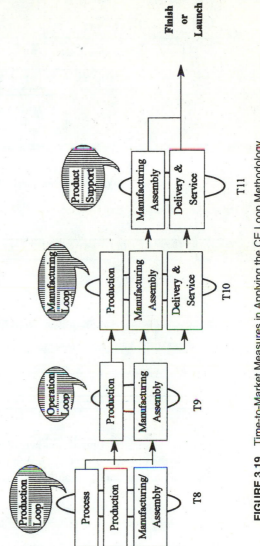

FIGURE 3.19 Time-to-Market Measures in Applying the CE Loop Methodology

157

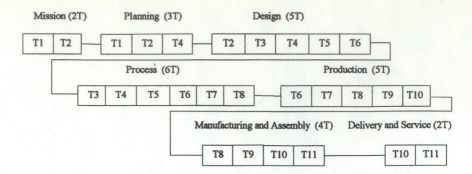

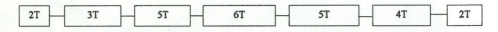

(a) Equivalent Serial Engineering Process Approach

| 2T | 3T | 5T | 6T | 5T | 4T | 2T |

Time-to-Market = 27T

(b) Simplified Serial Engineering Process when Ts are equal

FIGURE 3.20   Equivalent Measures in Applying a Serial Engineering Process

$$\text{TM}_{se} = 2T_1 + 3T_2 + 2T_3 + 3T_4 + 2T_5 + 3T_6 \\ + 2T_7 + 3T_8 + 2T_9 + 3T_{10} + 2T_{11} \tag{3.33}$$

where TM_{se} denotes the time-to-market for an equivalent serial engineering process.

Thus, the time-to-market savings in using a CE process as opposed to using a serial engineering process can be computed based on the following equation:

$$\text{Time-to-market Payoff} \equiv \frac{[\text{TM}_{se} - \text{TM}_{ce}]}{\text{TM}_{se}}. \tag{3.34}$$

An upper and lower bound on the percentage time-to-market payoff can be obtained by substituting

$\text{TM}_{2\text{-}T}$ or $\text{TM}_{3\text{-}T}$ values from Equations 3.30 or 3.31 in place of TM_{ce} into the equation 3.34.

The lower bound on savings will result when all the 2-T and 3-T loops operations are chosen in CE implementation. Thus,

$$\text{Saving (Lower Bound)} = \frac{(27T - 11T) \times 100}{(27T)} = \frac{(16) \times 100}{27} = 59\%. \tag{3.35}$$

The upper bound on savings will result when either 2-T or 3-T loops strategy is followed for a CE process,

$$\text{Saving (Upper Bound)} = \frac{(27T - 5T) \times 100}{(27T)} = \frac{(22) \times 100}{27} = 81\%. \qquad (3.36)$$

In computing the above upper bounds, it is assumed that each 3-T loop of the CE process can be completed in approximately the same time duration. That is,

$$T1 = T2 = T3 = T4 = \ldots = T11 = T \qquad (3.37)$$

The benefits of applying the CE process in terms of time-to-market savings thus range between 59% to 81%, depending upon the iterative path chosen for its implementation. The figures represent a remarkable gain in time-to-market objectives.

Figure 3.21 shows a comparison of a traditional development process and a CE development process as they relate to the product development expenditure and timing. Not only is the time between the start-up and the point of first product shipment shortened by 59% to 81% range, but the overall development cost over the life of the product is also reduced. The peak cost is reduced to a range of 25 to 40%.

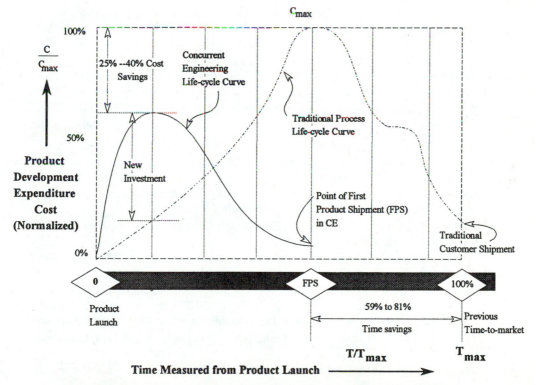

FIGURE 3.21   Comparison of a Traditional Design Process and a CE Design Process

REFERENCES

ASI. 1989. *Quality Function Deployment: Implementation Manual For Three Day QFD Workshop.* American Supplier Institute (ASI), Version 3.1.

BEGLEY, R.L. 1990. "Steering Column Concept Selection," Second Symposium on Quality Function Deployment, 1990.

CAGAN, J., and A.M. AGOGINO. 1987. "Innovative Design of Mechanical Structures from First Principles," *Artificial Intelligence for Engineering Design, Analysis and Manufacturing.* Volume 1, No. 3, pp. 169–189.

CAREY, W.R. 1992. *Tools for Today's Engineer Strategy for Achieving Engineering Excellence: Section 1: Quality Function Deployment.* SP-913, SAE International Congress and Exposition, Detroit, Michigan, USA (February 24–28) SAE Paper No. 920040.

CLAUSING, D.P., and J.R. HAUSER. 1988. "The House of Quality." *Harvard Business Review*, Volume 66, No. 3 (May–June) pp. 63–73.

CULLEN, J. 1987. "An Introduction to Taguchi Methods," *Quality Today.* (September).

DEMING, W.E. 1986. *Out of Crisis.* 2d ed. Cambridge MA: MIT Center for Advanced Engineering Study.

DEMING, W.E. 1993. *The New Economics.* Cambridge, MA: MIT Center for Advanced Engineering Study, November.

DIKA, R.J., and R.L. BEGLEY. 1991. "Concept Development Through Teamwork—Working for Quality, Cost, Weight and Investment," SAE Paper No. 910212, International Congress and Exposition, SAE (Feb. 25–March 1) 1991, pp. 1–12. Detroit, MI.

FREEZE, D.E., and H.B. AARON. 1990. "Customer Requirements Planning Process CRPII (beyond QFD)," SME, Paper No. MS90-03, Mid-America '90 Manufacturing Conference (April 30–May 3) 1990, Detroit, MI.

FEIGENBAUM, A.V. 1990. "America on the Threshold of Quality," *Quality.* (Jan. 1990) pp. 16–18.

FEIGENBAUM, A.V. 1991. *Total Quality Control.* 3d edition, Revised. New York: McGraw-Hill.

FOWLER, T.C. 1990. *Value Analysis in Design.* New York: Van Nostrand Reinhold.

GOLDRATT, E.M., and J. COX. 1986. *The Goal: A Process of Ongoing Improvement.* Croton-on-Hudson, NY: North River Press.

HARTLEY, J.R. 1992. *Concurrent Engineering—Shortening Lead Times, Raising Quality and Lowering Costs.* Cambridge, MA: Productivity Press, Inc.

ISHIKAWA, K., and D.J. LU. 1985. *What is Total Quality Control? The Japanese Way.* Translated by David J. Lu, Englewood Cliffs, NJ: Prentice Hall.

JURAN, J.M., and F.M. GRYNA. 1993. *The Quality Planning & Analysis.* 3d edition, New York: McGraw Hill, Inc.

MARGOLIAS, D.S., and M.H. O'CONNELL. 1990. "Part 3—Implementation of A Concurrent Engineering Architecture," SME Technical Papers No. MS90-205, Proceedings of the WESTEC '90, (March 26–29) 1990, Los Angeles, CA, Dearborn, MI: Society of Manufacturing Engineers.

MILES, L.D., 1982. *Techniques of Value Analysis.* New York: McGraw Hill Book Company Second Edition.

MARTIN, N., C.A. SAWYER, and M. SORGE. 1995. "Towards World-class," *Automotive Industries.* (September) pp. 84–89.

NEMOTO, M. and D.J. LU. 1987. *Total Quality Control for Management: Strategies and Techniques from Toyota and Toyoda Gosei.* Englewood Cliffs, N.J.: Prentice Hall.

OHNO, T. 1988. *Toyota Production System—Beyond Large-scale Production.* Cambridge, MA: Productivity Press.

PUGH, S. 1991. *Total Design—Integrating Methods for Successful Product Engineering.* Addison-Wesley Publishing Co. Inc.

SHINGO, S. 1989. *A Study of The Toyota Production System from an Industrial Engineering Viewpoint.* Revised Edition, Cambridge, MA: Productivity Press.

SME. 1989. NCSU Videotape (45 Minutes), *Simultaneous Engineering.* Society of Manufacturing Engineers (SME), One SME Drive, Dearborn, MI: SME.

TAGUCHI, G. 1986. *Introduction to Quality Engineering.* Tokyo, Japan: Asian Productivity Organization.

ZIAJA, H.J. 1990. "Total Product Quality Process Model." ASQC 44th Annual Quality Congress, (May 15).

WHEELWRIGHT, S.C., and K.B. CLARK. 1992. *Revolutionizing Product Development: Quantum Leaps in Speed, Efficiency, and Quality.* Cambridge, MA: Free Press.

WILSON, P.M., and J.G. GREAVES. 1990. "Forward Engineering—A Strategic Link between Design and Profit," *Mechatronic Systems Engineering*, Volume I, No. 1, pp. 53–64.

TEST PROBLEMS—TOTAL VALUE MANAGEMENT

3.1. Have you seen consumer preferences changing from quality to product values in the 1990s? If you believe in it, state your points with supporting examples or case histories.

3.2. How do you structure WBS for maximum value? How do you use a WBS system to tie work units to space and time dimensions?

3.3. What are the relationships between CE enablers and the 7Ts? How do you decide and gain consensus about which parts of your PD^3 process are in most need of improvements?

3.4. How do you select and empower a team with a motivation to add values and introduce necessary changes?

3.5. What roles do CE and TQM play in TVM? What are the TQM tools that can aid in quality management? How do these differ from quality tools such as continuous improvement tools?

3.6. Describe how value analysis can be used in process simplification. Discuss the usage and benefits of Value Analysis.

3.7. Why is value management important for a PD^3 process? What are the scientific thinking mechanisms for pursuit of constancy of purpose or common goals?

3.8. Explain with illustrations, how a typical production system works. What role does management play and should play in executing a true CE system? In a CE process where do you see *quality built-in*? What other alternatives do you see to ensure quality? Describe 8 TVM tools or concepts discussed in this chapter.

3.9. How does one assess or measure the overall value of a system? Can you quantify the quality or value associated with a product or service? List some items that you look for in an activity-plan.

3.10. What is meant by value Engineering? How can it be used in benchmarking? What benefits does value engineering offer?

3.11. What is a desired process of quality engineering operation? What is the basis of managing this operation?

3.12. Describe a multi-step concurrent process for TVM. How does this differ from Figure 3.18? What changes does TVM bring to a conventional PD^3 process?

3.13. What is DFV? Describe its steps. List the questions that will be asked at each DFV step to properly assess the value content. In design for value (DFV), how can you assess customer-perceived value? What is the relationship, if any, between DFV and life-cycle management?

3.14. How do you choose response functions that are insensitive to local variation in selected parameters?

3.15. When ranking individual design alternatives or concepts, should they be evaluated individually, or relative to each other? List five bottom-line measures to assess the goodness of a product or service.

3.16. In Figure 3.11, we measure savings associated with value Engineering, but an attempt to control costs in one area could lead to increased cost in another? What can you do to avoid that from happening?

3.17. Explain the Concurrent process of Quality Engineering (QE). What is the basis for quality improvement in this approach? How did Ford approach the 5 basics for Quality improvement in the 80's to cause a turnaround?

3.18. What is the design for robustness and why is it so important? Explain Taguchi Methodology. How does Taguchi's approach differ from a SPC based quality control technique?

3.19. What are the statistical methods to ensuring stable process? Explain Dr. Taguchi's six steps of Robust Design. How many of these are off-line variability reduction steps and how many are on-line steps?

3.20. What are the major factors in implementing TVM? What are the advantages associated with it? How can this be used in place of a TQM in the global business environment?

3.21. How does Taguchi's loss function contribute to changing the outlook of quality control practices in today's business environment? Is measure of quality based, at least in part, on how the product works to negate the noise factors?

3.22. What is design of experiment (DOE)? How can DOE techniques be used to improve product as well as processes? How do the number of factors and levels in Taguchi's method differ from the number of trials in DOE? What are the essential benefits of using Taguchi's robust design approach to produce a great product?

3.23. Compare the U.S. and Japanese plant approach to reducing defects and variability. Illustrate the differences from the published data in managing the variability of a Sony Color Television set's quality found in U.S. and Japanese plants. Discuss some examples of error-proofing as it pertains to product quality assessments.

3.24. What is Dr. Deming's philosophy regarding quality improvement? Describe his (Deming's) 14-point strategy. How would you apply this to TVM?

3.25. Why is quality loss function often approximated as a quadratic function? Give an example of a quantifiable quality measurement that is unlikely to follow a normal distribution. How do the zero defect and automatic adjustment loss functions behave?

3.26. What is the purpose of a control chart? What types of conclusions can be drawn from this type of chart in a production environment?

3.27. What QC activities are used during a conventional product realization process? What are some of the reasons most manufacturing companies do not employ QC methods during manufacturing?

3.28. What is TVM? Referring to the basis of value management in Figure 3.6, how does each of the outer layer sectors relate to the inner TVM hub?

3.29. If the assumptions made in the beginning of the product cycle are correct and appropriate measures are taken to DFX, what is the use of on-line quality techniques?

3.30. What are some of the advantages of using CE process of quality engineering over a Failure Reporting and Corrective Action System?

3.31. How do you identify areas in your process where improvements would bring the largest benefits? What is the danger of focusing too narrowly on the end cost, delivery and usefulness of the product and not much on the process or flexibility?

3.32. How do you balance company profits with customer and/or employee satisfaction? How do you introduce change in a way that balances new practices and goals with existing ones?

3.33. How do you use on-line process metrics to detect early errors in the quality engineering (QE) process and correct those errors with least disruption?

3.34 Describe an approach to ensure that changes have long-lasting effects. What is "one-inch wide but one mile deep" approach? How do you prioritize ideas for change to ensure the biggest rewards?

3.35. Describe a curve for profitability or cash flow for a PD^3 life-cycle. How do you determine a break-even point, beyond which negative cash flow changes sign? When is the optimum time to introduce a product to the market?

3.36. How do you estimates time-to-market savings in applying CE loop methodology described in Chapter 9 of Volume I? What is the potential range of savings that can be realized using CE taxonomic approach compared to an equivalent serial approach?

3.37. How do you qualify the impacts of a CE PD^3 process on (a) point of first production shipment, (b) amount of cost savings, and (c) the upstream life-cycle operations?

CHAPTER 4

PRODUCT DEVELOPMENT METHODOLOGY

4.0 INTRODUCTION

The product environment in modern manufacturing is very complex. It consists of many components of products, processes, and services, including information processing services (hardware and software). The design of an automobile, for example, involves 2000 to 3000 parts, and calls for thousands of engineers making millions of design decisions over its life cycle. None of these parts are designed and developed in isolation from each other. Figure 4.1 compares the process of product design, development, and delivery (PD^3) with a process of fluid flow through a maze of pipes. Each pipe of an assembly represents a part or an information build-up activity in a conventional PD^3 process. Serial engineering process involves a number of connected parts or repeated activities of an assembly, such as plan, redo, download, up-load, iteration, retrieve, store, etc., which must be performed in the proper sequences. The *fluid* flowing through the pipes denotes *information* flow of a PD^3 process. The *fluid pressure* is equivalent to needs for *information build-up*. The activities or parts to be designed are represented by straight pipes. The *cross-section* of each pipe represents the corresponding *design parameters*. A typical conventional decision making step is shown in Figure 4.1 by a pipe elbow or an end-coupling. Similar to how an end-coupling changes the direction of the fluid flow, decision making in the conventional serial process changes the steps or parts required for subsequent information build-up. The length of each pipe in the assembly denotes the time it takes to complete the step or build the necessary information for the next serial step of a PD^3 process. Each design decision is a trade-off affecting many other design parameters. Such a traditional breakdown of design tasks, even though it resembles a hierarchical pattern, is repetitive and inefficient. Decision making in the conventional process therefore can be very difficult and total lead time could be very large considering the

164

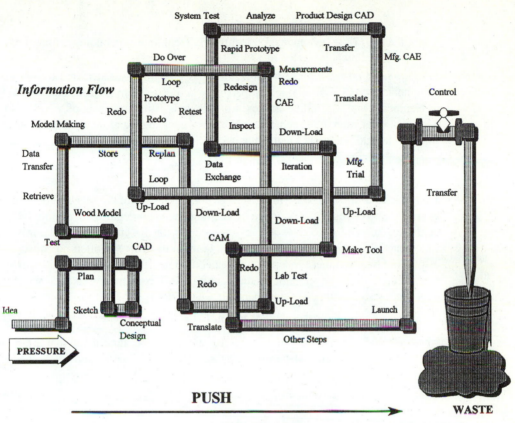

FIGURE 4.1   An Analogy for a Serial PD³ Process

magnitude and complexity of the products and processes that need to be addressed. These complexities are often compounded by the presence of the following four factors [Suri, 1988]:

- *Large Interconnected Components:* There is a high stake on decisions that must be made simultaneously. In modern manufacturing, where both parts and information move rapidly through the plant, a small change (say a material change) at design end of a PD³ process can have a significant impact on the production end. This is likely whether or not parts are stamped, machined, or injection molded. Most changes, static or dynamic, must be managed in real time.

$$\frac{\Delta \text{production}}{\Delta \text{design}} \Rightarrow \text{Large (A factor of 100 or more)} \tag{4.1}$$

- *Limited Resources:* Most modern manufacturers have down-sized their resources (7Ts, as shown in Figure 8.12 and Figure 4.1 of volume I) to a bare minimum. Resources are shared to contain costs. Repetitive demand of shared resources increases the burden of managing them efficiently. The limited resources can be mathematically expressed as:

$${T} \le {T_{\max}} \tag{4.2}$$

where the set ${T} \equiv$ [talents, tasks, teamwork, techniques, technology, time, tools]. ${T_{\max}}$ is the allowable stretch of ${T}$.

- **Geographical Distributions:** Manufacturing is global; it is distributed over a vast geographical areas. For example, a part may be designed in Detroit, manufactured in Kentucky, and assembled in Korea or Mexico. Thus, the costs of travel, transportation, relocation, communication, currency exchange, and labor agreements are some of the additional parameters that are factored into the cost equation.

$$\text{Cost} = C_{base} tf \text{ [Travel, Transportation, Relocation,} \\ \text{Communication}, \ldots, \text{etc.]} \tag{4.3}$$

- **Many Goals and Objectives:** In most large companies, there are sets of independent goals and objectives, F_{ui}, developed by each independent department or unit. Not all of these goals and objectives are in agreement with the enterprise goals, mission statements or its vision, F_{ej}, assuming the latter exits. There is often no constancy-of-purpose between these independently specified goals. The situation gets worse if there is more than one strategic business unit (SBU), each having its own set of independent vision or mission statements. If we define

$$\text{Goals that may be in conflict} = (F_{ui} \cup F_{ej}) \tag{4.4}$$

where u stands for a unit, and e stands for an enterprise, then goals that are good candidates for constancy-of-purpose are those for which intersections of F_{ui} and F_{ei} are nonzero.

$$\text{Constancy-of-purpose Goals} = (F_{ui} \cap F_{ei}) \tag{4.5}$$

The terms u and i in F_{ui} take a value of

$$1 \le u \le \text{number-of-SBUs, and}$$

$$1 \le i \le \text{number-of-unit-goals.}$$

The subscript j in F_{ej} takes a value of

$$1 \le j \le \text{number-of-enterprise-goals.}$$

A method is a specific way of capturing and displaying information concepts [Webster's New College Dictionary, 1988]. Various methods for CE were captured on the basis of taxonomy in the previous chapters. In Chapter 5 of volume I (Figure 5.2), key success factors for realizing team cooperation—7Cs (Collaboration, Commitment, Communications, Compromise, Consensus, Continuous Improvement, and Coordination) were described. Chapter 7 of volume I contained a general classification for information modeling, while Chapter 8 of volume I, described what constitutes a functional *whole system*. Chapter 9 of volume I discussed taxonomy for product realization.

$$\text{Product realization} = \cup \text{ (Planning, Design, Process, Production,} \\ \text{Manufacturing or Assembly, Delivery and Service)} \tag{4.6}$$

4.1   IPD PROCESS INVARIANT

In Chapter 7 of volume I, a set of five *model invariants* was described. They were essential elements of IPD that were considered constant or stationary (thus invariant) in the model dimension of IPD for a CE enterprise. In a similar way, *process invariants* are the essential elements that are constant or stationary (always present) in the process dimension of IPD. The taxonomic approach, described in Chapters 8 and 9, provided for the concurrent work groups a common and disciplined way of working together and communicating with each CE work group. The *process invariants* can be extracted from this taxonomy concept. If we analyze the taxonomic structure of the loop methodology, namely, 1-T, 2-T and 3-T loops as defined in Chapters 8 and 9, the following four *process invariants* can be extracted.

- *Specifications or Goals Structure:* This corresponds to a breakdown structure for specifications (inputs, requirements and constraints) from high level to its low level. It is independent of life-cycle aspects of the product design. This is discussed in sections 8.4.3 of volume I. Goal structure also deals with constancy of purpose goals throughout the enterprise (see section 8.4.4 of volume I) .
- *Process Structure:* This corresponds to a process breakdown structure (PsBS) of the manufacturing process to manufacture the product. This is described in section 4.3.4.2 of volume I.
- *Organ Structure:* The concept of organ fans out the structure of the whole product system using a decomposition technique. This normally corresponds to a 5-layer hierarchical breakdown structure of the product—(PtBS), namely, system, subsystem, components, parts, and material or form features. This is described in section 6.4 of volume I (Figure 6.3).
- *Work Structure:* This corresponds to activities, such as tasks and job contents, information flows, task dependencies, etc. The activities describe the structures associated with organization of work (WBS) and product development teams (PDTs). An example of WBS is shown in section 4.3.2 and Figure 4.7 of volume I.

The process invariants are vertical cross sections of the IPD realization process. *Model invariants* are horizontal cross sections of IPD realization process (see Figure 4.2). *Process invariants* are regarded as key contributors of the IPD realization process. The basic structure of these invariants and their interactions are shown in Figure 7.11 of volume 1. The invariants are linked by taxonomic relationships. The *specification or goals structure* describes the breakdown structure for inputs, requirements and constraints. The *process structure* determines the breakdown of manufacturing functions to be employed by the *organ structures*. *Organ structure* describes the five layer breakdown structures of the product (PtBS). The *work structure* describes the structure associated with organization and management of work tasks (WBS). The four *process invariants* are independent on the state of transformations as the product moves from a set of raw specifications to a real object (physical artifact). McMohan, Xianyi, Brown and Williams [1995] consider the de-

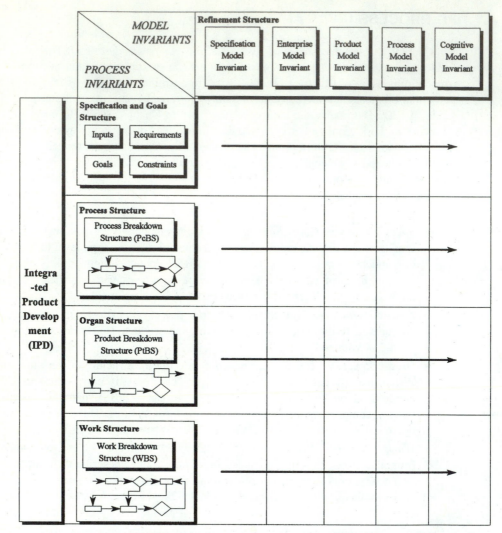

FIGURE 4.2   Invariants of the Taxonomy Dimensions

sign process a parallel development having transformation sequences consisting of several series of models representing different aspects of the emerging design. Two types of transformations are considered, equivalent and non-equivalent transformations:

- An *equivalent transformation* leaves the underlying design unchanged. It is used to develop model representations for the purposes of assessment and analysis.
- A *non-equivalent transformation*, by contrast, develops the underlying designs, for example, to reflect what was learned from the results of some previous analysis.

The loop methodology, discussed in Chapter 9, thus belongs to this non-traditional transformation class. Loop methodology describes the structure of the enrichment process as the product moves from one state of transformation to another while the initial specification transforms from its raw state to a physical form of an artifact.

The contents of each *model invariant* change from one product to another, but their structures remain the same. The *model invariants* determine the structure of the enrichment process, not its contents. Product realization is supported by the behaviors that are captured through the *model invariants*. Andreasen [1992] used a chromosome model to describe a modeling framework with genetic information, information about the origin of the design characteristics. The strengths of both approaches lie in the general well-defined terminology for describing the artifact. The *process invariants* are based on the process taxonomy dimension which has a strong link to a PD^3 process. The invariants are descriptive rather than prescriptive. They are intended to provide a common ground for representing enterprise or business-driven, product-driven, and process-driven works, activities, features, functions and decisions. The structure of the invariants should not be confused with a particular level of abstraction for the artifact or its degree of details. The loop methodology described in Chapter 9, volume I, supports the gradual emergence of design, and in particular the increasing completeness of specifications. In doing so, it reduces uncertainty in information and data, and enriches design content as design progresses through one loop to another. Any process invariant structure may make use of any of the five *model invariants* as discussed in Chapter 7, volume I, with varying degree of abstraction or level of details. For instance, in the *specifications and goal structure invariant* case, as the product moves closer to maturity, the content of specification takes the form of a physical shape. The structure and hierarchical breakdown of the specification do not change from one product to another product. Process invariant requires the *model invariants* (specification model, product model, process model, enterprise model and Cognitive model) to support the corresponding process function. The concept of invariant offers a disciplined way of thinking. This enables a design work group to tackle any PD^3 problem in a common way. Chapter 2, volume II, described CE metrics of measurement. Together, the information contained in the aforementioned chapters of the two volumes forms a coordinated pathway leading to the so called world-class product. *Process invariants* can be considered as an IPD framework to hang-in product, process, work and enterprise breakdown structures (decomposition) and to accommodate cross-functional decisions.

Each of the above *process invariants* corresponds to a well-defined aspect of a CE enterprise and involves several cross-functional teams and multi-disciplinary experts. In concurrent product and process design, it is not enough to have different sets of thinking and working possibilities. Each invariant is like a pathway (road, lane, highway, freeway, etc.) on a geographical atlas. In order to get from point A to point B, we need both good roads and good directions (a map of how to get there faster). In earlier chapters, we have mainly described roads or pathways. We have not discussed how to actually get to a world-class product. What would be an *operational* or a *solution* process, a good road map, associated with all these contingencies (classifications) combined? What would be a consistent set of goals for integrated product development (IPD), and what additional steps should be taken to achieve them? The key for efficient IPD is to identify the fundamental goals that aim for constancy-of-purpose, decompose or classify the problem, and attack the problem at the appropriate decomposited level. An effort for a sound resolution methodology should constitute one of the major works to uncover and understand the whole integrated PD^3 process.

Integrated product development is different in nature from general problem solving, even though IPD shares some common elements of problem solving. Product development is actually an open-ended problem. There is no correct answer, no unique array or line of transformation that would yield the best outcome. One of the reasons for most real-world problems being open-ended is that the given or available information is usually insufficient, incomplete, or uncertain. Work groups furnish the missing information based upon past experiences, product history, assumptions, and practices (3Ps) for the purpose of finding a good starting point.

Although there is no unique starting point, some solutions are better than others. The art of integrated product development is to choose a problem formulation (transformation sequences) that satisfies most of the specifications. The solution (a set of transformation sequences) a work group finds is not nearly as important as the logic behind the transformation and the rationale used during product design and development [Prasad, 1985]. As pointed out by Dr. Deming, any such rational plan, no matter how complex it may be, would merely be a prediction concerning conditions, behaviors, performance of the teams, procedures, equipment, or materials [Deming, 1993]. The team must decide what important measures of merits (MOMs) influence the outcome. For example, should the outcome emphasize safety, ergonomics, low cost, performance, aesthetics (appearances), design for X-ability (DFX), and so on? Reliability, maintainability and supportability (RMS) or other similar DFX measures are discussed in Chapter 2 of volume II. Most product designers do not experience problems in developing specific measures but they are short on creating an appropriate solution plan (an integrated methodology) to realize a complete constancy-of-purpose [Prasad and Emerson, 1984]. The solution plan is a coordinated set of 7Ts that helps ensure that quality solution is found within the constraints of a transformation domain. In Chapter 7, volume I, a basis to classify IPD information into five model classes or invariants was outlined.

1. Enterprise (business planning, information planning) model-class.
2. Specifications (inputs, requirements, and constraints) model-class.
3. Product (data, method and knowledge) model-class.
4. Process (project management, change management, configuration management, manufacturing, work, etc.) model-class.
5. Cognitive (or behavioral) model-class.

In section 7.9 of volume I, a cognitive model class based on how a human brain solves a problem was described. The rationale for reasoning was a cyclic process between different parts of the brain, namely, analysis, decision making, synthesis, and behavior (or relational). In the real world, this reasoning process remains pretty much the same. Due to the complexity of the problems that teams encounter today, human brains are incapable of processing them efficiently. As far as formulation of the problem is concerned, human brains do a relatively good job. If the tasks of analysis, decision making, synthesis, and behavior were supported by computer tools, the complexity would not matter as much. The computer-based activities for reasoning between virtual enterprise and cognitive tools would be very similar to the mental activities. The cognitive tools that replicate the mental activities constitute the virtual team environment. This environment may serve as a supplement to brain power during decision making. Personnel teams also go through the same mental process, but they process the information at a more abstract level, as discussed in section 7.9, volume I. Personnel (cognitive) and virtual teams may cycle independently or loop recurrently many times until all members of the two CE teams are happy with the outcome or have run out of time or budget. In Chapter 5, volume I, four cooperating CE teams were named (see Figure 5.7). The jobs of the PDT (Product Development Teams) leaders are to identify and concurrently schedule tasks among the CE work groups. A task is a set of steps required to support, change, or validate information build-up for a work or product related activity. In this chapter, a procedural methodology for making explicit high-level design trade-off is developed, based on more rigorous methods [e.g., Prasad, 1984a and 1984b] than just the heuristics.

Solving an open-ended product problem is not easy even for work groups. They must communicate, do some trade-off studies, and make collective decisions. The primary vehicle for IPD is the use of decision mechanisms, such as trade-off studies, involving DFX issues and VCM matrices as discussed in Chapter 2, volume II, as an integral part of the whole PD^3 process. The low level results can be combined with weighting factors to obtain the figures of merits for the higher (sub-system or system) levels. It is important, therefore, to establish a unified set of CE methodologies that would do a *lifecycle optimize* precisely but utilize more computer-supported decision tools than human-based decisions. IPD requires a good mix of information models, CE techniques and work groups, and a set of past proven 3Ps (policy, practices and procedures). A technique is a reusable pattern that consists of one or more forms and meanings. CE techniques are shown in Figure 4.3 to bridge the gaps between information models and CE teams. Five sets of such information models are discussed in Chapter 7 of volume I: enterprise, specifications, product, process, and cognitive (human) modeling. Many CE techniques are discussed in Chapter 4 of volume I. Four cooperating CE teams are discussed in Chapter 5 of volume I. This chapter addresses many of the integration issues and provides a systematic framework for carrying out these CE concepts. An integrated methodology for PD^3 integrates work classifications (WBS, team of teams), product classifications (PtBS), process classifications (PsBS), and process taxonomy (3Ps, 4Ms: models, methods, metrics and measures, and 7Ts). The outcome is an IPD system. Figure 4.3 shows the connections between several of the preceding terms leading to IPD. The ability to execute efficient and effective design revisions as a part of this integrated methodology is critical to achieving product quality, cost containment, and reduced delivery time [Chandrasekaran, 1989].

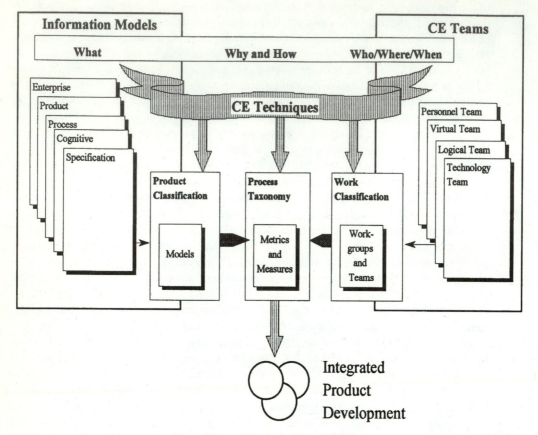

FIGURE 4.3   Integrated Product Development

4.2   INTEGRATED PRODUCT DEVELOPMENT PROCESS

Beyond concurrency, integration of distributed PPO (product, process and organization) seeks to offer better life-cycle alternatives and realization potentials. The taxonomy described in Chapter 9, volume I, establishes a methodology of systematically organizing the process necessary for new product realization [Berger, 1989] and for developing future product upgrades. Though this process taxonomy is useful for formulating and decomposing the system [Kusiak and Wang, 1993], it does not take the work groups to the next step, that is, to synthesize or optimize the design system with respect to the identified constraints. Taxonomy characterizes the design system problem into a well-structured set of decomposed tasks. This converts the IPD process into a topology of networks showing how tasks are interconnected. However, the taxonomy does not show how the network of tasks will be solved. Well-structured tasks are amenable to a variety of solution techniques such as analysis, simulation, sensitivity, optimization, mathematical programming and other weak numerical techniques [Prasad and Emerson, 1984]. There are two types of complexity in a design problem, one is structural complexity and the other is computational complexity [Prasad, 1984a and 1984b]. The two are not the same concept. Structural complexity is resolved through problem description, whereas computational complexity is resolved through problem solving. Models and structuring techniques help design work groups to reduce the structural complexity of the tasks and provide a structured road-map that accompanies problem solving steps [Prasad, 1985]. With predetermined computational models, work groups simply enter the specifications and the model forward-solves to provide the results. This process is very similar to a spreadsheet. In synthesis or design optimization, unlike the spreadsheet, the model or work group can back-solve, entering the desired result and making the synthesizer find suitable input specification values. AI based techniques are demonstrated to perform well in a closed environment (well-structured symbolic or rule-based domain) that uses weak methods of problem solving. There are few commercial tools that can accept a set of mathematical equations that are pre- or user-defined, and analyze or synthesize the problem on a need basis. However, most tools require the problems to be explicitly defined or have their characteristics explicitly known.

Most books on optimization, for example, concentrate on how to solve an optimization problem if it can be expressed in a mathematical form (e.g., a linear or a nonlinear function of design variables) [Prasad and Magee, 1984]. Such formulations are often of a closed type. The work groups often find no difficulty in arriving at a suitable solution since all the necessary information about the problem is given or known. Work groups also know when they are finished with the problem and generally know if it can be solved correctly. The most general statement of the optimization problem posed is to:

$$\text{Find a vector of design variables } v \in D \text{ that minimizes or maximizes} \tag{4.7}$$
$$\text{a set of value characteristics (objective functions), } VC_i(v),$$

$$\text{while satisfying a set of constraint equations,} \tag{4.8}$$
$$C_{ij}(v) \geq 0.$$

D is the design space in which the solution lies. The design space is actually an intersection of three sets: a set of requirements, a set of constraints and a set of design variables (see Figure 4.4). Most optimization problems have four parts.

1. A transformation system.
2. A set of objectives (a function or a criterion)—$VC_i(v)$.
3. A set of design variables, v_i.
4. A set of constraints, C_{ij}.

- **Transformation System:** In most product realization cases this is a part of the problem definition and, therefore, hidden. In books, this is often identified in explicit form,

$$[T]\underline{v} = [O] \tag{4.9}$$

where T stands for transformation and O stands for output. Both are characteristic matrices and v is a vector of design variables. Such explicit forms define the problem—how the objective functions and how the constraints are related to design variables.

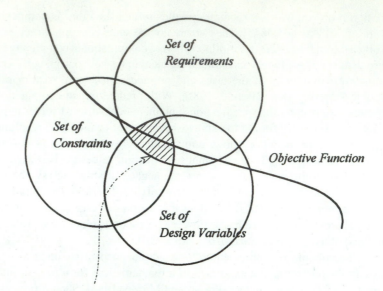

Intersection Set == Design Space

FIGURE 4.4   Definition of A Design Space

$$VC_i = f(\underline{v}) \tag{4.10}$$

$$C_{ij} = g(\underline{v}) \tag{4.11}$$

- *Objectives:* The objective is a function or criterion that characterizes the aspect of design to be improved. It is commonly posed as a single criterion problem, such as cost, weight, strength, stiffness, and so on. At times, the objectives are to optimize under the presence of a set of multiple criteria and task levels [Prasad and Emerson, 1984]. The technical merits are established by evaluating key performance, reliability, structural integrity and economy criteria and comparing these to the current best values forecast of these criteria [Agrell, 1994]. The problem of multi-criterion optimization is to minimize or maximize a set of merit functions, simultaneously:

$$VC_i(\underline{v}); \qquad \text{for } i = 1,2,\ldots \text{qc-max.} \tag{4.12}$$

For example, the design of a machine tool involves many aspects, such as transmission, control system, hydraulic components, power utility, and body frame. This is a case of configuration optimization. Configuration optimization is the process of first identifying the best from a group of configurations, and then embodying the configuration to provide the highest possible technical merit values, such as performance, reliability, durability and cost. If there are several performance attributes, they can be combined using multi-attribute utility theory [Keeny and Raiffa, 1976] to build a duty index. Similarly reliability attributes can be combined to obtain a reliability index and cost attributes can be combined to obtain a cost index. Each index can be normalized with respect to a carefully selected benchmark solution—

a datum value for comparison. Normalization scales the parameters and provides a better numerical stability and convergence.

- **Design Variables:** These represent those input parameters of a problem that are subject to change. The design variable is usually a vector, where

$$v = [\{v_{sizing}\}, \{v_{shape}\}, \{v_{topology}\}, \{v_{knowledge}\}]^T. \qquad (4.13)$$

There are four classes of design variables commonly used:

- *Sizing Variables*—$\{v_{sizing}\}$: These include variables like thicknesses (for thin-walled sections) and areas (for solid objects) that can be changed.
- *Shape Variables*—$\{v_{shape}\}$: These involve changing the configuration points or the geometry of the part that is represented such as length, width, height, coordinates, and so on.
- *Topology Variables*—$\{v_{topology}\}$: These define parameters that actually determine where material should or should not be removed. As long as the topology change can be represented parametrically in the CAD system, the model can be optimized. Topology optimization allows feature patterns, such as how many bolts are needed to hold down a given part, or how many ribs provide a given stiffness.
- *Process Variables*—$\{v_{process}\}$: These involve changing the rules concerning the part's forming or processing needs that have the effect on changing the part's size, shape, topology or functions themselves.

- **Constraints:** These are the response parameters (state variables) of the model used to evaluate the design based on the criteria that limit how it should function or behave. They are usually specified in an equality or inequality equation form.

$$C_{ij}(v) \geq 0; \text{ for } j = 1, 2, \dots \text{c-max.} \qquad (4.14)$$

Often such constraints also include limits on design variables:

$$\{v_{min}\} \leq \{v\} \leq \{v_{max}\} \qquad (4.15)$$

$\{v_{min}\}$ and $\{v_{max}\}$ vectors denote lower and upper bounds on design variables.

This completes the definition of an optimization formulation. The transformation system ties the optimization model of the product with objective functions, optimization constraints, and design variables in some mathematical or conceptual forms (see Figure 4.5). The objective functions are derived from performance type outputs of the transformation state. The set of constraints and the objective functions are derived from behavior type outputs. Examples of behavior type outputs include deflection, noise, vibration, frequency, stiffness, strength, and so on. Design variables are derived from specification attributes (inputs, requirements, and constraints). Note the shuttle differences between the input and output constraints. Transformation state constraints are the inputs to the baseline state whereas the optimization constraints are outputs from the baseline state. Books on optimization seldom focus on the transformation system. They assume it to be given or explicitly known. However, most physical problems cannot be modeled purely in an explicit form. For example, in a minimum structural design problem, the transformation system exists in a finite element model or a similar form. The relationship between constraints (such as stress and deflections) and objective functions (such as weight) is tied to

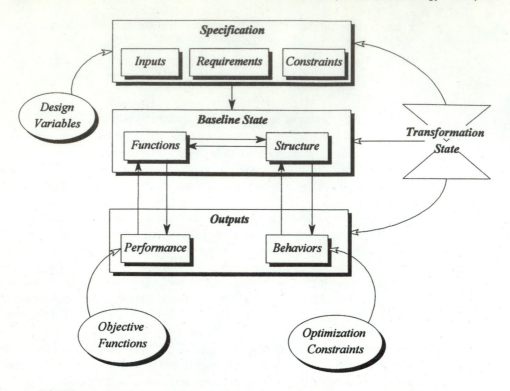

FIGURE 4.5   Relationship between a Transformation State Attributes and an Optimization Model Variables

the stiffness matrix of a FEA model. Consequently, in most cases, design trade-offs are made tacitly and implicitly. An implicit statement of a problem at a system level is shown in Figure 4.6. The solution of the problem is governed by:

1. Minimizing a set of functions.
2. Maximizing another set of functions, and, at the same time,
3. Desensitizing some parameters of the problem.

Most product designers in industry are not familiar with how to express the transformation system as an optimization model, explicit or implicit, so that the problem can be optimized. An open-ended optimization problem usually takes longer to solve in iteration than a closed form problem. Teams must evaluate the formulation, the validity of the assumptions, the credibility of MOMs and other criteria, the mechanics of analysis, and the reasonableness of decisions. Here, formulation goes beyond its mathematical sense to modeling of the elements of design variables, constraints, and objectives, as work groups move from one stage of product realization to the other. One of the important considerations in the product realization process is an output modeling. Performance modeling is one type of output modeling. Performance modeling provides the transformation from

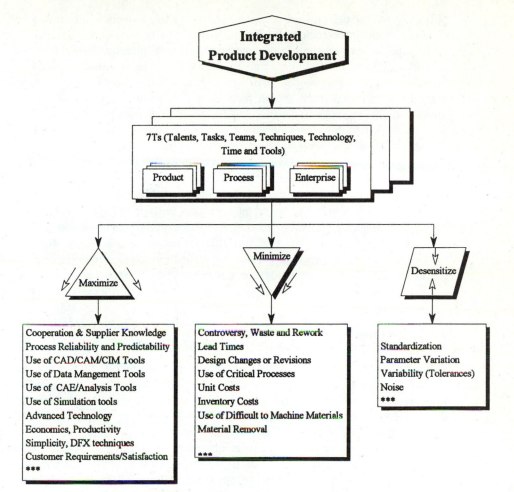

FIGURE 4.6   Definition of a Global IPD Solution

specifications to the outputs that are performance-based. One way to represent a performance-based transformation model is to use a set of symbol structures for the description of the incoming specifications and to use simulation or analyses to predict output behaviors or performance. In performance modeling, one may use performance metrics, (such as manufacturability, assemblability, reliability, etc.) to simulate, quantify or analyze the outcome. Analysis or simulation is not the only mechanism to capture outputs of a transformation system. Genetic algorithms are being explored with CAD systems to generate designs in a generate-and-test approach called *conceptual interpolation*. In conceptual interpolation a number of conceptual operators provide a genetic basis for generating interpolate designs. In genetic algorithms, interpolate designs exist in generations. Within each generation, designs can mate and produce offspring according to some measure of their value characteristics (a fitness function). The conceptual operators amplify the power of design work groups by allowing them to work at higher conceptual levels. Other techniques, such as the use of fuzzy set theory, AI approach, and simulation-based generate and test approach, are commonly used to alleviate formulation or trade-off difficulties [Saaty, 1978]. For example, if a set of fuzzy goals is modeled according to the life-cycle issues of the product, goals can be interpreted as a set of criteria. A premise of fuzzy set theory is that the overall preference of a product design alternative is represented well by aggregating the individual goals with respect to the criteria [Parsaei and Sullivan, 1993]. In this context, a variety of fuzzy set connectives can be used to form the framework on which the aggregation process of product realization goals can be based. Using this or similar methodology, future design evaluations are based on accumulated knowledge. If any of the elements of the fuzzy set or optimization model are incorrectly specified or are inappropriate, the resulting design would be incorrect, too. Thus, it is normally not sufficient to have a sound mathematical basis or to have the world's best algorithm. Formulating the transformation system and identifying a consistent fuzzy set or optimization model at each step of this transformation is critical to efficient product realization. A rational prediction to product realization requires building in a sound analytical or algorithmic methodology or a computer-based procedure at each step of this transformation.

4.2.1 IPD Methodology

A key requirement in a distributed environment is to provide a quantitative mechanism for integrating competing information from distributed agents, combining different opinions, resolving conflicts, and finding a feasible or optimal solution [Shumaker, 1990]. In this chapter, an IPD methodology has been developed to capture optimization formulation as part of the IPD process. It employs the conventions of Chapter 8, Figure 8.11, volume I, that is, inputs, requirement, constraints, and output, to reformulate the product realization problem in a decentralized decision integration setting. The purpose of this IPD methodology is to improve the performance characteristics of the product or process relative to customer needs and expectations. It builds the theory of knowledge through systematic revision and extension of the paradigms introduced in previous chapters. As shown in Figure 4.7, there are eight parts to an IPD methodology called building blocks. The first four blocks, 1 through 4, provide a conceptual framework for understanding the challenges and opportunities. The last four parts, 5 through 8, provide the building blocks for an analytical framework for decision making and improvements.

4.3 STEPS IN IPD METHODOLOGY

IPD is described as the process of going from a set of incomplete and inconsistent functional requirements to realizing a physical product. As stated earlier, this methodology has eight parts to it. Each part contributes to the overall effectiveness of the IPD process. These parts are listed here and shown graphically in Figure 4.7.

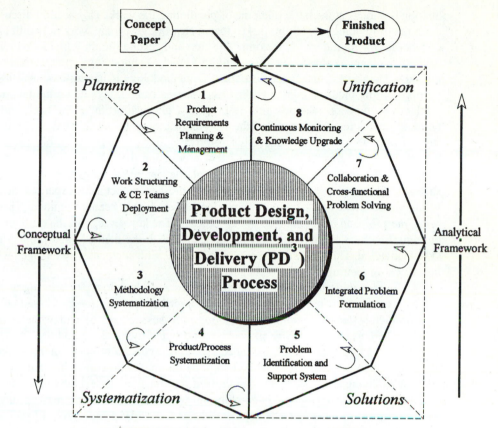

FIGURE 4.7   Building Blocks of an Integrated PD³ Process

1. Product Requirements Planning and Management.
2. Work Structuring and CE Team Deployment.
3. Methodology Systematization.
4. Product and Process Systematization.
5. Problem Identification and Support System.
6. Integrated Problem Formulation.
7. Collaboration and Cross-functional Problem Solving.
8. Continuous Monitoring and Knowledge Upgrade.

The first four blocks involve the construction and definition of problem statements to determine the artifact's functions, components, sub-functions and their interconnections. The first building block of IPD is determination of product requirements planning and management. The second building block is work structuring and CE team deployment. Structuring of work facilitates the integration of complementary engineering expertise.

Section 4.4 looks at the basic dilemma of paralleling the work. This section discusses the dilemma of dividing responsibilities so that each work group can work in parallel and yet collaborate for joint decision making. The first and second building block combined are called the planning phase. The third building block, Methodology systematization, outlines the basic conceptual framework for IPD methodology [Chandrasekaran, 1989]. The fourth building block, Product and Process systematization, lays down the taxonomy of product and process transformation leading to a physical artifact. The third and fourth building blocks together are called the systematization phase. This phase implies understanding the ability of the manufacturing process and communication of upstream and downstream concerns to produce consistent production parts. While the systematization phase of IPD may consume only 5–15% of the enterprise resources, it casts a long shadow. When this phase of IPD is complete, most manufacturing expenses, as well as value characteristics of the would be product, have already been committed. This represents, therefore, an important step in PD^3 process. The key strategy of the third and fourth building blocks is to predict problems associated with the methods and product/process systematization. This way the problems can be dealt with adequately or changed when the initial cost of their modifications is cheap.

The remaining four building blocks outline a methodology for arriving at an optimized or a consensus-based alternative. They describe solution strategies based on what is algorithmically feasible (from contributions of blocks 5 and 6) and what is possible through the consensus-building approach (contributed by blocks 7 and 8). The fifth and sixth building blocks combined are called *solutions*. In the sixth block, an integrated problem formulation approach, the methodology determines what constraints are violated, in addition to giving an alternate set of optimized solutions. In the seventh block, consensus-based approach, the methodology combines different opinions, which may or may not be analytically based. This resolves key conflicts and provides a set of possible solutions in addition to the conflicts that cannot be resolved. Collaboration means coordination of work group problem solving abilities. The last building block, continuous monitoring and knowledge upgrade, emphasizes the need for continuous improvement. The structuring of work and the need for redefinition of the methodology (product realization process) within the Concurrent Engineering framework (taxonomy) was discussed in section 8.4 of volume I at length. Section 8.4.5 of volume I presents conceptual frameworks that facilitate a comprehensive understanding of the system optimization methodology. Optimization is used here to indicate a process of finding interdependency from distributed agents, defining strategies, and finding a feasible or a consensus-based solution that resolves the conflicts. Interdependency can be classified as competitive, complimentary and cooperative. A consensus-based solution, then, entails a cooperation of multiple competing perspectives maneuvering through all considerations of the product's life-cycle values. Blocks 7 and 8 combined are called *unification*. Although this definition of unification will streamline the product realization process during the search for the best design, a solution cannot be arrived at without a good problem identification and support system (block 5) followed by integrated problem formulation (block 6). Therefore, all eight facets must be used within the context of a cross-functional team as a part of a simultaneous engineering effort. In reality all of the above tasks are inter-connected.

4.4   PRODUCT REQUIREMENTS PLANNING AND MANAGEMENT

Effective identification of design variables, constraints and objectives at each level of transformation requires a high level of intelligence. Human resources and expertise are needed to analyze adequately the transformation process and identify the optimization model from the prescribed set of inputs, requirements and constraints (RCs), and outputs. It is still beyond the current capabilities of modern computers to automate these functions. Various types of optimization models may be necessary to satisfy RCs, such as incorporating life-cycle values or assessments. These types of assessments are established through the metrics of measures (MOMs) in Chapter 2. The unknowns are: the priority, the distribution of these MOMs into objectives, and distribution of specifications into design variables and constraints. The latter may be looked upon as a type of transformation from a conceptual-model state to an optimization-model state. This is not an easy process. The model depends upon many factors: prior design history, life-cycle priorities, and other value implications. A team of work groups with different expertise may be required to formulate the optimization model that contains all essential elements of customer needs and RCs.

A systematic classification scheme for distributing RCs during product development must be designed. This is viewed in Figure 4.8 as a 3-D graphical pyramid. The pyramid

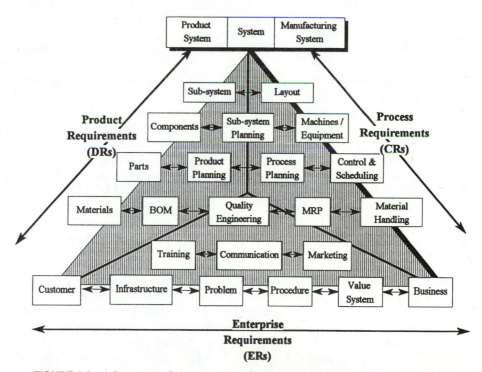

FIGURE 4.8   A Systematic Scheme for Classifying Requirements and Constraints (RCs)

has three inclined triangular ceilings and a hidden base. Each triangular ceiling is supported by three sides belonging to RCs for product, process, and enterprise, respectively. In modern manufacturing, product, process, and enterprise are constantly entwined. The base of the pyramid is the foundation—the CE philosophy of product realization on which the remaining three sides stand. The 7Cs (Collaboration, Commitment, Communications, Compromise, Consensus, Continuous Improvement, and Coordination), the 7Ts, 4Ms and the CE infrastructure provide strength to the hidden base of the pyramid. The foundation base touches the ceiling sides and supports them all. Corresponding to each side of the foundation, RCs are shown classified into three distinct forms corresponding to the three major sub-domains.

- Product requirements (DRs) for product design sub-domain.
- Process requirements (CRs) for production process sub-domain.
- Enterprise requirements (ERs) for customer and business sub-domains.

Along each base side, the RCs are further branched into independent block structures to generate the next level of transformation which can best serve the needs of different life-cycle sub-domains. The pyramid model shows perfect symmetry with equilateral sides, but it is unlikely that this will happen in all cases. There are some CRs, DRs, and ERs that are common, for example, manufacturing tolerances may appear at all places. The RCs are directly related to the geometrical form of an artifact that can be used to infer the functional form of a product at each level. In order to simplify the reasoning process and to better characterize the relationship between RCs and geometrical forms, these RCs can be represented as a hierarchical, tree-type structure. This hierarchical structure is based on the assumption that RCs can be independently propagated into a lower level RCs at each loop level (see Figures 8.16, and 8.17 of volume I). RCs at each loop level can then be mapped into their corresponding form features through a mapping process. By back-tracking (concatenating the RCs) as we move from the bottom up (similar to Figure 9.10 of volume I), the actual functional form of the artifact can be inferred.

At the beginning of the IPD process, initial specifications represent the highest level of abstractions representing a complete and consistent set of specifications for PPO (product, process and organization).

4.5 WORK STRUCTURING AND CE TEAM DEPLOYMENT

Work structuring defines a work breakdown structure (WBS). A WBS and its interrelationships allow for the structuring of tasks at different breakdown levels. This is useful in product realization because different alternatives can potentially be generated and evaluated at each level by using the system to execute WBS networks and by changing the 7Ts deployment and work structuring. Team or work group deployment defines a PDT organizational structure and a cooperative project team structure. Work structuring includes timing for the various transformations and loops applied to the product and process design cycles. Facilities to manage and organize the WBS at both computational and human levels is an important part of IPD. Figure 9.1 (Chapter 9, volume I) describes a program timing schedule for 1-T loop, 2-T loop, and 3-T loop transformations. A set of taxonomy for loop transformations has been thoroughly discussed in Chapter 9 of volume I. Methodology systematization, on the contrary, has nothing to do with program timing. Systematization sets a common methodology for problem solving that can be applied to any loop transformation.

4.6   METHODOLOGY SYSTEMATIZATION

A product realization process usually involves a large number of design and analysis activities that need to be managed. The quality or depth of information about a design solution evolve during this realization process. Without any systematization, it is not possible to process the design constraints of a sub-assembly or an individual part since, at that point, many of its details are unknown. It has long been recognized that problems, no matter what their size or complexity, can best be solved by working through a sequence of steps [Warfield and Hill, 1972]. Steps systematize the methodology of problem solving, which, in turn, helps to prevent adverse situations. Researchers now recognize systematization as a fundamental approach to understanding and controlling the interaction between the constituent elements [Ulrich and Eppinger, 1994]. Another important reason why designers work hierarchically is that an individual person or a work group is not able to process large numbers of constraints simultaneously. Systematization offers a simplification tool to reduce the inherent complexity of the problem domain. Methodology ensures that everything possible will be done to apply the 7Ts resources in the most effective manner. One such act of systematization in product design is to apply decomposition, branching the design into loops, activities and tasks. Stefik [1981] describes further motivations for decomposing design and manufacturing activities.

- The apparent structural and computational complexity of a design and manufacturing problem is often reduced as a result of decomposition
- If the decomposition is done with a view to minimizing interdependence while the activities are split into tasks, each discrete task can then be run in parallel.
- The problem is reduced to a series of self-contained smaller activities or tasks. For example, most of the details of a sub-system are irrelevant when the design problem is dealt with at the system level.
- The talent and expertise of designing sets in a decomposed problem can be divided among the area specialists. Each team of work groups can be assigned to work on each decomposed set concurrently.
- This enhances concurrency of the product realization process.

There are many levels of abstractions in a systematization. In problems as complex as IPD, systematization starts with product/process management (see Figure 4.9). In Chapter 8, the loop concept was introduced to manage the interactions between different life-cycle

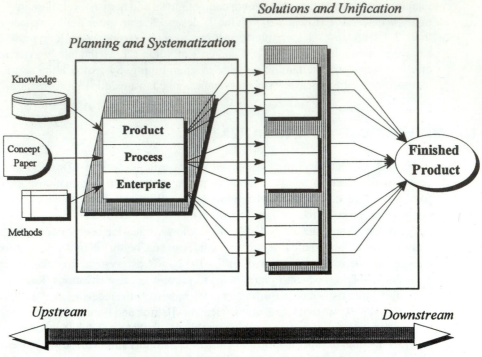

Solutions and Unification

Planning and Systematization

FIGURE 4.9   IPD Methodology Systematization

phases. Each loop consisted of five major components: a "baseline system," inputs, outputs, constraints, and requirements (see Figure 8.17 of volume I). In general, the solution to problems of this type can be approached by a four-stage systematization process as shown in Figures 4.7 and 4.9.

Stage 1.   ***Planning:*** This is the stage when system specification is defined. Partial ordering of intermediate goals is identified and a view to early determination of gross features is conceived.

Stage 2.   ***Systematization:*** There are two kinds of systematization: methodology systematization and Product and process systematization. Systematization is the:
- Systematic decomposition of the problem into discrete sub-problems.
- Decomposition of product, process, and enterprise specifications into different levels of abstractions.
- Description of this abstractions in terms of the functional and/or physical elements.
- Identification of the interactions that may occur between these elements.
- Aggregation of the discrete constituents back into their high level original definitions.

Stage 3.   ***Solutions:*** After stage 2 is completed, it is followed by solutions to the series of sub-problems, specifications, or constituents. In this stage, a number of al-

ternatives or options are obtained for each sub-problem, and the best solution is
selected based on the specified goals.

Stage 4.   ***Unification:*** The fourth stage is aggregation or reconstruction of an overall solution from the various solutions to the sub-problems. Unification is an important
step in solving IPD problems while using the decision making process. Unification may involve system optimization and may consist of a higher level of decision making and complex reasoning. Unification helps eliminate the constraint
violations and yields to refining the product for interface considerations.

The above four stage process highlights the major steps to be undertaken in tackling an
IPD problem. This process is very similar to decomposition and aggregation process used
during satisfying the integrity of the product as shown in Figures 9.10 and 9.19 of volume
I. The four stage process can be regarded as a continuous cycle. Systematization is an important step in yielding a faster and better solution. It reduces search and makes product
realization more efficient. If the decomposition is not discrete, the sub-problems would
neither be compatible nor necessarily complementary. The sub-problems will suffer from
excessive interdependence. There will be a large number of constraints and design variables that are common to these sub-problems. One or two iterations of the above four
stage process would not lead to a *good* solution. Several of the constraints would be in
conflict. It may require an excessively large number of iterations for convergence. In such
cases, decomposition may not have many real benefits. Real benefits are obtained when
the four stage process results in significant savings in time and effort, compared to solving the original system problem as a huge global optimization problem.

4.6.1 Branching and Bounding Methodology

A branching and bounding methodology has been used here first to branch the product and
process sub-domains into loops, and later bound these loops into a sub-domain and then en-
fold it back into an IPD system. The notion of branching allows for the exploration of possibilities, whereas bounding provides a way for the concurrent work groups to judge the outputs.

- ***Branching:*** Branching of each domain into loops is carried out uniformly (see Figures 8.16 of volume I and Figure 9.14 of volume I). A consistent representation is
  used throughout the branching and bounding process. In each group, the baseline
  system can be further branched into independent sub-units which can better serve
  the needs of these loops. The three loops, feasibility synthesis, design synthesis, and
  process planning synthesis, provide a basis for satisfying product-oriented requirements, giving rise to the so-called product-oriented loops (see Figure 4.10). The
  other three loops, process planning execution, production synthesis, and operation
  synthesis, provide a basis for satisfying process requirements, giving rise to the so-
  called process-oriented loops (Figure 8.17 of volume I). The second level and third
  level breakdowns of the product oriented loops are also illustrated in Figure 4.10.
  The constraints in each of the individual loops provide a basis for determining the
  goodness of a candidate baseline system.

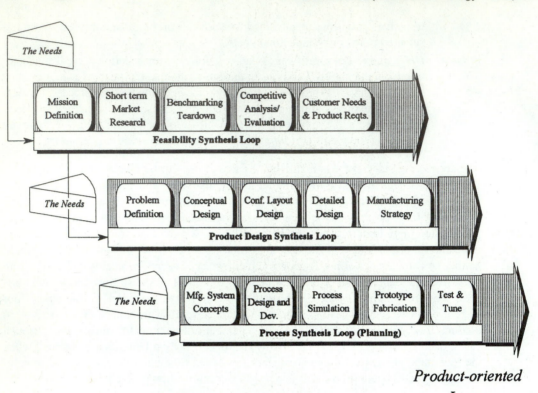

FIGURE 4.10   Concurrent Elements of Product-oriented Loops

- ***Bounding:*** This provides a mechanism to evaluate the goodness of a baseline candidate system with respect to meeting a common set of requirements. During the bounding phase, the system computes the *goodness value* based upon the constraints that are still not satisfied up to that stage. The common product and process constraints in the two half-domains provide a basis for determining the *goodness* of the total system (Figure 8.17 of volume I). The system value, a cumulative index on the measure of violations, envisions the trade-off possibilities or further exploration of the problem domain. This may require a comparison of each synthesis loop's outputs to the system's goals and objectives, refinement or reallocation of requirements, reevaluation of lower level objectives or reconfiguration of goals. Bounding occurs through the use of concurrent function deployment, or similar techniques (Chapter 1), for simultaneous consideration of a series of competing requirements and objectives during an IPD process.

The branching and bounding methodology provides the ability to refine successively the goodness or fitness of a baseline concept as one proceeds from one nested loop to the other (Figure 8.17, Chapter 8, volume I). The domain of the product realization process is, however, evolutionary. During a loop, the design or concept formulation is not fixed, rather it reflects CE teams' understanding of the design problem and the environments spanned by its (loop's) specification sets. As the satisfaction of the specification continues during a loop, the work groups learn more about the forthcoming baseline model and output (solution) state as new aspects of its behavior inherent in the formulation are revealed. As a result, work groups may gain new insight into the behavior of the model (and the solution output state) that may have an effect in reformation of a new set of specifications or changing the baseline system concept. This process of learning and reformation can continue until one or more of the following conditions is met:

- The incremental change in behavior of the baseline system concept due to change in inputs becomes insignificant (produce no change in outputs).
- The requirement sets are empty, no more requirement is left to be satisfied.
- The incremental satisfaction of the constraints become contradictory or result in a concept that is too costly or cannot be manufactured.

It is important to be able to use the prescribed evaluation criterion for each candidate baseline system in order to guide the redesign process as the product evolves from problem or customer needs to an artifact instance. Each branch and bound procedure inherent within each loop provides a mechanism to select an increasingly good design or concept. Any intermediate trial design based on such a bounding procedure is subject to iterative rework, or can potentially be discarded. This is quite natural. This has always been the case in the conventional process too when someone chooses to design a part from incomplete or uncertain data. So what is *different* in a taxonomy-based CE process? Without the taxonomy (ability to classify the PD^3 process), one is forced to instanciate all possible configurations and check the compliance with respect to all possible value characteristic requirements. It may not be possible to carry out this instanciation process every time, taking account of the inherent complexity of the product. Another alternative is to ignore many possible configurations, or consider a subset of inputs, requirements, or constraints one at a time. This often results in a series of design concepts that are sub-optimal in some way. A taxonomy-based CE process acknowledges the risk up front and provides a taxonomy-based procedure to manage this risk appropriately. The four stage methodology (see Figure 4.9) thus provides a built-in risk management technique. It balances the needed reduction in time-to-development against the risk of concept changes in the form of iterations and loops.

The four stage methodology can be applied to any new product introduction, or to any problem set having a deviation from its initial specifications. This can also be used to tackle a continuous improvement opportunity.

4.7 PRODUCT AND PROCESS SYSTEMATIZATION

Concurrency can be exploited by differentiation, followed by systematization—organizing the information in a hierarchical way (Figure 6.6 of volume I). In most product systems, there are complex interactions among many of its constituents: sub-systems, components, parts, materials, features, and so on. Product systematization is a technique of handling a larger class of problems, or a product, by decomposing and then concatenating the results of its behavior through a smaller set of problems or hierarchical organization. Examples of an automobile, an aircraft, and a helicopter are shown in Figures 6.8, 6.9, and 6.10 of volume I, respectively. Similarly, a process can be decomposed into activities. A group of activities aggregated into a high level activity group is called a scenario. By systematization of process in the early stages of product development, the work group can compare various scenarios. Concurrency can be affected by studying the dependency of the decomposed process set or scenario. If one is able to reduce the dependencies among the decomposed process sets or scenarios, concurrency can be increased. Concurrency can also be increased, and interdependency reduced, if one is able to maintain precedence between the consecutive decomposed process sets or scenarios. Since the effect of maintaining precedence between tasks is reduced interdependence, the degree to which concurrency can be affected depends upon the mode of product decomposition into constituents or process decomposition into activities. Figure 4.11 shows an example of two scenarios of the same process. The process shown in scenario X has been decomposed into five activity-groups, A through E, as shown in Figure 4.11. Such scenarios are said to be *serially decomposable*. The constraint equations between the scenario Y activities can be solved serially, yielding the value of one new activity for each constraint evaluation. When a set of constraint equations are not serially decomposable, other ordering methods are used [Navinchandra, Fox, and Gardner, 1993] to avoid solving a large set of equations simultaneously. Activities within a group can run in parallel. There are two parallel activities in group A, four in group B, and three in groups C, D, and E (see Figure 4.11). The activities within Groups A to D can be overlapped if the dependencies of the interfaces are not very strong. As differentiation proceeds, emphasis changes to interfaces between the decomposed sets: between system and subsystems, between subsystems and components, between components and parts, and between the constituents themselves (see Figure 9.10 of volume I). Besides some recent works on multi-level optimization and decomposition, the field of assembly, or system, is still new and growing. There are three possible ways in which product and process systematization can occur. These are described in sections 4.7.1 through 4.7.3.

4.7.1   Through Problem Structure Recognition

Many researchers look for commonalties between the structures of the organization or the structures of the problem, or both [Galbraith, 1973]. An engineering design can be divided into logically distinct modules that describe a portion of that design. This concept is known as Modular Design. Similarly, RCs for each module can be broken down into manageable components. The idea is similar to what is done in software design. The computer program is divided into sub-routines or procedures, which may be represented in pseudo code or an actual code. In modular design, we have modules of design that relate to each other. Alternate designs are created by giving values to their design modules. Each set of values given to a design module is called a version. Design session is a process of designing parts, and the version is an output of a design session. Versions of design modules can be related to one another in many ways. Several versions relate to a

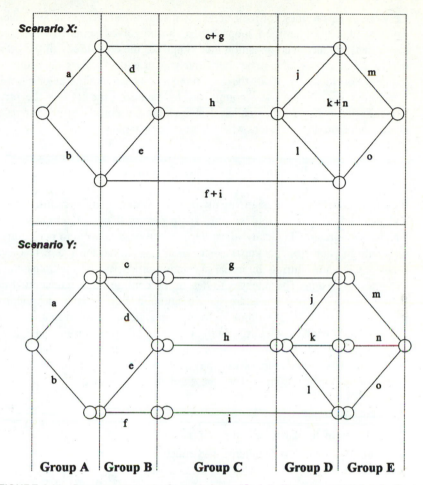

FIGURE 4.11   Decomposition of a Scenario into a "Serially Decomposable" Activity Groups.

perspective, several perspectives relate to an assembly, several assemblies relate to a product. For example, there may be versions of the same design module belonging to the same product, or describing the same assembly, but from a different perspective [Hardwick, et al., 1990].

4.7.2   Decomposition and Partitioning of Problems

Concurrency can be exploited by viewing the product as a complex system which can be decomposed or partitioned into a series of subproblems, each with its own set of requirements and constraints (RCs). Each subproblem can be solved in parallel with a subset of the RCs from the original problem and the results brought back to satisfy the coupling from the remainder of the RCs. The complexity of the products and the processes forces the work groups to look for their breakdown structures and exploit any inherent independence so that the identified subproblems can be solved somewhat independently from each other. Product decomposition thus simplifies the complexity of the original problem that would otherwise have to be alternatively dealt with. The original problem would have required consideration of all the requirements and constraints at one time. The use of smart modules alleviates this complexity by capturing the RCs in the form of product structure decomposed into smaller sizable chunks. It contains rules for reconfiguring or changing the product structure when there are new inputs.

4.7.3   Through Product or Process Organization

Products, in mechanical design, are divided into systems, sub-systems, components, parts, and materials (Figure 6.3 of volume I). In electrical design, circuits are divided into smaller sub-circuits. Typically, most designs can be divided into perspectives as well as into configurations (or mockups) as shown in Figures 6.6 through 6.12 of volume I. The focus of CE teams, during the initial realization process, is on the holistic elements of the PD^3 process design. The intent is to design the configuration (conceptual mockup) first, before the perspectives are addressed. In automobiles, airplanes or helicopters (see Figures 6.8, 6.9, and 6.10 of volume I), for example, perspectives might include aerodynamics, weight, strength, esthetics, vibration, noise, and so on. In mechanical design, a part might be described as a function or as a drawing; and in electrical design, a circuit might be described as a schematic or as a physical layout. It is important to note that such decompositions are not discrete. They are divided on the basis of

- Degree of independence.
- Degree of similarity or compatibility.
- Ability to provide complementary functions.

By decomposing a product into its constituents, and a process into its activities in such a way that their inherent dependencies are minimized, the level of concurrency can be maximized. Independence means that the strong interactions occur within an individual constituent or an activity itself. The weaker interactions occur across its decomposed constituents or activities. Pahl and Beitz [1991] and Suh [1990] use such concepts in modeling the functional requirements of product design. If the constituents or activities are inextricably related, such product or process divisions are not good. They are neither inherently compatible nor necessarily complementary.

4.7.4   Through Part/Activity Classifications

Classification is a basic concept. Once information is categorized, it can easily be found or tracked. A classic example is a periodic table. By categorizing the metals by their molecular weights, an understanding of physical laws of action and reaction is simplified. Another classical example of this is the library. All books in a library are categorized. This

TABLE 4.1   Organization by Table

Part Number	Cost	Price	Lead Time	BOM Routing	Planner Code
ABC123	$30.00	$23.00	2 Months	SAE1030	Drill, Machine
ABC456	$67.00	$45.00	1 Week	Titanium	EDM

makes it easy to find a book on any topic, even though there are millions of books stored on the shelves. Manufacturers commonly keep track of thousands, even hundreds of thousands of parts, yet the vast majority of companies do not have a classification system.

Classification is one of the basic scientific methods that can be used to understand a true product mix. The two most common methods of classifications are:

1. *Organization by Table:* In this method of classification, the variation of each item is captured in a tabular form (see Table 4.1). The advantage of this approach is that the table provides many different forms of sorting possibilities. For instance, a column of the table can be sorted based on a specific criteria. However, it results in a large number of possibilities (e.g., part number) if each item is varied independently.

2. *Organization by Properties:* Generic properties provide a logical group of common characteristics (e.g., size, color, finish, tint, etc.) of a part whose individual properties differ only in amount or types (e.g., cost, price, lead times, etc.) A template is defined for each property that uniquely defines its characteristics using a set of conditional rules or equations. Using this template, a specific selection or variation for a property is configured and evaluated as in Pugh [1991]. The template serves as a master database from which specific versions are generated on demand. Creators can control names, descriptions and placement of properties. In addition, one can link properties, thus allowing a way of grouping that makes most sense. There is no need to store all possible variations of the properties like the table in the previous example.

Once someone starts to understand the product-mix that a company possesses or markets, they can also begin standardizing the design and process plans. Group Technology (GT) classification provides an important organizational structure for process data. An article written by E. M. Fisher, in Datamation Magazine, included this quotation from B. Chandrasekharan, "Classification (of data) is perhaps the most universally applied generic task. By classifying, an expert relates a single situation to a larger group of like cases. Instead of remembering what to do in each individual instance, you need only to remember what to do in each class of situation." The two techniques for classifying objects are discussed in the following sub-sections.

4.7.4.1   Group Technology

Group Technology (GT) is one of the oldest, yet most powerful, classification techniques known to the manufacturing community. For manufacturers, GT categories are based upon a combination of the following:

- The part's design geometry (what the part looks like).
- Functional descriptions (what the part does).
- Manufacturing processes (how to make the part).

Classification techniques: Over the last two decades, many different techniques have been adopted in applying GT to manufacturing. The goals, however, have always been consistent. The basic methods of classifying parts are Monocodes, Polycodes, and Hybrids.

1. *Monocodes* are hierarchical systems that build relationships between features and attributes. For example, valves can be divided into sub-groups, or types of valves (i.e., butterfly, gate, etc.). These sub-groups (types) are related to the main group (valves).
2. *Polycodes* are not based on direct relationships. An example of this is an *attribute*, such as material, that can be applied to all groups regardless of what the part is— whether it be a valve or an electronic circuit card assembly.
3. *Hybrid* systems effectively combine Monocode and Polycode techniques. The team only has to enter, maintain, and update the logical relationships that are needed in an entire hierarchical system. Hence, it is easier and faster to implement or change a hybrid system.

Group technology classification and coding schemes can also be used as a basis for a design retrieval system. This way, a GT retrieval system can provide a unified classification scheme for both design and manufacturing situations. However, GT provides more benefits to the manufacturing community than what design standardization provides to the design community. With GT, both parts and assemblies can be categorized. It is easy for anyone in the company to retrieve and utilize the historical information originally used to make or assemble a part. The benefits are real. This is because, irrespective of the techniques used, productivity improvements have been realized by utilizing the company's existing information, even without building any database or employing an experienced work group.

4.7.4.2 Design Standardization

Standardization is another form of classification used by many design teams to improve the consistency of parts being designed. Like GT, with design standardization, work groups can frequently find and reuse existing designs rather than redesigning the same part. Even when the existing designs are not exactly right, design standardization allows work groups to locate similar designs that can be used as a starting point or for knowledge gathering. This can dramatically reduce the total design effort. Further, it is not necessary to recheck the design for all X-ability considerations since the standards ensure compliance with respect to many features that are common. Thus, design standardization allows a concurrent team to share quickly most of the proven design alternatives with other work groups and redesign the rest more quickly.

From a CE perspective, both approaches of classification provide a means of standardization. GT is a tool primarily known for manufacturing standardization. Parts, whose features are the same, need not be reanalyzed or redesigned. Similarly, if a manufacturing process plan is obtained for a part, no further work is required. The process plan stays the same for the entire group of similar parts. Such standardizations provide substantial benefits.

- Eliminates unnecessary duplication of parts.
- Controls excessive proliferation of new parts and manufacturing processes.
- Reduces setup time, work-in-process inventory.
- Reduces scraps while improving the quality of parts.

GT classification and design standardization are prerequisites for concurrent product design. With the help of these tools, the problem is broken down into a family of subproblems, taking all active design and manufacturing requirements into consideration. Breaking down the problem into subproblems makes the problem easier to solve, because it is easier for anyone to deal with and understand a set with hundred parts rather than thousands of parts.

4.8 PROBLEM IDENTIFICATION AND SOLVING METHODOLOGIES

There are two steps to solving a typical design problem: (1) developing a problem identification scheme [Agrell, 1994] and (2) identifying a suitable method of solution [Dowlatshahi, 1992]. The sub-steps involved in developing a problem identification scheme for a design concept are:

- Identification of parameters or design variables for input modifications.
- Identification of criteria for generation of objectives, requirements, and evaluation of constraints.
- Generation of modification.
- Selection of a suitable PPO—product, process, or organization—model and executing it.
- Reevaluation of constraints and further problem identification (e.g., sensitivity analysis).
- Selection of parameters based on a particular solution methodology (e.g., trade-off, sensitivity or optimization) that satisfies the imposed constraints and meets the stated objectives [Pugh, 1991].

Problem solving methodologies are discussed next.

In IPD schemes, work groups require multiple problem solving methodologies so that concurrent teams can move from one level of abstraction to another, or from one viewpoint to another, as the design concept evolves [Dowlatshasi, 1992]. This may consist of developing a number of alternate design schemes and possible alternate arrangements. Members of the CE work groups can employ one or more of the following methods of solutions depending upon the problem type at hand.

- *Performance Improvement:* Here the objective is to find a set of parameters that will improve some performance characteristics that are monitored. The goal is to move the product performance characteristics to lie within an acceptable range. The current design details and parameter values and bounds are analyzed or tested against acceptable criteria and to determine possible performance violations.

- *Parameter Design:* Here the objective is to minimize the impact of external uncontrolled processes or minimize the impact of process variations that cannot be controlled.

- *System Design:* Here the objective is to determine a design alternative, an alternate material, or a new method (process) which will provide a similar (close to previous values) or equivalent performance. The interactions and relationships between subparts of a product are measured with respect to a global set of system criteria.

- *Mathematical Programming:* Here the objective is to find a set of parameters that will satisfy a performance objective subject to a given set of constraints. The performance objective is a single function.

- *Multi-criterion Optimization:* Here the objective is to find a set of parameters that will satisfy a set of performance objectives, subject to a given set of constraints [Agrell, 1994]. An example is a min-max problem.

- *Heuristic-based Search:* This employs heuristic based rules to aid in estimating RCs. The objective is to narrow down the solution field by successive introduction of heuristic-based RCs until only one solution is left, or the solution converges to an interesting part of the design space.

Finding a product design or a concept that satisfies all the constraints is possible only when the constraint network represents all design alternatives, is complete and consistent, and results in a unique solution. These conditions are rarely, if ever, met. If the constraint network is over-constrained, no solution exists, and some constraints must be relaxed or a portion of the goals or objectives that are conflicting must be modified. If the network is under-constrained, many possible solutions exist, and additional constraints or goals must be added so that the resulting design or concept converges to some interesting domain.

4.9   INTEGRATED PROBLEM FORMULATION

Integrated problem formulation involves two steps, understanding the underlying structure or intent of the problem, and understanding the solution of the problem. In Chapter 9 of volume I, a taxonomy for product realization was established. A taxonomy plan driven by a complex knowledge-base can be used to carry out the above two steps. Part of identification of criteria is determined through the use of specifications. The inputs represent the source of information that is specified at every step during product enrichment. Another major source of criteria is a repertoire of CE metrics and measures discussed in Chapter 2, such as simplicity, use of standard parts, reduction of materials, and so on. PD^3 processes appear to be iterative at all life-cycle stages. The best results are obtained by generating, evaluating, and optimizing several alternative options at each stage of the PD^3 process. However, there appear to be numerous approaches to problem formulation and problem solution. Investigations to test the efficacy of various approaches has been done on embodiment, configuration, and structural (FEA/FEM) problem formulation using knowledge-based engineering (KBE), or smart models. However, little progress has been reported on types of solution techniques or modeling approaches. Coupling of FEA/FEM with optimization has been found useful for structural design problems. Coupling of enterprise modeling, simulation, and statistical analyses (mean values and statistical distributions, etc.) has been found useful for studying the organizational structures and for verifying the dynamic interactions of processes. For configuration and embodiment cases, bond graph based techniques have been found useful to represent and connect functional requirements. For process design, Taguchi's method is most common to minimize variations (design for robustness) and assign tolerance limits.

4.9.1   Determination of Dependency Index

The separation of tasks into parallel or serial mode is based upon the premise that the tasks are either dependent or independent. In general, however, each activity will not be strictly dependent or independent. If we consider d_{ij} or d_{ji} as a dependency index relating task i with task j, the value of d_{ij} and d_{ji} can be determined as follows:

    (a) $d_{ij} = 1$; if task i and j are completely dependent,

    (b) $d_{ij} = 0$; if task i and j are completely independent,   (4.16)

    (c) $d_{ij} = d_{ji}$; if the order of triggering a task is immaterial.

    (d)   If task i is neither completely dependent nor independent with respect to task j, then the range of value of d_{ij} will be:

$$0 < d_{ij} < 1. \qquad (4.17)$$

The integrated formulation must account for the following considerations.

4.9.1.1   Constraints Reformation

Constraints reformation is used to obtain a transformation matrix which is less complex and coupled. Three desirable characteristics of reformation include:

- Serial decomposability.
- Block decomposability.
- Skewness (over or under-constrained).

Figure 4.12 shows examples of some dependency matrices prearranged in a lower triangular form. No skewness is assumed to be present in the examples, that is, the matrix is symmetric. The rows represent the constraints and the columns are the unknown variables or parameters. The appearance of x in a cell indicates a dependency. The general relationship between a constraint t_i and variables p_i is:

$$\{t_i\}\, p_i \equiv o_i$$

or
$$[T]\, \underline{p} \equiv \underline{o} \qquad (4.18)$$

where T is a transformation matrix, $\underline{p}$ and $\underline{o}$ are vectors. The four matrices A, B, C and D are examples of T as shown in Figure 4.12. From this example work groups can compute the serial decomposability and block decomposability index of the matrices. The table underneath the examples shows their corresponding indices.

- *Serial decomposability* is simply the number of independent constraints (rows) for which there is at least one new variable.
- *Block decomposability* is the number of independent blocks of constraints that form smaller independent sub-problems. It is clear from Figure 4.12 that a better formulation is one that provides a low number of serial decomposability and a large number of block decomposability. If the block decomposability is equal to the size of the matrix (numbers of rows or columns), each constraint is independent.
- *Skewness Problem Pattern:* Skewness arises when a matrix is asymmetrical. Two cases are possible:
  1. rows are larger than columns (this leads to an over-constrained pattern) or
  2. columns are larger than the rows (this leads to an under-constrained problem pattern).

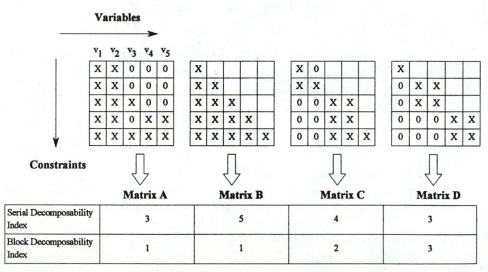

FIGURE 4.12   Constraints Reformation

Most design problems are an under-constrained (limited specifications, multiple outputs) problem type. They follow a pattern of solutions where several outputs that all meet the given specifications are produced. There are two major problems associated with this solution pattern: (a) how to find all feasible alternative designs that satisfy design requirements, and (b) how to select the best design among all feasible alternative candidate concepts.

4.9.1.2   Incremental Specifications of Inputs

Most traditional systems are single valued systems. They force the work groups to make premature commitments to value specifications, requiring costly iterations at later stages of PD^3 process. Intervals allow incremental availability of information, whether it is incomplete or uncertain. It does not force work groups to make a commitment for inputs. They provide a general framework for problem solving and keep some input values open for others to fill in, possibly avoiding costly iterations at later stages. Such a *least commitment* approach to problem solving is a useful and desirable characteristic of concurrent engineering.

4.9.1.3   Synchronization

This refers to time dependencies. If a process is synchronized, inputs will depend upon each other, whether complete or not. Synchronization helps improve coordination among engineering teams. The taxonomy need not be a complete continuous process. Discrete level decisions during product realization would not adversely affect the convergence.

4.9.1.4   Reasoning and Inferences

Reasoning explicitly about physical principles tends to reduce available options or bounding space of new design or concept alternatives. Reasoning with practical concerns in addition to physical principles, further narrows down the solution space of feasible designs/concepts. Reasoning can be qualitative or quantitative. Similarly, inference can be of a symbolic or graphic type. Conventional expert systems and AI tools emphasize symbolic processing and heuristic inferences. In situations where a quantitative goal (objective function) can be expressed, numerical optimization technique (e.g., mathematical programming methods) can be employed. Computer modeling and simulation is equally useful for analyzing the operating characteristics of existing decision making rules. It can be a powerful tool for evaluating and fine tuning new and existing processes and verifying the control strategies for automated systems. In contrast to optimization techniques, simulation provides a descriptive model predicting how a manufacturing system will behave under a given set of conditions. It can be used for studying the logistics and operations of material handling systems such as AGV (automated guided vehicles), conveyers and robots, as well as storage components, such as automated storage and retrieval systems.

4.9.2   Method of Solution

Problem solving depends not only on inferences (decision making ontology, such as algorithmic or heuristics types) but also on methods. Most numerical computational tools contain inferences that are algorithmic-types. In such tools, decision making is supported by some sensitivity or mathematical programming techniques, such as continuous or discrete optimization. Artificial intelligence, on the other hand, emphasizes symbolic processing and non-algorithmic inferences, such as rule-based, model-based or case-based reasoning. Most expert systems are heuristic-types. Real design problems require the use of both algorithmic and heuristic inferences. Combining an expert system approach (e.g., knowledge-based) with algorithmic inference (such as parametric, variational or feature-based CAD tools) can make these CAD tools more powerful for solving complex product realization problems. A problem solving algorithm is a form of representation, through which the method describes the dynamic knowledge that is applied to a static knowledge ontology. In section 6.5, two types of problem solving methods (in the form of constraint-based and knowledge-based) were discussed. Typical types of expert system software that result from using methods based on knowledge-based representation are diagnostics, designers, planners, configurers, classifiers, assistants, and so on. Problem solving methods are therefore useful in linking knowledge acquisition to problem-solving algorithms such as types of solutions for constraint-satisfaction problems (CSP). Knowledge acquisition is meant here to indicate some sort of template that is to be filled in during knowledge modeling.

4.10 COLLABORATION AND CROSS-FUNCTIONAL PROBLEM SOLVING

Problem solving is a process of finding a set of alternate solutions by applying the cross-functional teams and collaboratively improving the quality of decisions. Some simple methods of finding solutions are reasoning, calculations, table look-up, and graphics, as well as different types of knowledge representation, heuristics, algorithmic, and so on. Before we outline any strategy for solving problems, let us examine the cross-functional contributions. Most human beings, like any natural thing, choose a path of least resistance. Humans have a limited ability to entertain more than a few ideas at a time. For example, if several alternatives exist for satisfying a constraint, most people will satisfy rather than optimize. This observation in mechanical design is well noted by many [Stauffer and Slaughterbeck-Hyde, 1989; and Newell and Simon, 1972]. H. A. Simon was the first to coin the term "satisficing" to describe this phenomenon. This is illustrated in Figure 4.13 through a simple constraint diagram. Three straight lines are shown representing constraint boundaries for three constraint sets: g_1, g_2, and g_3. F_i represents a curve for design solutions. From this construction, it is clear that the region bounded by these three intersecting lines will define a feasible region where constraints will be satisfied. The symbol "x" marked along the design solution curve, in Figure 4.13, denotes a series of points at which solutions are "satisficing," and the symbol, @, denotes the optimal solution point. Pearl described "satisficing" as "discovering any qualified object with as little search effort as possible [Pearl, 1984]." Stauffer describes it as reducing the time and effort in searching for a solution by utilizing the first acceptable solution rather than searching for the best or optimal solution [Stauffer and Slaughterbeck-Hyde, 1989].

When faced with a number of alternatives, designers often employ a simplifying, "elimination by aspect" strategy, choosing only a few critical attributes that are the most heavily weighted. The above are some examples of ways a human mind reacts. Instead of fighting with these cognitive limitations of human beings, the strategy ought to minimize

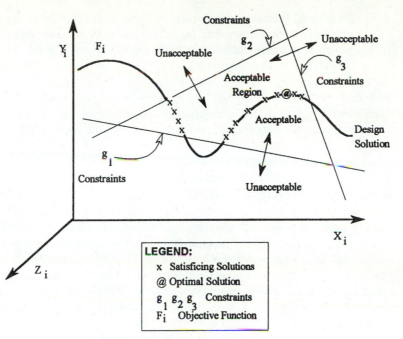

FIGURE 4.13   Feasible Vs. Optimal Solutions

its effect. The taxonomy could be such an strategy. Taxonomy streamlines the total PD^3 process into a limited number of loops. If the problem is classified in such a way that the entire problem is built from its constituents, then the problem reduces to applying *elimination by aspect* to each constituent-built, giving rise to *satisfaction by taxonomy*. By successive use of elimination by aspect, along with problem solving methodologies (see Figure 4.6), cross-functional teams are more likely to converge to a near-optimal solution. Depending upon the situation and the team focus, a number of alternatives may exist for problem formulation. Teams may use an analytical technique, such as a formal optimization, or a non linear programming technique, if the problem is well posed and can be quantified. If the tasks involve non-numerical or non-algorithmic information, expert system techniques can be used to represent domain expertise in terms of heuristics. For other problems, requiring both heuristics and algorithmic aspects, the two methods can be combined. The selection of the redesign method depends on three things.

Modeling Source: The types of individual analysis or simulation models chosen for a loop evaluation and their potential for relative impact on improving the product performance, such as product cost.

Time of Each Iteration: How long it takes to perform each iteration for a single configuration and how many configurations ought to be considered. The number of times a nested iterative sub-loop evaluation is performed depends on the chosen algorithm and its convergence criteria.

Number of Iterations: How many evaluations are performed per iteration and how many iterations are required for convergence. If the same redesign method, is repeated more than twice in a row to the same sub-loop, this may be a case of possible floundering in its strategy.

Alternatively, a work group may use a line of experimentation techniques if the problem is difficult to quantify. Many approaches for problem formulation are contained in this chapter. The solution techniques and tools are discussed in Chapter 7, volume II.

4.11 CONTINUOUS MONITORING AND KNOWLEDGE UPGRADE

In the world of global manufacturing a competitive advantage is often short-lived. As soon as a management becomes satisfied with the results of what process models have achieved, the company may again start losing ground to competitors. Competitors continually strive to improve their position in the world marketplace. In order to stay competitive, these process models must have a provision for upgrading knowledge as the process matures. An example of a typical process model for continuous knowledge monitoring and upgrade, CMKU, is shown in Figure 4.14.

$$CMKU = f \text{ [Monitor, Select, Analyze, Contain, Correct, and Prevent]} \quad (4.19)$$

where, f denotes the function. The CMKU process consists of the following six steps:

1. *Monitor:* During monitoring if discrepancies are noted, the suspected work groups are interviewed and *as-is* information is gathered.
2. *Select:* Here a portion of the as-is process is identified and the information is flow-charted.
3. *Analyze:* From the information flow charts, opportunities or bottlenecks are identified.
4. *Contain:* Alternative solutions are sought or evaluated to contain the problem.
5. *Correct:* A corrective action plan is determined and the work group's cooperation is sought.
6. *Prevent:* Changes are introduced in the process to prevent reoccurrence of the same or similar problems, concepts, or designs.

Normally during a process monitoring, there are two types of changes involved: scheduled changes (also referred as revisions) and unscheduled changes. Some changes may produce far reaching influences. Revisions introduced into the way a product definition is performed whether in the traditional sense or in the concurrent engineering mode, or even somewhere in between, is painstaking. As a company migrates from one life-cycle aspect to the other, it may induce unscheduled changes in other functions in the company. Through this process of continuous monitoring and upgrading it is possible to prevent some of these unscheduled changes.

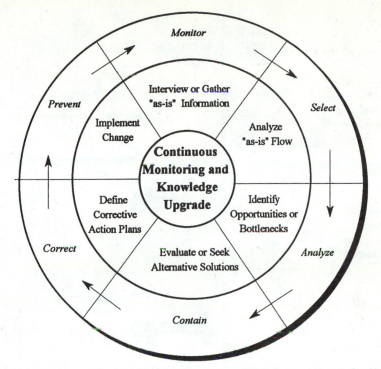

FIGURE 4.14   A Typical Process Model for Continuous Monitoring and Knowledge Upgrade

4.12 CONCURRENT IPD METHODOLOGY

In Figure 4.1, a fluid flow through a pipe analogy was shown for a serial process. The same analogy has been redesigned in Figure 4.15 for a concurrent PD^3 process. Instead of placing the control at the end of the pipe assembly as in serial engineering, the control is now placed at each loop level of the concurrent product realization process. This constitutes a *control by design* operation at each loop level. The configuration of the pipe connections and their relative positions along the vertical direction is governed by the taxonomy of product realization as discussed in Chapter 9 of volume I. The numerous bends and elbow-connections in Figure 4.1 have been replaced by loops tapped in at designated points (governed by the taxonomy) along the vertical tube. The fluid pressure in the pipe is created naturally due to gravity. In concurrent PD^3 process, this is equivalent to just-in-time information build-up for each loop. Five loops run concurrently as shown in Figure 4.15. The amount of information build-up at each loop is governed by a natural pull of the information due to gravity rather than a force *push* found in serial engineering. There are many advantages associated with the concurrent IPD methodology. The methodology recognizes that product and process data in the early stage of product development is fuzzy, incomplete and often uncertain [Wesner, Hiatt and Trimble, 1994]. Concurrent IPD provides a taxonomy-based CE process to sort through this fuzzy set of information to establish rationally what will work well and what will not. The methodology balances the needed reduction

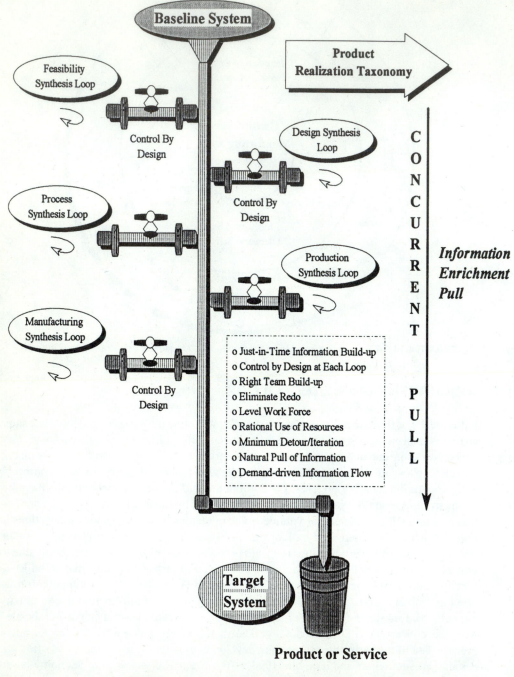

Baseline System

Product Realization Taxonomy

Feasibility Synthesis Loop

Control By Design

Design Synthesis Loop

Control By Design

Process Synthesis Loop

Production Synthesis Loop

Manufacturing Synthesis Loop

Control By Design

C O N C U R R E N T   P U L L

Information Enrichment Pull

o Just-in-Time Information Build-up
o Control by Design at Each Loop
o Right Team Build-up
o Eliminate Redo
o Level Work Force
o Rational Use of Resources
o Minimum Detour/Iteration
o Natural Pull of Information
o Demand-driven Information Flow

Target System

Product or Service

FIGURE 4.15   An analogy for a Concurrent IPD Process

in responsiveness (with respect to time-to-market) against the risk of design changes by using incomplete or uncertain information upfront in the taxonomy-based process. Control by design thus provides an integral mechanism to manage the risk appropriately. Concurrent IPD eliminates excessive redo, minimizes detour and iterations, and requires a leveled work force. The methodology is based on rationally utilizing the resources, and *demand driven* is the main accessing mechanism for pulling information.

REFERENCES

AGRELL, P.J. 1994. "Multicriteria Approach to Concurrent Engineering." *International Journal of Production Economics*, Volume 34, No. 1 (February) pp. 99–113.

ANDREASEN, M.M. 1992. "Designing on a Designer's Workbench (DWB)." Proceedings of the 9th WDK Workshop, Rigi, Swizerland.

BERGER, S. et al. 1989. "Towards a New Industrial America." *Scientific American* (June 1989) pp. 39–47.

CHANDRASEKARAN, B. 1989. "A Framework for Design Problem Solving," *Research in Engineering Design*, Volume 1, No. 2, pp. 75–86.

DEMING, W.E. 1993. *The New Economics*. Cambridge, MA: MIT Center for Advanced Engineering Study, (November).

DOWLATSHAHI, S. 1992. "Product Design in a Concurrent Engineering Environment: An Optimization Approach." *International Journal of Production Research*, Volume 30, No. 8 (August) pp. 1803–1818.

GALBRAITH, J.R. 1973. *Designing Complex Organizations*. Reading, MA: Addison-Wesley.

HARDWICK, M., et al. 1990. "ROSE: A Database System for Concurrent Engineering Applications." Proceedings of the Second National Symposium on Concurrent Engineering, Morgantown, West Virginia (February 7–9) pp. 33–65.

KEENY, R.L., and H. RAIFFA. 1976. *Decisions with Multiple Objectives*. Chichester: John Wiley.

KUSIAK, A., and J. WANG. 1993. "Decomposition of the Design Process," *Journal of Mechanical Design*. Volume 115, pp. 687–695.

McMOHAN, C.A., M. XIANYI, K.N. BROWN, and J.H.S. WILLIAMS. 1995. "A Parallel Multi-attribute Transformation Model of Design." Proceedings of the Design Engineering Technical Conferences, September 17–20, 1995, Boston, MA., edited by Jadaan, Ward, Fukuda, Feldy and Gadth, Volume 2, DE-Volume 83, pp. 341–350, New York: ASME Press.

NAVINCHANDRA, D., M.S. FOX, and E.S. GARDINER. 1993. "Constraint Management in Design Fusion." *Concurrent Engineering: Methodology and Applications*, edited by P. Gu and A. Kusiak, Elsevier Science Publishers B.V., 1993, pp. 1–30.

NEWELL, A., and H.A. SIMON. 1972. *Human Problem Solving*. Englewood Cliffs, NJ: Prentice-Hall.

PEARL, J. 1984, "Heuristics: Intelligent Search Strategies for Computer Problem Solving." New York: Addison-Wesley Publishing Company, Inc.

PAHL, G., and W. BEITZ. 1991. *Engineering Design: A Systematic Approach*, edited by K. Wallace, New York: Springer-Verlag.

PRASAD, B. 1985. "An Integrated System for Optimal Structural Synthesis and Remodeling," *Computers and Structures*, Volume 20, No. 5, pp. 827–839.

PRASAD, B. 1984a. "Novel Concepts for Constraint Treatments and Approximations in Efficient Structural Synthesis," *AIAA Journal*, Volume 22, No. 7 (July) pp. 957–966.

PRASAD, B. 1984b. "Explicit Constraint Approximation Forms in Structural Optimization—Part 2: Numerical Experiences," *Computer Methods in Applied Mechanics and Engineering*, Volume 46, No. 1, pp. 15–38.

PRASAD, B., and J.F. EMERSON, 1984. "Optimal Structural Remodeling of Multi-Objective Systems." *Computers and Structures*, Volume 18, No. 4, pp. 619–628.

PRASAD, B., and C.L. MAGEE. 1984. "Application of Optimization Techniques to Vehicle Design— A Review," *Recent Experiences in Multidisciplinary Analysis and Optimization*, NASA CP 2327, Part 1, pp. 147–171, NASA Langley Research Center VA (April).

PUGH, S. 1991. *Total Design—Integrated Methods for Successful Product Engineering*. Wokingham, England: Addison-Wesley Publishing Company Inc.

SAATY, T.L. 1978. "Exploring the Interface between Hierarchies, Multiple Objectives and Fuzzy Sets," *Fuzzy Sets and Systems*, Volume 1, pp. 57–68.

SHUMAKER, G.C. 1990. "Integrated Product Development Program Strategy." Concurrent Engineering Office, Manufacturing Technology Directorate, Wright Research and Development Center, Wright-Patterson AFB (July) p. 4.

STAUFFER, L.A., and R.A. SLAUGHTERBECK-HYDE. 1989. "The Nature of Constraints and Their Effect on Quality and Satisficing," DE-Vol. No. 17, Ist International Conference on Design Theory and Methodology, Sept. 17–21, ASME, pp. 1–7.

STEFIK, M. 1981. "Planning with Constraints (MOLGEN: Part I and Part II)." *Artificial Intelligence*, Volume 16, pp. 111–169.

SUH, W.P. 1990. *The Principles of Design.* Oxford, UK: Oxford University Press.

SURI, R. 1988. "A New Perspective on Manufacturing System," *Design and Analysis of Integrated Manufacturing Systems*, edited by W. Dale Compton, NAE. p. 118.

ULRICH, K.T., and S.D. EPPINGER. 1994. *Product Design and Development*, New York: McGraw-Hill.

WARFIELD, J., and J.D. HILL. 1972. *A Unified Systems Engineering Concept.* Battele Monograph, edited by B.B. Gordon, No. 1 (June).

WESNER, J.W., J.M. HIATT, and D.C. TRIMBLE. 1994. *Winning with Quality: Applying Quality Principles in Product Development*, (September) Reading, MA: Addison-Wesley Publishing Company, Inc.

TEST PROBLEMS—PRODUCT DEVELOPMENT METHODOLOGY

4.1. How do you adequately define an engineering process for product design? What is its relationship to decomposition and aggregation and why do the teams break that way?

4.2. How many (range) parts does an automobile have? How many decisions are normally made in the design and development of an automobile? What other factors cloud the inherent complexity associated with an automobile design?

4.3. What are the invariants of a PD^3 process? Are they intended to provide a common ground for representing enterprise or business-driven, product-driven, and process-driven works, activities, features, functions and decisions in a PD^3 process?

4.4. Describe the invariants of an IPD realization process. What are the process invariants and what relationships do they have with model invariants discussed in Chapter 6 of volume I? Does the basic structure of invariants change during a product realization process?

4.5. Is IPD an open ended problem? Is there a unique way of solving an IPD problem? If not, how does one go about solving an IPD problem? What is the primary vehicle for IPD?

4.6. What CE techniques shown in Figure 4.2 are important to bridge the gaps between information models and CE work groups?

4.7. What are the three major elements of a process taxonomy? How is it different from process classification or process breakdown structures (PsBS)?

4.8. Does process taxonomy synthesize or optimize a design system with respect to the identified constraints? If not, what does it do? Does the taxonomy show how a network of tasks will be solved?

4.9. What are the two types of complexity a work group encounters during a PD^3 process? What do you need to do to overcome them? Are there off-the-shelves tools available to solve such complexities?

4.10. What are the four parts of an explicit optimization problem? Which of these parts are implicit in an IPD formulation? How do you overcome a solution complexity? What types of relationships exist between transformation state attributes and optimization model variables?

4.11. What are the different types of design variables commonly found in a typical structural design problem? How do process variables impact which parts of the problem statements in a typical optimization sense?

4.12. What is an implicit statement of an IPD problem at a system level? What attributes govern its solution: maximize, minimize, or desensitize?

4.13. What are the building blocks of an integrated PD^3 process? How many parts does an IPD methodology have? What is the significance of each part? What constitutes a conceptual framework and an analytical framework and why?

4.14. What building blocks of an IPD methodology outline an optimized or consensus-based alternative? What is unification? Are these steps interconnected? If so, how is synergy accomplished between them? Why are education and training an important part of IPD?

4.15. Describe a pyramid way of classifying requirements and constraints. What constitutes the sides and its hidden base of this pyramid? Are all sides of this pyramid equal? How does one go about controlling specification creep during an IPD process?

4.16. State a production system from Figure 8.4 of volume I in terms of (linear/nonlinear programming) systems of equations. What is meant by a feasible set of solutions versus an optimal solution?

4.17. What are the differences between an IPD methodology and a conventional PD^3 process? How important do you think System optimization is to IPD? Illustrate your answers with scenarios.

4.18. What are the tools/techniques to control off-line and on-line quality of a product or a process? Who must be involved in the product cycle management to minimize product variation?

4.19. Why do tolerances required through design conflict with what is desired during a manufacturing process? What should work groups do to overcome these difficulties? How do work groups know when to stop (what is the right tolerance)?

4.20. How would you approach an overall solution to an IPD problem? What is the four-stage systematization process that work groups need to employ?

4.21. Why is methodology systematization critical to the success of an IPD solution? What is the branching and bounding methodology? How does that help to bound the solutions?

4.22. Do independence and concurrency increase the risks of providing multiple solutions? How do product and process systematization help in concurrency? What are the four ways to systematize a product or a process?

4.23. How do you decompose a scenario into a serially decomposable activity group? What impact does it have on the product or process concurrency?

4.24. How does group technology help in product and process systematization? What are the two ways to perform part/activity classifications? What is the significance of design standardization?

4.25. How do you examine an IPD methodology? How do you prioritize the PD^3 activities among the vital functional groups? What do you measure successes?

4.26. What are the two steps of an integrated problem formulation? How would you formulate your problem to identify those RCs or design variables that are missing or redundant?

4.27. Why do you require a number of problem solving methodologies during an IPD process? What are these potential methods of solution that one can employ during problem solving?

4.28. Why would you require continuous monitoring and knowledge upgrade as a part of an IPD methodology? If not, how would you detect (product or process) problems, perform related analysis, and take (long-term) corrective actions? Describe a process model to perform all of these functions.

4.29. Describe an analogy of a concurrent IPD. Compare this with a serial-based IPD analogy. Outline ten major benefits that are achieved by following a concurrent IPD methodology?

CHAPTER 5

FRAMEWORKS AND ARCHITECTURES

5.0 INTRODUCTION

Concurrent engineering requires teams, and teams require teamwork and cooperation. But cooperating teams need a platform through which the 7Ts and 7Cs activities (see Figures 4.1 and 5.1 of volume I) can be performed freely. No organization can hope to compete on a worldwide basis without a set of sound and modern technological frameworks and an open architecture to build the system.

A framework is actually a structure "serving to hold the parts of something together, or to support something constructed over or around it" [McKechnie, 1978]. There is no dearth of structures and frameworks. A large number of institutions are involved in developing many kinds of standards and frameworks, each serving a particular purpose. Architectures are designs of framework components erected on the principles of framework objectives. Frameworks and architectures together establish an environment. In a complex undertaking, such as concurrent engineering, a product or a system has to communicate with several work groups in addition to its core business units. Examples of environments include the office environment, engineering environment, design and manufacturing environment, product support environment, and so on. Appropriate architectures and frameworks are required to support such environments and their multi-level constructs. Constructs are parts of the environment. Each construct is supposed to be consistent with respect to how an individual product or a system operates or interfaces with other constructs. Examples of environments include consistent office environment (COe), consistent work group computing environment (CCe), and its constructs—consistent engineering environment, consistent design environment, consistent manufacturing environment, etc; consistent communication environment (CCe) and its constructs—Videoconferencing, E-mail, and so on. Division of concurrent engineering business operation into multi-level constructs of consistent environments allows orderly organization of tasks, systematic distribution of responsibilities, and efficient interaction between the CE work groups involved. Frameworks are the basic building blocks of concurrent engineering that make an enterprise more responsive. The idea of the architecture is to identify designs, framework features and metrics, so that product design, development, and delivery (PD^3) process can be managed and improved within a set of constraints imposed by an enterprise and its components (business, people, and customers). Appropriate selection of frameworks and architectures leads to a series of consistent environments for the intended concurrent business operations.

5.1   GENERAL ARCHITECTURE

Design and development of a product is a complex undertaking. During a PD^3 life cycle, a product goes through a number of changes. Some changes may be present as specifications. Some of these specifications may make the product robust while others are required to test the product under extreme operating conditions. Initial design specifications must accommodate the diverse environments that a product is subjected to during its normal life.

The specifications or characteristics associated with a particular environment of a product constitute a framework that dictates an initial phase of its design. A study of a complete product life cycle reveals varying operating conditions that must be accounted for during a product realization, PD^3, process. Frameworks are conceptual models of the operating conditions. All these frameworks together form an architecture. A general structure for developing a product is shown in Figure 7.12 of volume I. As the frameworks depend upon each other, so do the environments. The goal is to develop a series of building blocks, concurrent engineering frameworks, that would provide the CE work group members in a large dispersed organization the same freedom of interaction and information transfer as enjoyed by a small collocated team [Margolias and O'Connell, 1990].

5.1.1   Integration Schemes

There is an ongoing trend to integrate a product life-cycle activities into an intelligent computer-aided (such as CAD/CAM) environment to be used during an early design stage. The purpose is to shorten the cycle-time from ideas to production, increase productivity, and more importantly, come up with an optimum or near-optimum production plan to build the PD^3 system.

Various standard organizations, CAD/CAM/CAE/CIM vendors, and work groups are proposing comprehensive enterprise integration technologies and data models. Various approaches, such as the use of wrappers or translators, already exist, and more alternatives are being developed. Issues, such as the use of federation versus unification, pair-wise versus global translation, information loss in translating between different ontologies, significance of normative modeling languages, and so on, are yet to be explored.

Choosing an integration scheme can be very important in determining how efficient or how flexible (or general) a resulting architecture will turn out to be [Prasad, 1985]. The basic elements that need to be integrated during concurrent engineering are shown in Figure 5.1 through a context diagram. The elements are a variety of software, hardware, and database products that are either developed in-house or have been acquired by a company over the years. The elements may belong to a multi-level class structure. A two-level class structure is shown in Figure 5.1. The primary class consists of the following four elements: applications, tools, databases, and services. The tools may include commercial (third party) software for which source codes may not be available. Each of these primary class elements may have its own satellite class—a secondary (second level) class. This is shown in Figure 5.1 by a set of smaller ellipses surrounding a primary class (first level) ellipse. Information transfer takes place first at the primary level before it is passed on to the next level. In Figure 5.1 arrows originating in and out of the elements show the interactions between primary-and-primary class, secondary-and-primary class, or primary-and-secondary class elements.

In a typical organization, there exists a wide mix of computer programs and computer platforms of different brand names. No matter how disparate and fragmented their

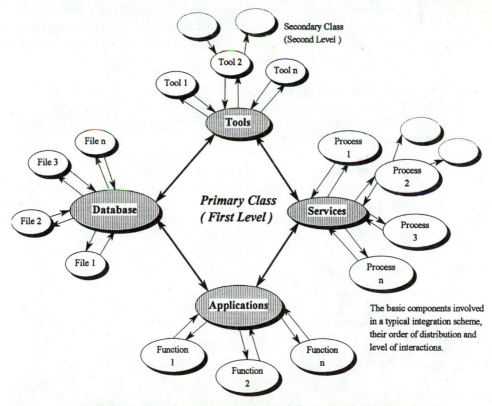

FIGURE 5.1   Basic Components involved in a Typical Integration Scheme

in-house programs may be, companies cannot be expected to abandon these tools and thereby lose the investments in these tools in order to implement a CE-based PD³ system. These tools must be allowed to coexist with any newly designed CE environment that may be deemed necessary. Organizations must be able to leverage the knowledge that has been gained over the years, as well as the investments in hardware and software that created it, in order to stay ahead of competitors. Many organizations wish to retain the best of what they already have. In addition they wish to complement that with the best of what a third party vendor in each class has to offer, or leverage with what a company can openly acquire or buy. This could mean a big-time integration.

There are many ways of integrating these components. Some schemes provide a higher degree of inter-operability than others. An ideal integration scheme should meet the following qualifications:

- *Transparent Mapping:* Each tool or a service, irrespective of the environment in which it was created for, is mapped into every other environment, where it appears as if it was written particularly for that environment. For instance, if you make a change to data in one tool, the change automatically occurs in the data used by all other tools.
- *Distributed Access:* Tools may reside both on stand alone computers or be shared through a network. Each owner of a database is able to define and manage the services in the environment most familiar to him. It then appears as if data was created originally for that environment.

Depending upon the components to be integrated and the choice of the integration scheme, a work group can achieve a varied degree of successes. The integration schemes can be characterized into four types: hard-wired, network-based, language-based, and data translators.

5.1.1.1   Hard-Wired

Figure 5.2a shows the inner workings of a simple minded hard-wired integration approach. Input data handling utilities, as for example in Engineering Analysis Language (EAL) [Whetstone, 1980], provide a major source for controlling program functions and for

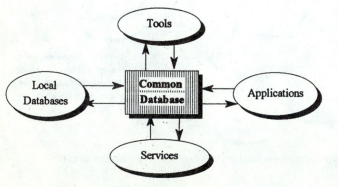

FIGURE 5.2a   Typical Model of a Hard-wired Integration Approach

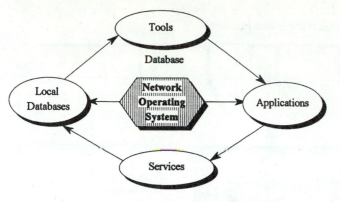

FIGURE 5.2b   Typical Model of A
Network-based Integration Approach

communicating with the common database in a most efficient way. The applications, services, tools and local databases are all linked programatically through a common database. The prime reason for integrating the components this way is efficiency. However, this approach does not allow much flexibility on the part of the work group for future upgrades. The scope of primary components remains what it was originally when the components were first hard-wired. At any future point of time, it is difficult to alter their scope, for example, to include other structures or constraints of interest.

5.1.1.2   Network-based

Network-based approach, as shown in Figure 5.2b, is more direct and is logically suited to connect systems with incompatible databases or structures. The approach is highly dependent upon the network communication mechanism and the infrastructure of the computing environment such as the operating system. The network establishes the order and the direction of the information flow between network components. New generation workstations are equipped with better network configuration and management facilities that enhance this computing environment tremendously. More and more of the computing environment is beginning to use interoperable tools that are interactively linked to a standard data management system [Donlin, 1991]. Figure 5.3 shows a set of four network traits that enable integration: open system traits, shared traits, client/server traits, and gateways and protocols. They are discussed in the following paragraphs.

Open System Traits: The architectures needed for CE have to be open, that is, the applications or tools from different vendors have to exhibit the following properties [Kahaner, Lu, Kimura, Kjellberg, Krause, and Wozny, 1993].

- *Adaptable:* The vendor applications or tools should have the ability to take advantage of new technologies and be implemented in changing environments. Flexibility also means additional functionality under the time constraint of parallelism.
- *Inter-operable:* Multiple tools will have access to the original data, not copies of the data. Tools should have the ability to work together over a network, be able to connect and share with heterogeneous database and information systems under local and distributed environments, as desired.

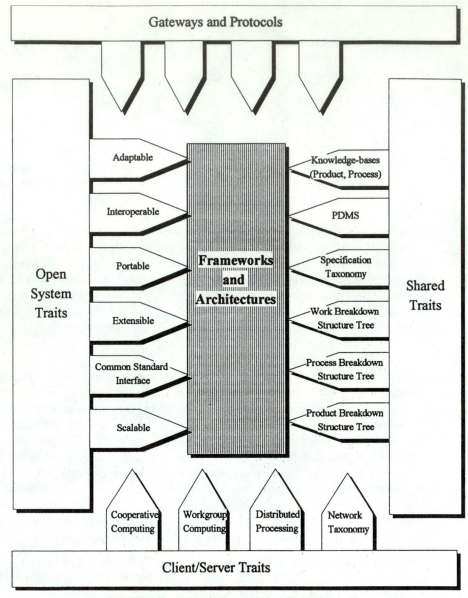

FIGURE 5.3 Network Features that Enable Integration

- *Extensible:* They should have the ability to enhance or support the business processes, accommodate in-house developed systems, or add new ones as they become available in near future.
- *Common Standard Interfaces:* The applications or tools should be easy to use. Standard interfaces should be intuitive and consistent in purpose and use. Tools will em-

ploy open emerging standards for data formats, operating system, application interfaces, and networks.

- *Scaleable:* The applications or tools should have the ability to migrate from a client to a server or to machines of greater or lesser power, depending upon the requirements, with little or no change to the underlying applications.

- *Portable:* The user should have the ability to run the code in a wide variety of platforms (hardware environments) regardless of their manufacturers or operating systems. Some operating system shell commands, for example, in UNIX, allow teams to interact with applications. In other cases, this is accomplished through a set of system-dependent commands. Unfortunately, in those cases, the program becomes dependent on its parent operating system and does not remain portable.

Shared Traits: The information generated in CE, such as product specifications, goals, enterprise data, process, and knowledge, and so on, must be shared among work groups [Kahaner, Lu, Kimura, Kjellberg, Krause, and Wozny, 1993]. It is useful to describe common information as shared resources so that each work group member can access this information irrespective of their tasks, sources, or place of work. Such shared resources may include product knowledge-bases, process knowledge-bases, PDMS, product specification taxonomy, process specification taxonomy, work breakdown structure (WBS), product breakdown structure (PtBS) tree, process breakdown structure (PsBS) tree, and network taxonomy (see section 6.4.1 of volume I). Shared resources can include the concurrent work group taxonomy (also referred as WBS) for both human and computer supported activities. Work group taxonomy can include distribution of work for each work group within each CE team, portable applications for distributed processing, common services or execution environment, and common shared data resources. Graphical user interface (GUI) mainly provides a friendly front-end for accomplishing the man-machine interactions. It is not a means to accomplish computational resource sharing.

Client/Server Traits: The client/server trait represents a software-defined model for computing. It is made out of two entities, a client and a server. The client requests services from a server, such as retrieval of data, or the printing of a document. The server processes the request, performs the service, and returns the results to the client. Establishing a streamlined communication involves configuring an open-system architecture that provides distributed computing in which data and resources can be shared easily. One type of architecture that provides such openness is a client/server model. A client-server approach is shown in Figure 5.4. It is generally characterized by a division of an application into components with the flexibility that each component could run on a different network or a computer. A transport layer provides the necessary communication protocol between the client-team interface and the server applications. There are features that are prerequisites of a good client-server model.

1. *Distributed Processing:* In this environment, concurrent teams tap the local processing power of "client" computers (personal computers and low-end workstations, for example) to "servers" (other high-end workstations and multi-user platforms). When clients request some information, servers respond by running an

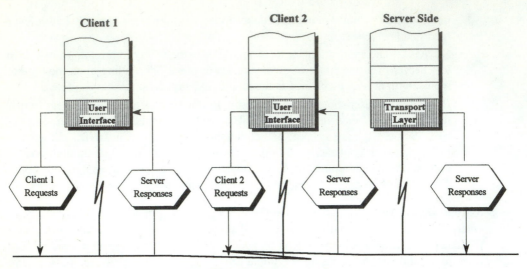

FIGURE 5.4   Client/Server Model

application locally and then delivering the requested information over the network (see Figure 5.4).

2. *Work-group Computing:* Globally interconnected clients and servers on standard-based networks can produce improved communication and greater access of data throughout an organization. The key benefit of the client/server form of integration lies in its ability to foster work group computing (WC). The WC implies a transparent access of data/system resources to the linked applications or work groups, independent of their locations, installations, processor hardware, operating systems, and programming languages. WC incorporates four different types of interactions between the work group members with respect to the dimensions of time and space. Figure 5.5 shows schematically these four shared computing environments between the work group members. These environments can be characterized as:

   • *Face-to-face:* Interaction occurs at the same time and at the same place. This environment supports face-to-face interactions among work group members regardless of temporal or geographical alignment with other group members. Examples include interactive meeting, collaboration laboratory, design reviews, and so on.

   • *Distributed Synchronous:* Interaction occurs at the same time, but at different places. Lotus Notes, for instance, supports a type of distributed synchronous interaction environment. Using Notes work group members can work on the same task regardless of their physical location or place of work. Video-conferencing is a groupware technology that enables geographically dispersed teams to conduct face-to-face meetings in real time, by combining interactive video, audio, and graphic/document display capability. Meeting on the network is a similar idea.

   • *Asynchronous:* Interaction occurs at different times, but at the same place. Examples include multi-purpose equipment (for cutting, milling, threading, etc.), and other resources installed in a plant, or database repository.

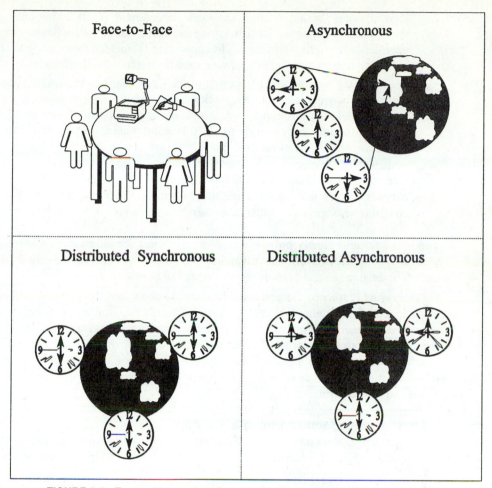

FIGURE 5.5   Types of Interactions Between Members of the Work Group Computing

- *Distributed Asynchronous:* Interaction occurs at different times and at different places. Distributed asynchronous is useful in providing functions such as computer-intensive processing, electronic-mail, automatic printing, and filing.

3. *Network Taxonomy:* For a client and server to communicate, a company must use a common network protocol, even though multiple protocols can be used by each individually. Examples of protocols include TCP/IP, DECnet, SNA, and so on. Many computer vendors, in order to optimize the use of different computers, support tools that allow client/server interactions for different protocols. Examples include the X-Window System, Distributed Name Service (DNS), Remote Procedure Call (RPC), etc. RPC is a procedural language mechanism for distributing application program procedures to remote network locations. The remote procedures become servers. The local client programs invoke the remote procedures as if they were local procedures.

X-Windows is also a network-based windowing system. Applications developed for the X-Windows system are hardware independent, that is, the application developer uses standard graphics protocols. The application is shielded from the details of individual implementation of the hardware-specific graphics display features.

4. *Cooperative Computing:* Another part of the CE solution is to provide cooperative computing between applications, work groups, and the computers using enterprise models or network taxonomy, as required. Work groups benefit because such models and taxonomy are common to all X-Window based workstations. Developers can move their applications from one UNIX platform to another or run the applications with relative ease, as long as both platforms support X-Window-based interface and are networked together. A work group member can run one application on one workstation and, at the same time, open an X-window interface on another computer belonging to another team-member and run his or her application.

The above features clearly provide a freedom of choice for information management as well as system and network administration to structure and cost-effectively manage the current and future (e.g., client/server) computational needs.

Gateways and Protocols: Gateways are functions to allow interfaces among different host environments (see Figure 5.6). Many of the protocols required to implement gateways now exist. A generic approach is to have applications request registered services that correspond to various platform services (real or advertised). How to index registered services to advertised services is a part of the name-space and binding problem. Gateways and protocols must be capable of addressing issues such as binding time, granularity of services, and state control.

5.1.1.3 Language-based Integration Approach

This approach, as shown in Figure 5.7, utilizes advanced programming concepts to build this system within a framework of a common data base. This programming concept is

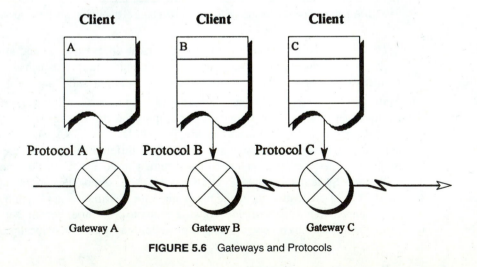

FIGURE 5.6 Gateways and Protocols

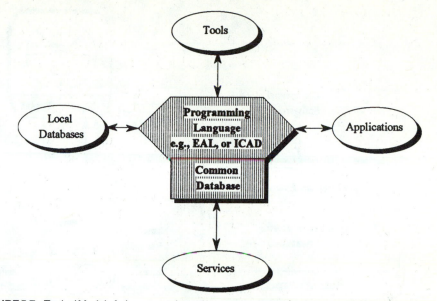

FIGURE 5.7   Typical Model of a Language-based Integration Approach (Independent of Operating System)

based on interweaving software and its multi-level language structure in its most efficient combinations.

   Wrapper Concept: A solution for integration under a new acronym called *wrapper concept* has recently emerged (see Figure 5.8). Though this idea is not very different from runstream-based and data-handling integration scheme, its application is new. The idea is to write pieces of software (like pre- and post-processor for an application) that encapsulate existing applications and map their internal protocols to a common standard. The common standard acts as a glue to allow communication between dissimilar systems. There are three parts to a wrapper concept: host wrapper, tool wrapper, and a Product, Process, and Organization (PPO) model [Lewis, 1990]. Host wrapper resides on the client and tool wrapper on the server. PPO responds to the host's named messages with values. A wrapper is required for each application. It is difficult, however, in this concept to communicate with applications for which no wrapper exits. This allows for the application to be supported and maintained independently. The wrapper piece is the only piece that needs to be changed if protocols of the original application are changed from the last time. This presents the need for regular maintenance at frequent intervals. However, the wrapper concept is an effective mechanism to integrate two dissimilar systems. To achieve this level of generality and efficiency, four levels of language structure are crucial.

1. *Attributes-based Features:* These include a high-level command language structure and some software capabilities to perform a number of input/output operations between team-mates, data bases, and the programs. Attributes-based features enable:
   - Definition of a real, integer, or alphanumeric string of names in terms of an array of registers, and storage of their assigned or computed values in their respective locations on the complex data base.

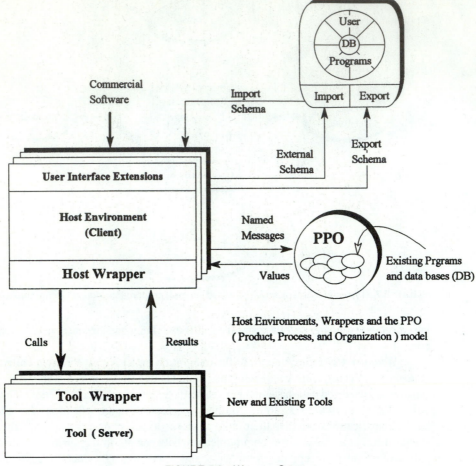

FIGURE 5.8   Wrapper Concept

- Performing control tasks such as let-stream collection, and sending messages to any element or variable object of a tree structure.
- Performing computational tasks.

Attribute-based features can automate routine processes and design custom forms and menus.

2. *Runstream-based Features:* These are more sophisticated capabilities than the attributes-based features. Runstream-based features let you treat a number of canned instructions as a unit and let you call it from an executable module. Typical examples are: retrieving the intrinsic stiffness or displacement matrices in finite element analysis, performing matrix or vector operations (addition, multiplication, inverse), and so on. Non-routine jobs may also be tackled in much the same way. For example, solution of a system of equations with modified (user perturbed) stiffness or mass matrices,

or retrieving pre-stored sets of control images from master libraries of generically applicable procedures.

3. *Data Handling Features:* These include programming facilities for creating part definitions, defining attributes, parts association, positions and orientations, and other functions. Data handling features consist of a library of efficient (Lisp, FORTRAN, C, or C++ callable) data handling and geometry manipulation routines that are crucial to the development of tasks from within a program unit. These tasks may include writing a new processor or modifying existing ones. This allows data communication from a common database to the work group processors in central memory, and vice versa. Sometimes, it may be necessary to reformat data to bridge the gap between incompatible systems.

4. *Object-based Data Handling Features:* Object-based data handling features let applications on computers scattered over the network behave as objects and communicate by sending messages to each other. An example of such a language capability is the MetaCourier language [Symbiotics, 1990]. This allows for quick integration between applications as diverse as FORTRAN programs, in-house analysis programs, databases and hard-copy devices. Using this approach, one could take applications that are not part of a common branch of CE models, inherit object class definitions, and quickly let the applications exhibit some object-based neat behavior.

The use of a higher degree of features (e.g., object-based data handling features) results in a more efficient system than the use of attributes-based or run-stream-based features. Nevertheless, they are all based on efficient data communication and dynamic core management techniques, so that for any problem application, no more than a minimum amount of memory and storage space are needed. These features become a seamless part of the CE environment.

5.1.1.4 Data Translators

There are two ways to translate data: indirect or neutral form translators, and direct point to point translators.

Indirect or Neutral Form Translators: Some translators convert data generated in one CAD environment into a neutral form and then read this neutral form data back into another CAD system. This is called indirect or neutral form translator, shown in Figure 5.9. Data translated by way of a neutral file has the ability to pass into any application environment as long as the latter has import schema, a built-in ability to read-in the neutral file format. Such import and export schema are shown in Figure 5.9 as separated from a CAD system. In actuality, import (identified as Translator A, Translator B and Translator C in Figure 5.9) and export schemes are usually an integral part of a CAD system.

In neutral-file format, one inward-translator and one outward-translator are needed for each application.

Thus, if there are n systems,

$$\text{No-of-indirect-translators} = 2 \times (n) \tag{5.1}$$

where n is the number of geometric applications (e.g., CAD/CAM systems) in use.

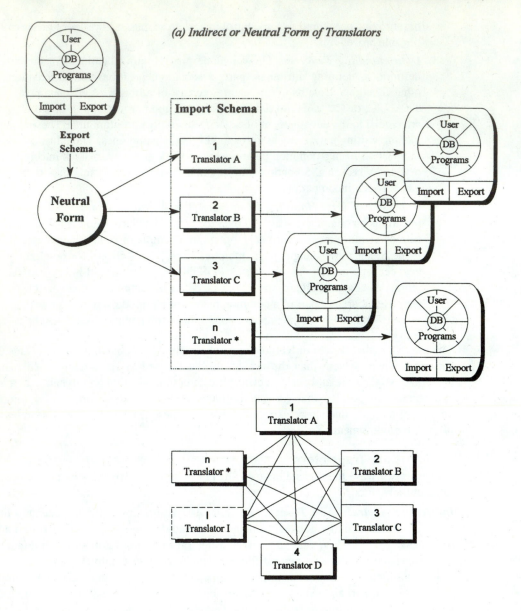

(a) Indirect or Neutral Form of Translators

(b) Direct (Point-to-point) Translators

FIGURE 5.9   Neutral Forms and Direct (Point-to-point) Translators

Direct (Point-to-point) Translators: A direct (i.e., point to point) translator, on the other hand, is often custom-designed specifically to translate data between two specific application environments. Unlike indirect translators, which go through an intermediate data conversion route (such as neutral form), direct translators are precisely engineered to handle a bi-directional data flow between each pair of CAD/CAM systems. If there are n systems between which data must be shared, in direct translator, a translator is required for each remaining $(n - 1)$ application. Evidently each translator (point-to-point) will require $2 \times (n - 1)$ translators. The number, $(n - 1)$ indicates the need to exchange data with all systems except itself, and the multiplication factor of 2 indicates that two-way interchange of data is generally required. Having written the translators for the first system, the second system would not require $2 \times (n - 1)$ data converters again. In fact, the second system would only require $2 \times (n - 2)$ data translators. One can extend this logic to the next to the last system, when only one bi-directional data translator is left to be written. Thus, the number of independent translators required to handle all conversion possibilities, in direct translator case, are

$$\text{No-of-direct-translators} = 2 \times (n - 1) + 2 \times (n - 2) + \ldots + 2 \qquad (5.2)$$

Rearranging the terms in the summation, this can be expressed as:

$$2 \times [1 + 2 + 3 + \ldots + (n - 2) + (n - 1)] \qquad (5.3)$$

The above represents a sum of $(n - 1)$ consecutive numbers. Using the formulae for the sums, it yields:

$$\text{No-of-direct-translators} = n \times (n - 1) \qquad (5.4)$$

Clearly, a data sharing approach based upon an intermediate neutral representation is preferable to one involving direct (point-to-point) translators. In neutral file format, however, it is not easy to control numerical accuracy. The inherent mathematical forms of the neutral file entities in most CAD applications are different or their internal representations are different. Reading or writing these to a neutral file introduces error possibilities. The standard neutral file translator, recognized by American National Standards Institute (ANSI) and International Standard Organization (ISO), is IGES (Initial Graphics Exchange Specification).

5.2   DISTRIBUTED COMPUTING

Most traditional architectures are based on mainframes and minicomputers connected through a centralized relational database system (RDMS). Most of the RDMS (e.g., CA-Ingres, Oracle, Sybase, DB2, etc.) provide a typical semblance of remote computing (see Figure 5.10a). However, their centralized nature of installation causes performance problems, storage retrieval delay, and maintenance problems. Personal computers, on the other hand, provide better access for doing routine tasks but are not designed for use by a group of people (such as work groups). They usually lack computing power and are not well con-

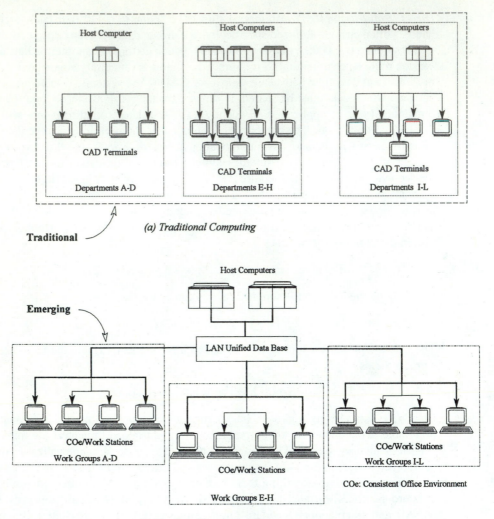

(a) Traditional Computing

Traditional

(b) Distributed Computing

FIGURE 5.10   Information Processing (a) Traditional Computing (b) Distributed Computing

nected. Local Area Network (LAN) or Wide Area Network (WAN) is usually tacked on as an ad-hoc or after-thought basis.

Distributed computing can provide a balanced solution for large scale multi-disciplinary computing if appropriately designed (see Figure 5.10b). Workstations and servers shared by a group of teams (such as work groups) are the two main ingredients for building a distributed computing network environment for concurrent engineering. The smallest unit of this network is called a work group network. A typical work group network unit comprises one or more client(s), consisting of powerful desktop workstations, interacting with one or more servers that store and manipulate information. Relevant programs are

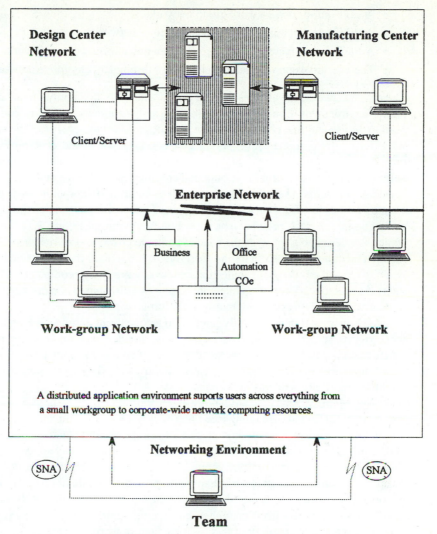

Design Center Network

Client/Server

Manufacturing Center Network

Client/Server

Enterprise Network

Business

Office Automation COe

Work-group Network

Work-group Network

A distributed application environment suports users across everything from a small workgroup to corporate-wide network computing resources.

Networking Environment

SNA

SNA

Team

FIGURE 5.11   Distributed Applications Environment

stored on the servers so that they can be accessed from any workstation on the network. This enables an efficient and timely sharing of data and resources between all the members of a work group connected to a network. Work group network is equipped by design to provide the right computing resources (capacity and power) for a specific group of concurrent teams to do their job efficiently. Depending upon the size of the work group, one or more work group networks may be needed and inter-linked. A distributed computing environment consists of a chain of interconnected networks linking work groups at various levels in a CE organization like a tree structure. The connections between a CE unit and the rest of organization are established in the following ways (see Figure 5.11). A CE team connects to:

- A PDT unit through a work group network.
- A PDT's design unit through a design center network.
- A PDT's manufacturing unit through a manufacturing center network.
- The executives on top management board through an enterprise network.

The above networks are all considered a part of a network taxonomy for a distributed computing environment. Center networks (design and manufacturing) may consist of a *center/unit server* and one or more of these linked *work group networks*. The enterprise network forms the corporate level server connection down to the center/unit networks. It provides work group access to organization level applications such as business and office automation tools. A distributed application environment is thus capable of supporting concurrent teams across every level from a small work group to corporate wide computing resources. Others in CE teams who need to access information can do so through the remote connection (e.g., SNA protocols) as shown at the bottom of Figure 5.11. Servers, at each center, are designed to provide the needed compute power and be responsive to individual runstream commands. Recent trends, therefore, are to replace the central database stored on mainframes and minis by distributed databases on server based computers. CORBA (Common Object Request Broker Architecture) provides a distributed objects' capability, an ability to invoke objects residing on multiple platforms.

The workstation is generally integral to a concurrent engineering since it empowers work groups, something which other systems are not designed to do. While personal computers automate tasks, they are intended to be used on a work alone basis. Similarly, while mainframes and minicomputers encourage multi-users, they lack interactive power and, therefore, mostly function as resource islands.

Sun Microsystems used an analogy of transportation to show how newer workstations are best for distributed computing (see Figure 5.12). One can compare mainframes to trains with a formal schedule, a very high degree of storage capacity, and useful in data security and control. Unfortunately, one cannot own a train personally; they are difficult to maintain and require a specialized infrastructure. Likewise, minicomputers can be compared to buses, which are more manageable than trains, provide more flexibility as to where one can go, operate in more flexible time-frames, and thus offer quicker service. But, due to declining cost per CPU power in the computer industry, minicomputers are under significant pressure from the top by mainframes and from the bottom by personal computers and workstations. They are, therefore, on a downhill path. Personal computers are like bicycles providing a lot of fun to the individual team member in getting around and completing one's specific tasks. Unfortunately, it is difficult to go very far on a bicycle. It is too difficult to travel too far, alone on a bike, away from your associates. There is a limit to what one can do all by oneself.

The workstation, of course, is the automobile. And like a fleet of vehicles comprising trains, buses, and bicycles, a workstation roughly corresponds to distributed computing (see Figure 5.12). The customers want to be able to drive to the train, the plane, the bus, and put their bike in the back. They want to own it personally so that they can control it, and they want to be able to carry more than one person. The workstation and work group computing as a whole are links between all of these previous platforms—mainframes, minicomputers, and personal computers. Workstations have all their benefits without many of the costs.

Comparision / *Sources* (E L E M E N T S O F D I S T R I B U T E D C O M P U T I N G)	*Analogy of Transportation*	*Distinguished Features*	*Major Hardware for Unix*
Personal Computer 	Bicycles 	Intended to be used on a work alone basis. Lot of fun but can't go very far on a bicycle.	
Mainframe Computer 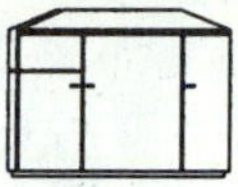	Train 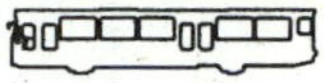	Encourage Multi-user but lack the interactive power. A high degree of storage Capacity and useful for data security & control.	AT&T Compac Data General DEC DELL HP IBM ICL
MiniComputer 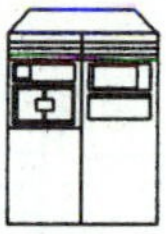	Buses 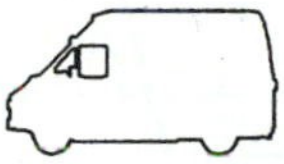	Decrease in Cost and Increase in Compute Power Multitasking and Multiuser	NCR RS/6000 Sequent Sun ****
Workstation 	Automobile 	Empowers Group Distributed Computing Multitasking and Multiuser Provides the Power of Minicomputer at about the same price	

FIGURE 5.12   An Analogy of Transportation to Power of Distributed Computing

5.3   WORK GROUP COMPUTING

In recent years, an architecture called work group computing (WC) has emerged due to the efforts of many workstation vendors competing for the CE market share (SUN, HP, DEC, IBM among others). WC provides a better integrated environment for CE compared to LAN based PC networks. With work group computing, computing tasks are distributed between clients, consisting of powerful desktop workstations and servers that store and manipulate information (see Figure 5.13). Work groups are provided a view into the computing complex through a window created by the client workstation. They may, however, be

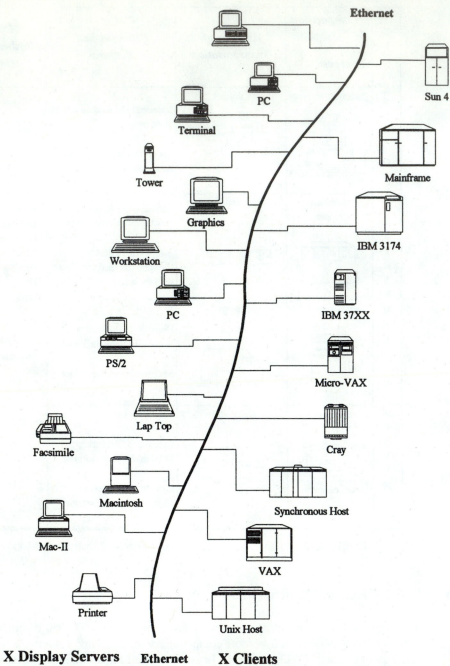

X Display Servers   **Ethernet**   **X Clients**

FIGURE 5.13   Distributed Computing

linked with virtually anyone else (see Figure 5.13). The X-display servers can range from a printer, to a facsimile, to a workstation.

And while mainframes, mini, and microcomputers will suffice for general needs, there will always be a need for more specialized machines. That is why work group computing is built around client-server architectures. In addition, modern databases can be distributed over many different machines, so work groups can create and execute applications locally on their own computers. These applications could have the same look and feel just like separate programs. Here, individual workstations, the clients, handle the local processing needs while the server has the power and capacities to access data from distributed databases beyond that of an individual workstation. The server may provide heavy-duty number crunching, distributed databases, and a link to the outside resources. When a work group runs the applications from any workstation, it draws on the resources of all the related applications and databases over the network no matter where they physically reside or are stored. This distributed concept maximizes the compute power needs of the work groups with the least amount of investment.

Work group computers are designed from the ground up. The intention is to help teams work together and automate group processes, including engineering, manufacturing, and other complex tasks. Since open systems are built around industry standards, they can be integrated easily with existing equipment from various vendors. Work group computing provides the power to perform individual tasks with ease while opening up the possibilities of information sharing between parallel work groups. Figure 5.14 illustrates the shift in key characteristics when moving from personal computing to work group comput-

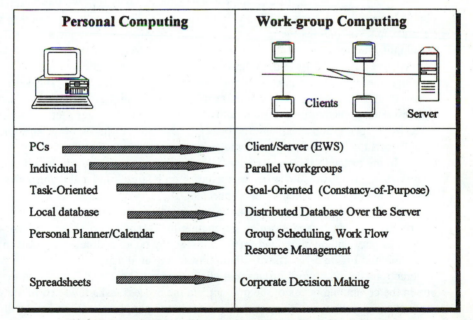

FIGURE 5.14   Shift from Personal Computing to Work Group Computing

ing. The shift pulls everyone, teams, computers, networks, transforming disparate computers into one flexible, easy-to-use client/server based system. The individual task-oriented environment of personal computing activities becomes a set of goal-oriented parallel work group activities. The use of distributed database over the network becomes the mainstream norm for file management as opposed to the local database residing in one's own personal PC (see Figure 5.14). The use of a personal planner and a calendar is replaced by a group scheduling system and an electronic work-flow resource management system. With work group computing, manufacturing organizations, small and large, can all benefit from concurrent engineering.

5.3.1   Parallel Work Groups

While implementing CE, a major challenge an organization faces is to provide a seamless connection between parallel work groups and computing machines. Work group computing provides a basis of distributing the work into cohesive parallel teams working in close association with each other. An example of such a distribution is shown in Figure 5.15. Here, four concurrent work groups, design, engineering, prototyping and manufacturing, are shown to be working together, each with its own work group computing system. The terminals represent the concurrent tasks that are being performed by a team within a work group. For example, in a engineering work group, different teams at any point may be working concurrently on tasks such as concept design, detail design, solid modeling, detailed analysis, drafting, and so on. Scheduling of work group tasks follows the IPD methodology as discussed in Chapter 4. Division of tasks follows the hierarchy of the breakdown structure trees (WBS, PtBS, PsBS, etc.) as discussed in Chapter 4 of volume I. Transparent communication and access to common databases provide a mode of constant communication and frequent interaction between the work groups. The network is represented in Figure 5.15 by a thick horizontal wavy line. It runs continuously and crosses through the work group partitions not shown in Figure 5.15. The vertical down-arrows connect the work group workstations, terminals, PCs, mainframe, minicomputers and the corresponding database server to a network. The network ensures that the messages created by a work group or a team member are passed on to the work groups and that the changes that affect the design outcome are propagated throughout the CE organization.

In the example shown in Figure 5.16, work groups are equipped with workstations with existing hardware and software, including both large (mini and mainframes) and personal computers, that need to be connected. There is no unique setup for work group computing. Applications can run on the server side or on the client side depending upon work group needs, tasks, priorities, computing loads, and so on. A typical distribution of activities between the client and server sides, at any point in time, may occur in the fashion shown in Figure 5.16. The distribution is merely an illustration to show the concurrent interactions between work groups and client/servers. The actual interaction may depend upon the computing needs of the product and the process tasks involved in the work session at a particular time. Many varieties of tools and peripheral equipment, facilities (plot-

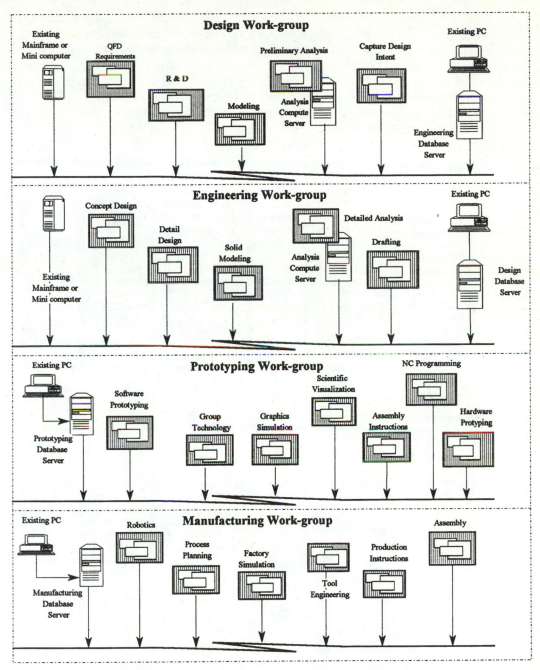

FIGURE 5.15   The Parallel Work Groups of Concurrent Engineering

ting, drafting, scanning), office automation machines (mechanical and electronic devices), computers, and networks usually form a major part of this computational environment. This is shown in Figure 5.16 by the unshaded blocks on the first column. Separating the two sides in the middle are the communication and network links, shown by a wavy horizontal remote line and a series of vertical remote lines. The horizontal remote line connects the work groups. It runs continuously and crosses through the work group partitions as shown in Figure 5.16. The series of vertical remote lines connect the server sides with their corresponding client sides.

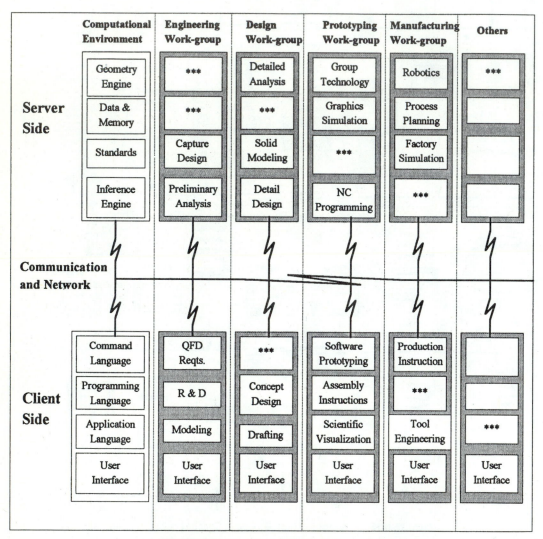

FIGURE 5.16 Client/Server Mode of Computing

5.4   PRODUCT INFORMATION MANAGEMENT (PIM)

The function of PIM is to improve the integrity of product definition data and make it accessible to all work group members. The information contained in the project's PIM system evolves from a high degree of abstraction to fully detailed parts as product development proceeds. Common problems encountered during product information (or data) management are:

- Heterogeneous data environment.
- Proliferation of data.
- Frequent data changes.
- Information not managed as a corporate asset.

A PIM solution is usually based on a set of hardware- and software-independence principles called freedom architecture. This is discussed later in this section.

5.4.1   Coordinating and Managing PIM Definition

The subject of enterprise ownership of the shared PIM definition was described as a necessary condition of the concurrent engineering process (see section 4.3.5 of volume I). This is a lot more extensive and difficult task than that of maintaining configuration control of a released design definition data. It entails the coordination and management of both working forms and released versions of the PIM definition, and therefore includes the capture and maintenance of all forms of data, process, and knowledge, including:

- Designs (CAD and non-CAD) for product, machining, NC programs, process planning tooling, product support and equipment.
- Bills-of-material for production and product support.
- Analyses (wire-harness, mass properties, structural, thermal, reliability, etc.), of specifications (functional, interfaces, materials and processes, test, and procurement).
- Manufacturing plans (materials handling, MRP, quality, procurement, inspection, fabrication, product support, and test).
- Quality Control (QC) programs for production and inspection.

Accordingly, current configuration and data management policies, practices, and procedures (3Ps), as well as traditional development process interfaces, must be modified as part of the concurrent engineering process development to account for this new PIM requirement [Donlin, 1991]. Such 3Ps must, of course, also provide for the distribution of PIM information to the various work group members.

5.4.2 PIM Requirements

There is no doubt that PIM plays a crucial role in managing and controlling the flow of information, data or knowledge through an organization. Its use in the concurrent engineering PD^3 environment, however, imposes some additional requirements on PIM. They are also listed in the following:

- *Adaptability:* To a varying degree, a CE-based PIM system typically requires customization before it is ready to roll. Customization sometimes means costly and constant reprogramming. The PIM system should be adaptive, that is, able to configure and reconfigure with little or no programming.

- *Intelligence:* The system should have built in CE policy, practices, and procedures (3Ps) in various standard forms with change provision for instant access by the work groups. It should be linked with a number of popular application softwares and be able to run across various platforms and operating systems. It must also work with distributed hardware when parts of the product design and manufacturing are done in different geographical locations.

- *Shared Access:* Multiple work groups should be able to process CAD or CAM data files, whether they are trying to access the same file or the corresponding storage area. To a work group, it should be just a matter of requesting an object (a data, a file, or a piece of knowledge) and the PIM should take care of the details such as:
  - Finding the object.
  - Bringing it up on the team member's workstations.
  - Maintaining version control.

- *Product Knowledge:* The PIM system should respond to innate knowledge of the product tree structure and its hierarchical breakdown (such as PtBS, PsBS, WBS) when dealing with work groups. The system should automatically generate bill-of-materials for components, assemblies, and systems. Teams should be able to locate and retrieve object/data simply by selecting items from the bill-of-materials (BOMs) listing. Teams should also be able to automatically retrieve associated objects/attributes from the family of parts' library.

- *Dynamic:* In concurrent engineering situations, PIM systems should link work groups and other life-cycle elements in a more informal manner, allowing frequent communication to take place. A concurrent team can access information at any time. When a team accesses a part, it brings in the latest part as being worked upon by the remaining work group members. Product information is enriched on a regular basis depending upon who is working on it. The process is dynamic in the sense that concurrent teams can do transactions on distinct portions many times a day instead of a limited number of times as in the static, rigid systems.

- *Unified Security and Control:* PIM needs changes as one proceeds from a work group to a department to an enterprise. The security and control needs are different depending on the following:

- *PIM Locations:* where the data moves from an individual work group to a department, and from a department to an enterprise.
- *Maturity Level:* when a design cycle is completed or when product data matures, meaning when it moves from one loop level to the other.
- *Team member:* who is using or modifying the information.

It is not clear whether or not these control needs can be translated into a unified release and control management at the PIM level. Nevertheless, all these releases and control mechanisms at each stage have to be seamlessly connected so that, from a work group point of view, there appears to be only one PIM. This is depicted in Figure 5.11 of volume I. The responsibilities can be distributed as follows:

- The PDT work group manager is responsible for change of product information,
- The PDT life-cycle manager is responsible for process data (e.g., 3Ps, policy, practices, and procedures),
- The lead product manager of the strategic business unit is responsible for enterprise or business data.
- ***Freedom architecture:*** PIM architecture should be configurable with a variety of tools and with different hardware and software options. Product information should be made available across diverse platforms in distributed computing environments and should accommodate and integrate C4 (CAD/CAM/CAE/CIM) tools from a variety of software vendors. PIM should also provide flexibility for tailoring to each installation's design practices and procedures. It should track any data type, including part lists, 3-D models, schematics, software applications, and CAD files residing in any media such as raster, paper-form, or microfiche.

5.4.3   Basic Components of PIM

There are five basic components of a PIM system.

1. Graphical User Interface (GUI).
2. Database Management Systems (DBMS).
3. Review and Release Management.
4. Project/Product Tracking & Management.
5. Network and Network Services.

5.4.3.1   Graphical User interfaces

User interfaces are emerging as key components in PIM systems. They help teams tap the power of PIM with little effort. Indeed, most team members usually do not work directly with the underlying DBMS. They merely fill out electronic forms. These forms are often made to closely mimic the paper, ones already in use by a company, easing the transition to PIM. Some PIMs also offer graphical user interfaces with point-and-click operation to further simplify the filling task. Some provide varying degrees of viewing and update privileges to individuals and work groups. User interfaces are important to both team members and their managers to allow managers to find the data they need and put it in a format they want. A good PIM system gives work groups a good user interface that minimizes the intrusion of PIM into their everyday work.

5.4.3.2 Database Management Systems (DBMS)

DBMS are an integral part of virtually all PIM packages. In fact, most PIM systems are general-purpose DBMS for engineering applications. PIM systems are mostly built on top of commercial relational DBMS such as INGRES or ORACLE. As a result, division of information between programs and database are quite strict, procedures go into programs and data goes into databases. This division has created problems when changes are made to the structure of the database. The effect is not reflected in the procedures. This also constitutes a violation of the good practices of modular object-oriented programming (OOP) that requires related procedures and data be packaged together [Taylor, 1993]. One way to overcome this violation and data incompatibility is to encapsulate related procedures and data together regardless of what applications they may serve or platform they run on. This is shown in Figure 5.17 using a network of circular dots—a dot represents a datum and a connecting line between two dots denotes the related procedures. The pattern

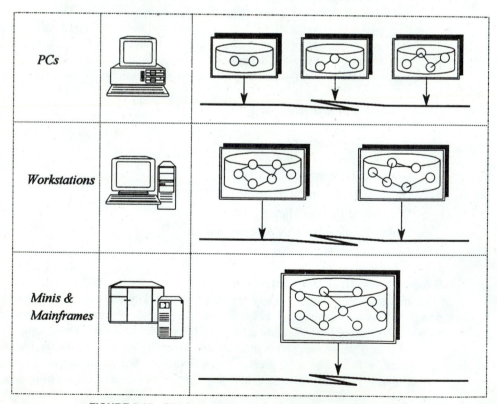

FIGURE 5.17 Distributed Applications in an Object-oriented DBMS

of encapsulation (two dots and a connecting line) is shown repeated no matter where the applications are run, on PCs, workstation, or mini and mainframes. This breaks the distinction between programs and databases. Newer PIM systems that are based on object-oriented DBMS could be the natural vehicles to bring about this merger. Object oriented philosophy allows methods to execute directly on the knowledge-base. Thus, it can deal with creating objects out of data as well as the processes that work on that data. The knowledge-base can contain the entire application including all their procedures as well as their data. Proponents of OOP claim this to be an ideal vehicle for CE applications since the process is often as important as the data itself.

Object-oriented DBMS are said to handle many different types of data in a PIM application more efficiently, and like other object-oriented software, they are reportedly easier to reconfigure or customize. Systems of this kind can be configured as a tree structure (in an object-oriented way) with standard templates as nodes of the tree for custom software modules. Reconfiguration of the tree can be used to cater to the unique team's work flows and to accommodate in-house standards and CE practices.

DBMS Functions: At the most basic level, DBMS help manage all files typically found in a concurrent engineering project. Key areas where DBMS is helpful include:

- *Access:* DBMS allows teams to access files by project, date, or serial number rather than arcane file names. Retrieval by meaningful engineering and manufacturing references, including SQL search and retrieval, is required for CE.

- *Security:* DBMS makes it easier for system managers to set up access privileges. They can ensure that only authorized teams access files and can make files available on a read only basis when needed.

- *Data transfer:* DBMS also greatly simplifies data transfer from one system to another, or from one program to another. While they do not usually have translation capabilities themselves, DBMS automates many needed commands to move data across the network. In doing so, DBMS ensures that data is in the right format and that it is transparent to the end users.

- *Archive:* DBMS tracks data no matter where it is stored, even if it has been moved to archival media such as tapes. In addition, they provide automated data archival and backup, keep an audit trail to record the movement of files to and fro, and also record the time of file access.

5.4.3.3   Review and Release Management

PIM plays a crucial role in management of data review and release process. Procedures and policies for review and release of design data at each stage of CAD refinement or design approval can be encapsulated into a single PIM system, helping to automate the CE process. Through network services such as electronic mail, PIM systems can automatically notify the changes, carry out design reviews, and notify post releases. Some PIM systems allow for document redlining or CAD annotation. The sign-off can be electronic as well. Moreover, the PIM system ensures that proper revision, annotations, and documentation procedures are used for design review, and version control is maintained for subsequent modifications.

Project Tracking and Control: Some PIM systems allow project or task tracking, and control features that tie into a resident project planning and management software. They provide functions such as milestone posting, electronic approval mechanism, and percentage complete estimating. Many PIM systems are also integrated with popular word processing software.

5.4.3.4   Project/Task Management

A product consists of assemblies, assemblies are made of components and sub-assemblies, and so forth. Typically, PIM systems allow project managers to set up such hierarchical product or process breakdown structures (PtBS, PsBS) as illustrated in section 4.3.2 of volume I, to bring order to the product realization process. Also included in this hierarchy are CAD geometry, analysis files, and most geometric or non-geometric specifications that describe the product. PDTs maintain that this hierarchical approach is ideally suited to design and manufacturing, and is easily understood by everyone in the organization. Moreover, this hierarchy fits well with the widely used bills-of-materials approach to pass product information from design to manufacturing. Many PIM systems use this hierarchical approach to create work breakdown structure (WBS), automate work process management or formalize relationships between various work groups or teams. These relationships can be defined with heuristics, mathematical formulas, or engineering rules.

PIM systems can assist CE managers in mapping out structures (e.g., PtBS, PsBS, WBS) for new products. The initial design of such structures could be just a skeleton unfilled tree. Later, when details become available, branches and nodes can be filled in. As the project progresses, this allows CE managers to check on its status. Some PIM systems even allow managers to create custom reports from such tree structures.

5.4.3.5   Network and Network Services

Network services are part of the foundation of PIM. They help to connect various systems and work groups in product development, regardless of what hardware or software they use. Network services include a wide range of open and proprietary networks, workstations, and personal computers as well as CAD outputs in drawings and model formats. The open system approach permits work groups to easily access and interchange historical and real-time product and process data across a variety of platforms. PIM electronic mail helps work groups to communicate with one another and automates several PIM tasks. Standard file formats such as NFS allow PIM files to be easily and transparently moved from one system to another, even among outside vendors.

5.4.4   Rewards of PIM

Figure 5.18 shows the relative affects of PIM on design changes and life-cycle time. Four regions are shown.

(a) A typical manufacturer (no PIM).
(b) World-class manufacturer (no PIM).
(c) World-class manufacturer with PIM and minimal process optimization.
(d) World-class manufacturer with PIM and an optimized process.

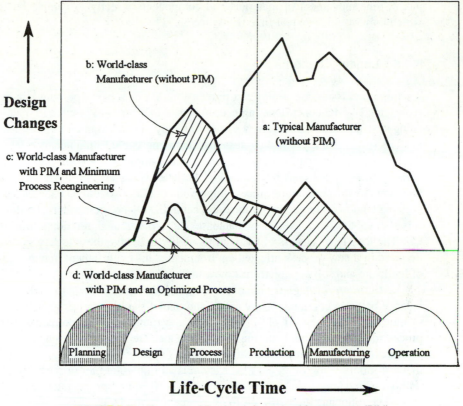

FIGURE 5.18   Rewards of Product Information Management (PIM)

They are identified in Figure 5.18 as a, b, c, and d regions. Clearly, PIM reduces the number of design changes and moves them further upstream, where it is less costly to implement. By leveraging downstream operations upstream, developmental teams can also make better use of existing facilities and have more freedom to make pertinent design changes, and ultimately shorten the product delivery time. The effects are more pronounced when the process is optimized. The latter may include:

- Object-oriented or modular programming techniques.
- Dynamic data linking (DDL) of applications, such as CAD/CAM with modeling and analysis packages.
- Dynamically managing parts in a tree structure, such as components in an assembly.

On the manufacturing side, PIM system solutions can help optimize supply-chain management by improving communication between manufacturers and suppliers. In most industrial applications, PIM employs electronic data interchange to transfer order information to vendors. It is not very far in the future when, through PIM, vendors could tie just-in-time movement of materials and components directly to vehicle build orders to which they would have real-time (EDI) access.

5.4.5 Legacy Systems

An important issue concerning the implementation of a new operating environment, as represented by the concurrent engineering based PD[3] process, is what to do about the existing system, the so called legacy systems already in use. Companies have made substantial investments in these legacy systems, including capital, maintenance, and personnel training costs. They will not easily give them up, nor are they willing to replace old systems with new open ones immediately. The latter could also be made obsolete within a few years by the continuous process improvement (CPI) system. Databases for traditional systems tend to be large, monolithic, and centralized—not quite what is required for CE. It is also not easy to convert these databases into a distributed, relational form required for CE. The long term success of CE is in its ability to evolve these legacy systems to function within a new process oriented environment. This environment may have a new slew of hardware and software platforms, new tools, or new networks.

Migration Strategies: One of the major challenges of CE is to find the appropriate balance between continuity and change, and one of the keys to establish that balance lies in migration strategies. If old PIM systems are found to be incompatible with the new process oriented environment, migration strategies are needed to re-engineer existing environments. Concurrent engineering developers must convince management that replacement or modification of an old PIM system is an important part of the migration strategy without which long-term success (productivity, efficiency, etc.) cannot be achieved.

The following are two ways to overcome this problem:

1. One way is to employ a migration strategy with a client-server software architecture that bridges the technology or competitive gap. A well-orchestrated client-server architecture can provide many of the advantages of an evolving open environment and could at the same time accommodate the old system. By maintaining newly created data in the new format, old systems can be phased out into a new open system environment where the shock of change is not so sudden. This route may require the following utilities:
   * A line of utilities (such as wrapper concept or window-based dynamic data linking (DDL) protocols) that encapsulate existing tools so that they can be embedded in the new environments with minimum change.
   * A line of smart utilities to reconfigure the distributed environment, including intelligent routing of existing software tools. The most effective way to implement this is through a client/server architecture based on open systems.

2. The second way is to run the two systems in parallel and slowly phase the old PIM system out. This is based on the premise that technology and automation enhance CE—they are not a necessary condition for concurrent engineering. Using this approach, one will find abandoning the old system a better option than compromising

the solutions by trying to force fit the old PIM system into the new one. Nevertheless, migration is still a problem that a company must address. The migration problem does not go away when implementing a new process in an ongoing operation.

An increasingly competitive business and economic environment, coupled with management desire to improve communication and maximize inter-operability, has driven the evolution of open systems into a PIM type functionality. PIM technology plays a pivotal role in re-engineering the migration process, at the same time bringing the manufacturers, suppliers, and customers closer together. PIM is very timely since its components form a strong basis for shaping CE architecture.

5.5   CE ARCHITECTURE

A compendium of abstractions (called frameworks) leads to a CE Architecture. These levels of abstractions are not progressive in nature (have no definite sequencing order) but are need-based (see Figure 5.19). Hence they are called frameworks. These frameworks are:

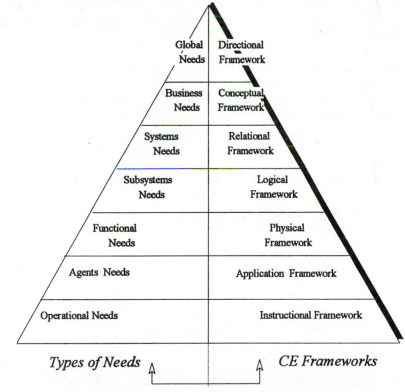

FIGURE 5.19   Types of Needs and CE Frameworks

- Directional Framework.
- Conceptual Framework.
- Relational Framework.
- Logical Framework.
- Physical Framework.
- Application Framework.
- Instructional Framework.

The following sections describe each in more detail.

5.5.1 Directional Framework

Directional Framework focuses on the enterprise's global needs (vision, mission, objectives, goals, etc.). A top level abstraction for a CE architecture defines the product vision in relation to external (sometime referred to as uncontrolled) environments such as vendors, suppliers, field support, marketing, and customers. Examples of system frameworks are the Air Force's Integrated Computer-Aided Manufacturing (ICAM) program, CASA/SME (Society of Manufacturing Engineers), CAM-I, and many others. The dominant means to meeting the challenges are global thinking, common tools, consistent standards, and uniform methodology applied across the enterprise.

CIMOSA Architecture: Very few methodologies cover the whole PD^3 development cycle from analysis to operation to maintenance of a CIM system. CIMOSA (Open System Architecture for CIM) is one of the most complete methodologies in terms of life-cycle coverage (see Figure 5.20). It was developed by the AMICE consortium in 1986 under the auspices of EC ESPRINT (European Strategic Program for Research and Development in Information Technology) project funding. The AMICE (European CIM Architecture—in reverse) consortium now comprises 15 companies and research institutes from eight European countries.

CIMOSA architecture is composed of the following three elements:

1. *Modeling Framework:* This provides modeling structures—how CIM systems should be modeled. It organizes the CIMOSA reference architecture into a generic and partial modeling level, each one supporting different views on the particular enterprise model. CIMOSA has defined four different modeling views: function, information, resource and organization. This set of views may be extended as well. This concept of views allows teams to work with the subset of the model rather than the complete model. Users can view only what they are interested in viewing, hiding the complexity from the particular area of concern.

2. *System Life Cycle:* This describes the operations or tasks to be used to generate CIMOSA Models. The architecture supports modeling of three life-cycle operations: requirements' definition, system specification definition, and implementation description. Sequence of modeling is optional, that is, modeling may start at any of the life-cycle phases and may be iterative as well.

3. *An Integration Infrastructure:* This provides a set of generic (system wide) information technology (IT), service entities, and resources to support the execution of
   CIMOSA Models in a heterogeneous environment. Types of service include: management entity, business entity, common entity, information entity, presentation entity; and resources include information technology resources and manufacturing resources. Control on execution of the implementation description model is provided
   by the business entity that receives the events and creates occurrences of the domain
   process and all its contents.

The above three together form a three-dimensional framework for the models, as shown
in Figure 5.20, also known as the CIMOSA CUBE. CIMOSA provides the basic framework for evolutionary enterprise modeling. CIMOSA is based on the object oriented concepts of inheritance, that is, structuring its constructs in recursive sets of object classes.

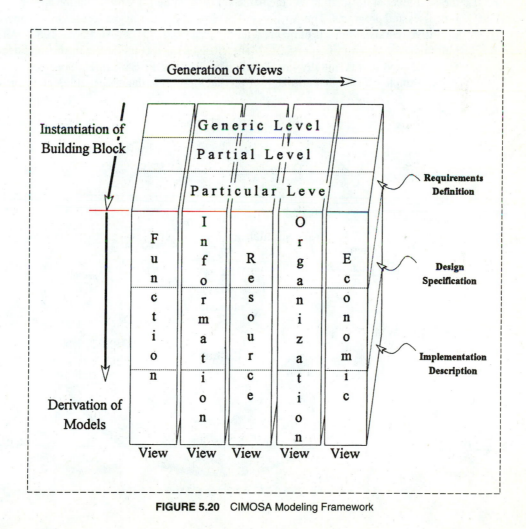

FIGURE 5.20   CIMOSA Modeling Framework

5.5.2   Conceptual Framework

Conceptual Framework addresses high level CE business perspectives (e.g., strategies, objectives/goals) for developing a PD^3 system [Margolias and O'Connell, 1990]. Examples and tools that fall in this category are multi-disciplinary or cross-functional team, work groups, quality function deployment (QFD) and its replacement concurrent function deployment (CFD), total quality management (TQM) and its replacement total value management (TVM), and so on.

5.5.3   Relational Framework

Relational Framework expands conceptual perspectives into system needs and then forms the next level of abstraction. As the name suggests, it describes the relationships between processes and programs. The Air Force ICAM program and its successors have generated extensive relational models of manufacturing organization [U.S. Air Force, 1982]. An initial model is shown in Figure 5.21. Many changes have come about in this architecture from the time it was initially introduced. The growing PDES/Express specification is a good example of a relational model for product definition databases, although Express is

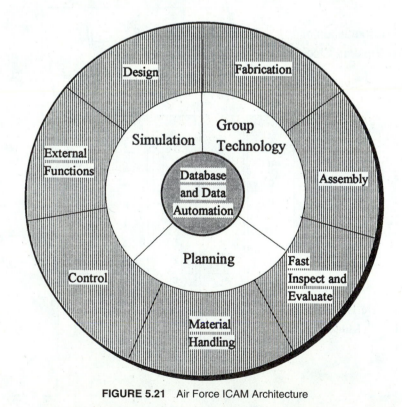

FIGURE 5.21   Air Force ICAM Architecture

also being used as a data definition mechanism. Less formally, the Commission of the European Communities (CEC)—ESPRIT (European Strategic Planning for Research in Information Technology)/CIM program—has generated a book-sized flow chart description of a typical product development organization [Yeomans, Choudry, and Hagen, 1985]. The key to meeting the CIM challenge is enterprise-wide system thinking on the relational framework (e.g., Figure 5.20) with standards being the glue to make it all stick.

5.5.4   Logical Framework

Logical Framework defines sub-system needs. It depicts logical design information with respect to a relational framework. It provides symbolic descriptions of the processes and programs to define the logical view of the system. The types of symbols that identify the structures of the framework are CAD, CAM, CAE, object oriented databases, and so on.

5.5.5   Physical Framework

Physical framework defines the functional needs for logical and relational frameworks so that they can work together. This level of abstraction defines:

- A computer system.
- A network that connects a collection of computers.
- A data representation, (e.g., data definition language (DDL), structured query language (SQL), etc.).
- A communication interface like Ethernet.
- A database that stores the underlying information.

An example of a typical physical framework for a computer system is shown in Figure 5.22. The operating system and utilities are shown to be the gate keepers for I/O transfers to CRT terminals, peripherals, and other computers.

5.5.6   Application Framework

Application Framework defines the agent's view. This defines the specifications of the physical framework components in relation to both hardware and software needs such as operating systems (DOS, UNIX, etc.), basic programming languages (PL1, UNIX, C, C++, FORTRAN, etc.), and hardware platforms (IBM RT's, SUN, HP, etc.).

5.5.7   Instructional Framework

This is the lowest level of abstraction and provides the basic instruction sets for computer programs or systems. This may include data encoding, character coding, data conversion, and program formats (real, alphanumeric, or floating point). While instructional frame-

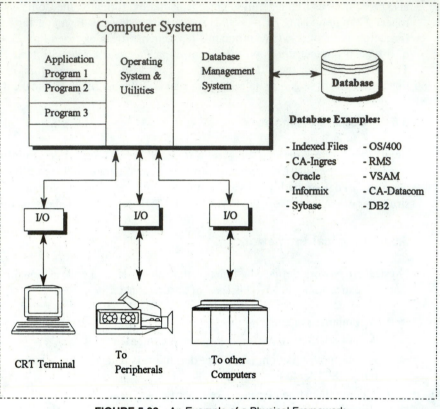

FIGURE 5.22   An Example of a Physical Framework

works have made considerable impact on physical applications, their lack of standardization is a well-known annoyance.

In the aforementioned sections 5.5, CE architecture was classified into 7 levels of abstractions. Figure 5.23 summarizes the salient features of each level with examples and tools in a tabular form. The examples and tools are only representative samples to show the main distinctions between these different levels. The boundary between two consecutive frameworks, such as application framework and instructional framework, or physical framework and application framework, is rather fuzzy. The first four frameworks: directional, conceptual, relational and logical are more distinct than the rest. Their models are more formally defined than the rest and, unlike the others, they are not directly executable by computer programs or systems. Logical and physical frameworks are closely related. Though not executable in the same environment, logical and physical frameworks share the same goal. They represent a major contribution to the success of the product life cycle and together they constitute a computational environment for a CE-based PD^3 system.

Type	Description	Examples and Tools
Directional	Enterprise global needs and directions (Vision, Mission, Objectives, Goals, etc.) High level reference models	Air Force ICAM Program, CIM-OSA, IBM CIM, CASA / SME, CAM-I, etc.
Conceptual	The high-level requirements or taxonomy for developing a product or a system. It addresses the major business needs	QFD, Multi disciplinary teams, Work-group, TQM, etc.
Relational	Relational descriptions of processes or programs (not executable). It addresses the major Systems needs	IDEF, PDES / Express, ESPRINT/Amice E-R models process, etc.
Logical	Symbolic descriptions of processes or programs to define the system. It defines the major subsystem needs.	Design, Object data bases, CAPP, CAE, CAD, CAM, etc.
Physical	It defines the functional needs for Relational and Logical frameworks to work together.	DDL, SQL, Ethernet, TCP/IP, and other communication Interfaces.
Application	Specifications of the physical framework components -- hardwares and Softwares	DOS, Unix,... ,etc. PL1, C, C++, Fortran, etc. IBM RT, SUN4, HP, etc.
Instructional	Basic instructional set -- Data descriptions, Information flow and formats	CISC / RISC Instruction sets, character codes, formats.

FIGURE 5.23   Salient Features of a CE Architecture

5.6   CE SUB-ARCHITECTURES

5.6.1   Sub-architectures for Logical Framework

The goal of a logical framework is to provide a flexible application development environment that shields end user applications from possible downstream changes. The logical framework is organized into a three-layered system. Figure 5.24 shows a logical view of this CE sub-architecture that forms the basis for the flexible CE environment described in this book. The lowest layer is the computing platform. The second-layer, intelligent interface, provides the primary programming interface to application developers. The top layer consists of end-user applications communicating among themselves (horizontally) and to the intelligent interface (vertically).

$$\int [\text{Compute Platform} \oplus \text{Intelligent Interface}] \times \Delta\text{Standards}$$
$$\Rightarrow \text{Long Life of End-user Applications} \tag{5.5}$$

When computing platforms with intelligent interface are integrated over a range of applicable standards, this results in a long life of the end-user applications shown on the top of Figure 5.24.

5.6.1.1   Computing Platform

This is the bottom layer of a CE sub-architecture. It consists of an actual hardware platform (such as Engineering Workstations (EWS), networks and standards), system enablers (such as operating system, languages, and standards), and core enablers. The core enablers are the core tools that enable transparent access to database and other system enablers. Core enablers have three engines.

1. *Object Engine:* This is used for managing objects supporting complex hierarchies of objects arranged into classes, subclasses, and instances.

2. *Rule Engine:* Because different engineering, manufacturing or business problems require different kinds of reasoning, a versatile rule engine is required that gives the work group members a choice of various rule-based reasoning techniques:
   • Forward chaining for data-driven reasoning.
   • Backward chaining for diagnostic-style reasoning.
   • Backtracking for intelligent searching and the management of iteration and looping.

3. *Data Access Engine*: for mapping data from a relational database onto objects.

5.6.1.2   Intelligent Interface

This layer consists of a strategic C4 (CAD/CAM/CAE/CIM) system, general tools, and application enablers (see Figure 5.24). The application enablers create and foster the adoption of a base set of enabling features, functions, and interfaces (called engines). The enabling base provides an environment for building applications utilizing the engines. The intelligent interface layer includes three main components:

1. *Strategic C4 System:* This is a set of C4 utilities, a set of graphical end-user interface construction facilities. It is employed to construct a common knowledge-base model. The common knowledge-base model is a common part/feature based model. This enables a view of the design that is common across all CE disciplines.

2. *General Tools:* Interactive C or C++ environment facilitates development of object-oriented (C or C++) source codes.

3. *Application Enablers:* This is a developer interface for developing object bases, rule sets, and database mapping in a graphical environment. Examples include workbench, class/function libraries, browsers, and so on. Browsers can be further subdivided into the following: application browser, module browser, entity browser, class browser, and error browser. Application enablers can be used to build model drivers. The model drivers are things like equation solvers, optimization, and visualization tools.

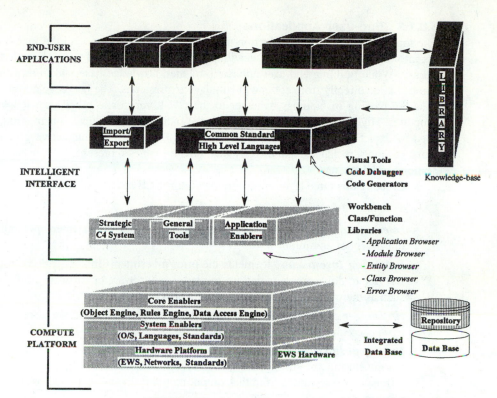

FIGURE 5.24   A Logical Framework for CE

The intelligent interface contains import/export facilities, common programming standards, and high-level language processors. High-level languages help in capturing business decision-making processes. Examples of such processes may include visual tools, code generators (CASE Tools), and a code debugger. Business logic, navigation, and database operations are triggered by visual-control. The application language combines rule, analysis, and procedure-based descriptions, along with sophisticated pattern-matching capability. The high level language is tightly integrated with the intelligent interface system. Access to the common knowledge-base model is achieved through the application (high level) language that allows the work group members to query the model in their own terms. High level language can be used in conjunction with, or even as an alternative to, the C++ language to write procedures for object system methods and monitors, rules, or even a command language. The language itself may be either interpreted or compiled in a single phase during development, thus providing one language, with a single syntax, for a variety of uses.

Intelligent interface with high level language provides a full application programming interface (API) to the various engines. This allows an application developer to function in any role within a broader application architecture including back-end server, intelligent front-end, and run-time library.

5.6.1.3 End-user Applications

The top layer contains the production-user interface that integrates all the functional layers into the so called end-user applications. This layer is equipped with a knowledge base library. While building end-user applications, the knowledge base and library functions can be called directly from command language expressions. The language compiler manages the passing of bindings, whether to the C^{++} library or to other knowledge library functions. High level language, along with the facility for graphical construction, can be used to depict the status of important events or actions. With interactive graphics, the work group users can see and directly manipulate images representing their business operations and decisions.

Creating this three layer logical framework for CE has the following benefits:

- *Inter-operability:* It provides the applications with better inter-operability in a multi-vendor environment.
- *Protecting Investments:* It shields the prior investments in applications from being lost due to downstream variations in platform hardware or software configurations, such as the winning brand names or standards of the month.
- *No Danger of Obsolescence:* It prevents the applications that are developed today from becoming the *legacy* of tomorrow.
- *Intelligent Interface Layer:* It allows quick generation of the applications, since development kits, or utilities, are placed on the common service layer (intelligent interface layer) separated from the computing platform and shared across multiple applications.
- *Less Programming:* It insulates the application developer from low level languages, platform, and network variations.

5.7 CE COMPUTATIONAL ARCHITECTURE

Computational Architecture (CA) is a 3-in-1 deal, a fusion of three framework components: relational, logical and physical. As a result, CA provides a uniform setting for collaboration, session management, data sharing and multimedia communication along with a power of numeric, symbolic, and graphics substrate. Because of their relevance to CE fields, this book focuses on computational architecture.

A CE developmental environment is actually an implementation of relational and logical objects into a particular physical architecture. An object is a particular design method, tool, or an advisor that is executed from within a computational environment. The object may be *extensible*, that is, other tools are accessible as native resources for an application, or *open*, that is, the facilities of that object are made available to other tools. A server is an object that provides facilities to other objects without a direct user interface of its own. An extensible environment that provides teams access to the facilities of one or more servers, work groups, and foreign codes is called an *open environment*.

5.7.1   Requirements for a Computational Architecture

In Chapter 4, section 4.5, the requirements for concurrency between different work group configurations and their degree of involvements were presented in a matrix form (see Figure 4.13 of volume I). While significant portions of any computational framework meeting these requirements must be included, concurrent engineering systems must also operate under other kinds of constraints related to infrastructure (see Figure 6.13 of volume I). A computational architecture for CE must exhibit the following qualities:

- *Accommodate distributed computing environment:* CA must support a distributed computing environment based on an open networking context to allow teams to work collaboratively, unhindered by types of compute platforms, operating systems, network protocols, and so on. The CA should provide seamless communication among work groups (with appropriate level of security).

- *Existing applications and product development procedures:* Because of large prior investment in tools and commitments to ongoing operations, most organizations will not be able to start with a clean slate. The CA should work temporarily with the legacy system without interrupting operations significantly while a new framework is being designed, developed, and installed.

- *Emerging frameworks and applications architectures:* Vendors and organizations developing software and hardware will continue to produce new frameworks/applications. This might dictate changes to the data or configurations of the computational framework. The computational architecture should allow all such tools to be integrated so that minimal effort is required to move data from one tool/application to the next.

- *Efficiency:* Methods, tools, and advisors within the CA framework should run at speeds comparable to their independent execution. System overhead must be commensurate with the benefits accruing from its use.

5.7.2   CE Developmental Environment

One of the purposes of computational architecture is to easily build new end-user applications that not only provide a new set of CE functionality but are integrated with the rest of CE applications developed earlier. As a development platform, the computational framework for CE can be specified in terms of the following nine layers tool-kit: user interface, command language, high-level application language, programming language, geometry engine, data and memory management, the inference engine, communication and networks, and finally, standards (see Figure 5.25). The specifications apply to the syntax, semantics, and representations of each of the above nine basic functional layers. Computational architectures range in complexity from time-sharing operating systems and conventional lanugages (FORTRAN like) to the newer object-oriented languages and data bases defined by organizations like the CAD Framework Initiative (CFI) and some of the system vendors. Today, many computational architectures (as a developmental tool-kit) seem to converge on some variation of the following nine-layer configuration (see Figure 5.25).

Facility	Description
1. User Interface	Mechanisim for user to activate and monitor application facilities (Window to the world)
2. Command Language	Language for defining user interfaces and capturing command sequences
3. High Level Application Language	Symbolic representations of attributes inputs/Outputs and other design and manufacturing parameters
4. Programming Language	Languages for expressing, developing and writing application functions
5. Geometry Engine	Languages, facility , library and utilities for modelling geometry of parts, family of designs and defining topology
6. Data and Memory Management	Languages, facilities, Library and utilities for managing information, sending messages, changing files, records and objects
7. Inference Engine	Chaining a set of instructions to infer results or an outcome of an event (values of an attribute)
8. Communication and Networks	Communication protocols facilities, library and utilities for communication amongst heterogenous h/w & s/w environments
9. Standards	Introduction of standardized neutral forms, languages and interchange formats

FIGURE 5.25   A Generic CE Developmental Environment (Nine-Layer Configuration)

1. *User Interface:* UI is often considered the *window to the world*, an external mechanism whereby a team or a work group can execute system functions and graphically view the results. Common UI structure allows the team members to move from one computer application tool to another with minimal learning. The "same look and feel" aspect of UI allows the teams to concentrate on the results and not on the interface. A simple example of UI is the pull-down menu in any window-based environment. The current emerging standard appears to be some variation of the X-Windows graphic windowing system with OSF/MOTIF libraries.

2. *Command Language:* This enables team members to issue interactive commands with programmatic modification capabilities in an object-oriented setting. The definition and interpretation of computational framework are done with the help of a command language. Using the command language, one can execute sequences of functions provided by an application. The Macintosh Hypercard language Hypertalk is a representative of this class of languages.

3. *High Level Application Languages:* As application languages evolve further, programming languages support more English-like syntax. At the 4GL level, one merely tells the application what is to be done, and the program automatically figures out how to do it. The term 4GL encompasses not only the particular programming languages but also the entire set of software-development tools that are associated with these languages. Such tools include:

- Debuggers.
- Text database editors.
- Application and menu generators.
- Forms and report writers.
- Screen layout.
- Presentation systems.
- Data manipulation languages.
- Data dictionary.

4. *Programming Language:* This includes the language and compiler (usually) for defining the functions needed by the application. While FORTRAN is still a popular language in engineering design, C language is dominating many new applications and C++ object oriented extension to C appears to be the choice of many electronic CAD system vendors.

5. *Geometry Engine:* The geometry engine has five principal elements: 3-D wireframe, constructive solid modeling, 3-D solids, rendering and scientific visualization, and interactive photorealistic rendering (see Figure 7.7 of volume I). Solid modeling provides a more natural understanding of proposed designs and makes it easier to discover the relationships among systems, structures, materials, and processes. Moreover, a solid model offers an unambiguous definition of geometry and topology, simplifies computation of physical properties, enables detection of interference, and supports determination of dimensions and tolerances.

Associativity helps to create 2-D drawings from 3-D solid models. Essentially 2-D views are taken off a 3-D solid model and automatically placed in a 2-D drawing. Small changes in the 3-D model can be automatically reflected in a 2-D drawing. A common math-based representation for solids can communicate seamlessly between various modules. For instance, a unified solid modeler can communicate with modules for production drafting, 3-D modeling (including wireframe, surface, and solid modeling), FEA, and NC programming.

6. *Data and Memory Management:* This includes utilities and libraries for managing collections of objects, files, and records in memory, for use by the application languages, or on permanent storage (disk). It can add functions, such as pre-drawn symbols, for other CAD/CAM packages. Relational data bases have been standardized by ANSI. Object oriented disk data-bases (the disk based equivalent of the memory management systems mentioned in Figure 5.10) combined with ionic desktop user interfaces seem to be the most likely technique for management of files in the future. A number of new data base products are emerging and existing relational data base vendors are extending their products with object oriented facilities such as clustering and rules. However, an early resolution in this area seems unlikely.

Gaining in popularity are object oriented data base extensions of memory object managers of languages like C^{++}. Soon, system vendors may provide language-independent object management systems and shared access to common object pools by independent applications. Some sort of run-time object manager will also be required because most existing object managers do not save enough class information for run-time dynamic objects. Object oriented data bases such as ROSE [Hardwick, et.al, 1990] can provide such layers. Because of considerable technical complexity, the development of standards will take some time in this area.

7. *Inference Engine:* Inference engine allows teams to define rules for capturing design intent and processing it in a demand-driven mode. Conventional AI tools emphasize symbolic processing and non-algorithmic inferences. In order that the resulting distributed intelligent environment is powerful, it is imperative that the inference engine has abilities to conduct both heuristic and algorithmic reasoning.

8. *Communications and Networks:* It must facilitate information sharing and enable organizations to coordinate their teams' activities and resources. The idea of an enterprise integration network is to provide electronic exchange of information and services both internally between work groups, teams, and departments of a company, and externally with customers, suppliers, and strategic partners. It must provide electronic access of information to the CE teams on technology, standards, methodology, and 3Ps. Teams should find meeting on the network useful to collaboratively guide and develop the necessary product ideas aimed towards a global product realization strategy.

9. *Standards:* Defining common semantics and schema for the objects are difficult propositions. With a number of viable options emerging, such as the PDES/Express language, some standardization in this area seems likely in the next few years.

A CE developmental environment must permit any application tool to access any product data in the PIM and process it at a workstation, whether or not any of these elements are inherently compatible. This is the underlying theme of a CA framework. A framework tool can be modified or updated any number of times, or replaced with a similar tool from another vendor, and it must still function without special preparation or custom programming. A CE development environment must support standard protocols (for example, DDL for windows that will make them "plug compatible"), allowing teams to request information and services from one another. The possible example tools for various layers of this developmental environment are shown in Figure 5.26

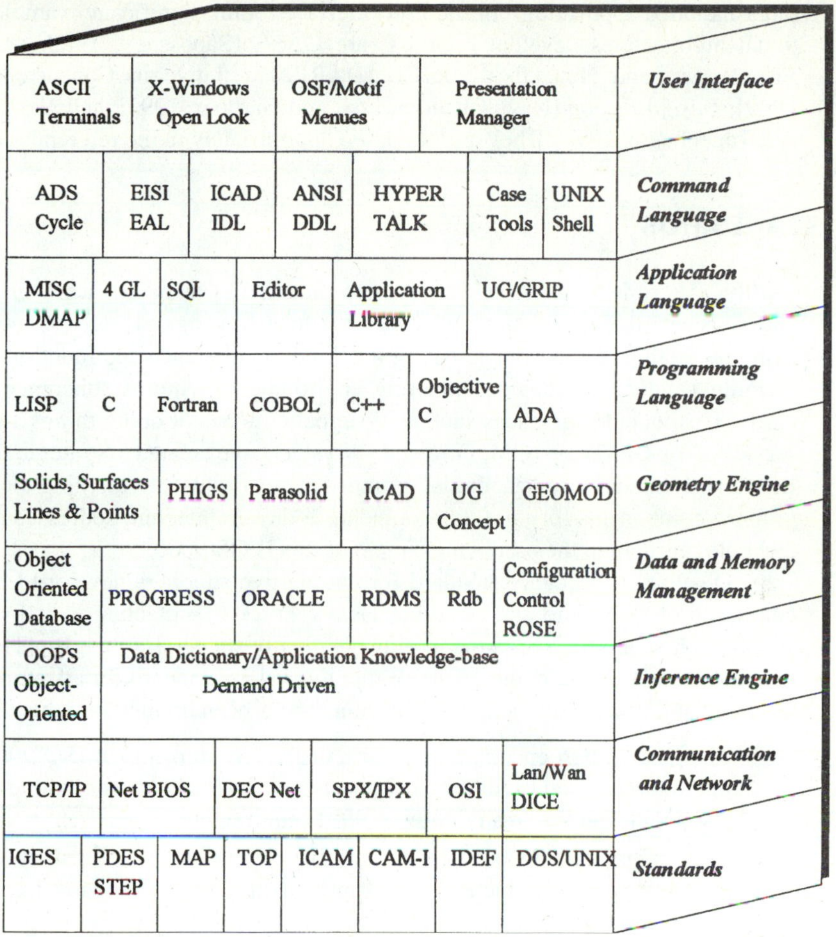

ASCII Terminals	X-Windows Open Look		OSF/Motif Menues		Presentation Manager			*User Interface*
ADS Cycle	EISI EAL	ICAD IDL	ANSI DDL	HYPER TALK	Case Tools	UNIX Shell		*Command Language*
MISC DMAP	4 GL	SQL	Editor	Application Library	UG/GRIP			*Application Language*
LISP	C	Fortran	COBOL	C++	Objective C	ADA		*Programming Language*
Solids, Surfaces Lines & Points	PHIGS	Parasolid	ICAD	UG Concept	GEOMOD			*Geometry Engine*
Object Oriented Database	PROGRESS	ORACLE	RDMS	Rdb	Configuration Control ROSE			*Data and Memory Management*
OOPS Object-Oriented	Data Dictionary/Application Knowledge-base Demand Driven							*Inference Engine*
TCP/IP	Net BIOS	DEC Net	SPX/IPX	OSI	Lan/Wan DICE			*Communication and Network*
IGES	PDES STEP	MAP	TOP	ICAM	CAM-I	IDEF	DOS/UNIX	*Standards*

FIGURE 5.26   Possible Example Tools for A CE Developmental Environment

5.7.3 Models of Computational Architecture

In the last few years many models of computational architecture have emerged. Each defines a consistent interface that has the same look and feel for all applications, thus allowing CE work group members to devote their creativity to perform PD^3 functions rather than learning the infrastructure. An overview of an approach proposed by Lewis [1990], which meets these constraints is shown in Figure 5.8. The basic idea is to implement most product development methods, tools, and advisors as independent servers or libraries. User interface or data management calls are not included as part of these libraries. The work groups and data interfaces are implemented through a collection of open environments based on one or two application frameworks described earlier. Consequently, individual applications will not require extensive modifications for integration into the rest of the open environments. Applications are thus insulated from changes in the frameworks, other included applications, or the user interface. Some significant examples of computational architectures developed for CE are CASE [Sapossnek, Talukdar, Elfes, Sedas, Eisenberger, and Hou, 1989], DICE [DARPA Initiative in Concurrent Engineering, 1989], *Design Fusion* [NavinChandra, Fox, and Gardner, 1993], and *Next-Cut* [Cutkosky and Tenenbaum, 1991]. They are discussed appropriately in the references cited.

5.8 STANDARDS

Ideally, everyone in the strategic business unit (SBU) team and work groups should be able to access the information, wherever it is located geographically, at any point of time from the established sources. Creating this kind of coherence and functionality in a world of multi-vendor, distributed computing environments requires adherence to some standards. This includes practices such as systematic modes of doing things and of providing services. The Big Three (GM, Ford and Chrysler) have created a Strategic Standardization Board to create uniform standards, promote compatibility of emerging technologies, and propose components for joint development. Today architecture addresses more and more technologies than in the past. An example of this is QS-9000.

QS-9000 is a quality standard for automotive suppliers developed by a task force composed of GM, Ford and Chrysler, Heavy Truck Manufacturers, and several tier-one suppliers. It supersedes all of the worldwide quality standards of the Big Three. The intent is to establish a single, comprehensive quality system standard for all suppliers of the automotive and truck manufacturers. The content has been divided into three basic sections.

- The *first section* encompasses 100% of the requirements of ISO 9001, which is an international quality standard. The OEMs have added automotive interpretations and supplements to many of these 9001 requirements.
- The *second section* outlines automotive requirements defined by the Big Three, which includes production part approval process, continuous improvement, and manufacturing capabilities.
- The *third section* identifies customer-specific requirements as defined by the task force. To a large extent, these requirements deal with safety, regulatory, and other special items. A major difference under QS-9000 is the required use of independent third-party registrars/auditors to verify compliance to the quality requirements.

Investments in technology are inevitably expensive. Automakers are hoping to eliminate wasteful duplication of research and development funds, thereby creating even greater advantages over offshore producers. Not very long ago, open systems were meant to signify UNIX and Transmission Control Protocol/Internet Protocol (TCP/IP). Management concerns for uniformity and cost-cutting have led to the adoption of many such standards in the development and implementation of their system solutions. The UNIX and TCP/IP approaches were initial attempts to protect large investments in existing systems, ranging from personal computers to minicomputers and mainframes. Attempts are now being made to provide facilities to move information from plant floor device outwardly to suppliers and customers. These new solutions must withstand the test of time and grow without costly system replacement as needs change.

The evolution in technology in the nineties has moved away from the fully integrated solutions (the so-called "bundled" approach) adopted in the eighties to an open system solution. Vendors are responding to the demands of the work groups by making their products more open so that the CE organization has a choice of buying the best of individual components and plugging them, independently, in any time order. A set of open, multi-vendor CA frameworks that are based on industry standards facilitates many of the above problems. Some of its notable characteristics are:

- *Portable:* An open CE developmental environment enables the development of distributed applications that are portable. *Portable* characterizes an application that can move from one piece of hardware to another without costly mainframe information system (MIS) conversions.
- *Scaleable:* Flexible processing power—as needs change, hardware and software grow without costly system replacement.
- *Same Look and Feel:* This necessitates a consistent user-interface to span disjointed systems. Example: Tuxedo System/T for distributed on-line transactional processing.
- *Common Interface:* It provides a single window to the world. Example: A personal workstation environment with an enhanced security operating system (like UNIX System V release 4.1 ES).
- *Inter-operability:* This includes easy and transparent information movement across various databases, operating systems, networks, and file formats. Example: Network File System (NFS) for distributed file management; or Digital's Network Application Services (NAS) for network integration.
- *Modular with Standard Interface:* This eases communication between plant-floor devices (for example, shop-floor control, purchasing, distribution, MRP) without complex interfaces with minicomputers, PCs, workstations, and mainframes. Example: Seven-layer OSI stack for advanced open systems networking, or integration of PCs into UNIX Networks.

Open systems allow work groups to implement a vast number of applications utilizing a common set of standards, facilitating inter-operability, and enabling hardware and software independence. This is illustrated by the matrix shown in Figure 5.27. Today non-standard proprietary technologies are linked to open standards via networks that are non-standard, yet open. Vendor specific technologies that are non-standard and proprietary, are shown at the lower left. At the upper right corner are the open standards, such as OSI/MAP, UNIX, X-window, PDES (Product Data Exchange using STEP) in the United States and their ISO counterpart STEP (Standard for the Exchange of Product Data), and other standards, such as Email/Internet, ISO/IGES, and so on. The lower right corner contains the other de-facto standards, such as TCP/IP, DECnet, SNA, and so on. These are de-facto standards adopted because of wide use or acceptance. TCP/IP, DECnet from Digital Equipment Corporation, and Systems Network Architecture (SNA) from IBM are examples of de-facto communication standards. Other examples include Windows, Adobe PostScript, and OSF/Motif. They enable computers from different vendors to co-exist on large networks and route data. Another is the Network File System (NFS), a protocol that lets files appear *local* to work group machines. In addition, a service named the

Sources / Nature	Migration of Standards		
	Vendor Proprietary	Company Proprietary	Open (Developed by Standard Setting Bodies)
Standard	IGES Flavors X/Open, Motif SUN's RPC HP's NCS HP's TaskBroker DEC's NAS DEC's PowerFrame VMS, MS-DOS IBM OS/2, PS/2 VM, MVS	ULTRIX IBM/UNIX GM/Infranet GM/MAP GM/C4	OSI/MAP UNIX X-Window PDES/STEP Email/Internet ISO/IGES
Non-standard	Vendor Specific Technology	Company Specific Technology	TCP/IP DECnet SNA (Defacto Standards)

FIGURE 5.27   Migration of Standards

Remote Procedure Call (RPC) permits a program running on one system to use computing resources elsewhere on the network. Vendor proprietary network versions (shown on the top left corner) include Sun's RPC and HP's Network Computing System and its (HP) TaskBroker program; Digital's Network Application Services (NAS) and its PowerFrame program. Vendors' Operating system environment includes, VMS, ULTRIX, MS-DOS, Macintosh, OS/2, VM, MVS, and PS/2. Between these two groups are company-proprietary standard and non-standard flavors of open systems customized to their unique operating conditions, product-lines, and environments, such as ULTRIX, IBM/Unix, GM/Infranet, GM/MAP and GM/C4.

As time passes, many of the underlying standards for software, hardware, and communications will take root. Standard-setting bodies and organizations have already defined and/or are in the process of defining services in the following areas:

• Windows.
• Forms.
• Terminals.
• Graphics.
• Application control.
• Compound document architectures.
• File sharing systems.
• Repositories.

- Printing services.
- Operating system shells.
- Message services.
- Networks.

The American standard bodies include ISO, PDES, American National Standard Institute (ANSI), IEEE, NIST, OSF, UNIX International (UI-Atlas) framework, and X/Open. International bodies include: British Standard Institution (BSI), French Association for Standardization (AFNOR), German Industrial Standards Institute (DIN), and Japanese Industrial Standards Committee (JISC). Applications developed for the X-Window system are hardware independent, that is, the application can ignore the details of graphics display hardware. X-Window is designed specifically as a network-based windowing system. An application can be run on one computer and the X-window interface on another computer connected to the same network. X11 is a common graphics-oriented user interface for applications running on different computers in a network. Standards for tying applications together are X11 and ATIS (A Tools' Integration Standard). Other examples of vendor and company proprietary emerging standards are contained in Table 5.1.

Some industry compliant specification setting bodies have emerged to encourage use of these standards. Examples include the Continuous Acquisition and Life-cycle Support (CALS- earlier known as Computer-Aided Acquisition and Logistics Support) initiative, GOSIP, and the PDES/standard for the exchange of product model data (STEP) for geometry exchange. Unlike the Initial Graphics Exchange Specification (IGES), which covers only the geometric view of product data, STEP addresses both product design and manufacturing data. PDES/STEP goes well beyond previous geometry definitions to include information about features, versions, tolerance specifications, surface finish, and topology.

IGES is a graphics-oriented and file-based exchange specification adopted as an ANSI standard in 1981. Today, it is the world's most widely used product data exchange standard. Its structure and configuration center around the geometrical representation of a part instead of that part's function, tolerances, material make-up, and other relevant data.

TABLE 5.1   Examples of Vendor and Company Proprietary Emerging Standards

Service	Example Standards
Windows	X.11 and Motif
E-mail	X.400
Portable applications across operating systems	POSIX
Distributing computing	Remote Procedure Call (RPC)
File share systems	NFS and AFS
Relational database query	SQL
Electronic commerce communications or EDI	ANSI electronic commerce message sets, known as ANSI X.12
3-D graphics programming	PHIGS (Programmers' Hierarchical Interactive Graphic Standard) and GKS
Plant floor device communications	MMS and MAP

STEP, on the other hand, is written to allow access to data across a product's entire life-cycle concerns and includes database sharing in addition to file-based exchange. Another important evolving standard is PEX. PEX is an extension to the X11 windowing standard in support of PHIGS. Probabilities are high that the standard setting process will continue to churn. Those who are waiting for a *true* open system will fall further behind their manufacturing competitors who have already employed these in their CE process.

REFERENCES

CUTKOSKY, M.R., and J.M. TENENBAUM. 1991. "Providing Computational Support for Concurrent Engineering," *International Journal of Systems Automation: Research and Applications.* Volume 1, No. 3, 1991, pp. 239–261.

DICE, DARPA. Initiative in Concurrent Engineering. 1989. *Red Book of Functional Specifications for the DICE Architecture.* Technical Report (February 28) Morgantown, WV: Concurrent Engineering Research Center, West Virginia University.

DONLIN, M. 1991. "Data Management Tools—the Framework to Concurrent Engineering," *Computer Design*, Volume 30, (April), p. 28.

HARDWICK, M., ET. AL. 1990. "ROSE: A Database System for Concurrent Engineering Applications," Proceedings of the Second National Symposium on Concurrent Engineering, February 7–9, Morgantown, West Virginia: Concurrent Engineering Research Center, pp. 33–65.

KAHANER, D., S. LU, F. KIMURA, T. KJELLBERG, F. KRAUSE, and M. WOZNY. 1993. "First CIRP International Workshop on Concurrent Engineering for Product Realization—Workshop Results," *Concurrent Engineering Research in Review*, Volume 5, pp. 6–14.

LEWIS, J.W. 1990. "An Approach to Applications Integration for Concurrent Engineering," Proceedings of the Second National Symposium on Concurrent Engineering, Morgantown, WV, February 7–9, 1990, pp. 141–154.

MARGOLIAS, D.S., and M.H. O'CONNELL. 1990. "Part 3—Implementation of A Concurrent Engineering Architecture." SME, Paper No. MS90-205, WESTEC'90, Los Angeles, CA, March 1990.

MCKECHNIE, J.L. 1978. *Webster's New Twentieth Century Dictionary*, Collins World.

NAVINCHANDRA, D., M.S. FOX, and E.S. GARDNER. 1993. "Constraint Management in Design Fusion," *Concurrent Engineering: Methodology & Applications*, edited by P. Gu and A. Kusiak, pp. 1–30, Elsevier Science Publishers, B.V.

PRASAD, B. 1985. "An Integrated System for Optimal Structural Synthesis and Remodeling," *Computers and Structures*, Volume 20, No. 5, 1985, pp. 827-839.

SAPOSSNEK, M., S. TALUKDAR, A. ELFES, S. SEDAS, M. EISENBERGER, and L. HOU. 1989. "Design Critics in the Computer-Aided Simultaneous Engineering (CASE) Project." *Concurrent Product and Process Design*, DE-Vol. 21, PED-Vol. 36, edited by Chao and Lu, pp. 137–141, Proceedings of the Winter Annual Meeting of the American Society of Mechanical Engineers (ASME), December 10–15, 1989, San Francisco, CA: ASME

SYMBIOTICS. 1990. *MetaCourier Users Manual*, Symbiotics, Inc.

TAYLOR, D.A. 1993. *Object-Oriented Technology—A Manager's Guide.* Reading, MA: Addison-Wesley Publishing Company.

U.S. AIR FORCE. 1982. *ICAM Project Reports*. Dayton, Ohio: Air Force Wright Patterson Lab (AFWAL).

YEOMANS, R.W., A. CHOUDRY, and P.J. W TEN HAGEN, editors, *Design Rules for a CIM System*, 1985, First Edition, North-Holland, The Netherlands: Elsevier Science Publishing Company.

WHETSTONE, W.D. 1980. "EISI-EAL:Engineering Analysis Language," Proceedings of the Second Conference on Computing In Civil Engineering, New York: American Society of Civil Engineering (ASCE), pp. 276–285.

TEST PROBLEMS—FRAMEWORKS AND ARCHITECTURES

5.1. What is a framework or an architecture? Why do we need a set of frameworks and architectures for CE?

5.2. Does framework constitute an environment in which a product exists or process operates? If not, what is it? What are the relationships between a frame and an architecture?

5.3. What are some basic components that are involved in a typical integration scheme? Describe four schemes for integrating these basic components. Which of these are more flexible and why?

5.4. Describe a set of four network traits that enable integration. Out of these which one is perhaps the most popular and why? What are some network features that enable integration?

5.5. What are some features that are prerequisites for a client/server model?

5.6. Describe a set of four types of interactions that take place between work group members with respect to dimensions of time and space? Which of these are the most popular in a collaborative engineering situation?

5.7. What is a typical approach used in a language-based integration? Describe two approaches to language-based integration. What properties do you require the language structure to possess to enable you to create an integrated, flexible, and efficient language-based system?

5.8. What are two common ways to translate data between two dissimilar systems? How many translators do you need to translate data among "n systems" in each case? Which one is more efficient and why? Which one is less reliable and why?

5.9. Describe two types of information processing. Why is distributed processing more popular than traditional computing? Describe a scenario of distributed computing environment where work group networks are linked with design center networks, manufacturing center networks, and also to an enterprise network.

5.10. What constitutes a distributed application environment? Describe an analogy of transportation to various types of computing platforms used in distributed computing?

5.11. What is a work group computing? How is it different from a distributed computing? Describe a shift in key characteristics when moving from a personal computing to a work group computing. Does the shift pull everyone, teams, computers, networks, into one flexible easy to use client/server based system?

5.12. How are companies meeting the challenge in providing a seamless connection between parallel work groups and computing machines (brand names)? Is work group computing an answer to this challenge? If not, what is?

5.13. Describe a scenario when four work groups, engineering, design, prototyping and manufacturing, could work together. How would they be able to communicate with each other, each with their own work group computing system? How are the changes that affect the design outcome are propagated throughout the CE organization?

5.14. Describe a typical distribution of activities between a client and a server side that may take place in CE-based cooperating work group sessions. Assume that a variety of tools and peripheral equipment, facilities, office automation machines, computers and networks are part of this CE computational environment.

5.15. What is a product information management (PIM) system? What is its function? What are some typical problems encountered during a PIM session? How do you overcome these problems?

5.16. What is a set of PIM requirements? What are some basic components of PIM? Describe each briefly one by one.

5.17. Describe how OOPS helps encapsulate procedures and data together in an object oriented data base management system. Why is this an ideal vehicle for a CE application?

CHAPTER 6

CAPTURING LIFE-CYCLE INTENT

6.0 INTRODUCTION

Concurrent engineering is sometimes looked upon primarily as a management-based product design, development, and delivery (PD^3) philosophy as described in Chapter 5 of volume I. However, the management philosophy of CE cannot, by itself, yield the needed benefits in terms of time to market or responsiveness. The computer-aided environment of CE is definitely the most important approach to realizing its (PD^3) full potential. One of the purposes of CE is to capture the life-cycle intent, that is, the knowledge about the PD^3 process. Knowing how a product is designed, how it functions, and how it will be delivered is necessary so that the work groups can leverage this knowledge of life-cycle concerns to upgrade the design as the product moves from one domain to the other.

Except in a few rare cases, products are now so complex that it is extremely difficult to *capture* correctly their life-cycle intent the first time, no matter what C4 (CAD/CAM/CAE/CIM) tools, productivity gadgets, or automation widgets are used. Traditionally, CAD tools have been primarily used for activities that occur at the end of the design process. Such usage of CAD tools, for instance, during detailing a geometry of an artifact, is in generating a production drawing, or in documenting geometry in a digitized form (see Figure 6.1a). CAM systems are conventionally used to program machining or cutting instructions on the NC machines for a part whose mock-up design, clay, or plaster prototype may already exist. CAE systems are used to check the integrity of the designed artifact (such as structural analysis for stress, thermal, etc.), when most of the critical design decisions have already been made. Studies have revealed that 75% of the eventual cost of a product is determined before any full scale development or a CAD tool usage actually begins [Nevins and Whitney, 1989].

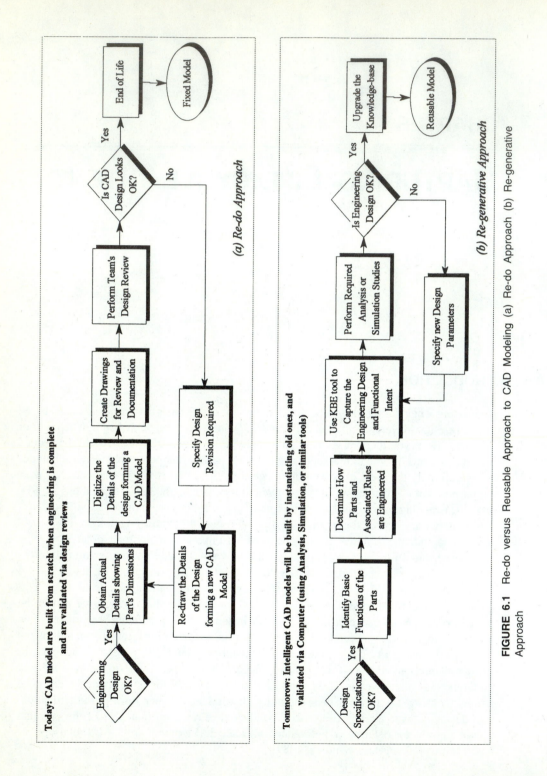

Today: CAD model are built from scratch when engineering is complete and are validated via design reviews

(a) Re-do Approach

Tommorow: Intelligent CAD models will be built by instantiating old ones, and validated via Computer (using Analysis, Simulation, or similar tools)

(b) Re-generative Approach

FIGURE 6.1 Re-do versus Reusable Approach to CAD Modeling (a) Re-do Approach (b) Re-generative Approach

Most C4 tools in use today are not really *capture* tools. The need for a *change* after a design model is initially *built* is all but inevitable. Today, CAD models are built from scratch only when engineering activity is complete and outcomes are validated via a series of design reviews (see Figure 6.1a). Design work groups typically document the design through a CAD software only after the completion of major engineering processes and only after resolving all of the pressing engineering issues. A work group captures the geometry in a static form, such as lines and surfaces. Static representation is actually a *documentation* that tells a designer what the final design looks like but not how *it has come to be*. If changes are required in the design, a new CAD model is recreated (see Figure 6.1a) using some types of computer-aided redo or backtracking methods. Such CAD methods of activating change or modification (e.g., a redo or a backtracking) can be extremely time-consuming and costly that late in the life-cycle process. In such static representations of geometry, configuration changes cannot be handled easily, particularly when parts and dimensions are linked. In addition to the actual process that led to the final design, most of the useful lessons learned along the way are also lost. In the absence of the latter, such efforts have resulted in loss of configuration control, proliferation of changes to fix the errors caused by other changes, and sometimes ambiguous designs. Hence, in recent days, during a PD^3 process, emphasis is often placed on the methods used for capturing the life-cycle intent with ease of modifications in mind. The power of a capture tool comes from the methods used in capturing the design intent initially so that if warranted the anticipated changes can be made easily and quickly. By capturing a design intent as opposed to a static geometry, configuration changes could be made and controlled more effectively using the power of the computer than through the traditional CAD attributes (such as lines and surfaces). *Life-cycle capture* refers to the definition of a physical object and its environment in some generic form [Kulkarni, Prasad, and Emerson, 1981]. *Life-cycle intent* means representing a life-cycle capture in a form that can be modified and iterated until all the life-cycle specifications for the product are fully satisfied. *Design-capture* likewise refers to the design definitions of the physical objects and its surroundings. *Design-intent* means representing the *design capture* in a form (e.g., a parametric or a variational scheme) that can be iterated. Design in this case means one of the life-cycle functions (see Figure 4.2 of volume I). In the future, CAD models will be reusable. The new models will be built by instanciating the old ones and validating them via computer (using simulation, analysis, sensitivity, optimization, etc., see Figure 6.1b). Such models will have some level of intelligence built into them [Finger, Fox, Navinchandra, Prinz, and Rinderle, 1988].

6.1   DESIGN CLASSIFICATION

Notwithstanding the life-cycle capture, applying knowledge uniformly across various design descriptions is at best a difficult task. The way in which one designs a product changes as one's knowledge about that product is enhanced. The most common design descriptions, however, have not changed for a long time [Finger and Dixon, 1989]. Following works of Pahl and Beitz [1992], one may classify design processes as follows:

- Routine or fixed principal design.
- Variant design.
- Adaptive or innovative design.
- Original or creative design.

Figure 6.2 shows, through a Venn diagram, the existence of the above classifications in a design space.

6.1.1 Routine or Fixed Principle Design

These are design situations where the generic aspects of the design are well understood [Kulkarni, Prasad, and Emerson, 1981]. All the design variables and their possible ranges are known and the problem is one of instantiation. Instantiation involves setting values for the attributes of the design that have been identified previously as variables. Setting values involves selection of relevant design variables of a *given subtype*, selection of relevant design criteria and subsequent determination of their values based on the set criteria. New designs are created by a series of instantiations and criteria evaluations. No negotiation is necessary in this process. A similar concept was earlier proposed around the notion of Primitive Generic Components (PGCs) [Kulkarni, Prasad, and Emerson, 1981]. PGCs

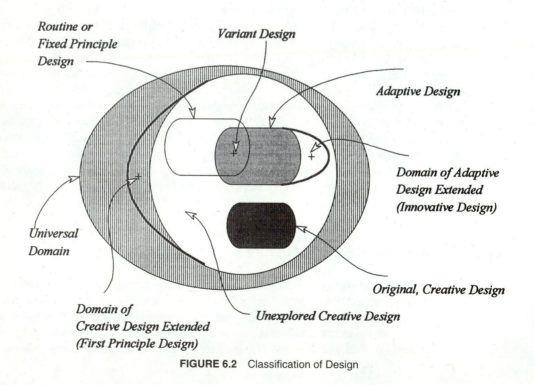

FIGURE 6.2   Classification of Design

represent design principles, such as the principle of a beam, at a certain level of abstraction. They can be considered as building blocks of a routine or fixed principle design. Design descriptions built from PGCs integrate a variety of views of interest coming from different members of a CE work group.

6.1.2   Variant Design

For this type of design, new designs are created by combining existing parts or features of previous routine designs in some way. In this case, the basis for common understanding about information and processes among different work groups has not changed. The need to re-negotiate, in this case, is minimal and is due only to the extent of changes made to the design variables or design criteria previously imposed on the routine designs. Detailing involves the retrieval of some *specific design subtype* based on the closest (or at least good) match between actual and required behavior attributes. Most variant design is *by mutation*—extrapolating the domain of design beyond its routine environment (beyond a given range of design variables or design criteria). The range of values for the existing design variables is either extended, or the original criteria on which basis a routine design outcome was adjudged, is changed, or varied. The instantiation process is repeated with slightly varied design variables or criteria to get a variation of a routine design. The rest of the functional logic for tying the criteria and the design variables remains intact. The existing design approach, as well as specific parts of previous designs, are kept the same.

6.1.3   Adaptive or Innovative Design

Adaptive or innovative design involves the generation of new design subtypes by adding new parameters (design variables) or adding theory to the original set spanned by routine design. In this case, each new design is *an adaptation of an existing design.* The process is one of identifying an adaptable aspect of the original design and some new aspects (e.g. new design variables) that is different from the routine or variant design. Since new design variables or criteria are added here, in this case, the space of known solutions is extended by making new adaptation to the variant design or subtype. That is, designs that were not previously known are produced although there is no departure in kind from previous designs of the same type [Rosenman and Gero, 1993]. The team may choose to enhance (e.g. increase the range of the design variables) or slightly refine the information (e.g., change a few criteria that were previously employed for the routine design) or the process used to produce the product. The CE work groups may also generalize the design in such a way that it can be applied to a wider variety of applications. In any case, adaptive design involves a systematic refinement of the design criteria, design processes defined by some set of transformation and their parameters. Since the basis for common understanding about information and processes among diverse work groups can change in this situation, there is a need to re-negotiate amongst the work groups.

6.1.4 Original, Creative or First Principle Design

Creative design involves *generation of entirely new configuration subtypes* (no relationship to the old types). Creative design incorporates innovative design but involves creation of products that have little obvious relationship to an existing line of products. Creative design is the creation of new structure in response to functional requirements [Rosenman and Gero, 1993]. Creativity is concerned with exploration within a space that is only partially defined or known. Product specification information is somewhat sketchy, existing product design information is very sparse, and the design process is not very well understood. The processes are obscure (e.g., intuitive or creative) and the domain knowledge is mostly in implicit forms. The concurrent team must be concerned with issues such as defining necessary information and developing procedures to produce this information. Once this information is produced, the work group members must have a way of assessing original assumptions to see if they still apply and be ready to iterate on these assumptions if necessary. In addition, the design team must negotiate with other work group members to ensure that the information and procedures that the design team has developed so far are consistent with other work groups and the rest of the life-cycle.

The above four design classifications are captured in Figure 6.3. As noted, routine

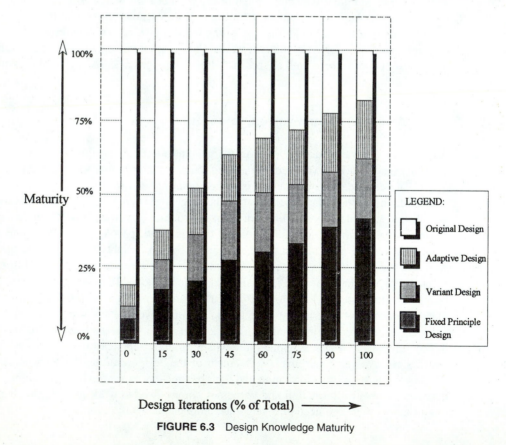

FIGURE 6.3 Design Knowledge Maturity

or fixed principle design is the most suited for automation. Original design (or first principle design) is best suited for knowledge capture. Adaptive and variant designs help to solidify the generality of a design and bridge the gap between original and fixed principle.

Routine or fixed principle design has more potential for automation since it contains stable design information such as features and parameters, and because design processes (in fixed principle) are well defined. However, in order to achieve a better return on investments from automation, some level of life-cycle capture must start during the original (or first principle) design and must be refined during adaptive and variant designs. During this refinement process, a design description can be considered as a set of inputs, outputs, requirements, and constraints. It is necessary to have specific tools available to the CE design teams for each type of design, and these tools should reflect the specific nature within each design type. Bowen and Bahler [1993], for example, have looked into developing a multi-domain tool using a constraint language, Galileo3, using those design classifications and their respective descriptions that can be modeled. Galileo3 supports an interactive design process in which designers can add new constraints. The language allows the generation of a series of progressive design views for the CE work groups.

6.2   LIFE-CYCLE CAPTURE

Design intent is a relatively simple concept, but the process of capturing life-cycle rationale is complex [Klein, 1993]. There are many known ways of capturing the life-cycle intent of a manufacturable product. As shown earlier, design and analysis languages [Whetstone, 1980] provide some useful tools to capture not just the design (documentation or detailing of a physical part) but its life-cycle rationale—the logic, relationships and the knowledge behind the functional intent of the product. Examples of such commercial language tools are Engineering Analysis Language (EAL) [Whetstone, 1980], Patran Command Language (PCL) [PDA, 1992], and Intelligent Design Language (ICAD/IDL) [Rosenfeld, 1989; ICAD, 1995]. The important elements of a life-cycle intent tool are:

1. Flexibility in capturing the decision tree structure in terms of spec-features (input-parameters, requirements, and constraints).
2. Flexibility in relating major keywords with user actions to define the life-cycle specifications. This is shown in Figure 6.4 by a five-step process. These steps are:
   Definition of Decision Tree Structure.
   Definition of Features.
   Definition of Physical Objects.
   Definition of Functional Intent.
   Definition of Symbolic Relationships.

6.2.1   Definition of Decision Tree Structure

Hierarchical or decision tree structure has been known to the process planning community since the 1970s. Brigham Young University pioneered the use of this concept for generative process planning (GPP) in 1978. GPP is a highly structured process very similar to

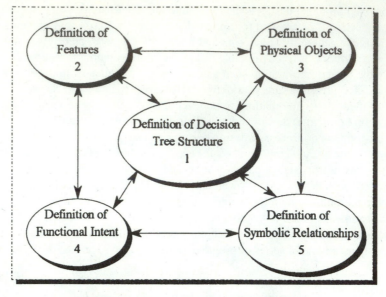

FIGURE 6.4   Steps in the Life-cycle Intent Capture

writing a software program, but with a higher level of granularity. In GPP, a user structures the control flow of the decision rules. The same technique can be applied to set up generically a decision tree logic for each grouping of parts (see Figure 6.5a). This technique is most easily applicable to products with a small number of parts and a reasonably small number of operations per part. One can establish and test a decision tree relatively quickly. As design progresses from early stages to detail design, the decision tree logic grows not just in volume but also in complexity. Many manufacturing companies have successfully implemented complex decision tree logic. With increasing product complexity, users are often required to perform a multitude of dependent decision steps. Calls to external programs are common, along with data storage to an external database.

6.2.2   Definition of Features

Features are named entities that encapsulate how geometry behaves using rules and attributes in addition to their defining constraints and geometry. The most practical definition of features is provided by Dixon and Cunningham [Dixon, 1988; Cunningham and Dixon, 1988]. They define a feature as "any geometric form or entity that is used in reasoning one or more design or manufacturing activities" and as "an entity with both form and function." Features mimic the way engineers think, allowing concurrent teams to communicate with the CAD system using terms familiar to the work groups. Definition of features leads to capturing the life-cycle intent and making the description of the problems more user-friendly. It helps effectively organize both geometric and non-geometric product data. Features may

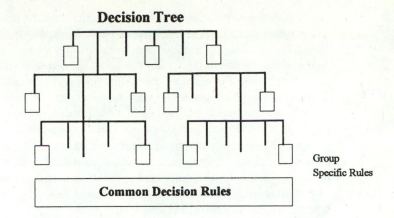

Decision Tree

Group
Specific Rules

Common Decision Rules

(a) Definition of Decision Tree Structure

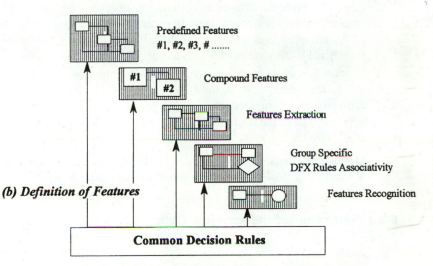

Predefined Features
#1, #2, #3, # ....

#1 **#2**   Compound Features

Features Extraction

Group Specific
DFX Rules Associativity

(b) Definition of Features

Features Recognition

Common Decision Rules

FIGURE 6.5   Definition of (a) Decision Tree Structure (b) Features

also have attribute information that can be used by other life-cycle processes. For example, addition of form feature information to a product model can provide a mechanism to establish a mapping method to relate product geometry (form features) to manufacturing processes [Henderson, 1986]. Features are identified in a number of ways.

- *Predetermined Features:* One way to ensure the availability of features during design is to parse the original object with a set of its previously defined features. This is often known as a *design with feature* approach. This allows the team to build an object with simple operators (such as add, delete, modify, position, move, etc.) [see Cunningham and Dixon, 1988; Dixon, 1988; Shah, 1991; and Finger and Dixon, 1989]. The advantage of this approach is that features are explicitly available for DFX type evaluations where DFX rules can be associated with the attributes of features (see Figure 6.5b). For example, both blind-hole and through-hole features could be defined using a predefined Boolean difference operation between a part and a cylinder. Some limitations may result from the fact that freedom or flexibility of creating different shapes and sizes is restricted by the availability of predefined features in the library.

- *Compound Features:* Compound features are obtained by combining some existing predefined features. However, when one does that, it is difficult to associate DFX rules. It is relatively straightforward to combine existing features, but since the resulting combined features are not a predefined feature, it is difficult to associate with them a manufacturing evaluation rule.

- *Dynamic Feature Extraction:* This allows the teams to define the design more artistically and later extract the desired features. Dynamic feature extraction allows the identification and selection of topology of complex details like bosses and pockets as features during the design process, without, a priori, having created these as features. However, the science of feature extraction has not advanced to a point that ambiguities between common features can be eliminated. Comprehensive reviews of feature extraction techniques can be found in the paper by Joshi and Chang [1990] and in the state-of-the-art report by Chen and Miller [1990].

- *Feature Recognition:* Feature recognition is primarily a process of pattern matching. In feature recognition, the system scans the points, lines, surfaces or solids information and compares them with a set of primitive feature library. Depending upon the type of the decision rules, one class of features might be better suited than others (see Figure 6.5b).

$$\text{Methods of Features Identification} = \cup \ [\textit{Predetermined Features, Dynamic Feature Extraction, Feature Recognition}] \tag{6.1}$$

6.2.3 Definition of the Physical Objects

This includes capturing the definition of the product in terms of the following objects:

Parent class.

Physical attributes.

Form features.

Children parts.

Query attributes.

Rules or relationships.

Initial default values.

Inputs, part type.

Referencing chain messages.

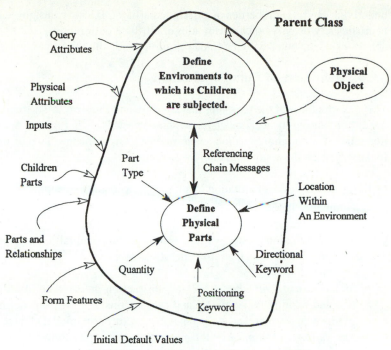

FIGURE 6.6 Definition of the "Physical Object"

Other information (see Figure 6.6).

Environment to which its children are subjected:

• Part's physical configuration.
• Topology.
• Children's attributes.
• Positioning keywords for placing or assembling parts.
• Directional keywords for orienting the parts within an assembly.

Examples of "physical attributes" include parameters and constraints, such as geometric relations, engineering equations, positions (top, bottom, front, rear, left, right, etc.), orientations (e.g., translate, rotate, transform, etc.), and non-geometric relations. An object oriented programming (OOP) approach is most flexible from the point of view of the product's instanciation and code reuse, as well as future modification and maintenance. A OOP developer comes up with a set of reusable components that exist physically and defines how this new physical object class will differ from the old class. The object of capturing the definition of a product using OOP is to conserve corporate memory and facilitate design reuse. In traditional programming, "commonalty among processes" is usually an afterthought instead of a planned activity in the PD^3 process. Attributes in OOP can present form-feature objects. A form-feature object is actually a pseudo-part of a formed object that is physically differentiable from the rest of the attributes but adds certain functional details [Henderson, 1986]. Features are hierarchical and are composed of primitive features, constraints, or the underlying physical properties of a designed artifact. Because design standards and expert heuristics rely heavily on characteristics of attribute definitions, an OOP approach must be able to share attributes with the CAD primitives and with the form-features' library.

6.2.4   Definition of Functional Intent

This includes three of the five steps that are shown in Figure 6.7. The first two steps, including definitions of physical objects, were shown previously in Figure 6.6. Definition of functional intent is commonly called an alternative design change. Two of the main strategies for speeding the design of a product are to reduce:

1. Cycle-time of each iteration in a life-cycle process.
2. Number of iterations.

Capturing the functional intent of a product has the potential to contribute in both areas. The corresponding actions that affect iterations are:

- *Model Abstraction Change:* Model abstraction involves defining a design perspective and relating its keywords and user actions to a previous definition—definition of physical objects. Design perspective captures the model change in terms of parameters (e.g., design variables, methods, attribute examiner, query report, graphic display, output specifications, etc.). It captures the intent information at a functional (meaning concept) level, rather than source code, since the latter is firmware or tool dependent. Function is the key concept to integrate object modeling with process modeling. Capturing involves adhering to a certain protocol or structure while capturing the format or information. This also allows members of the work groups to describe their work as it develops in logical terms. If a candidate model of a previously working design could be recorded in a generic form, then information can be reused to eliminate iterations in future designs involving similar works [Kulkarni, Prasad, and Emerson, 1981].
- *Individual Parameter Change:* Individual parameter change involves capture of the iterative process itself. It is based on an understanding of existing good engineering practices and rules of thumbs including work habits. If some critical trade-off analysis and rules occur in the design process repeatedly and that leads to a converged solution or acceptable behavior, iterations can proceed faster if such procedures can be captured. This is accomplished by a three-step design procedure.
  - *First*, create a set of trials or iterations (with goals to be satisfied) for design modifications (perturbations).
  - *Second*, set up test criteria, convergence rules, trial attributes, etc.
  - *Third*, define the initial values of the trial attributes so that iteration can start from this point. A trial-iteration with trial attributes is set up to perturb the design and make it available for other applications. Trial is repeated, new trial values are assigned, and conditional checks performed.

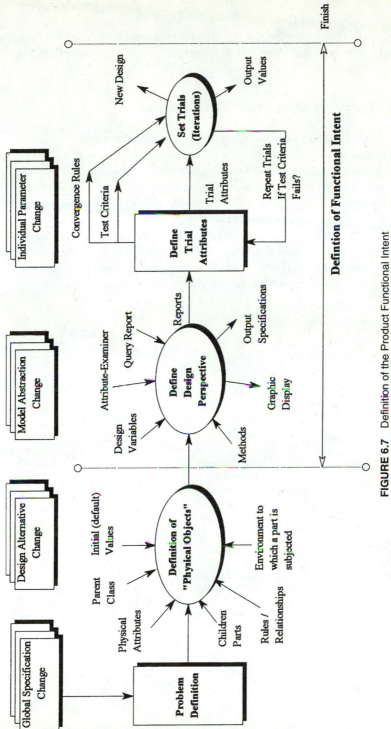

FIGURE 6.7 Definition of the Product Functional Intent

273

The process stops when the trial criteria or the convergence criteria are satisfied (see Figure 6.7). The resulting output is a converged set of attribute values for an improved design.

6.2.5   Definition of Symbolic Relationships

In the previous two definitions, relationship is captured through symbolic referencing and instanciation techniques. Defining symbolic relationships provides for maintenance of the aforementioned definitions and variables. Figure 6.8 shows a list of user-defined parts and primitive parts available with a command language library. Examples of user-defined parts are points, lines, vectors, curves, surfaces, quantified parts, and so on. Primitive parts are broken into wire-frame, feature primitives, parts assemblies, and volumetric assemblies. Ribs, fillets, chamfers, holes, slots, and bosses are just some of the feature primitives that can be defined. Other forms of primitive parts are volumetric primitives (box, cylinder, cone, etc.) and wire-frame primitives (points, lines, polylines, arcs, circles, ellipses, etc.) A part assembly results from combining two or more of these primitive

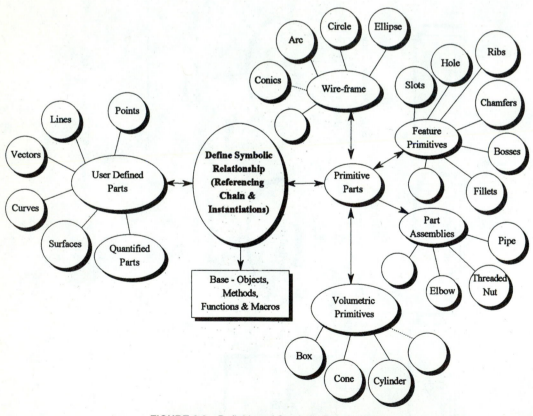

FIGURE 6.8   Definition of Symbolic Relationships

parts. For example, a pipe is obtained by revolving a hollow cylinder along a 3-D curve arc. A threaded nut is obtained by combining some of the feature primitives with volumetric primitive parts. A filleted knot is obtained by filleting a cylinder with a sphere. The symbolic relationships have intelligence enabling primitive parts to be placed and positioned using parameters and mating constraints. Such mating constraints may include horizontal and vertical leveling, parallelism, pependicularity, tangency, concentricity, coincidence, and so on. Constraints ensure associativity. When a parametric model changes, the parts stay connected to the model in a predefined way. For instance, a rib knows it must stay connected to the sides when the sides move. A definition of any part within a primitive parts family can be retrieved by sending a message through an instanciation process.

Using the five-step process discussed above (see also Figure 6.4), a concurrent team can capture the complex life-cycle intent of a product in a *generic* form. Such life-cycle manufacture knowledge of its constituent parts can be captured in terms of a generic description called *generative constituents*. The knowledge may be in the form of attributes, symbolic referencing, or parent-child relationships. A set of *generative constituents* can then be used to make-up the entire product by collecting the required information and the constituent product relationships. The engineering and manufacturing knowledge required to translate a set of *generic constituents* into a specific product instance is embedded in the *total product model* (TPM). Information may include product knowledge, such as geometry, material type and performance, and/or process knowledge—the processes by which the product is analyzed, optimized, manufactured, assembled and tested. Once required information has been collected and stored in a total product model, the real advantages of a TPM become apparent. Design work groups can generate and evaluate new designs quickly and easily by changing the input specifications in a TPM. Generic TPM is useful since different perspectives of designs can then be replicated merely by specifying a new set of values for its specifications [Kulkarni, Prasad, and Emerson, 1981]. One of the biggest challenges is to identity and to define a generic TPM that can provide the greatest flexibility in capturing all possible product and process variations with the least amount of reprogramming or computer recoding.

6.3   LANGUAGE FOR LIFE-CYCLE CAPTURE

Languages are means of capturing the knowledge for the design and development of a product. Models are the results of such knowledge capture. The primary goal of a knowledge-capture formalism is to provide a means of defining ontology. An ontology is a set of basic attributes and relations comprising the vocabulary of the product realization domain as well as rules for combining the attributes and relations. Engineering Analysis Language (EAL), for example, provides a means of creating analysis or design models as run-streams. Later, they form the basis for iterative analysis and design [Whetstone, 1980]. ICAD/IDL, on the other hand, captures the knowledge about the process of designing and developing a product [Rosenfeld, 1989; ICAD, 1995]. There are three types of languages that can be employed to capture a life-cycle intent.

1. Geometry-based language.
2. Constraint-based language.
3. Knowledge-based language.

Figure 6.9 shows an evolution of languages for capturing knowledge over a thirty year time period. These are C4 (CAD/CAM/CIM/CAE) specification languages for product engineers or designers to define configurations of parts and assemblies. They are not computer languages for software programmers (such as C, or C++). During this thirty year period, there had been tremendous innovation.

- The *first generation* of C4 languages, first introduced during 1960, only dealt with 2-D drafting and 2-D wire-frame design.
- The *second generation* of C4 languages dealt with surfaces and 3-D solids.

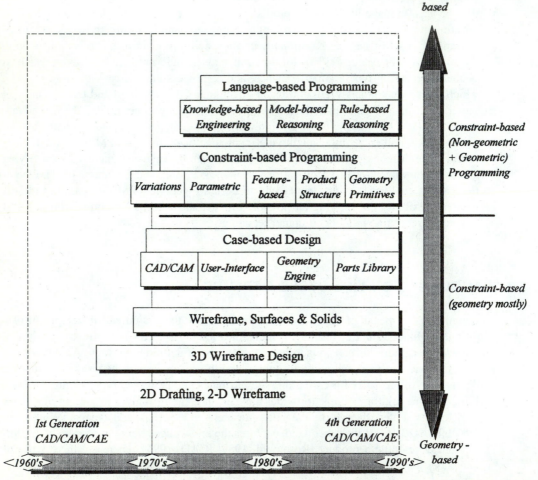

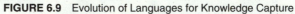

FIGURE 6.9   Evolution of Languages for Knowledge Capture

- The *third generation* of C4 languages was constraint-based but mostly dealt with geometry. Examples include case-based design, parametric scheme, variational scheme, and so on. From 1980 onward, there was a history of development in making C4 codes more user friendly, use of a solid-based geometry engine (CSG versus B-Rep), and introduction of part library concept. There was also a flurry of activity in the use of techniques, such as parametric schemes, variational schemes, featured-based concepts for creating product structure and for defining geometric primitives [Finger, Fox, Navinchandra, Prinz, and Rinderle, 1988].

- Then came the *fourth generation of languages.* Today is the age of fourth generation C4 languages, quite different from past languages. Fourth generation languages are knowledge-based techniques giving CE design work groups the ability to capture both geometric and non-geometric information.

$$\text{Languages for Life-cycle Capture} = \cup \text{ [Geometry-based language,} \atop \text{Constraint-based language, Knowledge-based language]} \tag{6.2}$$

Knowledge-based Systems (KBS) are software programs designed to capture and apply domain-specific knowledge and expertise in order to facilitate solving problems. Languages can be used as a means to build KBS. Knowledge-based Engineering (KBE) deals with processing of knowledge. There are many ways to capture knowledge to control its processing. KBE is a process of implementing knowledge-based systems in which domain-specific knowledge about a part or a process is stored along with other attributes (geometry, form features, etc.). A computational architecture for a typical knowledge-based system is shown in Figure 6.10. It consists of five layers, each layer supports the others.

- *Environment:* The *first layer* is an environment that provides a foundation for the rest of the layers. A typical environment consists of a slew of operating systems, standards, and compute platforms (workstations, hardware, etc.).

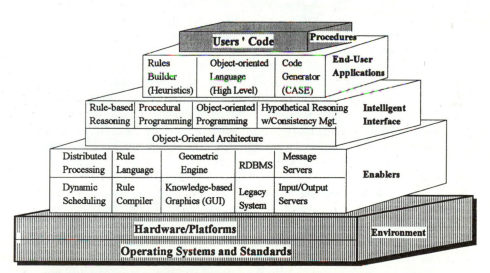

FIGURE 6.10   A Computational Architecture for a Typical Knowledge-based System

- *Enablers:* The *second layer* consists of core enablers. Some of the key enablers included at this layer are: distributed processing, dynamic scheduling, real time rule language, rule compiler, knowledge-based graphics, GUI, geometry engine, relational database management system (RDBMS), legacy system, input/output servers, message servers, and so on.

- *Intelligent Interface:* The *third layer* adds intelligence to the enabling tools (second layer) and gives a programming interface to build the end-user applications. Most knowledge-based engineering tools encompass five critical technologies within an object-oriented architecture:
  - Rule-based reasoning.
  - Procedural programming.
  - Object-oriented programming.
  - Hypothetical reasoning with consistency management.
  - Case-based reasoning.

- *End-user applications:* The *fourth layer* is composed of end-user applications. It consists of a high level object-oriented language, a rule builder, and a code generator similar to a CASE tool.

- *Procedures:* The *topmost layer* embodies the procedures for the users' code.

6.3.1   Geometry-based Language

In the past, knowledge about products was mainly present in the form of geometry. Now 3-D solid geometry, as opposed to surfaces, wire-frames, and other forms of geometry, is being used more often. Most traditional languages are geometry-based. They capture the attributes of solid primitives including lines and curves of a modeled object and their relationships to each other. Some high-end languages also capture information about the space inhabited by an object or about its enclosure (e.g., Constructive Solid Geometry—solids). Some modelers develop complex solids by adding an extension to the traditional Boolean (join, intersect, and subtract) operations. For example, a combined solid can be driven by a 2-D sketch. As illustrated in the IDEAS Master series [1994], the sketches can also be driven by geometric elements of other solids. Most solid modelers, however, fail to draw on knowledge about what the object is, its relationship to other objects or components, or its life-cycle aspects. Constraint-based CAD programs speed the design-change process by controlling and constraining object relationships based on dimensions (size, orientations, etc.), positioning, or geometrical inputs. However, such programs still focus on the geometrical aspects of the product development, not the knowledge about its life-cycle manufacture.

6.3.2   Constraint-based Language

Constraint-based language provides facilities for defining constraints. Most constraint-based languages provide means of incorporating arithmetic, logical functions, and mathematical expressions within a procedure. Such constraints may have a simple, linear algebraic relationship between entities to control shape (e.g., the length of line A is twice the length of line B) or geometry. The examples of geometric relationships include horizontal and vertical leveling, parallelism, perpendicularity, tangency, concentricity, coincidence, and so on. Some constraints provide means to define and solve a system of linking equations that constitute a set of necessary design constraints and bounds. Finite element analysis and sensitivity analysis are some of the options that are generally considered an integral part of a constraint-based language. Such languages encompass command structures, symbol substitution, user-written macros, control branching, matrix analysis functions, engineering data base manager, and user interface to integrate complex multi-disciplinary analysis, design, and pre- and post-processing work tasks.

Finite element systems usually consist of:

(a) A set of preprocessors through which a team defines finite element meshes, applied loads, constraints, etc.

(b) A central program that primarily performs numerical computation.

(c) A set of post-processors for displaying the results.

They do not provide instantiation needs.

An issue often encountered in a conventionally structured program is how can CE work groups add additional capabilities that go beyond what the FEA programs provide. If it is necessary to perform functions that are not explicit capabilities of a program, the only recourse available to the teams is a very expensive and time-consuming one. The product developer has to write a new module (in a conventional language, such as FORTRAN, C or C^{++}) that operates on an output data file produced by the finite element analysis program. The constraint-based language largely eliminates this difficulty, making it very easy for teams to integrate complex and highly specialized analysis and design tasks, and to create specialized input formats and output displays. Another class of constraint-based languages that use AI techniques is based on solving a constraint satisfaction problem (CSP). Formally a CSP is defined as follows [Kumar, 1992]: "Given a set of n variables each with an associated domain and a set of constraining relations each involving a subset of the variables, find an n-tuple that is an instantiation of the n variables satisfying the relations." In the CSP approach, most of the efforts are in the area of solving a constraint satisfaction problem automatically. Most design problems, on the other hand, are open ended problems. They are evolutionary in nature requiring a series of frequent model updates and user interactions, such as what is commonly encountered during a loop and track methodology (discussed in section 9, volume I). How to manage such team interactions in the CSP approach has been the topic of research in the AI community for some time.

6.3.3 Knowledge-based Language

In a knowledge-based engineering (KBE) system, work group members use a design language to build a smart model of the product. A formalism for defining smart models is a knowledge-based representation scheme for describing the life-cycle domain knowledge.

KBE languages go well beyond parametric, variational, or feature-based geometry capture mode to a knowledge-based life-cycle capture mode. A design work group does not just design parts. Work groups design products—a collection of functional parts placed in an assembly to form a finished (that is, a functional) product. A KBE language provides ways to capture geometry and non-geometric attributes, and to write the rules that describe the process to create the assembly. Such rules might include design stresses, resultant volume, or other parts' positioning, mating, and orientation rules. These rules form a part of an intelligent planning procedure derived from a domain specific knowledge. The use of intelligent planning procedure in KBE replaces an exhaustive enumeration of all feasible assembly plans that would have been needed otherwise. Most of the present KBE languages use object-oriented techniques. Unlike constraint programming languages that define procedures for the manipulation of objects and entities, knowledge-based languages define classes of objects, and their characteristics and behaviors that possess built-in manipulation capabilities. KBE languages capture the totality of the functions and relationships between model elements.

The following are some of the characteristics of a knowledge-based language.

- *Object Symbols or Attributes:* In a KBE language, object symbols or attributes are the backbone of the system. Attributes describe an object geometry, overall physical parts, its environments, location of the parts within that environment, and any other characteristics that are required. Figure 6.6 shows a definition of a physical object and a few examples of some associated attributes. Some attributes that are fixed are defined as constant-attributes. Variable attributes are design specifications whose values change. Inputs and children of an object are considered as variable attributes. Most of the physical attributes are fixed attributes. KBE languages allow a team member to define attributes in any order but they are internally recognized as *keywords*. Directional and positioning keywords are keyword examples shown in Figure 6.6. Keywords enable demand-driven-operations to take place [Rosenfeld, 1989; ICAD, 1995].

- *Demand-driven Operations:* In this mode the system determines the *order* and the *necessity* to evaluate an attribute. If a value for an attribute is demanded for the first time, the system computes the value and remembers it. In subsequent operations of the same attribute, when its value is required, the system returns the cached value instead of recomputing it. This method of evaluation, called demand-driven evaluation [Rosenfeld, 1989], is considered an important property for recalculating the value each time it is demanded. This type of operation relieves the programmer from assigning the order in which to evaluate the attributes. This makes programming in KBE languages significantly easier than in other languages.

- *Frame Structures with Rules:* Frames are object-oriented structures that allow for the storage of attribute information as object hierarchies. Frame representation is, thus, convenient for the storage of geometric dimensional and quantitative knowledge. Rules are used to implement the procedural expressions. The combination of an object-oriented frame structure with rules results in an adequate framework for capturing life-cycle knowledge.

- **Symbolic logic:** Symbolic logic is an underlying logic theory used in KBE to let knowledge engineers represent and manipulate the various types of knowledge required in CE. Symbolic logics are composed of object symbols (attributes), predicates, frame structures with rules, classes and instances, and kind-of inheritances. Simple logic statements can be connected using logical connectives to form compound logic statements. The set of such logic statements, simple or compound, is commonly called logic theory. A full accounting for how objects and relations in the real world map onto the logic symbols forms an interpretation of this logic theory [Winston, 1992].

- **Classes and Instances:** Most KBE languages allow definition of classes and instances. Classes are generic descriptions of objects, and instances are specific outcomes of an object-class. An object is a software packet that contains a set of related data and procedures. An object's procedures are called its methods. Objects communicate by sending messages to other objects requesting that they perform one of their methods. *Object* is an occurrence, or instance of a class. KBE languages often provide tools, such as browser, to represent objects and review instances, both graphically and non-graphically.

- **Kind-of inheritance:** The language allows definition of a new class from an old class where the new class is derived from the old class with some *same but except* characteristics. A new class is said to inherit a portion of definitions from an existing class. Users only define the *except* changes, for example, *square* is a kind of inheritance from a rectangle object class. Inheritance allows the developer to define generalized behavior classes that can be used by multiple, slightly different subclasses. It also allows existing classes to be extended and modified without changing the source code. This is accomplished by overriding previous methods at the subclass levels. It supports the creation of object models by allowing object designers or programmers to specify class hierarchies through selection of methods. The resulting object is maintained in a storage-independent form.

- **Generic Parts:** A generic part is an object-oriented structure that includes engineering rules, methods, attributes, and references to the children of sub-parts. The KBE language provides options for specifying the parts' attributes as variable attributes with no initial values specified. Other attributes are defined as a function of the variable attributes. Generic parts receive their attribute-values by means of inputs at run-time. The concept is useful since the generic parts' family can be replicated or instantiated at run-time merely by specifying the required inputs for each part throughout an assembly. The generic parts can encapsulate other sub-parts or contain positioning or assembly information.

- **Referencing-chain:** Referencing-chain is a useful concept to access an object or an attribute of a tree from any other place in the tree. It is often used to define dependencies that exist, or are desired, between children. An access is permitted by identifying a path that leads to the object whose attribute descriptions are required. In the definition of a physical object, shown in Figure 6.6, a referencing chain is shown connecting an environment definition with physical parts definitions. The concept is useful since attributes or parts can be retrieved by passing messages without actually replicating the source code, reasoning, or logic behind the definition of the parts or the attributes.

$$\text{Methods of Capturing Life-cycle Intent} = \cup \; [\textit{Attribute definition,}$$
$$\textit{Demand-driven Operations, Frame Structures with Rules, Symbolic logic,} \quad (6.3)$$
$$\textit{Classes and Instances, Kind-of inheritance, Generic Parts, Referencing-chain}]$$

Depending upon the available library of primitive parts, some languages are easier than others.

6.4 CAPTURE PRODUCT MODELS

The first step in developing an integrated product design, development, and delivery (PD^3) process capability is to capture a product life-cycle intent. This includes its underlying structure, function, geometry, topology, history, and the assembly configuration of a product. Figure 6.11 represents a mathematical replica of a physical product and its environments. It captures a model of the product, not an instance of the product itself. Central to the definition of the product model is the smart model concept. The smart model is composed of function definitions, geometry definitions, product hierarchical structures, design principles, engineering constraints, library of parts, mating and positioning information, empirical knowledge, generic parts, and so on. CE knowledge teams create smart model definitions based on *functional evolution* on one hand and *standards* on the other. *Functional evolution* means product functions get gradually refined as product and process specifications are enriched in a representation of a model object, leading to the realization of a physical artifact. Recording of functional evolution of a design process is critical to preserving a product's life-cycle intent. Product or process specifications can be present as CAD data, tables, schedules, and so on. *Standards* can be present as tooling standards, standard parts, or standard procedures, such as data bases, design handbook, technical memory, and so on. The geometry definition is composed of part feature definition, parts assembly sequence, relationship among features, and specifications and knowledge about its constituent design. The relationships among features are represented within a product model by a hierarchical tree structure based on parent child dependencies between the identified feature objects as shown in Figure 6.7. Each node of the tree is either a parent for the descending nodes, a child for the ascending nodes, or both. This tree structure supports the concept of inheritance and association. The total product model (TPM) is decomposed into a number of relatively independent loosely coupled models (e.g., analysis models, design models, process models, etc.) that may be performed in parallel. TPM often uses sensitivity analysis, optimization or some form of empirical knowledge as a part of an intelligent decision support system. This is also why TPM is referred to as the smart model. By changing in-

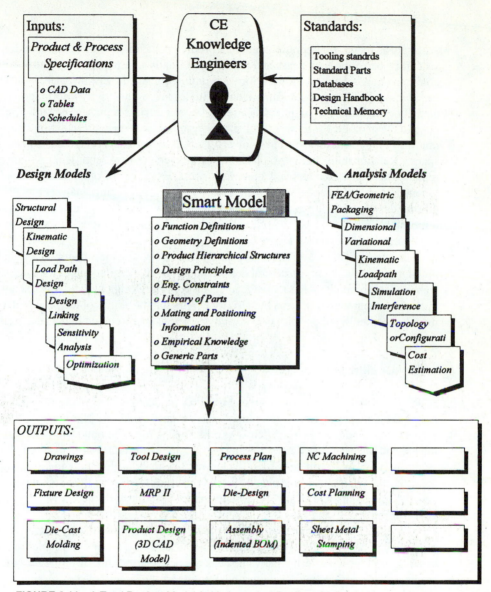

FIGURE 6.11   A Total Product Model (a Mathematical Replica of a Product and its Environments)

puts to a TPM, any number of alternate designs that satisfy the specifications and standards may be produced. Typical outputs of a TPM consist of product design (3-D CAD model), fixture design, tool design, die-design, assembly, indented BOM, die-cast molding, process plan, MRP-II, sheet metal stamping, NC machining, exploded-view assembly drawing, cost planning, and so on (see Figure 6.11).

6.4.1   Analysis Models

An analysis model is a representation of a product behavior so that a corresponding analysis can be performed with ease and flexibility. Usually a runstream procedure written in a command language is established for each required analysis. For instance, in a finite element analysis (FEA) model, finite element nodes and connectivity information (grid and element description) are required as a set of key parameters representing structured geometry. In the geometry analysis model, a wire-frame (boundary representation type) configuration model is defined in terms of a set of characteristic points and lengths in 3-D-space. Relationships between different parameters (points and lengths) are usually required to ensure compatible geometrical forms. Examples of other analysis models are: kinematics analysis model, load path analysis model, packaging analysis model, simulation model, interference analysis model, metal forming analysis model, cost estimation, dimensional analysis, topology or configuration analysis, variational analysis (for fits and clearances), and so on (see Figure 6.11).

   Design models are different from analysis models. The most commonly found example of a design model is a structural design model. In an integrated PD^3, many types of design models are required to satisfy the integrity of parts. Design models are used not just for evaluating the integrity of parts but also for evaluating the ability of the proposed design to meet a specified set of performance objectives. These are discussed in the next section.

6.4.2   Design Models

A design model may consist of several analysis models coupled with sensitivity analysis and optimization procedures to optimize the product behavior. The goal of these design models is to produce a product that is *best* in some sense. The bounds on the design parameters and the established relationships are considered as constraint equations. Each constraint may be described by a set of linear, nonlinear, or arithmetic equations, depending upon the established relationships between constraint forms and their linking parameters. Logical statements and looping can be introduced similarly through a programming language or through if-and-then instructions.

   The advent of design language has allowed the creation of very detailed and complex mathematical design models. Using such design models, it is possible for members of a CE work group to establish what steps are necessary to optimize the design. For example, with a design optimization model, the work group establishes an objective function that is chosen to characterize the *best* design. The goal is then to minimize or maximize this objective function (weight, cost, shape, minimum gages, and so on.). Mass of the structure is often chosen as the quantity (objective function) to be minimized. The design variables are the physical quantities, a design work group can control, such as thickness, material properties, and so on. Sensitivity analysis is the relative comparison of a design in terms of how a design variable affects the integrity of the design. This is achieved by computing design sensitivity coefficients or by measuring the rate of change of a constraint with respect to a change in the attributes of a part.

In many design situations, linking of design models is necessary to satisfy multiple objectives for optimization. Individual design models (e.g., structural design, kinematics design, load path design model) separately address individual performance needs such as structural, kinematics, loads, and so on. Linking of the design models allows teams to design products that can simultaneously satisfy multi-disciplinary constraints.

6.4.3   Process Models

Geometry-based programs (e.g., a CAD/CAM system) cannot be used to design complex non-geometric objects like processes. For example, a new chemical plant certainly has a geometry, but geometry does not capture the dynamics of the process itself. KBE systems are particularly suited to process designs that have little to do with geometry. Process models incorporate rules about the business, the chemical phenomena, the manufacturing facilities, and other important elements of the manufacturing process. Typical examples of manufacturing processes include stamping, casting, metal forming, roll forming, injection molding, and so on. Like product models, by changing inputs to a process model, any number of alternate processes can be designed, simulated, or verified.

- *Training new users or employees:* The process modules can be used to train new work group members. The experienced employees need not take valuable time from their busy schedules.
- *Hazardous applications:* The captured process knowledge can serve as an alternative to an on-site, in-person inspection and monitoring. Knowledge capturing is more attractive if the environments are hazardous and contaminated (nuclear waste, etc.).

6.5   CREATION OF SMART OR INTELLIGENT MODELS

An important component of a smart or an intelligent model is the ability to define geometry in terms of parameters and constraints. Constraints are rules about dimensions, geometric relationships, or algebraic relationships. A smart model is a reusable conceptualization of an application domain. The models contain the knowledge (attributes, rules, or relations) of the application domains forming the basis for future problem solving. Rules define how design entities behave, for example, whether a hole-feature is through or blind. In the case of a blind-feature, if the part becomes thicker and the cylinder is not long enough, the hole will become a blind hole. Unlike the blind-hole feature, a through-hole feature understands the rule that the cylinder must pass completely through the part and will do so no matter how the part changes. There are two ways of formulating a problem that leads to a smart model.

1. Constraint-based programming.
2. Knowledge-based programming.

Constraint-based modeling (CBM) or programming yields constraint-based models. Knowledge-based programming results in knowledge-based models. The major differences between the two model types (constraint-based modeling and language-based modeling), in contrast to conventional modeling, (a traditional CAD/CAM system) are shown in Figure 6.12. In conventional modeling, the geometry is captured using a static representation of wire-frames, surfaces, or solids, that is, the geometry is captured in digitized (fixed value) form. The modeling process is largely interactive. In constraint-based modeling, since the mechanism of geometry capture is through parametric, variational, or feature-based techniques, each model represents an instantiated (or dynamic) geometry.

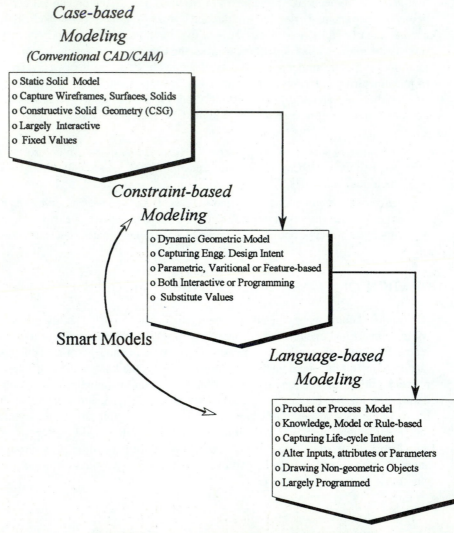

Case-based
Modeling
(Conventional CAD/CAM)

o Static Solid Model
o Capture Wireframes, Surfaces, Solids
o Constructive Solid Geometry (CSG)
o Largely Interactive
o Fixed Values

Constraint-based
Modeling

o Dynamic Geometric Model
o Capturing Engg. Design Intent
o Parametric, Varitional or Feature-based
o Both Interactive or Programming
o Substitute Values

Smart Models

Language-based
Modeling

o Product or Process Model
o Knowledge, Model or Rule-based
o Capturing Life-cycle Intent
o Alter Inputs, attributes or Parameters
o Drawing Non-geometric Objects
o Largely Programmed

FIGURE 6.12   Properties of Conventional and Smart Models

Thus, by setting new values to the CBM attributes, several instances of the geometry can be obtained. Knowledge-based modeling (KBM) is similar to constraint-based when it comes down to capturing the geometry. However, because of its (KBM) abilities to capture non-geometric information and to associate rules with attributes, KBM is also suited for capturing life-cycle intent. Other contrasting features of CBM and KBM are listed in Figure 6.12 and further explained in the following section.

6.5.1   Constraint Based programming

Constraint-based Programming (CBP) is a concept of formulating a problem in terms of constraints, which may be a part of a product definition, a process definition, or an environment for the problem definition. No distinction is made between types of constraints or their sources. A spreadsheet program is a simple example of a constraint-based programming. Here equations representing constraint relationships are input to cells in a spreadsheet program. There is a close resemblance between design rationales (DRs) [Klein, 1993] and spreadsheets. Equations that are entered into cells of a spreadsheet are analogous to capturing DRs, and computed cell values in the spreadsheet program are analogous to identifications of DRs. The cells themselves constitute the *knowledge* of CBP. If a smart model is thought of as a series of spreadsheets for a concurrent team, *programming* in CBP is analogous to specifying the relationships between the cells of a spreadsheet.

The following are some typical constraint parameters that can be employed during a constraint-based programming.

- Design specifications.
- Design criteria.
- Subjective qualifications.
- Design constraints.
- Manufacturing constraints and tolerances.
- Material properties.
- Geometry.
- Sectional properties.
- Configuration and topology.
- Heuristics or rules.
- Historical data.
- Performance requirements.
- Test specifications and data.

Figure 6.13 shows a three-way comparison between a set of key characteristics of smart models (created using constraint-based and knowledge-based programming approaches) and conventional models. Three categories employed for comparison are types of representations, types of relationships between parameters and constraints, and the method of solving the constraint satisfaction problem. They are shown in Figure 6.13 as three

	Representation of Parameters and Constraints	Type of Relationships between Parameters & Constraints	Method of Solving the Constraint Satisfaction Problem
Conventional Models	Non Interpretative (Fixed Dimensions)	Fixed Attributes: Points, Lines and Surfaces	Computational Geometry B-Splines, NURBS
	Interpretative (Variable Dimensions)	Variable Attributes: Points, Lines & Surfaces	Computational Geometry B-Splines, NURBS, Linear Algebra
Constraint-based Programming (Smart Models)	Parametric/Symbolic	Explicit/Algebraic	Equation Solver, Variational Geometry/Analysis
	Features/Forms	Explicit/Algorithmic	Linear/Non-Linear Programming, Optimization
	Mixed	Explicit/Algebraic + Algorithmic	Equation Solver/Linear Algebra, Multi-Criterian Optimizatiion, Optimal Remodeling
Knowledge-based Programming (Smart Models)	Rule-based	Implicit/Heuristics	Inference Engine (Object-Oriented Programming -- OOP)
	Model-based Reasoning	Implicit/Heuristics	Inference Engine (OOP)
	Case-based Reasoning	Implicit/Heuristics	Inference Engine (OOP)

FIGURE 6.13   Comparison between the Conventional and Smart Modeling Approaches

columns of a matrix. The modeling categories (conventional and smart models) are listed as rows. A cell of the matrix shows the differences in the approaches used in each modeling category. In the conventional models, the geometry compatibility (e.g., line and arc constraints) and consistency issues are resolved through computational geometry, linear algebra, B-spline, and NURBS techniques. In CBP, a product design problem is defined in terms of constraints and the inter-relationships that exist between them (see Figure 6.14). Constraints may be a part of the following sets:

- Product definition, such as geometry, materials, size, etc.
- Process definition, such as assembly constraints, tolerances, fits/clearances, etc.
- Product environment such as loads, performance, test results, etc.

In constraint-based modeling, the constraints are of explicit/algorithmic or algebraic types. They are resolved through a set of linear and nonlinear programming, optimization, and optimal remodeling techniques (see Figure 6.14). Such methods help develop a set of individualized criteria for life-cycle design improvement (more than what it generally ap-

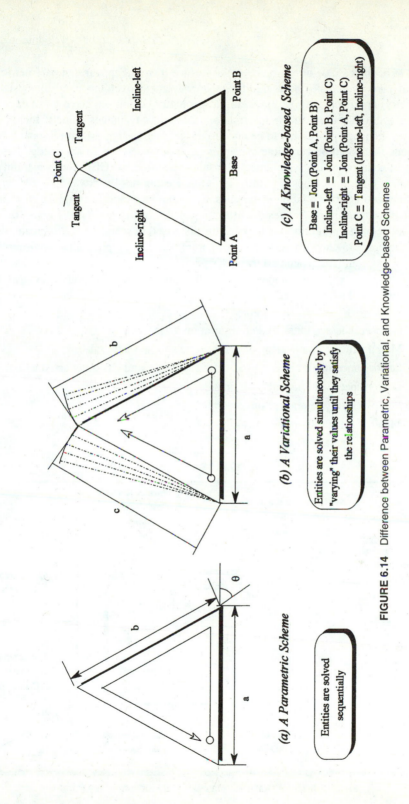

(a) A Parametric Scheme

Entities are solved sequentially

(b) A Variational Scheme

Entities are solved simultaneously by "varying" their values until they satisfy the relationships

(c) A Knowledge-based Scheme

Base = Join (Point A, Point B)
Incline-left = Join (Point B, Point C)
Incline-right = Join (Point A, Point C)
Point C = Tangent (Incline-left, Incline-right)

FIGURE 6.14 Difference between Parametric, Variational, and Knowledge-based Schemes

pears to be the case). For example, on the surface it would appear that parametric generation of parts would not be a significant improvement over the fixed-dimension (static
geometry) approach. In both cases, initial geometry definitions have to be captured and
shared with other users of the information. The real advantages come if there is a large
amount of change processing to be done to the original design. In a traditional CAD environment, this could be very time consuming and cumbersome. In knowledge-based modeling, in addition to the types of relationships specified for CBP, the constraint set also
contains implicit/heuristics type of rules. Inference engines (backward and forward chaining) or constraint propagation techniques are commonly used in conjunction with object-
oriented programming to resolve the imposed constraints (see Figure 6.14). The
knowledge-based programming approach is discussed in section 6.5.2 in greater depth.

Figure 6.15 shows a sample set of parameters for a typical CBP environment. The
types of environment depend on the descriptions of the functional intent behind the product or the process that are modeled. The following are the key parameters surrounding a
constraint-based smart model:

- Geometrical, sectional, and configuration variables.
- Manufacturing engineering design criteria and heuristic rules.
- Performance requirements, cost, efficiency data, and customer satisfaction.
- Design specifications or constraints.
- Manufacturing tolerances or constraints.
- Material selection or qualification.

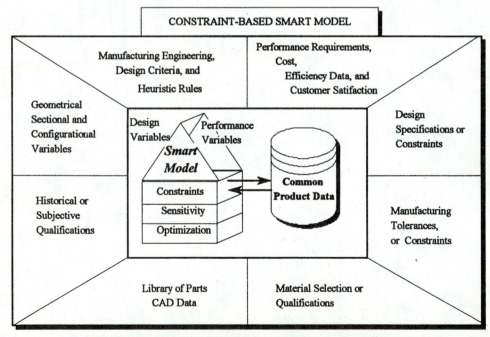

FIGURE 6.15   A Constraint-based Smart Model with Key Parameters

- Library of parts, CAD data.
- Historical or subjective qualifications.

In constraint-based programming, algorithmic tools such as analysis, simulation, generic modeling, sensitivity, and optimization are often used as an integral part of the design process. The original set of parameters is grouped into three categories: design variables, v_i, performance functions, f_i, and design constraints, c_i. The dependencies between performance and design constraints with respect to design variables are controlled through sensitivity analysis:

$$\text{Sensitivity} = \frac{\partial f_i}{\partial v_i} \qquad (6.4)$$

Additional discussion of problem formulation can be found in section 4.8 of Chapter 4.

6.5.1.1   Types of Representations

There are four modeling modes or types of representations that can be used in a constraint-based programming [Pabon, Young, and Keirouz, 1992].

1. Parametric or symbolic representation.
2. Variational representation.
3. Feature-based representation.
4. Mixed representation.

These modeling modes possess some common, overlapping characteristics. Each mode captures the design intent during product definition by enumerating those relationships that constrain the design. These representations, however, can all be classified as *variable-driven modelers* [Kurland, 1994]. That is, each mode allows the part to be readily edited by simply changing those relationships, thus allowing team members to efficiently explore design alternatives or perform revisions. Figure 6.16 shows a relative comparison of the salient features of three variable-driven modelers. The quality of each representation is shown in a tabular format. Fourteen key basic characteristics are outlined in general terms to understand their major differences or similarities. Actual implementation of each representation in a CAD program may differ greatly. For instance, most CAD vendors often put their own spins and twists to distinguish their software from others. The mixed representation indicates such an attempt in the use of one or more of the following techniques to build a new hybrid representation.

Parametric or Symbolic Representation: The use of parametric or symbolic representation with CAD primitives results in a parametric design. Parametric design is referred to as a method for determining a geometry (e.g., size and orientation of geometrical elements), whereas relationships are defined using a cyclic graph and related geometric (say, line or arc) constraints governing the system. Here disclosure of design rationales (DRs) is made through design plans and design constraints. Capturing of design rationale consists of problem formulations and modeling of the design descriptions in parametric forms. The parameters are used as representations for DRs. The parametric approach solves equations sequentially. Computational determination of design descriptions based on parametric formulation is analogous to sequential identification of DRs.

Modeling Modes	Parametric or Symbolic Rep.	Variational Representation	Feature-based Representation
Method of Encapsulating Associativity	Variable Dimension Modeling (VDM) (change of dimensions results in changes to the model) + Geometry Constructs	Variable Dimension Modeling + Engineering Equations	Variable Dimension Modeling + Named Entities that encapsulate how geometry behaves, when it is a part of a larger topology
Key Parameters	Size and Orientation of Geometric Elements	Symbols and Attributes	Attributes, Behavioral rules and Constraints
Key Constraints	Dimensional and geometric type constraints	Dimensions , geometric & algebraic relationships	Geometric, Algebraic and Positioning Rules
Methods to Relate Keywords and Actions of the user to Variables	Geometric-specific: A predefined set of actions to specify what constraints are known, find a valid solution and bring a design to a fully constrained state	Problem-specific: A pre-defined set of problem domain, which comes with a set of problem-specific equations.	Context specific: Library of features as a part of a CAD modifiable macro geometry primitives
Mathematical Representation	A set of acyclic directed graph which interactively link parameters with constraints	A set of nonlinear equations which declaratiely represent the set of constraints.	A set of boolean operations with behavioral rules to control the output behaviors as consistently desired
Constraint solving	Procedural or Programatic Manner	Simultaneous Equation Solver	Rule-based Boolean Operations
Editing Features	Construct-centered : Captures a transaction history of construction operations and their variables	Attribute -biased: Free-form sketching with gradual constraining of attributes.	Rule -biased : Rules define how commonplace design feature behave, such as through or blind holes
Method of Problem Formulation	Interactive Build-up: ability to interactively buildup a under constrained geometry to a fully constraint state	Programmatic Definitions: ability to solve under-constrained problem when geometry is not fully defined	Macro definitions: Library of macro-features can be built which is a composite of a user-defined set
Constraint Dependency	Order Dependent: parametric relationships & construct operations are captured in an order dependent or sequencial manner	Order Independent: User designs without being concerned about the order in which constraints are placed or solved	Rule –dependent: user controls the occurences through rule-sets. Rules can define both order dependency and independency as needed
Definition of 2 Parallel Lines: Line A and Line B	Either line A or Line B can be moved depending upon which is defined in relation to which . If line A is defined parallel to line B, line B may be moved causing line A to move.	Either to be moved while maintaing the constraint equation -- which states " Lines A and B are Parallel"	Parallel lines, if specified as a rule, can mimic either situation depending upon what rules were encapsulated
Change Propagation Mechanism	Chained: Preceeding steps dictate intermeciate results, which in turn drives the relatipnship for succeding steps. In bidirectional setup, the information can flow	Simultaneous: The addition or removal of an attribute in a problem domain changes the structure or content of the equations requiring simultaneous solutions of the entire set , not just its subset.	Iterative: The behavior can be chained or simultaneous depending upon the appropriateness of the features
Associative Mechanism	Transaction History: Requires compiling a transation history of editing operations performed by the user	Equations linking: Requires a known problem domain and generic definition of equations and their associativity with line and arc constraints	Rules linking: Requires linking the known behavior of the features to the "part of the whole" positioning rules.
Degrees of Freedom	Reasons about dimensions, actions, and geometry	Reasons about dimensions, equations (algebraic), actions and geometry	Reasons about dimensions, rules, positioning, actions and geometry
Re-Ordering Mechanism	History List Reordering: Creates a new sequence or updates the old ones parametrically	Equations Re-ordering: Allows a new set of equations to be redefined or modifies the old set programmatically	Rules Reordering: Allows a new set of rules to be created programmatically to ensure "a part to the whole" associativity

FIGURE 6.16 Comparison between Parametric, Variational, and Feature-based Representations

In parametric or symbolic representation, a smart model is created using a set of predetermined geometric constraints, parameters, and output formats. The parametric-based smart model (PSM) does the following:

- PSM employs generic modeling, analysis, and optimization as an integral part of the model methods. In this way, the system knows the identity and behavior of the individual object as well as the environment in which it fits (or is subjected to). All information resides symbolically in a unified data base. This goes beyond the conventional CAD method, which captures the geometry in terms of basic attributes (points, lines, arcs, and surfaces).
- PSM captures the transaction history of construction operations, their variables (e.g., geometric primitives), and team actions (editing operations performed by the work group parametrically) in terms of the symbolic attributes' data and model objects.
- PSM captures the information into a sequential or orderly transactional history file containing the real-world relationships between data, model objects, objects' physical characteristics, and the objects' environments.

The major advantage of a parametric definition is its solution speed and its major disadvantage is the flexibility. If the users decide to alter the product geometry to a different class, a majority of problem parameters and relationships need to be redefined.

Variational Representation: The use of variational representation with CAD results in variational design. Variational design is a method for determining geometry (e.g., size and orientation of geometric elements or primitives) and non-geometrical relations by simultaneously solving a set of nonlinear equations. The equations declaratively represent the set of constraints governing the system. The constraints are dimensions, geometric relationships, algebraic and engineering equations. Editing a design with such relationships is done simply by changing these relationships programmatically. Variational design allows the work group members to design without being concerned about the order in which constraints are placed or solved (see Figure 6.16). Another key benefit of a variational design is the ability to deal with under-constrained geometry, that is, geometry not yet fully defined. This allows the CE work groups to define only those constraints that are known at any point. It is important to note, however, if any variables of the equation set are not completely defined, or are changed in such a way that they alter the problem definition, the variational system is forced to make assumptions about the undefined constraint set. Through an incremental process of adding new constraints and finding a corresponding solution, the design can be brought to a fully constrained state. Under-constrained sections force the system to create missing equations by making assumptions about the designer's intent. These assumptions fall into two categories.

1. *Assumptions made to facilitate the sketching process:* This results in increased sketching and modeling performance (an example includes snapping vertices to be coincident). Such assumptions force the users to completely and accurately define the design, and guarantees that any changes will yield predictable results.

2. *Dimensional assumptions made due to lack of critical design information:* This is worse, since under the banner of ease of use, these assumptions are never made known to the users. The system simply regenerates the design and the user continues to model the part. At a later point, it is not certain how the model will react when later imparted to the design during a downstream process.

Two elements are essential to applying a variational representation.

1. *Known Problem Domain:* Variational modelers require a known problem domain, or a solution domain close to one being searched, since equations that are generated are problem-specific. It primarily involves free-form sketching with gradual constraining of attributes rather than recording a set of construction steps as in parametric modeling. It is an attribute-based approach rather than construct-centered (see Figure 6.16). A simple example is the definition of two parallel lines. A variational definition would merely state that Lines A and B are parallel, allowing either to be moved while maintaining the parallel intent. In parametric modeling, either Line A would have to be defined as a line parallel to Line B, or vice versa. This makes one's movement dependent upon the other's. Generation of constraints and simultaneous solutions of those equations are the backbone of any variational system. Another example is the third point of a triangle (see Figure 6.14).

2. *Equation solver:* An equation solver in a variational program forms the underlying *calculator*. Adding a module is analogous to adding a set of predetermined macros to a spreadsheet program, the team simply plugs in the requirements and the system takes over. A variational program executes a sequence of steps where order is unimportant and finally returns the results in the designated variable. Executing the single step commands, for example, the equation volume = length × width × height, represents the volume of a rectangular prism. If the variables were assigned independent cells as in a spreadsheet and their values were specified, calculating volume through a variational program, parametric programming, or a spreadsheet would be very similar. However, unlike a spreadsheet, variational programs can back-solve. One can enter a desired volume, guess on the magnitudes and the constraints, and the system would find the corresponding right input values. For example, a financial model might include an equation for calculating a loan payment and interest rate. Using the back-solving features, the team can enter the desired payment and compute the interest or vice versa through a single equation definition. In a spreadsheet, different equations would be necessary, one with the loan payment as an output and the other with interest as an output. Depending upon the desired functions, back-solving is handled in variational programming by a direct, or a list solver. The direct solver computes a single result while the list solver can solve for multiple results (multi-valued functions). Non-linear equations are examples of multi-valued functions since many solutions are possible. Work groups can input initial estimates as starting input parameters to aid in solving this set of nonlinear equations for the nearest values.

Feature-based Representation: The use of feature-based representation with CAD results in a feature-based design. Features are named entities that encapsulate how a design behaves through rules and attributes along with the defining constraints and geometry governing the system. Related information, such as design attributes, manufacturing semantics, and attribute dependencies, are attached to each feature in the product model. Feature-based modeling in geometric terms is very similar to rule-based Boolean operations (see Figure 6.16). It views the geometry of a product as a set of interrelated form features. For example, a through hole is a standard difference operation between a part and a cylinder. In reality, if the part became thicker than the cylinder length, the hole would become a blind hole. In contrast, if through hole was a feature-based representation, it will understand that it has to pass completely through the part no matter how the part changes.

During the last decade, noticeable research has been attempted in the area of design by features [Cunningham and Dixon, 1988; Shah, 1991]. There are many definitions of the term *features*. In general, features are defined as the representation of object characteristics, knowledge, or actions that have significance in problem solving. One of the original arguments of feature-based design is that features (such as geometrical shape) can better represent the function of a product, compared to using the lines and surfaces. However, in the desired CAD systems there are a number of standard features that can be identified in mechanically discrete parts that constrain a design to function in some way. Similarly, there are constraints associated with downstream production activities that express their functions in relation to some form features. These can be characterized as [Hummel and Brown, 1989]:

- *Design Features*: These features result from defining objects through 2-D or 3-D primitives, or by constructing a 2-D profile shape and sweeping that profile along a trajectory. Most of the design features are defined during modeling a part geometry and hence they are also called *geometry or volume* features.

- *Machining Features*: These features contain process related information. For example, the process knowledge or actions can be used to remove materials from the initial stock to obtain a final part. A machining feature includes knowledge about each distinct machining step, such as cutting tool, setup, orientation, and machinability information.

- *Inspection Features*: These features identify inspection type features of the part that could be useful in verifying the design dimensions and tolerances. Such features are generally represented by topological faces and edges of the parts.

- *Deburring Features*: These features identify knowledge of parts' deburring including common techniques of surface and edge preparations. Most features of this kind deal with convex edges of parts where burrs are often present.

Table 6.1 shows typical examples of the classification of features. The list is not complete but indicates a dilemma faced by concurrent teams in characterizing an object in different contexts.

Clearly, features are useful in identifying the complexity of design and process related operations. Under certain conditions, some product design features can be mapped directly to process design features. The existence of the latter is useful to provide process feedback. In a majority of cases, the product or process constraints can be determined from the mapping process between the functions and features. This led to the idea that features

TABLE 6.1   Feature Class Examples

Classification of Features	Examples
Design Features	Points, Lines, Vectors, Curves, Circular-curves, Elliptical-curves Quantified parts, Conics, Arc, Surfaces Circle, Ellipse, Polygon, Spiral , Notch Slots, Hole, Ribs, Chamfers, Bosses, Fillets Cylinder, Hollow-cylinder, Truncated-cylinder Box, Cones, Hollow-cones, Frustum Pipe, Threaded Nut, Elbow, Tube, Hollow-tube Parallelogram, Prism, etc.
Machining Features	Slab, Step, Notch, Step-notch Pocket, Wedge, Pocket-wedge Holes, through-holes, Counter-bore, Blind-holes, Step-through-holes, Counter-sunk hole, etc. Profile, Profile-wedge, Profile-notch, Profile-step-notch, etc.
Inspection Features	Planar Faces, Holes, Planar Hole-Faces Datum A, Profile-group, Circular or Round Hole Rectangular or Square Hole Rectangular faces, Cylindrical Faces, Cylindrical Surfaces Datum B, Datum-B Hole, Datum B Profile Group Datum C, Datum C Planar Faces, Datum C Holes Cylindrical and Planar faces, etc.
Debarring Features	Open-planar-path, Confined-planar-path, Confined-linear-edges Blind-holes, Through-holes Open-planar-path(s), Blended-planar-path Open-linear-edges, Blended-sculptured-path, etc.
Derived Features	Offset features, Compound Features, Macro features Features Extraction, Features Recognition Combined Features, etc.

can serve as a common link for considering the process requirements at the design stage. This has also led many researchers to use features as a representation scheme to drive the concurrent product and process design. Features contain the following characteristics:

- *Feature Attributes or parameters:* A feature has attributes or parameters that describe its characteristics and behavior. The attributes can be geometrical (i.e., height, width, thickness, diameter, etc.) and non-geometrical (material properties, type, yield stress, form properties, etc.) entities, rules, or relationships. The positioning constraints and values, in addition to dimensions, geometric definition of features, non-geometric entities, and algebraic constraints, make up a large portion of information useful to do CE.

- *Feature Structure Definitions:* A hierarchy of features library can be built by creating a macro feature structure, which, in turn, is composed of two or more micro features. This compounding can occur parallel to product or process decomposition hierarchy. Sometimes, features can be superfluous in a particular situation or a process. Feature suppression (temporary removal from the model) is an useful compounding feature.

- *Feature Class Definitions:* Hierarchy of product and process feature class definitions can be built parallel to product and process decomposition. Their relationships are automatically established due to the fact that the objects themselves are hierarchically decomposed. The key is that these definitions are set before the commencement of actual design activity.

- *Rules for Boolean operations and the intelligence:* Features contain rules for Boolean operations and the intelligence (behavior rules) to be a part of their definition. Features have the ability to reason. Features incorporate behavior rules and continue to observe these rules when applied in diverse situations. For example, a through hole continues to be a through hole even when the part shape is changed.

- *Encapsulation:* Features encapsulate their definitions. Once applied, the topology must continue to be recognized as a feature (hole, slot, etc.). The features allow their defining parameters to be changed (diameter, depth, draft, etc.) and hence the new feature continue to observe these changes as a part of the whole.

- *Automation of interface functions:* Designing with features enables automation of interface functions between design and other life-cycle aspects. For example, feature representation can be used to automate the model preparation for engineering analysis, such as FEM. Similarly, by establishing techniques to relate product geometry to manufacturing processes (as in attribute mapping) one can automate functions in manufacturing, such as process planning, numerical control cut, material removal, tolerance, and so on.

- *Common design and manufacturing characteristics:* Features can be used to model design characteristics that exhibit desired properties such as geometrical compatibility or technical concurrence with downstream operations. Also the understanding of manufacturing semantics captured by form features paves the way for efficient on-line X-ability evaluation of a product design at varying levels of design details.

However, there are situations where constraints cannot be defined between any of the standardized discrete parts and the identified form features. The problem is complicated further by the fact that features are usually context dependent, that is, there can be two features that geometrically look identical but may exhibit distinct and useful problem-solving knowledge. A mixed representation is often found useful in those situations.

Mixed Representation: In this mode, a constraint is represented using either a parametric, a variational, or a feature, or a combination of the above, depending upon the simplicity of representation or the flexibility of use. The use of parameterized *macro features* in feature based CAD systems exemplifies this scenario. With such mixed representation, it is possible to instantiate features that differ in dimensions while maintaining their basic properties. In another scenario, Hummel and Brown [1989] employed a basic parametric feature representation to develop two counter-bore feature representations.

A smart model of this type contains duel ("parametric and feature" or "variational and feature") representation of the problem's knowledge base (design criteria, constraints, etc.) with appropriate facts and logic (implicit, explicit, or derived). The model is supported by one or more of the decision support tools such as variational geometry, analysis, simulation, sensitivity, optimization, and so on. Constraint-style smart models, based upon object oriented programming tools or techniques (such as AI shells, objective C, C++, etc.), are found to contain better software engineering qualities.

6.5.2 Knowledge-based Programming

Knowledge-based Programming is another way of creating a smart model as shown in Figure 6.17. In order to develop an integrated view of PD^3, which is rich and comprehensive, it is necessary to include a variety of knowledge sources and representations. Knowledge-based representations deal with three types of knowledge: *explicit* knowledge, *implicit* knowledge, and *derived* knowledge.

- *Explicit Knowledge:* Statements of explicit knowledge are available in product or process domains as retrievable information like CAD data, procedures, industrial practice, computer programs, theory, and so on. They can be found both within and outside of a company. The explicit knowledge can be present as a set of engineering attributes, rules, relations, or requirements. Inside knowledge deals with observations and experiences of the concurrent work groups. Outside knowledge sources include papers, journals, books, and other product design and standard literature.

- *Implicit Knowledge:* Statements of implicit knowledge are mainly available as process details, such as memory of past designs, personal experience, intuition, myth, what worked, what did not, and so on. Difficulties arise when such processes are obscure, for example, intuitive or creative. Implicit knowledge includes the skills and abilities of the work groups in application tasks and problem-solving methods. Implicit knowledge that is found outside a work group circle is mostly in case studies and discussion dialogues. Difficulties arise when such implicit knowledge has not been articulated in a form that allows ease of use and transfer.

- *Derived Knowledge:* Statements of derived knowledge are those that are discovered only by running external programs, such as analyses, simulation, and so on. Derived knowledge is like extrapolation or interpolation of the current domain for which explicit knowledge is missing or incomplete. Undiscovered knowledge has been the driving force of most research and development organizations.

Knowledge-based programming (KBP) provides an environment for storing explicit and implicit knowledge, as well as for capturing derived knowledge. When these sets of knowledge are combined into a total product model (TPM), it can generate a family of designs, tooling, or process plans automatically. Unlike traditional computer-assisted drafting (e.g., a typical CAD) programs that capture geometric information only, knowledge-based programming captures the complete intent behind the design of a product—*HOWs* and *WHYs*, in addition to the *WHATs* of the design. Besides design intent there is other knowledge (such as materials, design for X-ability, 3Ps, process rules) that too must be captured.

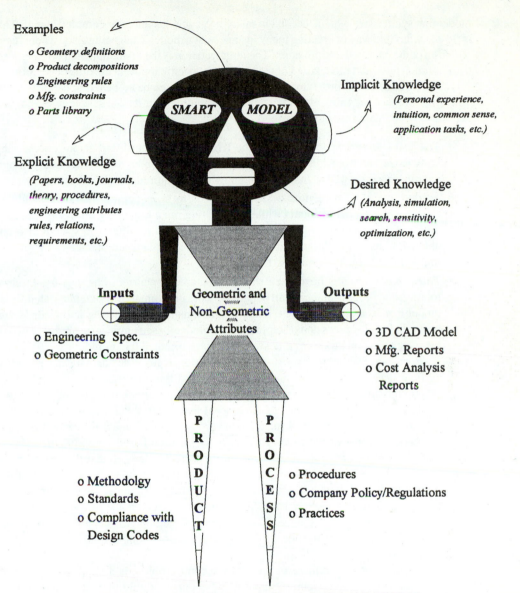

Examples

   o *Geomtery definitions*
   o *Product decompositions*
   o *Engineering rules*
   o *Mfg. constraints*
   o *Parts library*

Implicit Knowledge

   (Personal experience,
   intuition, common sense,
   application tasks, etc.)

Explicit Knowledge

(Papers, books, journals,
theory, procedures,
engineering attributes
rules, relations,
requirements, etc.)

Desired Knowledge

   (Analysis, simulation,
   search, sensitivity,
   optimization, etc.)

Inputs

**Geometric and
Non-Geometric
Attributes**

Outputs

   o Engineering Spec.
   o Geometric Constraints

   o 3D CAD Model
   o Mfg. Reports
   o Cost Analysis
      Reports

P R O D U C T

P R O C E S S

   o Methodolgy
   o Standards
   o Compliance with
      Design Codes

   o Procedures
   o Company Policy/Regulations
   o Practices

FIGURE 6.17   Salient Features of a Knowledge-based Smart Model

Knowledge-based engineering (KBE) is an implementation paradigm in which complete knowledge about an object (such as a part) is stored along with its geometry. Later when the part is instantiated, the captured knowledge is utilized to verify the manufacturability, processabilty and other X-ability concerns of the part. One important aspect of knowledge-based engineering is the ability to generate quickly many sets of consistent designs instead of just capturing a single case in a digitized CAD form that cannot be easily changed. Knowledge-based programming technology encourages development of a *generic* smart model that synthesizes what is needed in many life-cycle instances in complete detail. This is the most flexible way of creating many instances of a model, each being a consistent interpretation of the captured design intent. The interpretation is the result of acting on rules captured through the smart models and of feeding in the specific inputs at the time of the request.

Knowledge-based programming software offers three basic benefits: capture of engineering knowledge, quick alterations of the product within its acceptable gyration, and facilitation of concurrent engineering.

- The *first* strategic benefit is due to the KBE system capturing precious engineering knowledge electronically. This capture allows companies to leverage scarce engineering expertise and to build on the knowledge acquired slowly over time.

- *Second*, the system permits design variations to be generated rapidly by quickly changing a long list of inputs while maintaining model integrity. Products designed on the KBE system can practically design their own tooling. The system also enables designs to rough out their own macro process plans automatically by retrieving on knowledge for similar GT parts.

- *Third*, KBE systems have been shown to enable concurrent engineering. Design, tooling, and process planning all benefit by working from a common integrated smart model that is able to represent, retrieve, and integrate engineering knowledge from many different sources and disciplines. Knowledge-based engineering reduces two of the most common problems that arise with team-oriented CE, boredom and time.

  1. *Boredom:* Boredom crops into most traditional processes as part and parcel of their detail. Work group members do not find it attractive to check hundreds of common drudgery details that occur every time a new design is obtained (from checking its specifications to tolerances) as part of a PD^3 cycle. The idea is to capture those design and manufacturing issues that impact the design of most products, most of the time. This action is justified based on *8020 rule*. This is commonly called the 80:20 heuristic or Pareto's law of distribution of costs [Dieter, 1983]. In a typical situation, 80% of the assignments are routine or detail works and only 20% are creative (Figure 6.18). Pareto's law states that while only 20% of the possible issues are creative, they generally consume about 80% of the work group resources. The 80% of the assignments that are routine do not seem to cause any significant product problem or seem to consume as much in resources.

  2. *Shortage of time and resources:* The second problem is a shortage of 7Ts and resources. Many concurrent team members are not able to find enough time to devote to the actual design process due to their heavy involvement with other time-demanding chores such as staff-meetings, E-mail notes, management briefings, technical-walk-throughs, design reviews, and so on.

KBE reduces boredom by attending to the 3Ps details in ways that reflect design procedures, standards, company policies, and compliance with design and manufacturing codes and regulations. By packaging the life-cycle behaviors into a smart model, KBE improves productivity and automation. The 80% of routine tasks is reduced to a 20% level (see Figure 6.18). The CE work groups spend more time adding functional value to the design (during

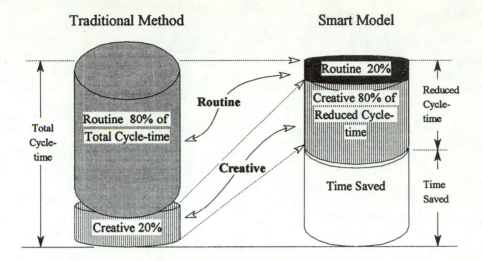

Time Saved = Total Cycle-time - Reduced Cycle-time

FIGURE 6.18   Key Benefits of Knowledge-based Engineering

the saved time period) rather than repeating engineering calculations for each case or recreating existing design items. If the traditional cycle-time were reduced from an initial total of
30 months to 20 months using smart models, the following calculation applies (see
Table 6.2).

$$\text{Time Saved} = \text{Total Cycle-time} - \text{Reduced Cycle-time} \qquad (6.5)$$

The work group is able to concentrate more on satisfying the creative tasks up to a
maximum of 80% of the reduced cycle time. They are freed from worrying about meeting
the drudgery details (routine tasks) that previously took up 80% of the total cycle time in
the traditional method. Thus, even after spending 40% of the work group engineers' time
more on creative tasks, there was a surplus of two months (30-20-8) using smart models
compared to the time it took following traditional methods.

6.6   SMART OR INTELLIGENT MODELS

Depending upon the frequency and need for creating modified designs, it is desirable to
weigh the benefits of smart models. Alternatives are either to apply an additional effort and
the time required to develop the smart models or to carry on the design in a traditional way.
Parametric development of design may not be worth the effort if each design is a unique design and significant changes to the product over its life-cycle are not expected. However,
this perspective changes quickly if one attempts to view the development worth with respect
to overall company performance. Techniques like parametric, feature-based, and

TABLE 6.2   Comparison of Savings

Actions or Tasks	Traditional Methods	Smart Models	Remarks
Total time taken to finish the tasks	30 months	20 months	Assumptions
Ratio of time spent doing routine/creative tasks in months (percentage)	24/6 months (80%/20%)	4/16 months (20%/80%)	Following the definitions of smart and traditional models
If 40% more time is spent doing creative tasks, then number of extra months needed	12 months (40%)	8 months (40%)	This shows how you can do more with less.
Difference in time compared to traditional method	12 months (deficit)	2 months (surplus)	

knowledge-based models all facilitate CE. Some techniques offer better programming capabilities than others. For example, parametric or feature-based techniques can change the geometry of the design very quickly. However, in doing so, design work groups no longer have the assurance of knowing if all, or indeed, any of the non-geometric (e.g., engineering and manufacturing rules) rules have been violated. Through knowledge-based engineering, one can capture, besides parametric geometry, the engineering and manufacturing rules for geometric modifications. When the specifications demand a new geometric design, the corresponding rules are automatically engaged to meet the engineering and manufacturing requirements and to achieve the best possible compromise. An interesting aspect of this approach is that one can capture and build trial processes (e.g., levels of analysis iterations, sensitivity, optimization, etc.) into a TPM in order to establish the best design. The effect of these trials is to automatically run thousands of analysis iterations in the background before the final design is selected. All of this can be transparent to the work group members, who simply want to feed in specifications and are interested in reviewing the outcome that works. Most of the smart techniques provide some form of an electronic control over the design process that can afford a dramatic reduction in change processing, such as design revisions, design changes, and so on. The term *automatic generation i*s construed as computer-aided generation of outputs, such as design renderings, process plans, bill-of-materials, numerical control instructions, software prototyping, machining, replacement parts, product illustrations, and so on. If an electronic capture of design intent is extended to include automatic or partially automatic design generation, the *worth* of electronic capture of design increases substantially. Furthermore, if such generation is done in 3-D solids that do not require assembly information in downstream processing, electronic capture becomes almost an irrefutable requirement of the product definition process.

There are four main constituents of a smart model (see Figure 6.19).

1. *Objects, independent problem domains and tasks:* Objects and the system are partitioned into a number of relatively independent loosely coupled problem domains so that the decomposed tasks are of manageable size. Each sub-problem can be solved somewhat independently from other domains and each can be performed in parallel. This is discussed in Chapter 4, Product Development Methodology.

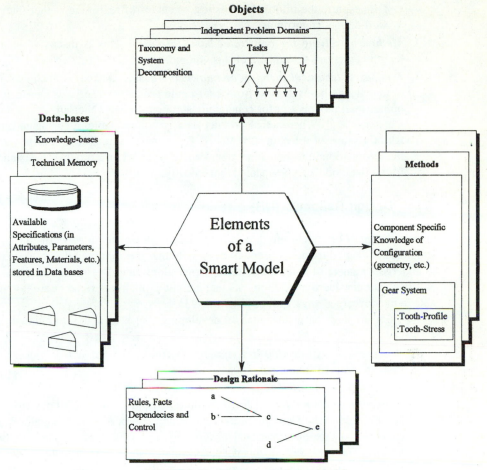

FIGURE 6.19   Elements of a Smart Model

2. ***Design rationales (DRs):*** Irrespective of the modeler used for creating smart models (knowledge-based modeler or constraint-based modeler), the modeler splits the data into *rule-sheets* and *frame-sheets*. The rule-sheets contain the equations, and the variables are automatically placed in the frame-sheets. Additional sheets can be defined as needed, such as data settings, constraints-sheet and tables where results are tabulated, and plot-sheets that define how results are plotted. This gives the original design problem a symbolic structure rather than a numerical processing structure. There are other advantages of using design rationales [Klein, 1993].

   • Definition of objectives and constraints are in terms of problem parameters, design rules, or intent with possibly incomplete data.
   • DRs establish a set of relationships (explicit or implicit) between parameters and constraints.

- DRs capture the informal heuristics, or chains of reasoning, rather than a set of well-defined algorithms.

3. ***Method of solving the constraint based problem:*** This is discussed at a greater length in sections 4.1 and 4.2 of this volume.

4. ***Databases, technical memory or knowledge-bases:*** Databases, technical memory or knowledge-bases derived from a surrogate product can serve as a basis for developing smart models and for conducting strategic studies [Nielsen, Dixon, and Zinsmeister, 1991]. When actual product data is not available, a surrogate object can take the place of a technical memory. Later, when actual data becomes available, the information in technical memory can be replaced on a modular basis and the information model can be updated dynamically.

6.6.1   Design Rationale (DR)

One of the goals of concurrent design is to identify downstream concerns during an upstream stage (e.g., during a conceptual design) so that the approved design does not have to undergo a number of unnecessary changes/iterations later. Design rationale can be used as a basis for considering requirements that are not explicitly defined or are not yet available to the concerned work groups [Klein, 1993].

There are three basic attributes for developing a successful design rationale (DR).

1. *Representation of explicit descriptions:* Definition of product or process descriptions in terms of functional, physical or geometrical features, attributes, or parameters so that the representations can be explicitly defined.

2. *Capturing functional intent:* Intent behind the product or process descriptions must be disclosed to the team, that is, functions, logic, procedures, and methods must be captured so that the functional *design intent* can be made explicit to the work groups. This includes dependencies and relationships among features and parts of the product so that changes to one part or to a feature propagate throughout all features and parts that depend on it.

3. *Identification of rest of product descriptions:* This consists of identifying and interpreting the rest of the product descriptions or their functions that are not explicitly defined. This is usually obtained through a series of numerical calculations or analyses, applying (1) and (2) successively. Examples include engineering rules for design optimization, decision criteria for extracting information from external data bases, simulation, and so on. Selection of standard parts from catalogs, parts from feature-based design libraries, and selection of rules for using geometric design data imported from CAD systems are some additional examples not covered in (1) and (2) above.

The above three basic attributes of design rationale constitutes an electronic memory or captured knowledge of the product and process design configurations. Later, when these DRs are employed during an actual execution of a concurrent process, the captured memory can provide a valuable resource for carrying out concurrent design activity.

Application of design rationale in PD^3 introduces the following benefits:

- *Integration:* DR helps integrate product with process synthesis loops, since DR representations can be used to satisfy requirements that are not part of a current activity. This enables process work groups to interpret product design requirements, and vice versa, as a regular part of a PD^3 process.
- *Design Identification:* When the design is incomplete or information is missing, DR could assist in identifying a design alternative whose descriptions closely match the stated specifications.
- *Design Consistency or Validity:* DR can check the consistency or validity of the work-in-progress design or the applicability of a design rationale. For example, using DR non-manufacturable features can be detected early in the design phase and not after the design is frozen or material has been committed to production.
- *Explanations:* DR can provide explanations for design descriptions using the basic attributes captured as a part of an electronic memory.
- *Mediation:* When constraints are violated and conflicts occur during a product realization loop, design rationale can help mediate conflicting goals.
- *Parameter Identification:* It can assist in impact evaluation due to an anticipated revision. If a revision is mandated across realization loops, DR can guide the redesign process by identifying the parameters that need to be changed.

The above merits clearly support the use of DRs as a valuable resource to facilitate a number of concurrent engineering tasks.

6.6.2   Capturing Life-cycle Intent—Applications

Recent work by Jaguar, Lotus Engineering, and others has made significant forward strides in using knowledge-based engineering for interior car design [Wilson, 1995]. They have developed a product model called smart mannequin that provides a good example of what a KBE and CE system can do. On the surface, the KBE mannequin appears to be no different than a CAD mannequin with dimensions specified by published ergonomics data tables. Like a CAD mannequin, the smart mannequin must be positioned by a user within a chosen vehicle package. This is where the resemblance ends. The smart mannequin can actually interact with the car interior environment. It can perform instructions such as adjust the seat, touch the radio, and push the clutch pedal or brake or gas pedal. Smart mannequin does this because it has captured the knowledge of the following fields:

- *Physiology*—how to configure a mannequin's limbs to perform a particular task.
- *Comfort*—given a choice of limb angles, what is most comfortable.
- *Geometric symbolism*—what is a radio and where do you find it?
- *Vision*—where are the mannequin's eyes and how do you relate what the mannequin *sees* to what a human would see?

The smart model was able to do all the above because it was programmed to perform such tasks intelligently. Effectively, the system is able to understand a set of high level instructions (e.g., touch steering wheel) and then turn the instructions into a more basic instruction set that defines the kinematics of limbs and comfort actions [Wilson, 1995]. In functional applications, it is not essential that all instructions be complex. Complex smart models can be based on combining a number of simplistic objects. Such simplistic examples of smart objects that can be created using KBE are:

* *Design intent model creation:* This involves capturing the key parameters, processes, rules, and methods of design that a design work group uses during decision making and redesigning parts.

* *Automatic geometry generation:* This involves capturing the geometry of axisymmetric parts, such as automobile wheels, bearings, steering wheels, crank shaft, connecting rod, piston, bearing assembly, brake master cylinder, etc.

* *Manufacturing form features generation:* This involves generating manufacturing form features, such as slotted-holes, center-drilled hole, counter-bore, fillet, rectangular pocket, nuts and bolts, etc.

* *Automatic drawing generation:* This involves generating drawing views, such as 2-D isometric, plan view, projections, etc.

* *Associative geometry creation:* This involves creating associative geometry for parts, such as cam shaft, rack and pinion assembly, linked geometry, etc.

* *Generic analysis model creation:* This includes creating the following from the applicable 3-D geometry objects: finite element mesh, FEA input data, analysis control data, pre-processor data, post-processor data, etc.

* *Design for assembly:* This involves creating 3-D macros for assembly, for example, association between 2-D and 3-D, interference checks, fits and clearances, etc.

* *Create and manage assembly of parts:* In conventional modeling, assembly of parts can be a laborious and repetitive task. Design work groups are required to insert actual copies of the model at appropriate locations and orientations. In smart model for assembly, one or more references to a part model only need to be identified. When a part model is changed, all instances of a part model in the assembly are automatically updated. For example, in a car model, to scale-up the diameters of the wheel from 13 inches to 15 inches, a team needs to reset the referenced parameter. Instantly all the wheels in the car assembly are enlarged from 13 inch to 15 inch wheels. Another example is a blow-up of a product assembly with the parts positioned in the same order in which they will be assembled as a unit.

* *Cost and producibility computation:* This involves capturing information on new design cost, "same as old design except" cost, new materials cost, cost of processing, operation cost, machinability cost, etc.

* *Design alternatives or what if Analyses:* This involves performing spreadsheet calculations, such as what are the effects of changing parameters (e.g., spring constants, number of turns, bore size, etc).

- *Design Sensitivity or Optimization:* This is a way to find an optimal trade-off. For example, what sets of design parameters will have the best pressure distribution and will require the least materials.

- *Scheduling, process, and layout Planning:* Such applications require access to information on manufacturing entities such as parts, assemblies, machines, tools, and transports and their various relationships to each other. It also requires knowledge about alternative machines and tools that can be used to convert one part to the other.

- *Family of parts modeling:* Variable dimensional modeling [Kurland, 1994] and macro languages facilitate obtaining variants of a generic family of parts. For example, work groups can write a generic program to generate variants of a beam design for a prescribed deflection, Δ based on a set of input and output parameters. From the principles of strength of materials, we know that maximum deflection, Δ, under a concentrated load is given by:

Δ (for a cantilever beam) = $\qquad PL^3/(3 \times E \times I)$

Δ (for a simply supported beam) = $\qquad PL^3/(48 \times E \times I)$ (6.6)

Δ (for a fixed-fixed beam) = $\qquad PL^3/(3 \times E \times I)$

Where P is the applied force, L is the length of the beam, E is the modulus of elasticity, and I the second moment of inertia about the neutral axis. Additional constraint, such as maximum bending stress occurring at the outermost fiber (σ_{max}), can be added to the above problem. The latter requires adding more equations derived from strength of materials, engineering mechanics, and variational geometry in addition to the deflection.

$$\sigma_{max} = Mc/I \qquad (6.7)$$

where M is the maximum moment, which is usually a function of load P and length L:

$$M = F(P, L) \qquad (6.8)$$

For instance, $M = PL$ (for a cantilever beam) and c is the distance from the neutral axis.

For a constant rectangular cross-section

$$c = h/2 \qquad (6.9)$$

and

$$I = bh^3/12. \qquad (6.10)$$

The above relationships between parameters can be represented as a design diagram as shown in Figure 6.20. In the design diagram, the like parameters are blocked into a set of convenient groups enclosed by dashed-line rectangles. The identifying labels signify the grouping of like parameters:

- L, I, and c are geometric parameters,
- E is the material parameter, and
- P is the load parameter.

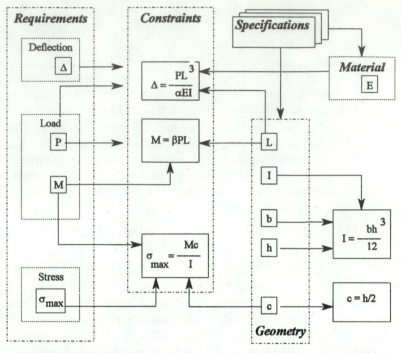

FIGURE 6.20   Design Diagram for Stress and Deflection in A Parametric Beam

- *I* and *c* can be further broken into lower level geometric parameters, *b* and *h*, as shown in Figure 6.20.
- Load, *P*, deflection, Δ, and stress, σ, are the given requirements. This is shown in Figure 6.20 by a dashed rectangle at its right.

The equations representing the relationships between the requirements and the parameters are the constraints. Constraints are shown in Figure 6.20 as a dashed rectangle in the middle. The arrows represent the flow of information. The above set of relationships represents a simple example of a parametric model where most of the requirements and constraints are explicitly defined. When a problem is complex, these relationships exist in some analyses (such as structural or finite element analysis (FEA)) or simulation form. The computations of deflection and stresses may require finite element modeling and running a FEA program from inside a generative program.

In such applications, the relationships contain the knowledge required to create a perturbed design from a baseline model and a set of specifications rather than the relationship between the parameter sets themselves. The programs are frequently called parametric models because their output, such as a variant of a generic 3-D model, is controlled by specifications. Different values of the specification set yield different designs of the same family. Creating such models can be difficult and time-consuming, but once it is written, obtaining variations of the design is a matter of familiarity. It is useful to create such parametric programs for parts that are changed more frequently several times in a model year than for the ones that remain fixed over a long period of time.

REFERENCES

Bowen J., and D. Bahler. 1993. "Constraint-based Software for Concurrent Engineering." *IEEE Computer*, Special Issue on Computer Support for Concurrent Engineering, Vol. 26 (January) pp. 66–68.

Chen, Y-M., and R.A. Miller. 1990. "Feature Extraction and Part Representation State-of-the-art," Report No. ERC/NSM-1-90-31, Columbus, OH: NSF Engineering Research Center.

Cunningham, J.J., and J.R. Dixon. 1988. "Designing With Features: The Origin of Features." ASME Computers in Engineering, Proceedings of the International Computers in Engineering Conference, Volume I, pp. 237–243, July 31–August 3, 1988, San Francisco, CA: New York: ASME Press.

Dieter, G.E. 1983. *Engineering Design: A Materials and Processing Approach*, New York: McGraw Hill.

Dixon, J.R. 1988. "Designing with Features: Building Manufacturing Knowledge into More Intelligent CAD Systems." Proceedings of the ASME Manufacturing International '88, April 17–20, 1988, ASME Press.

Finger, S., M.S. Fox, D. Navinchandra, F.B. Prinz, and J.R. Rinderle. 1988. "Design Fusion: A Product Life-cycle View for Engineering Designs." Second *IFIP WG 5.2 Workshop on Intelligent CAD*, September 19–22, 1988, University of Cambridge, Cambridge, UK.

Finger, S., and J.R. Dixon. 1989. "A Review of Research in Mechanical Engineering Design, Part II: Representations, Analysis and Design for the Life-cycle," *Research in Engineering Design*, Volume I, No. 2, pp. 121–137.

Henderson, M. 1986. "Extraction and Organization of Form-Features," Proceedings of the Vth International IFIP/IFAC Conference, pp. 547–557.

Hummel, K.E., and C.W. Brown. 1989. "The Role of Features in the Implementation of Concurrent Product and Process Design," Proceedings on the Winter Annual Meeting of the ASME, December 10–15, San Francisco, CA, *Concurrent Product and Process Design*, edited by Chao and Lu, DE-Vol. 21, PED-Vol. 36, pp. 1–8, New York: ASME Press,.

ICAD. 1995. *Understanding the ICAD Systems*, Internal Report, Concentra Corporation, Burlington, MA.

IDEAS Master Series. 1994. Structural Dynamics Research Corporation, SDRC IDEAS Master Series Solid Modeler, Internal Report, Version 6, 1994.

Joshi, S., and T.-C. Chang. 1990. "Feature-Extraction and Feature-based Design Approaches in the Development of Design Interface for Process Planning," *Journal of Intelligent Manufacturing*, Volume 1, 1990, pp. 1–15.

Klein, M. 1993. "Capturing Design Rationale in Concurrent Engineering Teams." *IEEE Computer* (Jan. 1993) pp. 39–47.

Kulkarni, H.T., B. Prasad, and J.F. Emerson. 1981. "Generic Modeling Procedure for Complex Component Design," SAE Paper 811320, Proceedings of the Fourth International Conference on Vehicle Structural Mechanics, Detroit, Warrendale, PA: SAE.

KUMAR V. 1992. "Algorithms for Constraint-Satisfaction Problems: A Survey." *AI Magazine*, Volume 13, No. 1.

KURLAND, R.H. 1994. "Understanding Variable-driven Modeling." *Computer-Aided Engineering* (January 1994) pp. 38–43.

LUBY, S.C., J.R. DIXON, and M.K. SIMMONS. 1986. "Designing With Features: Creating and Using a Features Data Base for Evaluation of Manufacturability of Casting." Proceedings of the ASME Computers in Engineering, Volume 1, pp. 285–292, ASME.

NIELSEN, E.H., J.R. DIXON, and G.E. ZINSMEISTER. 1991. "Capturing and Using Designer Intent in a Design-With-Features System," Proceedings, 1991 ASME Design Technical Conferences, 2nd International Conference on Design Theory and Methodology, Miami, FL.

NEVINS, J.L., and D.E. WHITNEY, eds. 1989. *Concurrent Design of Products and Processes*, New York: McGraw Hill.

PABON, J., R. YOUNG, and W. KEIROUZ. 1992. "Integrating Parametric Geometry, Features, and Variational Modeling for Conceptual Design." *International Journal of Systems Automation: Research and Applications (SARA)*, Volume 2, No. 1, pp. 17–36.

PAHL, G., and W. BEITZ. 1992. *Engineering Design: A Systematic Approach,* edited by Ken Wallace, New York: Springer-Verlag.

PDA ENGINEERING. 1992. "What's Needed for Competitive Manufacturing—Concurrent Thinking: Simplifying the Process," *Concurrent Engineering Supplement to Computer Aided Engineering,* (October 1992) pp. S1–S15.

ROSENFELD, L.W. 1989. "Using Knowledge-based Engineering." *Production* (November) pp. 74–76.

ROSENMAN, M.A., and J.S. GERO. 1993. "Creativity in Design Using a Design Prototype Approach." Chapter 6 of *Modeling Creativity and Knowledge-based Creative Design*, pp. 111–138, edited by Gero and Maher, Hillsdale, New Jersey: Lawrence Erlbaum Associates, Publishers.

SHAH, J. 1991. "Assessment of Features Technology." *Computer Aided Design*, Volume 23, No. 5 (June 1991) pp. 331–343.

SRIRAM, D., G. STEPHANOPOULOS, R. LOGCHER, D. GOSSARD, N. GROLEAU, D. SERRANO, and D. NAVINCHANDRA. 1989. "Knowledge-based System Applications in Engineering Design: Research at MIT," AI Magazine (Fall 1989) Volume 10, No. 3, pp. 79–96, AAAI Press.

WHETSTONE, W.D. 1980. "EISI-EAL:Engineering Analysis Language," Proceedings of the Second Conference on Computing In Civil Engineering, pp. 276–285. New York: American Society of Civil Engineering (ASCE).

WILSON, J. 1995. "Knowledge-based Engineering—A New Tool in Car Design." *Automotive Interiors International*.

WINSTON, P.H. 1992. *Artificial Intelligence,* 3rd Edition, New York: Addison-Wesley Publishing Company, Inc.

TEST PROBLEMS—CAPTURING LIFE-CYCLE INTENT

6.1. What percentage of the eventual cost of a product is determined (during a PD^3 process) before any full scale development or a CAD tool usage actually begins?

6.2. What is the static representation of a design? Why is this representation inadequate for handling configuration type design changes? What are other significant drawbacks of static representations?

6.3. What is meant by capture of a design intent? Why is this *intent capture* considered a better representation than a *static capture*?

6.4. What is the difference between a life-cycle capture and a design-intent capture? How would you capture a life-cycle intent? Does a conventional tool provide such functionality. If not, why?

6.5. How do you classify the various processes for performing a design? Show, using a Venn diagram, the existence of these classifications in a design space. Which of this class is more suited for automation and why?

6.6. Describe a progression of knowledge maturity with time for the classes of design that were discussed in exercise 6.5. How does product maturity relate to knowledge-capture techniques and what criteria determine whether to computerize, automate, or leave the process alone?

6.7. What are the various ways of capturing a life-cycle intent of a manufacturable product? What are the steps in capturing a life-cycle intent using a language-based tool? How could design intent be communicated along with other design details?

6.8. How do you define a decision tree structure? What is a definition of a feature? What are the different ways to identify a feature? How can the use of a decision tree make the solutions obvious?

6.9. How do you define a physical object? How do you capture its definition? What are its physical attributes?

6.10. How do you capture a functional intent of a product? What are the benefits of such a capture? How does it affect iterations? Does the function intent capture include a definition of its iterative process? If not, why?

6.11. How do you capture the relationships between various parts? What process do you use? How do you express the definition of a part in terms of its primitives? Give some examples of primitive parts.

6.12. How do you obtain an assembly of parts? How do you place and position primitive parts? What are the mating constraints?

6.13. How do you capture the knowledge of constituent parts? How do you represent this knowledge? What is a generic total product model (TPM)? What constitutes a product knowledge and a process knowledge? What are the rationales of capturing the WBS, PtBS, and PsBS functions through a network of intelligent models?

6.14. Describe an evolution of languages for capturing intent knowledge. Describe three types of languages that are common for capturing a life-cycle intent.

6.15. What is a knowledge-based system (KBS)? How is KBS built? What is the process for building KBS? How does an effective KBS system minimize resource utilization and time?

6.16. Describe a computational architecture of a typical KBS. What are its components? Describe the inter-relationships of these components. Which of these is a foundation layer?

6.17. What is the significance of an intelligent interface layer in KBS? What are some critical technologies employed within this layer?

6.18. What are the differences between geometry-based, constraint-based, and a language-based engineering system? Describe the characteristics of a knowledge-based language.

6.19. Why are some languages easier than others? Which of the languages provide a way to capture both geometry and non-geometric attributes?

6.20. Describe the importance of knowledge capture. What is meant by capturing a model of a product not an instance of a product itself? What is a smart model?

6.21. Describe a composition of a smart TPM. What constitutes an analysis model? What constitutes a design model? What constitutes a process model? How does the advent of design languages help creation of a detailed complex series of models?

6.22. What gives a smart or intelligent model the ability to define geometry, parameters, and constraints? How do you formulate a problem that leads to a smart model? Are smart models the results of good knowledge acquisition team, good knowledge capture method, good knowledge content, or all of the above?

6.23. What is constraint-based programming (CBP)? What are some of its constraint properties that can be employed during CBP?

6.24. What are the key characteristics of a smart model? How do they differ from a conventional model? What differentiates a constraint-based smart model from a knowledge-based smart model? What are the key parameters surrounding a constraint-based smart model? How do you compute sensitivities in CBP?

6.25. What are the four modeling modes or types of representations that can be used in CBP? Discuss the differences and similarities between them.

6.26. How does a feature-based representation differ from a parametric or a symbolic representation? What are some of the standard features that can be identified in a mechanical discrete part? How do you characterize these features?

6.27. How do you classify various characteristics of features? How useful are features in identifying the complexity of design and process related operations? In what situation is a mixed representation useful?

6.28. What are the different forms of knowledge logic that is employed in knowledge-based programming (KBP)? How are these sets of knowledge combined into a smart TPM? What are the benefits of KBP?

6.29. How does KBE improve productivity and automation? How do you establish productivity targets in life-cycle capture and automation? Is the purpose of automating processes to assure consistent quality? What is the percentage of routine tasks that are usually affected?

6.30. What are the four main elements of a smart model? How can you capture and build trial process into a TPM in order to establish a pathway to the best design?

6.31. Building smart models are not always advantageous. How would you determine when to build a smart model and when not to build?

6.32. One of the goals of concurrent designs is to identify downstream concerns in upstream stages. What are the three basic attributes to developing a successful design rationale (DR)?

6.33. What are the key benefits of KBE? What are the benefits of capturing design rationales? How does DR facilitate CE? Describe some examples of smart models that can be created using KBE.

6.34. What is a single product model (SPM) concept? What does a typical SPM model include? Why is coupling product and process constraints with one common geometry model critical for CE success?

CHAPTER 7

DECISION SUPPORT SYSTEMS

7.0 INTRODUCTION

Design decisions are a complex array of diverse and often contradictory cognitive activities. In CE, most work groups get involved directly in decision making and consulting activities. There are probably thousands of design decisions that are made in an ordinary kind of product. Almost every one of such decisions involves some sort of trade off—performance against cost, what one team (or person) wants against another team (or person), organizational issues against technical issues, and so on. Design decisions differ with each new piece of added information, new person, or new issue discovered. Design issues continually change and evolve during every step of the design. This is because design is an open-ended problem. Ordinarily, many solutions to a design problem are possible. The outcome is determined largely by the extent to which a design problem is understood by the work groups and by the process that is applied to solve them, including 3Ps (policy, practices, and procedures). In many cases, decisions can spread over multiple teams and organizations. If some of these organizations were not part of the original PDT tree structure, this can delay an agreeable outcome. It is hard to resolve the issues early with only a partial team present, since an agreed plan requires everyone's inputs. It is also hard to kick the issues to their superiors, if some teams were not part of an original product development team (PDT) or a CE structure.

Decisions made in early stages of design processes have profound affects on later stages as explained in Chapter 2 of volume I. Right decisions, if timely folded in, can produce tremendous savings in the life-cycle cost of a product. Conversely, the price paid for late or wrong decisions can be devastating. Early right decisions can ensure business success by enabling the production of better performing, more robust, and more reliable products. This requires early determination of some key part characteristics (KPCs) that make a product robust and reliable. In traditional organization, however, information on such characteristics (KPCs) is obtained only after a great delay, by which time the product concept is mostly frozen. It is too late then to make any major product design modification decisions. Teams are required to collaborate over long distances, often cross-nationally. Fortunately, distance normally has not been a problem for delay in decision making. Today, remote collaboration can take place with startling accuracy and speed through electronic networks. Decision support environment (DSE) provides a virtual framework for allowing concurrent team members to establish communication with each other early during a PD^3 (product design, development and delivery) process. This can have a major impact on the design if critical decisions about major product modification strategies have to be made quickly and cooperatively.

Technical communication in DSE is much more demanding, even more so than the airline reservation system or the transactions in banking business. Unfortunately, the technical communication and other software capabilities in DSE have not yet caught up. The transactions in airline and banking businesses are fairly routine, quantitative communications. Other communication tools such as video conferencing and the like are in use today, but unfortunately they are unsuitable for engineering collaboration. They are appropriate mainly for conducting business meetings. In engineering collaboration, the work groups need to know, in real time basis, how a set of parameters proposed by a work group affects other sets of parameters whether or not he or she is directly responsible for them. For instance, if a work group is trying to design a mechanism, it is not enough that the mechanism functions kinematically. Someone may like to know the sweep volume or the trajectory traced if this has to fit into somebody's assembly. If a work group is designing an automobile door panel, it is not enough for the design work group to design it for aesthetic consideration alone. Someone from the variation analysis work group may want to check to see how the parts fit with the body. Someone from the processing work group may want to check for the sheet metal formability of the part. The DFM work group may want to check for DFMA rules, and the structural work group may want to insure integrity of the design from strength and stiffness perspectives.

In a similar manner, an evolving product design goes through a number of revisions, a number of CAD modifications, a number of alternative proposals, prototypes, and so on. One then has to sort through the accompanying electronic or CAD file versions to determine the right output (design) from those alternatives that were evaluated. There is a lack of adequate collaborative tools useful for making early product trade-off during DSE. Most existing DSE tools, for example, are not equipped to compare alternative designs (i.e., to identify a good design from a bad design), or to compute design sensitivities. In the absence of such capabilities, and with tight manufacturing schedules, most trade-off studies, in the beginning of a design cycle, are left incomplete when a product is passed on to the next work group. Decision support tools include both manual and computer techniques to aid decision making, analysis, simulation, documentation, sensitivity, optimization, and control of everything in an enterprise. Though there are a number of developments in CAD/CAM and CAE arenas, their DSE capabilities are still inadequate to enable CE. In the present form, most C4 (CAD/CAM/CAE/CIM) systems are mainly suitable for analyzing a problem or for capturing an explicit, static geometric representation of an existing part. They are not suited for altering a part geometry, say using variable dimensions, or capturing its engineering design intent. Some recent C4 systems use parametric modeling, variational design, adaptive modeling, feature-based modeling or knowledge-based techniques to capture a part life-cycle intent. Such developments are dynamic in nature when it comes down to managing changes. These new C4 systems use the intent-driven techniques to generically capture a product's life-cycle values.

In recent years, more and more emphasis is being placed on the use of such intent-driven techniques for decision support during concurrent engineering [Talukdar and Fenves, 1989; and Sapossnek, Talukdar, Elfes, Sedas, Eisenberger and Hou, 1989]. Alternatively, many companies are developing specialized KBE systems (often called expert systems) targeted towards providing decision support for a particular product line or a line of product family. Automated Simultaneous Engineering (ASE) is an example of a prototype critic-based system, an expandable library of autonomous programs called critics. ASE was a joint research project between Carnegie Mellon's Engineering Design Research Center and General Motors' Inland Fisher Guide Division, now called Delphi-I [Sapossnek, Talukdar, Elfes, Sedas, Eisenberger and Hou, 1989]. The initial domain of ASE was a window regulator design. ASE consists of four components, a synthesis system and three critics (a tolerance critic, a mechanical strength critic, and a kinematics critic). Flexible Organization (FORS) [Papanikolopoulos, 1988] was employed as a framework for integrating critics. A constraint based design language called Design Objects and Constraints (DOC) [Sapossnek, 1989] was used to create a system for a window regulator synthesis.

7.1   BASIS OF DECISION MAKING

Decision making can be viewed as a process of creating an artifact that performs what is expected (specified as a set of requirements) in the presence of all sorts of constraints and operating environments that govern its behavior. The constraints can be pre-specified or can evolve during a design (or PD^3) process. The concepts of concurrency as described in volume I and the decision support system that enables concurrency constantly interact, each pushes the other to ever greater heights. The types of decisions that engineers make today to solve design problems are bounded by a spectrum with the cognitive aspect at one end of this spectrum and the progressive aspect at the other end [Finger and Dixon, 1989]. In cognitive-type situations, cognitive knowledge about the problem and its environment helps the problem solver. A CE team identifies an outcome, a pattern, or an hypothesis from a finite set of possible outcomes that any team-member has experienced, which is closer to the functions of the product. In progressive-type situations, however, information about the product and its behavior is unknown. The problem solver is required to follow an explicit method for approaching the solution.

7.1.1 Cognitive Decision Models

Knowledge of designs certainly plays a very important role in coming up with a suitable artifact. Depending upon the cognitive knowledge about a product available to a decision maker, design decision may range from cognitive (that can be adapted) to progressive (having less cognitive knowledge). It is known that experienced designers can create better designs, and in a shorter time-frame compared to a novice. What is not known is how the cognitive process in such cases works and how it helps in accelerating the decision making process. For example, why don't individual designers seem to be as creative and productive as a team of designers? Even if a vast amount of information is available about a product, it is hard to replace the cognitive knowledge the designers have. Most experienced designers can explore alternatives even with missing information and lack of data because their knowledge is broader, more general, and abstract. Less experienced designers are not able to explore alternatives as well. Instead, they try to extend their initial ideas for the new cases. AI techniques such as rule-based inferencing and expert systems that incorporate and capture the expert's knowledge of the experienced designers, can aid the novice (or less experienced) in decision making.

Commonly, a cognitive model is structured with four pieces of information: alternatives, criteria, knowledge, and belief [Herling, Ullman and D'Ambrosio, 1995] as shown in Figure 7.1.

1. **Criteria:** Criteria measure the quality of the proposed solution. The complexity of the stated problem is defined in terms of its constituents—inputs, requirements and constraints. In the cognitive decision model, the criteria are chosen from the inputs, requirements, or constraints, since outputs are unknowns.

$$\{\text{Criteria}\} = f\,[\{\text{Inputs}\}, \{\text{Requirements}\}, \text{and } \{\text{Constraints}\}] \qquad (7.1)$$

2. **Alternatives:** Alternatives are derived from a set of proposed solutions to the original problem. By specifying alternatives as solutions to the problem, the CE teams begin populating the solution space with design alternatives. Population of alternatives is based on the knowledge of the expected solution and the degree of belief (or confidence) that this alternative will meet the stated requirements or would satisfy the imposed constraints. Arguments for and against alternatives are stated in terms of the knowledge and belief. The stated values for belief and knowledge assert how well an alternative satisfies or will satisfy each of the criteria. In the context of parameters defined earlier, the following relationship holds good:

$$\{\text{Alternatives}\} = f\,\{\text{Anticipated Outputs or Solutions}\} \qquad (7.2)$$

3. **Knowledge:** In the cognitive decision model, knowledge is captured from a participant's personal experience, and his profound knowledge about the level of goodness of an alternative relative to a chosen criterion. The profound knowledge comes from working with other similar parts and his previous knowledge and experience about the alternative's function. Ten example levels are chosen as shown in Table 7.1 to rank the level of knowledge the participants possess.

4. **Belief:** Belief quantifies the level of surety about an assertion that an alternative will satisfy the criterion. Ten levels are chosen as shown in Table 7.1 to grade this belief. The belief that an alternative will or will not satisfy the criterion is assumed independent of the participant's knowledge.

$$\text{Measure of goodness of } \{\text{Alternatives}\} = f\left[\{\text{Belief}\}, \{\text{Knowledge}\}\right]$$
$$\text{or in short } f\left[\{B\}, \{K\}\right] \tag{7.3}$$

where B is the degree of belief (or confidence) and K is the level of knowledge (or expertise). Table 7.1 gives the numerical values for B and K that a participant can assign. The values depend on the level of confidence and the knowledge that one possesses about the alternative and whether or not the alternative would be able to satisfy the criterion.

7.1.1.1 Steps of a Cognitive Decision Model

In the following, a procedure similar to a QFD matrix is followed to obtain the cognitive decision model (CDM) matrix. The CDM matrix is similar to a QFD matrix. Like the QFD matrix, CDM has eight rooms, four line vectors and four 2-D matrices. The line vector corresponding to *WHATs* row contains the list of *criteria*. The *HOWs* column lists the *alterna-*

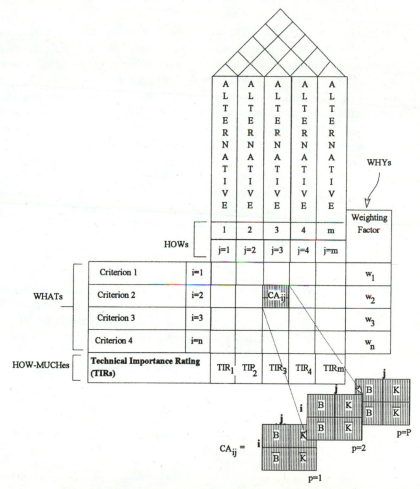

FIGURE 7.1   Cognitive Model Diagram for Decision Making

TABLE 7.1 Quantified Values for Several Degrees of Knowledge and Belief

Belief (B)		Knowledge (K)	
Word Description	**Value**	**Word Description**	**Value**
Full Confidence (Definite)	1.0	Full knowledge (Perfect)	1.0
Certain	0.9	Expert	0.9
Very Likely	0.8	Experienced	0.8
Likely	0.7	Informed	0.7
Potential	0.6	Amateur	0.6
Average	0.5	Normal	0.5
Questionable	0.4	Weak	0.4
Unlikely	0.3	Uninformed	0.3
Very Unlikely	0.2	Inexperienced	0.2
Uncertain	0.1	Novice	0.1
No Confidence (Indefinite)	0.0	No Knowledge (Empty)	0.0

tives that need to be assessed. The *WHYs* column contains the weighting factors—the factor assigned by each participant showing the relative importance indicator of each criterion. The values may range from an equal weighting to an extreme weighting, when a single criterion dominates. The *HOW-MUCHes* row contains the computed values for technical importance rating (TIRs). Technical importance ratings for each alternative can be computed based on the Bayes Equation proposed in Herling, Ullman and D'Ambrosio [1995].

  Step 1: Filling the Correlation Matrix: As in QFD the correlation matrix [*WHATs* versus *HOWs*] contains the results of surveys from each participant. Two pieces of information are elicited from each of the participants towards completion of this step. The first piece is an identification of belief and knowledge in terms of a set of word-pairs selected from a Table. A belief word is paired with each knowledge word. Let us assume person 1 chose "very likely, normal" and person 2 chose "likely, amateur" pairs. A number of such knowledge words and belief words is shown in Table 7.1.

  The second piece of information is obtained through translating these word-pairs into an equivalent numerical value pair. For example, as shown in Table 7.1, likely belief is equivalent to 0.7 and an amateur knowledge is equivalent to 0.6. Since two locations are required to store values for *B* and *K*, each (i, j) shell of the *WHATs-HOWs* matrix is further divided into four parts as shown in Figure 7.1. This allows the storage of the four sets of numbers: *B, K, $\overline{B}$,* and *$\overline{K}$* in the divided shells corresponding to each (i, j) location. The second row is derived from the first row (see Table 7.2). The terms are defined as follows:

  B = The degree of a participant's belief (or confidence) that the *j*th alternative would satisfy the *i*th criterion.

$$\overline{B} = (1 - B). \tag{7.4}$$

  K = The degree of a participant's knowledge (or expertise) that the *j*th alternative would satisfy the *i*th criterion.

$$\overline{K} = (1 - K). \tag{7.5}$$

TABLE 7.2 CA_{ij} Shell of a Criterion-alternative (CA) Matrix

	*j*th Alternative	
*i*th shell	$\dfrac{B}{\bar{B}}$	$\dfrac{K}{\bar{K}}$

The belief word B in step 1 and its associated number from Table 7.1 quantifies the participant's confidence in meeting the criterion. The two words and the corresponding numerical value pair (B, K) together quantify the participant's belief and the knowledge level that jth alternative will satisfy the ith criterion. The other two numbers $\bar{B}$ and $\bar{K}$ indicate the opposites or complements of B and K, respectively.

Using this and the Bayes Equation [Herling, Ullman and D'Ambrosio, 1995] for example, one can express the acceptability of an alternative by all involved participants (p) as

$$\text{Acceptability of the alternative for a particular set of criteria} = \Pi_p\,[BK + \bar{B}\bar{K}] \quad (7.6)$$

where Π indicates a product symbol taken over all participants.

Step 2: Filling in the WHYs column: This is similar to the *WHYs* column of an extended QFD matrix. The participants fill in the weighting factors associated with the criteria indicating the relative importance of the criteria. If there is more than one participant involved in the survey, this column lists the values after averaging the survey numbers from all participants. If the participants have assigned no value to a criterion then its normalized value is zero.

W_i = weighting factors—the normalized factor showing the relative importance indicator for each criterion. Normalized is meant to indicate a scalar length of unity.

$$\sum_{i=1}^{n} W_i^2 = 1.0 \quad (7.7)$$

Step 3: Computation of CA_{ij}: This amounts to filling in the computed values in the $(i,j)^{\text{th}}$ shell of the Criterion-Alternative matrix based on the procedure shown in Figure 7.1. The magnitudes of B, K, $\bar{B}$, and $\bar{K}$ assigned by each participant in step 1 is used in the computation.

Where p denotes the number of participants, $p = 1, 2, 3, \ldots, P$,
P is the total number of participants.
CA_{ij} indicates the $(i, j)^{\text{th}}$ location of the criterion-alternative matrix.

It contains the computed value of all the participants in the cognitive survey as shown in equation (7.8).

$$CA_{ij} = \left[\frac{\Pi_p\,[BK + \bar{B}\bar{K}]_p^{ij}}{\Pi_p\,[BK + \bar{B}\bar{K}]_p^{ij} + \Pi_p\,[B\bar{K} + \bar{B}K]_p^{ij}} \right] \quad (7.8)$$

where Π_p is a symbol of product, which is taken over all participants ($p = 1$ through P) providing the pair (B,K) information.

n = the number of criteria present and

m = the number of alternatives or concepts selected for evaluation.

Step 4: Computation of TIRs: This is similar to technical importance ratings for QFD. The numbers in the (i,j)th location of the criterion-alternative matrix is multiplied with the weighting factors stored in the *WHYs* column and the results are added. This is based on equation 7.9.

$$\text{TIR}_j = \sum_{i=1}^{n} W_i \times CA_{ij}; \qquad j = 1, m \qquad (7.9)$$

The TIR result for each *j*th alternative is placed in the *HOW* versus *HOW-MUCHes* of the CDM matrix, similar to QFD case, as shown in Figure 7.1.

Table 7.3 defines the equivalence between QFD and CDM matrices.

The other basis of decision making, which is not so dependent upon cognitive knowledge, is progressive modeling. This is discussed next.

7.1.2   Progressive Decision Models

The basis for decision making in progressive models is through experimentation of procedures or through analysis of the process of design. Progressive models describe and substantiate the elements, properties, sequences, and effects of ideal and actually observed design processes in their socio-technical context, including all aspects of a company, an organization and leadership. Many people have written about the real process of conducting design [Finger and Dixon, 1989, Pahl and Beitz, 1991; Suh, 1990]. Though there is little actual research to support the outcomes, two models seem to emerge in this progressive category: the German method and the axiomatic method.

7.1.2.1   German Method
German method is perhaps the most popular progressive model of design. Pahl and Beitz [1984 or 1991] in their book titled *Engineering Design* describe the thought process behind this method. The German method delineates three distinct stages of design: conceptual, embodiment, and detailed. The book describes explicit tasks associated with each of these three stages. The concept behind distribution is the key to the success of this method. There is no set pattern as to how to subdivide the design process. Others, who

TABLE 7.3   Equivalence Between QFD and CDM

QFD	CDM
WHATs	Criterion
HOWs	Alternatives
WHYs	Weighting Factor
WHATs versus *HOWs* Matrix (WH_{ij})	Criterion vs. Alternative Matrix (CA_{ij})
WHATs versus *WHYs* Column (W_i)	Criterion Weighting Factor Column (W_i)
Technical Importance Rating (TIR) for a *j*th *HOW* to satisfy all the stated *WHATs*	Technical Importance Rating (TIR) for a *j*th alternative to satisfy all the stated criteria.

consider design in the life-cycle sense, view the design as a five-step process: problem definition, conceptual design, configuration/layout design, detailed design, and manufacturing design. Design assessment, value redesign, simulation/analysis/prototyping, test/tune and evaluation are considered parts of a detailed design. German methodology is quite progressive. It states what to do and when to do it. In recent years, many researchers in the United States are paying attention to this method [Dixon, Guenette, Irani, Neilson, Orelup and Welch, 1989; Finger and Dixon, 1989] and many German manufacturing industries have achieved considerable success. Figure 9.22 of volume I shows a breakdown of tasks embodying the various stages of design.

7.1.2.2   Axiomatic Method

Axiomatic Design is a term used by Nam Suh in his book *The Principles of Design* [Suh, 1990] to describe a process by which high quality designs can be achieved by adhering to a small number of axioms. It is based upon the belief that fundamental principles, well accepted truths (axioms) and corollaries of good design exist in real life and their use in guiding and evaluating design decisions can lead to good designs. Analysis and refinement of initial attempts have shown that good design embodies two basic concepts. The first of these is based on the idea that functional requirements of a product can be satisfied independently by some aspect, feature, or component of the design. This concept, to a large extent, reduces or eliminates functional coupling between its constituent parts. The second basic concept is that good designs maximize simplicity, that is, constituent parts could provide the required functions with minimal complexity. The two concepts thus keep the constituent parts simple with less functional couplings and associations. These two concepts have been formalized by Albano and Suh [1992] as the Independence Axiom and Information Axiom.

- *Independence Axiom*: In good design, the independence of functional requirements is maintained.
- *Information Axiom*: Among the designs that satisfy the Independence axiom, the best design is the one that has the minimum information content.

Gebala and Shu [1992] expanded on the axiomatic approach and applied it to specific industrial design strategies. Though the axiomatic approach is shown to be useful, application of design axioms to the analysis and design of products and manufacturing systems is not always easy. It requires considerable practice and ample on-the-job experience [Dixon, Guenette, Irani, Neilson, Orelup and Welch, 1989]. In Chapter 4 of this volume, an alternative plan is proposed, which accounts for these two axioms. The proposed plan follows a four stage IPD systematization process discussed in section 4.6 of volume II. The four stages are planning, systematization, solution, and unification. The combination of axioms and IPD systematization leads to the following four step procedure (see also Figure 4.7 of volume II).

Step 1: The first step is to identify the functional requirements (FRs) for each cycle and for each loop within a cycle. There are two cycles and five loops (see Chapter 9, volume I). Each FR should be distributed into the loops such that FRs are neither redundant nor inconsistent. Similar to loops, FRs are also decomposed into corresponding levels of specifications. This is explained in section 4.3, volume I. The decomposition of FRs into loops proceeds simultaneously with the product decomposition process. When an FR is decomposed into corresponding loops, redundancies can occur. Redundancies can be removed by first searching the current list of loop level FRs, and then performing a dependency check according to the following rules:

- If an FR is found to be the same as a member of the loop list, then work-group deletes the current FR.
- If an FR is implied by another FR, then work group deletes one of the FRs.
- If an FR is found to be similar to an old FR, then the work group decomposes both FRs, if they can be decomposed further.

By following the above three steps (with its alternate decomposition and dependency check of FRs), the first axiom can be satisfied implicitly. After the dependency check of FRs, a condensed form of the functional requirements that represent an independent set of FRs for each loop is generated.

Step 2: The second step is to order the FRs in some hierarchical structure. In section 4.6, a system-based 4-stage hierarchy was proposed. In each hierarchical structure, the FRs are arranged in preferential order, from high importance to low importance ones. After the decomposition and dependency check of functional requirements, the final functional structure contains the minimum set of functional requirements, and all functional requirements in the lowest level are independent from one another.

Step 3: The third step is to identify the functional constraints that result from the unification stage (discussed in section 4.6) but need to be considered. This step is similar to Step 1 except that this time a unification cycle is followed. Instead of working with FRs, teams work with FCs and do unification and distributions of the FCs. These will form constraints for each loop. After the dependency check of all constraints, a condensed form of functional constraints that represent an independent set of FCs for each loop is generated.

Step 4: The fourth step is to apply the functional requirements (FRs) originating from the decomposition stage and functional constraints (FCs) imposed by the unification stage (discussed in section 4.6) so that the product is evolved through different levels of decisions. As work groups move from feasibility to manufacturing loop (represented as a part of five steps loops methodology discussed in section 9.2 of volume 1), FRs and FCs are simultaneously met.

Beginning with the conceptual design and following through each realization loop in ascending order, the above four-step procedure can be used to provide insight into design problems. The procedure forms the basis for decision making, and leads the way to good design. The steps must support each decision and satisfy all the constraints. The increased complexity of the modern information age, the continual need for change, and the constant emergence of new materials and technology place ever increasing demands on proper and complete understanding of the design problem. The broad spectrum of such issues needs to be addressed when applying this axiomatic approach.

7.2 TYPICAL PROGRESSIVE MODELS

Fundamental approaches to progressive design stem from different design views and methodologies used in solving a problem. Any decision making situation can be viewed as a series of interactions between inputs, outputs, requirements, and constraints as shown in Figure 7.2. Solution logic may be driven by types and sources of the information that describe the design process. This section contains in its extended sense all formal and ideal aids (tools and methods) that the work groups can use in order to think out possible design models to calculate, analyze, or to evaluate design alternatives, or to come up with a design of new or revised products. These have been categorized into seven types of progressive models: (a) Model-based, (b) Manual-based, (c) Experimental-based, (d) Analytic-based, (e) Heuristic-based, (f) Algorithm-based, and (g) Hybrid-based.

7.2.1 Model-based System

Model-based approach is a structured method of decision making aided by techniques such as parametrics, feature-based, or knowledge-based engineering as shown in Figure 7.2. These techniques are also called computer-based because of the use of computers in dealing with and capturing parameters, features, and knowledge. The model-based system captures models (design methods) or processes of design through a series of cooperating modules. The model-based techniques, in turn, provide an experimental apparatus for studying design methods in much the same way that an experimental physical apparatus helps to query a physical behavior. The model-based approach is a trade-off process where a method is set to modify the initial model as the design proceeds. Need for modification of models may come from multiple sources due to the uncertainties inherent with system design, the need to address many design criteria arising from many perspectives of design, and the needs to alter the criteria arising from the assumptions used to model the process. Assumptions may deal with the level of detail used in modeling the system or in interpreting the outputs. Independent of the sources of needs, feedback mechanisms actually drive the change process. As the design process continues, the feedback resulting from a synthesis part of the model provides knowledge to a work group (a human team, a virtual team (computer program) or a decision-maker). The work group, in turn, adjusts the problem sources by reducing the level of uncertainty, tightening the assumptions, or expanding the perspectives. The model provides the foundation for the solution process. The model-based approach can be looked upon as a multi-criterion problem (such as minimax approach), where a series of successive model refinements are obtained in which each successive new design satisfies more objectives than the old ones. The model-based system finally converges to a solution that satisfies most of the design objectives in the best manner. Problem complexity is dealt with and managed by using captured knowledge and computer tools. The model-based system involves the choice of the following strategies:

- Careful consideration of a PD^3 process that is based on a concurrent product realization method (such as loop methodology discussed in Chapter 9 of volume I).

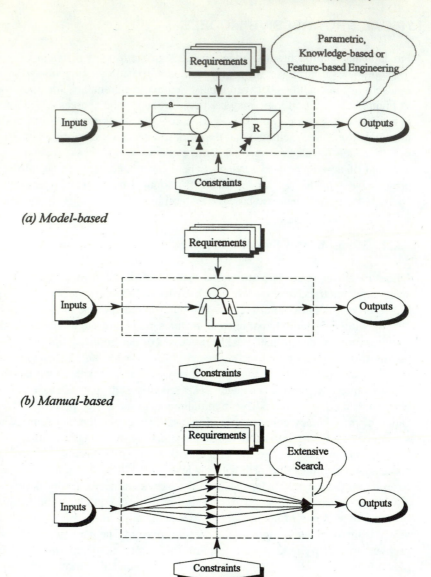

(a) Model-based

(b) Manual-based

(c) Experimental-based

FIGURE 7.2   Types of Progressive Models: (a) Model-based, (b) Manual-based, and (c) Experimental-based

- A coordinated application of appropriate modeling tools and programs at each stage of product development. This includes knowledge about the general strategic approach to designing and knowledge about tactics and methods for designing.
- A coordinated application of *groupware tools* to aid in problem solving. The *groupware tools* include all equipment, apparatus, multi-media computers and machines at the workplace, which directly support the concurrent team in their problem solving. Other collaborative tools are WBS structure definition, search for solution, decision making, and other analysis tools.
- Aspects of the PD^3 process that are computationally intensive are dealt with using analytical models within the confines of product realization loops. Those aspects that are knowledge intensive are dealt with by one of the four CE teams.

The advantages of the model-based system are many. The user interfaces are much friendlier (same look and feel). The model logic can be structured with less dependencies than in procedural systems if Object-oriented programming techniques (OOPS) are included as a part of the development environment. Many of the core modules can be linked to an existing relational or OOPS database. These factors allow model-based systems to be developed, implemented, and maintained by the cooperating CE work groups. A model-based system that is well integrated with a manufacturing execution system (MES) on a factory floor can be very powerful for product manufacturers. A simple model-based system can allow someone to build, for example, a process plan by selecting standard object operations and inserting values into these standard object descriptions. Time standard formulas and costing data can easily be calculated.

7.2.2   Manual-based Approach

In a manual-based design approach, as shown in Figure 7.2(b), the work group members play a key role in decision making. When an input information comes to a team member, the work group evaluates the baseline design, controlling the design development and interpreting the solution. The output reflects the range of objectives and constraints included in the baseline system that the work groups choose to include.

7.2.3   Experimental-based Approach

The experimental-based approach to design relies on matching design attributes to the objectives of a design process. As shown in Figure 7.2(c), this approach employs a series of possible searches to arrive at a solution of the design problem. An experiment is not intended to solve the design problem, but to seek a point along the solution path. Each solution point identifies an outcome on the solution path in which the performance and parameters change. The purpose of such an experiment-based approach is to reduce or control variations in a product or process. Taguchi's methods of design for experiment is an example of this type of approach. Depending upon the complexity and stage of design, there could be a large number of iterations required. Experimenting with all those possibilities may be difficult from a practical standpoint. Apparent inefficiency, due to the need for conducting a large number of experiments, makes this approach less attractive. However, when the behavior of the system is unknown or the system is new, experimental-based approach is often the only recourse available. The following are some of the common experimental techniques.

- *Test Plan*: The most common test plan is to evaluate the affect of one parameter at a time on product performance. Thus, for n parameters, n results are evaluated. In order for the data to make any sense, parameters are varied one at a time while the rest are kept constant. This gives the sensitivity of performance on design variables.
- *Taguchi's Loss Function*: The loss function quantifies the need to understand which of the design factors cause the most, or above average, variation in the performance characteristics of a product or a process. This is frequently referred to as *quality loss function*. It is defined as the amount of functional variation in a product plus all possible negative effects, such as environmental damages and operational costs (see Figure 3.17, volume II). This is discussed at length in section 3.7.3 of volume II.
- *Taguchi's Experiment*: The major steps of design for experiments are:
  - Selection of factors or interactions to be evaluated.
  - Selection of number of levels for the factors.
  - Selection of an appropriate orthogonal array.
  - Assignment of factors and/or interactions to columns.
  - Performing tests.
  - Analyzing results.
  - Verifying results (run a final test).
  - Eliminating parameters and repeating the above steps.
  - Continuous improvement.

  The method is explained in section 3.7.3 of volume II (see Figure 3.15).

7.2.4   Analytic-based Approach

Analytic-based approach describes, through mathematical models, the knowledge about properties and constituents of socio-technical systems and their elements. This knowledge is correlated with the available theory of properties (e.g., strength of materials). This assumes that a precise objective and a set of specifications for the problems are given or known as shown in Figure 7.3d. In reality, most physical problems cannot be modeled into an analytical basis since many of their behaviors cannot be quantified. Some ad-hoc analysis methods include formulae and graphs, hand-books, exact mathematics (ordinary and partial differentiation equations), and numerical methods (finite difference, finite elements, or boundary element methods). The use of the analytical approach is therefore limited to simple shapes and parts and the assumptions, if approximate methods are employed. Simulations are approximate methods to analyze a phenomena in great detail and are typically evaluative in nature. Such approximation methods are included in the analytical approach.

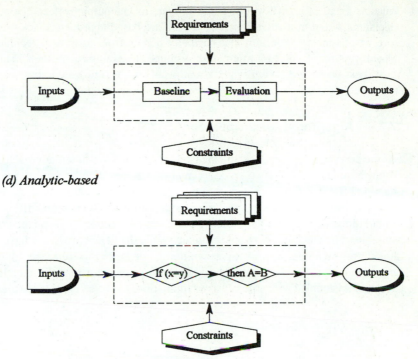

(d) Analytic-based

(e) Heuristic-based or Expert system

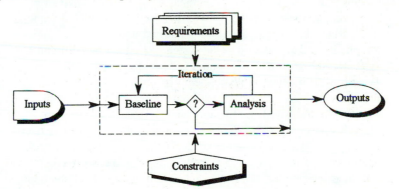

(f) Algorithmic-based or Iterative

FIGURE 7.3   Types of Progressive Models: (d) Analytic-based, (e) Heuristic-based, and (f) Algorithmic-based

7.2.5   Heuristic-based Approach

Heuristic-based approach is very similar to model-based approach except the needed knowledge is derived from heuristics (see Figure 7.3e). Heuristics are simple rules of thumb that many organizations have developed from their experiences or from intuition of what has been found to work well and what has not. The approach computerizes these rules as a part of a DSE for the baseline system. The output reflects the accuracy of such decisions stored in the database. Most heuristic-based approaches lack the continuum nature of parameter definition that normally comes from analytical techniques or model-based systems. Expert or rule-based systems are examples of such heuristic-based approaches where an extensive taxonomy of rules (knowledge-base), along with an inference mechanism, provides the basis for decision making.

7.2.6 Algorithm-based Approach

Algorithm-based approach is an extension of the analytic-based approach, where an algorithm-based feedback loop (in place of analysis) is present as an integral part of the decision making. The iteration is shown in Figure 7.3f. Compared to the analytic-based approach, this is an iterative process where design modification is based on sensitivity of performance calculations with respect to problem parameters. A feedback loop can be created using any suitable search techniques to automatically guide the PD^3 process. Such techniques include linear programming, nonlinear programming, integer or mixed integer, dynamic programming, and so on. For this reason, most classical optimization models, if applied to a baseline system and coupled with an analysis or simulation, can be categorized as algorithmic based.

7.2.7 Hybrid Approach

In real practice, however, individual approaches (sections 7.2.2 through 7.2.6) are not enough to model every level of complexity in solving difficult multi-criterion problems. For instance, most physical problems cannot be modeled using analytical techniques alone. Many behaviors cannot be quantified and are, at best, approximated. Due to the complexity and uncertainty of the design path, it is often difficult to specify design objectives precisely. Since specifications are not quantitative, they do not lend themselves to the analytical method. A combination of progressive approaches is generally needed to aid in reaching the solution. For example, when the problem is ill-conditioned, a combination of heuristic-based and manual-based or combined (analytic-based + manual-based + heuristics-based) models are employed in reaching an effective decision. However, a vast majority of literature recommends using the analytic-based approach as the primary design method. There is a need for further research to explore cross-fertilization possibilities with other progressive approaches. A recent trend in heuristic approach (expert systems) has been its integration with analytic- and algorithmic-based systems (in an object-oriented sense) to make the applicable source of knowledge transparent.

$$\text{Progressive Model} \ni \{\text{Model-based, Manual-based, Experimental-based,} \\ \text{Analytic-based, Heuristic-based, Algorithm-based, and Hybrid-based}\} \quad (7.10)$$

7.3 INTELLIGENT MODELS

Most of the CAD/CAM, CAE, FEA/FEM, mold flow, costing, value engineering, optimization, analysis, and simulation tools serve only a limited PD^3 purpose. Independently they are incapable of addressing the challenges that are listed in the next section. For example, the traditional CAD approach offers a *static* (explicit geometry capture or visual aid capability) to the work groups for the documentation of a preconceived part or an assembly. CAD/CAM tools provide a slew of functions for defining geometry (such as surfaces, curves, and boundaries) explicitly. The traditional FEA program (e.g. Nastran) checks the structural integrity of a preconceived part to withstand a given static or dynamic load. Traditional optimization programs (e.g., NASTRAN-OPTI), for instance, are suitable for minimum weight design of structures. If the objectives are different from minimum weight design, companies cannot use such packaged programs. Most expert systems can process symbolic information and conduct heuristic inferences, but their problem-solving ability depends mainly on inferences not on algorithms, such as what is commonly found in most optimization tools.

There are many similar issues that limit the applicability of most C4 tools. For example, structural performance is one of the many possible design criteria used. Performance parameters and criteria differ from one line of product to another. Some of the criteria are rules of thumb or heuristics, others are algorithm-based. When the behavior of the system is unknown, experimental data, instead of analysis or simulation, are often used as alternatives. In any company, over time, a vast amount of knowledge about key products is available. However, useful data are in proprietary forms and are fragmented across various work groups. With knowledge-based systems, especially in a workstation-based client/server computing environment, it is possible to search for designs across the vast engineering repositories. The design search may spread across many cooperative work groups or repositories that could meet functional specifications. In intelligent models, the rules come from such divergent sources as company policy manuals or standard design handbooks. Anything of importance to the company, the work groups, the customers, or any of the concerned parties may be captured as rules and made available over an electronic network. The automation world today presents a variety of standard and nonstandard (proprietary) options for the creation of texts, graphics, and translators. They are becoming more and more reliable especially for multimedia documents (text, line-art, photographs, audio, and video) and CAD media outputs.

7.3.1   Major Challenges

The major challenges faced during a typical PD^3 process are how to accomplish the following tasks. The challenges are how to:

- Modify easily and quickly an existing design (or a CAD model) with minimum human intervention.
- Determine rapidly the effects of desired changes on performance.
- Evaluate several design alternatives.
- Limit the use of (or number of) prototype testing for decision making.
- Obtain cost and producibility information.
- Obtain sensitivity information.

- Reduce the number of hardware prototypes.
- Visualize quickly the effects of the design changes.
- Obtain a feasible or an optimum design.
- Define any problem parameter as a design variable.
- Use constraints and performance parameters interchangeably in an optimization formulation.
- Perform *what if* design iterations.
- Consider manufacturing constraints up-front, e.g., during a conceptual design stage of a PD^3 process.

Many CAD/CAM vendors have started, or are in the process of, developing tools based on *dynamic* techniques rather than *static* techniques to manage more and more of the above challenges. Often, such tools employ variable-driven modeling, symbolic processing, and heuristic inference methods to capture the design and development rules, geometry, and a taxonomy for a product realization.

7.3.2   Modes of Decision Making

There are two modes in which decision making can take place during a PD^3 process—serial and parallel (see Figure 7.4). In a serial mode, decision making is an integral part of an activity. An activity not only entails information build-up but could also include decision making. A work group may need to carry this serial process when it is desirable to sub-optimize the design *one activity at a time* as teams go along meeting the specifications gradually. This may be all right if the activities are dependent or if no feedback across the activities is necessary for product realization. In parallel decision making, all the major outputs of activities are brought together. This way the activities can run in parallel allowing them to be processed concurrently. The decision making then involves determining on a cumulative basis the effects of the resultant outputs (see Figure 7.4).

7.3.3   Capturing Geometry

Most conventional CAD/CAM systems supply a rich array of model building tools geared towards expediting the interactive manual tasks. Most of these are directed towards capturing the *as-is,* that is, a *static* representation of a given part. Since the most laborious task in 3-D modeling is capturing the *as-is* geometry, this is quite helpful if the design that was captured does not change. Among the tools that aid this static process are:

Surface Generation Tools: Most CAD/CAM systems supply a rich array of model building tools for creating surfaces. The most primitive are the tools to create:

- *Analytic surfaces*—These tools are defined by specifying a simple geometric entity, such as cube, sphere, cylinder or cone, whose surface can be described by a mathematical equation.

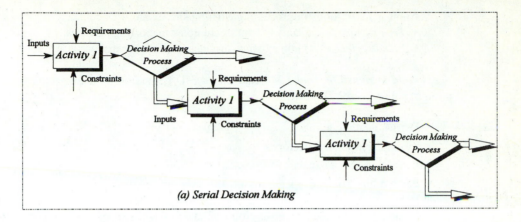

(a) Serial Decision Making

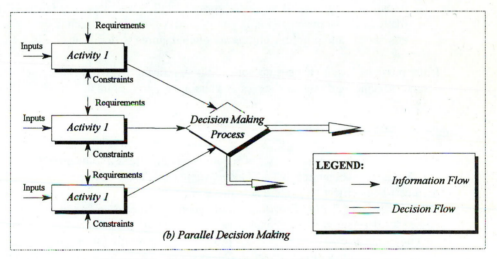

(b) Parallel Decision Making

FIGURE 7.4   Decision Making Modes (a) Serial and (b) Parallel

- *Surface of revolution*—revolving a profile around an axis.
- *Swept surfaces*—sweeping a surface along a path in space.
- *Fitted surfaces*—specifying an array of points and curves in space through which a surface is fitted. Another class of tools that are available for building complex shapes are:
  - *Skin surfaces*—stretching a surface over a series of profiles.
  - *Freedom surfaces*—sweeping and revolving a profile simultaneously. Skin surfaces are useful in defining objects such as boat hulls or wings whose cross-sections vary. Freedom surfaces are useful in defining helical objects, such as screws, allowing sweep and rotate profiles along an arbitrary path in space and for modeling bent objects such as pipes.

Another set of tools for processing existing shapes are:
- *Trim surfaces*—capacity to trim or cut holes in existing surfaces.
- *Blend surfaces*—to join existing surfaces to eliminate discontinuity, gaps, etc.

Solid Generation Tools: Most advanced 3-D geometric modeling tools allow creation of solids from 2-D profiles. These tools might consist of:

- Extrusion along a line.
- Sweeping along a trajectory.
- Revolving around an axis of symmetry.
- Performing Boolean operations on the existing solid shapes. The resulting solid stays *connected* to its 2-D shape and can thus be altered by altering its 2-D shape. Some class of tools convert a group of surfaces (skin, swept, freedom, etc.) into a solid model. Some CAD/CAM systems propose to bypass 2-D profile routes by automatically parameterizing some types of explicit 3-D geometry (also called automatic 3-D parameterization). Variable-driven modeling techniques, topological constraints, and primitive objects are often required to define and build a 3-D solid.

If the possible design changes do not exhibit significant diversion from a product initial geometry, static representation serves as a simple way to capture and build a 3-D solid.

7.3.4 Beyond Modeling

Static modeling is a minimum set of requirements that form a foundation for a DSE system. Manipulating features that are *beyond static modeling* are more advantageous in supporting 3-D modification goals. These include:

Record and Replay Provision: Many geometric modeling systems that have no facilities for defining variable geometry as a set of controlling parameters entail recreating the entire model. Some have a limited provision for permitting user modification of their geometry once a model is completed. Others do provide facilities to store the interactive steps leading to the generation of a finished shape. To facilitate changes, the system maintains a log file that records the steps used by the design work groups. This script can be edited and rerun to create a modified version of the CAD model. Though this *record and play* provision automates generation of the model geometry and thus saves design work group time, it is a cumbersome process since even a small change requires rerunning the entire script. This can be very time consuming computationally depending upon the complexity of the part and the number of steps involved. There is also no guarantee that the modified part will be compatible with other parts of the model.

Tweaking: Some geometrical modeling systems provide a limited set of tweaking features to speed up the model editing process. These operations expedite insertions of local changes in a finished model without its recreation. Examples of such changes are moving a face, deleting, replacing, tapering and subdividing 2-D curves, 3-D surfaces, or 3-D objects.

Work planes: Many geometric modeling systems allow a design work group to define 2-D work planes at any location and orientation. This facilitates generation of all geometric constructions confined to that work plane. It constrains all corners to have the same depth, eliminating the possibility of inadvertently inserting the wrong depth for one of the corners. In this way, the design work group is assured that all geometry that should be co-planar is co-planar. However, the process is temporary and user controlled. There is no guarantee that, in the future, if a work group changes one of the corners, the other corner points will adjust automatically.

Copy and Move Features: When an assembly of a component is designed, it requires multiple instances of the same part. For example, for creating an automobile assembly, four identical wheels are required. One way of creating such an assembly would be to individually create four instances of the parts. This could be very time consuming and unproductive. At a later date, if changes are needed to any of the wheels, each wheel has to be individually altered. Some CAD/CAM systems allow insertion of parts without duplications. Work groups need to identify the locations and orientations where parts are to be inserted. Though this process saves the work group time to recreate all the individual instances of parts, some systems actually create all these instances and store them in a database occupying a large disk storage space.

Associativity: Associativity means design consistency to a larger extent. When work groups make a change to a part, they generally want the rest of the design to be updated automatically. This is analogous to a spreadsheet where every value change is followed by the rest of the spreadsheet reflecting this change. Most general associativity is bi-directional, that is, a change at any PD^3 level propagates both ways—from model-to-drawing if an attribute is changed, or from drawing-to-model if the drawing is changed.

Variable Dimension Modeling: Variable dimension modeling (VDM) is a type of associative geometric modeling in which changes to a geometric dimension result in changes to the CAD model. Previously, CAD vendors used to provide VDM capability through their macro programming language structures. Today, most CAD/CAM systems claim to provide some type of this VDM capability. Most typical of these are parametric modeling, variational modeling, adaptive modeling, dynamic modeling, or feature-based modeling. Key differences among them are with respect to the completeness of the associative mechanisms, mechanics of design-intent capture, geometric versus non-geometric modeling, speed of updates, ease of use, and error trapping.

7.3.5   Capturing Smarts

The concept of parallelism and feedback control required for concurrent engineering cannot be exploited easily using conventional techniques. Smart tools and concepts, based on variable modeling techniques, symbolic processing and hybrid inferencing (heuristics + algorithmic) techniques, are required to achieve this parallelism and feedback control. *Software Prototyping* is a concept of virtual modeling through software programming. Each software module captures one aspect of the functional knowledge toward a PD^3 life-cycle process. The software prototyping concept, besides employing many integration and automation methods during the capturing process, capitalizes on dynamic modeling techniques. Careful planning ensures that these prototype models will have the characteristics that provide some form of dynamic (e.g., variable-driven modeling), as opposed to static (e.g., explicit modeling), environment. Variable-driven modeling is an example of the kind of characteristics that are needed for a sound decision support environment (DSE). Each model of the DSE must be based on sound characteristics. Each model must:

- Identify and capture consensus on problem or product parameters.
- Utilize an early design evaluation philosophy to bring all engineering up front at the conceptual design stage.
- Consider analysis or simulation an integral part of a concurrent PD^3 process.
- Employ generic modeling techniques. Employ principles of geometrical similarity for modeling a family of parts. Establish *library of parts* families.
- Utilize a *simultaneous engineering* philosophy. This is ensured by making provision for simultaneous (as opposed to sequential) treatment of problem parameters, inputs, requirements and constraints.
- Exploit the basic characteristics of a products' life-cycle functions that are of *generic* nature. Classify all designs by form features, such as ribs, fillets, chamfers, holes, slots, bosses, etc.
- Establish a library of standard features such as button, hole, thread, punch, warp, etc.
- Employ a solid modeling design approach that creates a 3-D data base that mathematically describes 100% of the design information.
- Employ numerical sensitivity for trend determinations.
- Employ hybrid inferencing techniques (optimization techniques for algorithmic computation and decision making; and an expert system approach for symbolic processing and heuristic reasoning).
- Establish a library of standard design practices.
- Utilize model based reasoning and other AI techniques to create the knowledge-base and to capture the expertise of senior designers and planners if some knowledge cannot be obtained analytically.

Smart models of a PD^3 system enable independent SBU sub-groups to work in parallel teams, provide the required electronic feedback to the interfacing groups, and share the results with project managers at marked check points. This system substantially reduces the time required for the completion of the design-to-manufacturing (PD^3) life-cycle.

7.4 SMART REGENERATIVE SYSTEM

Smart Regenerative System is obtained by combining progressive as well as cognitive aspects of product life-cycle functions. Cognitive aspects capture the knowledge and progressive aspects add a systematic structure for the new design evolution.

The *smart* here means a knowledge-based engineering (KBE) powered application or a computer module. Several types of rules go into creating knowledge in a KBE environment (see Figure 7.5). Captured knowledge typically includes:

- The product structure (or something like a bill-of-materials).
- Rules for reconfiguring or changing the product structure when there are new inputs.
- Dependencies and relationships among features and parts of the product so changes to one part or feature automatically change those parts and features that depend on it.
- Functional, physical and geometric attributes.
- Engineering rules from contributory engineering disciplines (e.g., manufacturability, structural analysis), engineering rules for design optimization.
- Decision criteria for extracting information from external data bases, including selection of standard parts from company catalogs and parts from feature-based (CAD) design libraries.
- Rules for using geometric design data imported from CAD systems (see Figure 7.5).

Some of the rules about a product might involve calculating values; for example, fabrication materials might be evaluated based upon the stress the component is subjected to. They combine with each other to create a design optimized for current inputs, requirements, and constraints at hand. Obtaining a smart regenerative system depends largely on the problem set that work groups are trying to solve. The more structured knowledge the problem has, the easier it is to implement a regenerative system. If a work group uses GT-classification and analysis tools in the PD^3 process, the most benefiting factor is the ease of machining and maintaining the logic.

KBE regenerative systems can work with external systems. Rules can call on knowledge in external databases like the preferred parts' list or supplier parts catalog. The latter is easier to maintain than the embedded rules. An example of an embedded rule would be when a company uses some new equipment for a part manufacture or uses a new manufacturing process for its manufacture. The company would have to investigate every hierarchical process to locate a part and every hierarchical decision tree to locate the equipment or the process. If a match is found, the appropriate changes can be made, tested, and documented. Until this preferred parts list is updated in the KBE internally, the process planners will continue to make parts using the old machines and old processes. With an external parts in KBE system list this can be fixed in no time. Another external action could be to communicate design parameters. If the task is to develop a new design that has certain strength or stiffness, the information must come from an external FEA program (NASTRAN, PATRAN, etc.) An input to the FEA and an output from FEA to the KBE system need to be established for the two to work properly. Such procedural steps cannot be eliminated completely if the goal is to achieve 100% automation. This makes the resultant KBE system dependent. To reduce dependency, work groups need to separate the logic from the data and document the reasoning for decisions in a manner that can be reviewed on-line.

Many CE teams who want regenerative product design forget that when the knowledge engine in a KBE system is done creating a design layout, work groups still need a de-

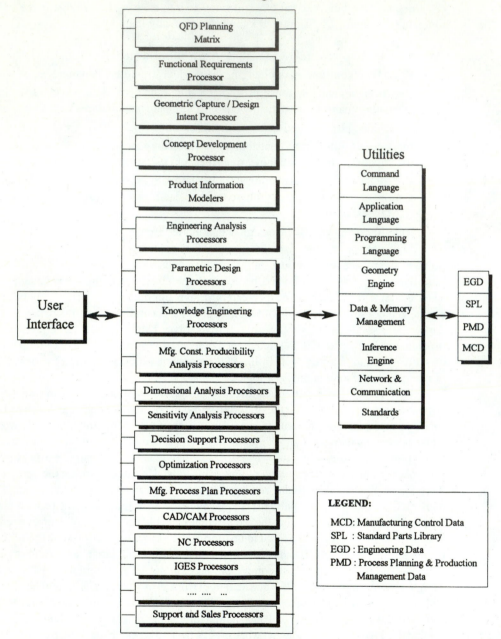

FIGURE 7.5   Decision Support Environment (DSE)—Regenerative Product Design

cision support system to maintain control of the information while it gets refined, formatted, approved, revised, and eventually built. It would be ideal to have an integrated (variant and a regenerative) product design system. Work groups must set reasonable goals for the company. It is a considerable, and nearly impossible, job for some companies to get 100% of their parts automated with a regenerative system. A more easily attainable goal is to get the majority of parts design done automatically, and for the unusual parts, let the regenerative system complete as much of the part as it can be and let the work groups complete the rest with a decision support system. It is a great accomplishment to become 80% regenerative (see Figure 6.17, volume II). Work groups should continuously look at the cost/benefit curve to determine when they are reaching an optimal mix of regenerative and variant product designs. If a company does not introduce many new parts in a year, or if new parts are similar to the old parts, then there are more opportunities to become regenerative.

Management must understand that the enterprise's rule-base or knowledge-base needs to be periodically reviewed, analyzed, and updated. Most work groups that build a smart regenerative system like this, often become dependent on its ability to provide intelligent designs. Most companies change their product mix (new part introductions) and technology approximately 20% annually. This means that their smart regenerative system needs to keep up. If the new parts closely resemble existing parts, then the regenerative system can generate the new parts with little or no additional logic. For instance, a ball bearing company will have no problem producing a design and a process plan for a slightly different size ball bearing, even though the company may not have made that exact part before. Whether such a part proliferation would be a desirable thing to do is another matter. Many companies have built-in ability to search the available (corporate) parts database and provide a short list of those manufactured parts that are in some way similar to the part being redesigned.

Smart regenerative system for assemblies creates a complete set of constituent bill-of-materials, part list, and so on. For each part, the associated DSE has the knowledge about the relationship of the part to its sub-assembly. Smart regenerative system evaluates how the parts will fit together (with respect to tolerance, clearance for other parts, part dimensions, etc.). This adds complexity to the regenerative system because each module has to deal with each set of assembly information. Smart regenerative system should be able to address most aspects of the life-cycle design. This is discussed in section 7.5.

7.4.1   Technical Memory or Knowledge

Smart models result when concurrent work groups employ knowledge about the products or the processes to predict their behaviors in an extended fashion. Tasks in manufacturing tend to be hybrid in nature. A large portion of knowledge used by the work groups comprises a mixture of well-defined algorithms, and a deep knowledge of engineering and manufacturing principles embedded in a set of casual and structural models. On a small scale, work groups also depend on superfluous or qualitative models based on shallow knowledge and experiences that are hard to quantify. One logical approach that has proved effective is to quantify these models using a manufacturing technical memory derived from an existing product line. The knowledge extracted from such sources ought to be verified by the various CE factions before placing them in the technical memory. It is important that the data points be used as relative pointers (in relation to each other) rather than as accurate or absolute values. It is also not essential that the outputs represent the new product precisely. There are many techniques used to implement technical memory representations of which frame structure and production rules are the most common. There are AI concepts which result in the so-called expert systems. Besides AI concepts there are procedural ways to use this knowledge. The term smart model is used here to include both these classes of knowledge-based developments. Such knowledge of engineering design exists in the following forms [Boyle, 1989]:

- *Imperative Knowledge:* This includes control of information. Imperative knowledge is categorized as formal in functions and explicable procedures, and informal in heuristic procedures. A procedural representation is one in which the control mechanisms are embedded within the knowledge.
- *Declarative Knowledge:* A declarative representation is one in which the knowledge is specified but the control structure to use the knowledge is not given. Declarative knowledge is categorized as formal when represented by attributes, and informal when represented by principles and background theory.
- *Innate Knowledge:* This contains common sense, learning, and unstructured reasoning.

Most manufacturing problems possess imperative or declarative knowledge. Clearly many manufacturing applications for KBE systems address decision-support aspects more than constant advisory roles. This requires effective integration of the KBE system with other models—analyses, simulation or decision aids.

7.5 LIFE-CYCLE VALUES

Life-cycle design is generally recognized as a PD^3 methodology of incorporating life-cycle values of a product at an early stage of design [Barkan, 1988; Suh, 1990]. These values include not only functional requirements, but also those related to design for X-ability (DFX) (e.g., assembly, testability, serviceability, transportability, recyclability etc.). Perhaps the most mature DFX methodology is Design for Assembly (DFA). Boothroyd and Dewhurst [1983] and many others have proven that DFA can bring significant cost savings in part production. Very often, but not always, DFA leads to more reliable and serviceable designs primarily due to the reduced number of actual parts. Unfortunately, DFA may also lead to designs that are very costly to develop, produce, and service. Some parts designed through DFA may be impossible to test or very costly to replace on the field. The lack of pertinent parts in a product assembly may degrade its overall performance. On the another hand, the increased cost of each repair could be offset by enhanced reliability due to DFA and modular designs, that is reduced service frequency. However, manufacturable design without a thorough consideration for all life-cycle values could lead to an unexpected increase in servicing and warranty costs that could off-set one's competitive advantage. This could also have an impact on customer satisfaction.

Many companies have compiled comprehensive guidelines for DFX values. The guidelines may address the following DFX options:

- Provisions to detect DFX needs.
- Redesign features to enhance easing of the phenomena.
- Compute estimated life-cycle cost.

However, the strong push for some life-cycle values beyond a justified level, such as manufacturability, assembly, modularity, and so on, can compromise the consideration of other values (serviceability, reliability). For instance, in a rush to reduce parts using DFMA, mainly by combining several parts into one, designers can easily move away from standard, off-the-shelf, high volume parts. Such new designs in turn may require complex tooling and longer lead time. It is possible, therefore, that by combing many parts into one, a design work group may have ignored ease of service in favor of reducing parts. What is required is a methodology that effectively deploys the knowledge compiled in pertinent DFX values as guidelines during the early stages of PD^3 process. This may require a system that assists design work groups to check the calculated values against the recommended DFX values in the proposed design. DFX value checks together with QFD could enhance the life-cycle quality of the product by increasing reliability, reducing service cost, and sustaining performance. The DSE for product modularity is an advanced topic of research towards this and involves the use of design for value (DFV), quality function deployment (QFD), and service mode analysis (SMA) [Ishii, Eubanks and Marks, 1993].

7.5.1   Understanding What's Critical

In order to shorten the PD^3 cycle-time, it is important to understand what is critical. The goal of an *as-is* process study is to develop a critical path model of the process. The critical path method (CPM) is a general technique for describing and analyzing project schedules. CPM identifies activities that determine how long it will take to complete a project and estimate the total completion time.

There are a variety of ways in which advanced technology can be applied in reducing the critical path needed to bring a new product from a marketing concept to successful production. For example, if prototype construction is a critical path activity, eliminating one or more prototypes through the use of better analysis or simulation can cut down time-to-market. If manufacture of tooling for certain items lies on the critical path, then reducing manufacturing time by supplying a CAD (NC tool path) model to tooling suppliers may improve time-to-market. Firms which furnish custom or semi-custom products may find that cutting lead-time needed to respond to a customer inquiry or bid invitation is critical to winning more business. Where know-hows or cost estimates are critical to making a proposal, appropriate intelligent CAD technology, such as knowledge-based automation or bill-of-material, may help bring new business.

The point here is that only those activities that fall on the critical path of a PD^3 life-cycle process need to be automated. To apply expensive CAD/CAM technology or to automate products or parts whose lead times are not critical may improve quality but will not decrease time to market.

7.5.2 Cost Estimating

Cutting life-cycle costs may be deemed critical to winning market share (or holding on to the current share). However, to cut PD^3 cost, work groups have to identify what the major cost drivers are during the life of a product. There is little point in slashing the cost of drafting labor if drafting labor makes up a mere 1 percent of the cost of doing business. Similarly, there is no point in applying new technology in an area where it can have only a small effect on reducing costs. Knowing the source of costs and which of these costs have a major impact on the profit potential of the company is important. This, however, does not refer to product price alone as is often assumed, especially if some of its parts are purchased from an outside source. Customers want high quality products at a cost-competitive price. Understanding what are critical life-cycle costs, will yield areas within a company where new technology could be applied to reduce PD^3 cost.

7.5.2.1 Direct Labor
Direct labor refers to the actual labor used to manufacture and assemble the products. In many manufacturing companies, direct labor makes up a small fraction of the total labor cost. However, when direct labor savings are important, they can be achieved most effectively by designing products that require less labor to build. Boothroyd and Dewhurst have developed some computer programs that can be used to analyze assembly costs very early in the PD^3 process. These programs let design work groups make trade-offs between material and direct labor costs early in the product development—before any detailed drawing or CAD model is created. Assembly by robots can also reduce direct labor cost. Off-line robot programming systems can make robot programming faster and less expensive.

7.5.2.2 Capital Equipment and Tooling
Capital equipment and tooling includes depreciation costs of manufacturing equipment, facilities, special tools needed to make the products, and the cost of procuring capital needed to purchase these items. Most tooling manufacturers are capable of using modern CAD/CAM technology and downloading part information to graphics-based numerically controlled (NC) milling systems that can cut, mold, and die cavities. Today 3-D CAD systems are being used for jig and fixture design and to provide data for coordinate measuring machines (CMM). If outside suppliers are producing tooling, it might be advantageous to supply them with 3-D CAD models to expedite tooling manufacture and to reduce machining errors. As more outside shops acquire graphics-based NC systems, competitive pressures will force the CE firms to deliver good CAD data to tooling sup-

pliers through EDI. This not only reduces tool delivery time, but also saves data processing expenses to OEMs, customers, and suppliers.

7.5.2.3 Indirect Labor

Indirect Labor refers to the labor not directly charged to a job or to a customer, for example, labor for service, repair, testing, technical support, sales, and administration. Too many firms start with the assumption that they must make every decision with the most advanced CAD system they can find in the marketplace. Firms that take this approach generally find that the cost of low-productivity CAD work eats up the savings from high-productivity work. A more effective approach is to apply high-cost CAD systems to high-productivity applications. Automated design and variable dimension modeling (VDM) design can be applied to families of topologically similar products or to those products that are modified repeatedly. Low-cost design aids, such as PC-based CAD systems, should be applied to low-productivity, one-of-a-kind job. Both types of applications may utilize products from the same CAD firmware to minimize waste in data processing and translation.

In most manufacturing firms, the cost of design and drafting labor is a small part of total indirect labor. A large part is spent on sales, marketing, testing, cost estimating, field service, and all other *white collar* activities in the business. Better quality engineering can save these expenses if carefully applied across the board. Here are some examples:

Mistake/Cost Avoidance—A complete 3-D model of an assembly can assure that parts fit correctly before they are made. They also allow stylists and marketing work groups to get a better sense of what the product will look like before firms commit to expensive prototypes. Automated bill-of-material capability can reduce mistakes caused by manual material handling and assure that only the right material is delivered to the production area. The following are the costs avoidance savings due to early detection of a mistake.

- *Material Costs*—the costs of the materials, features, or parts used in the products.
- *Scrap Costs*—the costs of the materials that must be discarded because they are made incorrectly or are quality deficient, or parts that have unacceptable tolerances, finish, etc.
- *Rework and Warranty Service Costs*—the cost of fixing defective products. Products made incorrectly can become a major cost burden. Quality control expert Armand Feigenbaum [1990] says that in many factories, fifteen to forty percent of plant capacity is devoted to reworking defective parts, replacing scrapped parts, and retesting or reinspecting defective units. In those cases, reducing mistakes before they occur can yield big dividends.

Serviceability—Products that are difficult to service because they cannot be disassembled easily require increased field labor and high service costs. Better serviceability considerations allow field service work groups and managers an opportunity to review designs for maintainability before details are cast in concrete.

Cost Estimating—Estimators and subcontractors need drawings or pictures of parts and subassemblies before they bid on them. CAD/CAM technology with a delivery system such as Lotus Notes can be used to distribute images electronically instead of printing on paper. Automated drafting can speed the production of bid and proposal drawings.

Inventory Carrying Charges—there is a significant cost penalty associated with carrying finished products, partially finished products, or materials in inventory. Work-in-progress inventories become burdensome if parts in stock are turned over very slowly. However, manufacturers of made-to-order equipment can enjoy significant inventory reduction and major cost savings. Such firms can use CAD and graphics based NC tools to help program the machining centers to manufacture parts in small lot sizes with one setup per cast. Inventory savings alone can pay back the cost of acquiring such technologies.

7.5.2.4   Unit Production Cost

The production cost of a manufacturing system depends on the manufacturing output and the transformation processes that are in place to produce the part. A free body diagram of a simple transformation process is shown in Figure 7.6. Although there are many elements that make up production costs, the two primary elements are the design and processing costs.

Design costs are the inherent variable costs of the materials, parts, components, and features of the constituents that provide product function for the consumer. Sometimes, these factors are documented on production drawings and in product specifications.

Processing costs include both the direct and indirect costs of running a production line, tooling plus the initial capital investment necessary to purchase the machines and equipment. This method primarily looks at manufacturing costs from the knowledge of the costs associated with the transformation process of a manufacturing system [Singh and Sushil, 1990]. Processing costs include technology elements, such as quality (e.g., cost-to-correct-quality and cost-to-ensure-quality) and events, that need to be considered to control cost.

Let T_{jk} denote a transformation process of a multi-stage manufacturing system in which

j represents a stage of a production system, and

k represents the number of alternate technologies that are employed at the jth stage.

T_{jk} is assumed to convert an input (resource, materials, etc.) to achieve a well-desired output (changes in the physical, chemical, technological, biological, or functional characteristics of the part). Figure 7.6 represents a conservation model of this transformation process, T_{jk}, corresponding to the jth stage, when subjected to the kth alternative technology. Besides output, the transformation model contains two other variables, namely, waste and rework. The model is very simple, that is, it is assumed that there is one type of input and one type of output with a simple disposal mechanism. The output units that do not meet the specification can be of two types (a) those units that can be reprocessed and salvaged in some way, and (b) those which have to be scrapped as waste or that may not be feasible to recycle.

Consider a production system with m stages, $j = 1, 2, \ldots, m$,

and each jth stage having n_j number of alternate technologies,
$k = 1, 2, \ldots n_j$, for the jth stage.

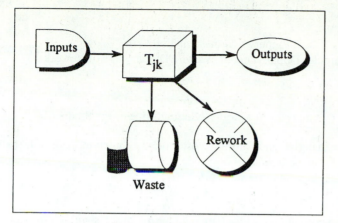

FIGURE 7.6 Free Body Diagram of a
Simple Production Unit

The multi-stage manufacturing system in this case is represented as

$$\text{Production System} = \underset{j=1}{\overset{j=m}{U}}\ T_{jk*} \tag{7.11}$$

where U denotes the union and

T_{jk*} denotes the jth stage transformation with $k*$ technology used for the jth stage,

where $k* \in (1, \ldots, n_j)$

If a product is required to undergo three operations ($j = 3$), for example, turning, milling and drilling, they represent the stages of T_{jk}.

The operations, turning, milling, and drilling, each may have a number of alternate technologies. Examples of alternate technologies for turning are: engine lathe, turret lathe, and single spindle automatic lathe; for milling: Universal Milling Machine—ordinary grade I, subgrade II or supergrade III; and for drilling: Pillar Drilling Machine and Radial Drilling Machine, and so on.

If a manufacturing system consists of three stages and each has 3, 2, and 4 sets of alternate technological options, respectively, to produce it, the manufacturing cost corresponding to technology selection ($k*$, $l*$, and $m*$) can be expressed as

$$C_p = U\ \{T_{1k*}, T_{21*}, T_{3m*}\} \tag{7.12}$$

where

for $j = 1$, $k* \in (1, 2, 3)$
for $j = 2$, $1* \in (1, 2)$
for $j = 3$, $m* \in (1, 2, 3, 4)$

7.5.2.5 Transformation Cost

The transformation cost of a part can be computed from the principle of conservation model shown in Figure 7.6. From the conservation of materials, at the transformation process T_{jk}, we get

Input = Waste + Rework + Output.

In terms of the cost, it can be expressed as

$$\text{Dollar value of inputs + Transformation Cost}$$
$$= \text{Dollar Value of Output + Dollar value of Waste + Dollar value of Rework} \quad (7.13)$$

where

$$\text{Dollar Value of the Output = Profit Margin + Selling Price of the Product.}$$

Using the notations of M for material, and C for the cost terms as shown in Table 7.4, the previous expression eq (7.13) can be written as

$$M_i C_i + C_{trj} = M_o C_o + M_w C_w + M_r C_r \quad (7.14)$$

where C_{trj} is the jth stage transformation cost for a manufacturing process with $k*$ technology selection.

TABLE 7.4   Definitions of Material, Cost, Labor, Machine and Time Parameters of Process T_{jk}

	Material		Cost per unit		Labor Rate per Unit of Material		Machine Hourly Rate per unit of material		Time it takes to Process a Specific Task	
Inputs	M_i	Material input	C_i	Unit cost per unit material input	L_i	Labor rate per unit material input	m_i	Machine hourly rate per unit material input at T_{jk}	T_i	Time it takes to process input
Waste	M_w	Material waste	C_w	Unit cost of waste of process T_{jk}	L_w	Labor rate per unit waste of process T_{jk}	m_w	Machine hourly rate per unit waste of process T_{jk}	T_w	Time it takes to process waste
Rework	M_r	Material rework	C_r	Unit cost of rework of process T_{jk}	L_r	Labor rate per unit rework of process T_{jk}	m_r	Machine hourly rate per unit rework of process T_{jk}	T_r	Time it takes to process rework
Outputs	M_o	Material output	C_o	Unit cost of output of process T_{jk}	L_o	Labor rate per unit output of process T_{jk}	m_o	Machine hourly rate per unit output of process T_{jk}	T_o	Time it takes to process output
Cost			C_{tr}	Transformation Cost of process T_{jk}	C_{lp}	Labor Processing Cost at T_{jk}	C_{eo}	Equipment Operation Cost of process T_{jk}		

This is because only one technology (e.g., machine) can be selected at each stage (turning, milling, or drilling) of material transformation.

or
$$C_{trj} = [M_o C_o + M_w C_w + M_r C_r - M_i C_i] \qquad (7.15)$$

The indices i, w, r, and o stand for inputs, waste, rework and outputs, respectively.

$$\text{Total Transformation Cost } (C_{tr}) = \sum_{j=1}^{j=m} C_{trj} \qquad (7.16)$$

There is a processing cost associated with this transformation process, discussed in section 7.5.2.6.

7.5.2.6   Unit Processing Cost

The processing cost for transforming a raw material into a useful product, which a consumer can buy or want, can be computed using the same conservation model, equation 7.13.

Using the notations L for the labor rates, T for the time, and M for the quantity of material, the processing cost can be expressed as

$$C_{lpj} + T_i M_i L_i = M_o T_o L_o + T_w M_w L_w + M_r T_r L_r$$

or
$$C_{lpj} = M_o T_o L_o + T_w M_w L_w + M_r T_r L_r - M_i L_i T_i \qquad (7.17)$$

$$\text{Labor Processing Cost } (C_{lp}) = \sum_{j=1}^{j=m} C_{lpj} \qquad (7.18)$$

The indices i, w, r, and o stand for inputs, waste, rework, and outputs, respectively.

7.5.2.7   Equipment Operating Cost

Let us assume that each technological alternative for obtaining the useful output is carried out on a separate machine and the cost of running a particular operation is measured in terms of the machine usage. The cost of the machine is depreciated over the working life of the machine.

Let us assume

$$C_{eo} = \text{Equipment Operation Cost}$$

and using the notation T for the time (hours) the machine takes in processing a specific task, prorated for any idle time,

then operation cost of manufacturing the part is

$$C_{eo} + m_i M_i T_i = m_w M_w T_w + m_o M_o T_o + m_r M_r T_r$$

or
$$C_{eo} = m_w M_w T_w + m_o M_o T_o + m_r M_r T_r - m_i M_i T_i \qquad (7.19)$$

$$\text{Equipment Operating Cost } (C_{eo}) = \sum_{j=1}^{j=m} C_{eoj} \qquad (7.20)$$

The indices i, w, r, and o stand for inputs, waste, rework, and outputs, respectively.

7.5.2.8   Unit Material Cost

The material cost of the part can be computed based upon the material consumed in the transformation of raw material into a useful product.

$$C_{mj} = C_i M_i \qquad (7.21)$$

where C_{mj} = The material cost of the part.

C_i = Cost/Unit material input.

M_i = Quantity of Material input at process T_{jk}.

7.5.2.9 Unit Tooling Cost

The tooling cost for the part, C_{to}, can be written as the sum of die construction cost (C_{dc}) and die material cost (C_{dm}). Then

$$C_{toj} = (C_{dc} + C_{dm})/N \qquad (7.22)$$

where N = the production volume of the part.

7.6 TOTAL LIFE-CYCLE COST

Many types of savings or benefits of concurrent engineering have been projected in various sections of this book, such as shortened time to market, reduced costs, improved quality, and so on. These benefits or savings are the result of implementing various tools and technologies such as those discussed earlier. It is difficult, however, to attach relative values to these goals without knowing their impact on the business bottom line. In this section, we will look at ways to quantify and measure the savings.

7.6.1 Total Production Cost

The total production cost of a part can be expressed as the sum of the material cost, transformation cost, tooling cost, labor processing cost, and the equipment operating cost of the part:

$$C_{tp} = \sum_{j=1}^{j=m} [C_{mj} + C_{trj} + C_{toj} + C_{lpj} + C_{eoj}] \qquad (7.23)$$

where C_{tp} denotes the total production cost. The expressions for these costs are discussed earlier.

7.6.2 Assembly Cost

The assembly cost of the part depends upon the number of parts, n. If C_{ai} represents the cost of assembling the ith part for a n-component assembly, then the total assembly cost, C_{ta}, can be expressed as

$$C_{ta} = \sum_{i=1}^{i=n} C_{ai} \cdot \qquad (7.24)$$

If the variable cost of assembly, C_{ai}, is constant for each part assembled (say C_a), then C_{ta} can be expressed as $C_{ta} = n.C_a$.

Reducing assembly cost obviously requires limiting the number of parts in the assembly. The number of parts can be reduced either by eliminating parts, for example, replacing screws, nuts, and washers, with press (snap fits) or by combining two or more individual parts into one injection molded, die-cast, or stamped part. When combining parts it is necessary to consider whether the increased tooling, processing, and material costs for the redesigned part offset the savings in assembly cost of the old part [Poli, Escudero, and Fernandez, 1988]. The production cost for a one-part composite assembly would be less than the production costs of an *n*-part assembly, if the following expression is true.

$$(C_{tp*} + C_{ai*}) < \sum_{i=1}^{i=n} [C_{tpi} + C_{ai}] \tag{7.25}$$

where C_{tp*} and C_{ai*} are the total production cost and assembly cost of a one-part composite assembly.

C_{tpi} = the total production cost of an *i*th component of *n*-part assembly.

7.6.3   Manufacturing Planning Cost

From an operational planning and control point of view, the loading problem in flexible manufacturing systems poses an interesting challenge. Manufacturing planning cost is often characterized by the presence of the following complexities:

- Multiple networks.
- Multiple parts, products, and machines.
- Multiple machining operations that each part is required to undergo for the completion of the pertinent manufacturing process.
- Complexity introduced by machine failure or downtime, repair rates and time loss, presence of location and size of buffer, mix and availability of pallets, etc.
- Uncertainty in determining the best process plan for manufacturing the product.
- Types of the machines, multi-purpose or single purpose, introduce variations in cost of processing, setup time requirements, and accuracy of the machining operation performed.

The situation is further complicated by the fact that there are a large number of parts, each part perhaps requiring a different sequence of operations, and the tools magazine capacity is limited.

7.6.4   Service Costs

Serviceability is a measure of the ease of performing all service-related operations. There are many factors that affect serviceability [Gershenson and Ishii, 1991]; and [Ishii, Eubanks and Marks, 1993]:

- Reliability of components and sub-systems
- Labor cost
- Inventory cost
- Accessibility of components to be serviced
- Availability of necessary parts, tools, and anything else needed for service
- Mechanic training
- Customer preferences
- Location of service
- Length of warranty
- Diagnosability, maintainability, malfunction repairability, and crash repairability

Service modes are the ways in which a system may be serviced. Service mode analysis (SMA) is the method of describing which service modes will impact a particular design and in what manner. There are two ways of looking at service modes and computing related service costs.

7.6.4.1 Component-based Service Cost (CBSC)

The component-based service costs are the results of malfunctions directly attributed to a particular component, for example, a broken lock rod. This approach pinpoints the malfunctioning component in the system and then computes the cost of repair. For component-based service modes:

$$\text{CBSC} = \sum_{i=1}^{n} C_i R_i \qquad (7.26)$$

where CBSC = Component based service cost

C_i = Cost for repairing an ith component.

R_i = Reliability of an ith component.

n = total number of components in the system.

The general function for the component based service cost C_i is:

$$C_i = f_i (L_t, L_r, F_t, F_e, C_{rp}, A_{rp}) \qquad \text{for } i=1,2,\ldots n \qquad (7.27)$$

where L_t = Labor time, the total time it takes to repair the malfunction.

L_r = Labor rate, the current average cost per hour for a mechanic.

F_t = Necessary tools factor $(0 < F_t < 1)$—the tools needed to perform the repairs and, therefore, the work groups who would own them (availability of tools).

F_e = Necessary mechanic training factor $(0 < F_e < 1)$—the level of training needed to accomplish the labor operation.

C_{rp} = Cost of replacement parts factor $(0 < C_{rp} < 1)$—the cost of any parts that are replaced during the repair.

A_{rp} = Availability factor $(0 < A_{rp} < 1)$—where the replacement part discussed above can be purchased (accessibility).

Each of the variables has an effect upon the serviceability cost of the particular service mode phenomenon.

7.6.4.2   Symptom-based Service Cost (SBSC)

Symptom-based service costs are the results of malfunctions of the system. They are exactly what the customer sees, the failure of the product to perform a function in an adequate manner. For example, a typical customer does not come to a mechanic and say "my lock rod is cracked". They come in and say "my door does not lock". This is the symptom side of the malfunction.

For symptom-based service modes, the cost is:

$$\text{SBSC} = \sum_{j=1}^{m} [S_j R_j] \tag{7.28}$$

where SBSC = Symptom-based service cost.

S_j = Cost of repairing a symptom or a phenomenon.

R_j = Reliability of a component to perform a function with a jth symptom present.

m = the number of symptoms that are detected over its life-cycle.

Symptom based service cost, S_j, is the cost to correct a *j*th service symptom.

When implementing a total life-cycle service cost (LCSC) based on a component-based SMA, several companies use:

$$\text{LCSC} = \sum_{i=1}^{i=n} [\{(L_t \times L_r) \times F_t \times F_e + (C_{rp} \times A_{rp})\} \times f_i] \tag{7.29}$$

where LCSC = Life-cycle service cost.

L_t = Labor time.

L_r = Labor cost per hour.

F_t = Necessary tool factors $(0 < F_t < 1)$.

F_e = Necessary mechanic training factor $(0 < F_e < 1)$.

C_{rp} = Replacement part cost.

A_{rp} = Replacement part availability factor.

f_i = Frequency range or the number of times an *i*th part is replaced.

n = Total number of components in the system.

The sums are taken over all the labor operations for the parts replaced. Another form expressing LCSC using the symptom-based SMA is

$$\text{LCSC} = \sum_{j=1}^{m} [\{(F_{lt} \times F_{lc}) + F_{rp} + F_t + F_e\} \times f_j] \tag{7.30}$$

where

F_{lt} = Labor time factor in each operation.

F_{lc} = Labor cost per hour factor in each operation.

F_{rp} = Replacement part cost factor in each operation.

F_t and F_e are defined earlier with equations 7.27 and 7.29.

f_j = Frequency range or the number of times a symptom may repeat for each operation.

m = Total number of labor operations performed to fix the symptoms over its life cycle.

Component-based SMA service mode analysis does not indicate which components will have the biggest impact upon serviceability cost. This means work groups must take the sum of all the components in the system. There are many problems with this approach. The frequency and causes of failure are scarce and inaccurate at best. All that is known is unreliable data on how often components are replaced. It is commonly believed that some service shops replace many components that do not need servicing.

Comparing the two approaches, Gershenson and Ishii [1991] find the symptom based approach more suitable for application to life cycle. In the component based case, one needs to calculate C_i for every component and include unreliable failure frequency data. The symptom based approach requires fewer calculations with more reliable data. Gershenson and Ishii [1991] argue that there may be one hundred possible malfunctions in the system, but often the top four or five malfunctions represent over eighty percent of the normal service cost.

7.7 COMPATIBILITY ANALYSIS

Over the past several years, many companies have focused their attention on modeling the *design review* process that commonly involves multi-disciplinary participants with different expertise, such as design, tooling, process, assembly, etc. Ishii and Barkan [1987] have led the method of Design Compatibility Analysis (DCA). Based on a set of compatibility information, that is, good, poor, and bad examples of a product design or concept, DCA evaluates a candidate design or concept with respect to given specifications and constraints. In addition DCA provides an overall assessment, and suggests improvements. DCA has proven effective in many domains, component selection [Ishii, 1991]; design for injection molding [Ishii, Hornberger and Liou, 1989]; forging product and process design [Maloney, Ishii and Miller, 1989]; and serviceability design of personal computers [Makino, Barkan, Reynolds and Pfaff, 1989].

DCA uses compatibility knowledge made up of good and bad examples of design case histories, which are called compatibility data or C-data. Each compatibility datum indicates various components, configurations, and situations to which the datum applies. DCA allows the work group to essentially search through thousands of guidelines by answering only a few questions. If a candidate design satisfies these conditions, a comment is triggered.

Each C-data has a qualitative design rating (good, bad, excellent, etc.) which informs a work group member of the importance of the process and whether the design process warrants improvement. For example, a poor process does not have the same effect on serviceability as a process that is totally incompatible. C-data also contains an explanation which gives reasons for the incompatibility in a textual and visual form. Experience indicates that visual justification and suggestions are considerably more effective than textual explanations.

7.8 SENSITIVITY ANALYSIS

The goal of any design program is to produce a product that is *best* in some sense. The advent of digital computers has allowed the creation of very detailed and complex mathematical models of a product design. Thus, the sophistication of design models has increased to the point where it has become very difficult for the concurrent work groups to comprehend what steps are necessary to optimize the design. Using computer tools for design optimization, a work group establishes a measure of merit, or objective function that is to characterize the so called *best* design. The goal then is to minimize or maximize this objective or *technical merit* function (weight, cost, shape, maintainability) subject to a set of constraints (stress, buckling, minimum gages, etc.). By changing design variables (physical quantities), the design team can control parameters such as thickness, material properties, etc. Technical merit is the measure of the proximity of a technical system to the limitations of the current technology, defined in terms of performance, reliability, economy, mass, etc. For example, in elastic structural optimization, the mass of the structure is chosen as the quantity (a merit function) to be minimized.

   Sensitivity analysis is the analysis of a product design in terms of how a problem parameter or design variable affects the objective or merit functions. This is achieved by computing design sensitivity coefficients that measure the rate of change of the constraint with respect to a change in a design variable. For example, the rate of change in stress with respect to a change in the thickness of a part is an example of a design sensitivity. A list of pertinent tools is given in section 8.4 of volume II (also in Figure 8.14).

7.9 LIFE-CYCLE RANKING OR RATING SCHEME (LCRS)

Measure of merits can be chosen in terms of several life-cycle ratings.

- Life-cycle index.
- Serviceability index.
- Spider or Amoeba chart (or Polygon Graph).

7.9.1 Life-Cycle Index

The life-cycle index (LCI) is a measure of design compatibility when a number of cost drivers is varied at once, for example, piece cost, tooling cost, and service cost. This is a normalized measure of the design or how a concept stands as a whole in terms of a life-cycle cost. LCI is the normalization of all the individual indices, as follows:

$$\text{LCI} = \sum_{i=1}^{N_c} (CI_i)/N_c \qquad (7.31)$$

where LCI = life-cycle index.

CI_i = Individual normalized index in ith cost driver.

N_c = Number of cost drivers taken into account in evaluating the life-cycle index.

7.9.2 Serviceability Index

Serviceability index (SI) gives the work groups an idea of how well they have incorporated serviceability. Work groups can use SI to compare two designs in terms of serviceability or view the effect of some design changes on serviceability. The index is calculated by taking into account all of the factors mentioned earlier for the major malfunctions in section 7.6.4. The index corresponds to the estimate of the life-cycle service cost and can be expressed as [Ishii, Eubanks and Marks, 1993]:

$$SI_\kappa = (\text{LCSC})_k/L \qquad (7.32)$$

where k indicates a kth version of design or kth design alternative, and L depends upon the method used for cost analysis.

$L = n$; if LCSC is computed using component-based SMA,

$L = m$; if LCSC is computed using symptom-based SMA.

The overall serviceability index can be scaled from 0–10;

10 being the best rating and 0 being the worst. In order to obtain a rating of 10, the repairs must incur no costs at all. If no serviceability information is known on the product, it receives a rating of 5. This is analogous to no one saying anything at a design review. The normalized index can be expressed as:

$$\text{Normalized } SI = (SI_k) \times 10/SIL, \qquad 0 \le NSI \le 10, \qquad (7.33)$$

where
$$SIL = [\sum_{k=1}^{L} SI_k^2]^{1/2}$$

7.9.3 Spider or Amoeba Chart (or Polygon Graph)

The spider or Amoeba chart demonstrates the tradeoffs between performance (e.g., serviceability index) and other cost life-cycle indices. Reducing cost in one focus area often hurts the product design in another focus area. By displaying the eight focus areas (or cost indices) in normalized terms on a single chart, the work groups can view all of the repercussions of changing a design parameter (see Figure 1.11 in volume 1).

As shown in Figure 7.7, an exponential function can be used to normalize the performance and cost indicator values with the following equations.

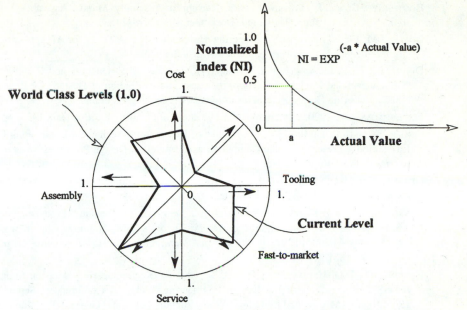

FIGURE 7.7   Mapping of Actual Values into Spider, Amoeba Chart (or Polygon Graph)

$$\text{Normalized Index (NI)} = \text{Exp}^{(-a \times \text{unnormalized actual value})} \qquad (7.34)$$

$$a = (1/b) \ln (0.5) \qquad (7.35)$$

and b is a chosen unnormalized *reference* value of the actual cost in that focus area.

Spider charts are ways to solve a specific problem that a work group may have created. Suggestions for redesign are accompanied by pictures visually showing the poor design and the good design side by side and noting the differences. The suggestions, along with the comments, help to make up the compatibility data that is used in DCA [Gershenson and Ishii, 1991].

In summary, the LCRS ranges from a quantitative estimate of life-cycle cost (LCI index) to a qualitative assessment (design, comments, and suggestions). The quantitative measure gives the work group engineers a comparison table and a tool to justify their designs or concepts. The qualitative comments simulate a design review where teams of work groups offer comments.

REFERENCES

ALBANO, L.D., and N.P. SUH. 1992. "Axiomatic Approach to Structural Design." *Research in Engineering Design,* Volume 4, pp. 171–183.

BARKAN, P. 1988. "Simultaneous Engineering: AI Helps Balance Production and Bottom Lines," *Design News*, Volume 44, No. 5, March 7, 1988, p. A30.

BOOTHROYD, G., and P. DEWHURST. 1983. "Design for Assembly: Manual Assembly." *Machine Design*, (November 10) pp. 94–98 and (December 8) pp. 140–145.

BOYLE, J.M. 1989. "Interactive Engineering Systems Design: A Study for AI Applications." *Artificial Intelligence in Engineering*, Volume 4, No. 2, pp. 58–69.

DIXON, J.R., M.J. GUENETTE, R.K. IRANI, E.H. NEILSON, M.F. ORELUP, and R.V. WELCH. 1989. "Computer-based Models of Design Processes: The Evaluation of Designs for Redesign." 1989 NSF Engineering Design Research Conference Proceedings, Volume 1, pp. 491–506.

FEIGENBAUM, A.V. 1990. "America on the Threshold of Quality." *Quality* (Jan. 1990) pp. 16–18.

FINGER, S., and J.R. DIXON. 1989. "A Review of Research in Mechanical Engineering Design, Part I: Descriptive Prescriptive, and Computer-based Models of Design Processes." *Research in Engineering Design*, Volume I, No. 1, pp. 51–67.

GEBALA, D.A., AND N.P. SUH. 1992. "An Application of Axiomatic Design." *Researching in Engineering Design,* Volume 3, pp. 149–162.

GERSHENSON, J., and K. ISHII. 1991. "Life-cycle Serviceability Design." Proceedings of the 1991 ASME Design Technical Conferences, Design Theory and Methodology Conference, September 22–25, 1991, Miami, FL, *Design Theory and Methodology* 1991, DE-Volume 31, pp. 127–134, New York: ASME Press.

HERLING, D., D.G. ULLMAN, and B D'AMBROSIO. 1995. "Engineering Decision Support System (EDSS)." Proceedings of the Design Engineering Technical Conferences, September 17–20, 1995, Boston, MA., edited by Jadaan, Ward, Fukuda, Feldy and Gadth, DE-Volume 83, Volume 2, pp. 619–626, New York: ASME Press.

ISHII, K., and P. BARKAN. 1987. "Design Compatibility Analysis—A Framework for Expert Systems in Mechanical System Design." *Computers in Engineering*, Volume I, pp. 95–102, New York: ASME Press.

ISHII, K., 1991. "Role of Computerized Compatibility Analysis in Life-cycle Design." *International Journal Of Systems Automation: Research and Applications*, Volume 1, No. 4, pp. 325–345.

ISHII, K., C.F. EUBANKS, and M. MARKS. 1993. "Evaluation Methodology for Post-manufacturing Issues in Life-cycle Design." *Concurrent Engineering: Research and Applications—An International Journal*, Volume 1, No. 1, pp. 61–69.

ISHII, K., L. HORNBERGER, and M. LIOU. 1989. "Compatibility-based Design of Injection Molding," Proceedings of the 1989 ASME Winter Annual Meeting, San Francisco, CA, December 10–15, 1989; *Concurrent Product and Process Design*, edited by Chao and Lu, DE-Volume 21, PED-Volume 36, pp. 153–160, New York: ASME Press.

MALONEY, L., K. ISHII, and R.A. MILLER. 1989. "Compatibility-based Selection of Forging Machines and Processes." Proceedings of the Winter Annual Meeting of American Society of Mechanical Engineers, San Francisco, CA, December 10–15, 1989: *Concurrent Product and Process Design*, edited by Chao and Lu, DE-Volume 21, PED-Volume 36, pp. 161–168, New York: ASME Press.

MAKINO, A., P. BARKAN, L. REYNOLDS, and E. PFAFF. 1989. "Design for Serviceability Expert System." Proceedings of the Winter Annual Meeting of the ASME, San Francisco, CA, December 10–15, 1989; *Concurrent Product and Process Design*, edited by Chao and Lu, DE-Volume 21, PED-Volume 36, pp. 213–218, New York: ASME Press.

PAPANIKOLOPOULOUS, N. 1988. "FORS: Flexible Organizations." Master Thesis, Carnegie-Mellon University, Pittsburgh, PA, December 1988.

PAHL, G., and W. BEITZ. 1992. *Engineering Design: A Systematic Approach*, edited by K. Wallace, New York: Springer-Verlag.

POLI, C., J. ESCUDERO, and F. FERNANDEZ. 1988. "How Part Design Affects Injection Molding Tool Costs." *Machine Design*, November 24, 1988.

SAPOSSNEK, M. 1989. "Research on Constraint-based Design System." Proceedings of the Fourth International Conference on Applications of AI to Engineering, Cambridge, England, July 1989.

SAPOSSNEK, M., S. TALUKDAR, A. ELFES, S. SEDAS, M. EISENBERGER, and L. HOU. 1989. "Design Critics in the Computer-aided Simultaneous Engineering (CASE) Project." Proceedings of the Winter Annual Meeting of ASME, San Francisco, CA, December 10–15, 1989; *Concurrent Product and Process Design*, edited by Chao and Lu, DE-Volume 21, PED-Volume 36, pp. 137–142, New York: ASME Press.

SINGH, N., and SUSHIL. 1990. "A Physical System Theory Framework for Modeling Manufacturing Systems." *International Journal of Production Research*, Volume 28, No. 6, pp. 1067–1082.

SUH, N.P., 1990, *The Principes of Design*. New York. Oxford University Press.

TALUKDAR, S., and S.J. FENVES. 1989. "Towards a Framework for Concurrent Design." Proceedings of the Winter Annual Meeting of the ASME, San Francisco, CA, December 10–15, 1989, *Concurrent Product & Process Design*, edited by Chao and Lu, DE-Volume 21, PED-Volume 36, pp. 35–40, New York: ASME Press.

TEST PROBLEMS—DECISION SUPPORT SYSTEMS

7.1. Apply axiomatic design in designing a child's car seat that can be used in both large and small size cars and that can protect the occupant in a 35 mph head-on collision with another car of equal mass. How would you design this using progressive methods?

7.2. Using the design of a kitchen blender and mixer, apply several of the progressive methods to the product design and development and explain how the PD^3 process for the kitchen blender would benefit?

7.3. How do you introduce the IPD optimization concepts to the CE teams to make design decisions? How do you generally overcome conflicts and satisfy constraints in order to come up with a working design?

7.4. How are design decisions made in a conventional PD^3 process? Why is it too late to make any major design changes in the latter stages of the process?

7.5. Why do decisions made in the early stages have the potential of tremendous savings in the life-cycle cost? What is the price for a late or incorrect decision? How do you recognize and deal with *dirty dozen* design and manufacturing *quality killers*?

7.6. Why is video conferencing unsuitable for engineering collaboration? What are the available tools for early product trade-off during a design support environment (DSE)? What are its limitations?

7.7. What features do work groups require in new emerging C4 tools so that they are better equipped to help in a DSE? Why is it important to establish measurable goals for a Decision Support System (DSS)?

7.8. What is the span of decision making? Does it span from a cognitive aspect to a progressive aspect? If not, how does it span?

7.9. Describe a cognitive model of decision making. What are the four pieces of information required to create this model? Describe some of the steps of a cognitive decision model (CDM)?

7.10. Describe some of the CDM features that are borrowed from QFD. Explain the similarities of CDM and QFD. How do work groups compute technical importance ratings (for decision making)? What are the QFD equivalence of *WHATs, HOWs,* and *WHYs* (in CDM)?

7.11. Describe a basis of decision making in progressive models. Describe two models that falls in this category.

7.12. What is the axiomatic method of design? What are the two axioms on which this method is founded?

7.13. What results when a four-stage process systematization is combined with Shu's two axioms? Discuss the combined procedure that incorporates the two Shu axioms.

7.14. What are the fundamental approaches to progressive design? Describe seven progressive models that stem from employing design views and methodologies. Which of these is the most popular and why?

7.15. In spite of a number of available progressive models, why would one still need a hybrid approach to creating a progressive model? Describe how one could obtain such a hybrid model and its significance?

7.16. What are the major challenges faced during a typical PD^3 process? How does a work group plan to overcome them?

7.17. Why do most current C4 systems serve only a limited purpose? What can be done to enable them to be used widely? Does an intelligent model provide a solution that bridges the knowledge and interface gaps? What is being done today in this area?

7.18. What are the two modes of decision making that work groups can use. Which one is better and why? How is the performance level of the designed product affected by the CE teams involvement during a planning process?

7.19. Why is manipulating features *beyond modeling* considered more advantageous in facilitating modification goals? What are those features?

7.20. When do you characterize a feature as a smart feature? What are some bases of building intelligence into a smart model? What do you call a software prototyping?

7.21. What characteristics are needed for a sound DSE system? How do you capture those characteristics in a smart model? How do you build flexibility into the configuration of a Design Support System (DSS)? What is a smart regenerative system? What types of knowledge go into creating this system. List some examples of typical knowledge used.

7.22. What is the significance of a smart regenerative system? What types of rules does it contain? Does this work with external systems? What percentage of outputs is regenerative and what percentage is creative? Do JIT and TQM fit into the smart model concept? If so, how?

7.23. How do you maintain an enterprise rule-base in a smart regenerative system? What are the types of knowledge required to create a product assembly? How do you select parts, features, and materials to be included in designing an effective assembly plan?

7.24. How do you capture technical memory and knowledge about the products and processes? What is the purpose of capturing such knowledge? What types of concepts yield an expert type regenerative system? Who is the owner for technical memory or knowledge in a KBE system? What are the keys to a successful DSS operation?

7.25. Give some examples of the three types of knowledge employed in building a KBE application? Why does a KBE application sometimes provide a decision support role and sometimes an advisory role? What causes the system to act this way?

7.26. How do you obtain a life-cycle design? With what life-cycle values must the regenerative system comply in various stages of PD^3 process. What are the seven evaluation criteria to help you select the best design?

7.27. How does your company compile a list of DFX values? How is this stored, retrieved, and maintained to support the ongoing product development needs. What are the nine specific actions you can take to save bogged down derailed CE projects?

7.28. What are the different ways advanced technology can be used to reduce time-to-market. Why does it help to know which activities are on the critical path? What do you achieve if you automate product or part activities that are not on the critical path?

7.29. How do you estimate costs? What are the benefits of understanding what critical life-cycle costs are? What are the various costs that a product sustains in a PD^3 process?

7.30. What are direct and indirect costs? How do you minimize these costs? How does quality engineering save indirect costs?

7.31. How do you compute unit production, unit processing, and transformation costs? What are the two elements that make up a production cost? Do these costs follow any conservation model? If so, explain what is it?

7.32. How do you compute the various aspects of a total life-cycle cost, such as total production cost, assembly cost, manufacturing planning cost, service cost, etc. What are the differences between a component-based service cost and a symptom-based service cost?

7.33. What is a design compatibility analysis (DCA)? What types of knowledge are required for DCA? How do you use CE metric analysis to target parts best suited for an IPD?

7.34. What is a design sensitivity analysis (DSA)? What are the benefits that sensitivity analysis offers for a design trade-off? How do computer simulation and sensitivity analyses help in decision making processes?

7.35. What are the life-cycle ranking or rating schemes commonly used in a PD^3 process. How do you graphically display this information on a spider chart? What are the reasons for using a normalized rating?

CHAPTER 8

INTELLIGENT INFORMATION SYSTEM

8.0 INTRODUCTION

The product development environment typically suffers from a number of shortcomings. Some are partly due to the lack of integrated tools that CE work groups has to deal with, while others are partly due to the diverse nature of an enterprise's business operations. Too often, tool related shortcomings are caused by inappropriate or inadequate computer group-ware or aids. Such aids range from hardware and software tools to technologies and standardization. Technology is used here in a generalized sense similar to its definition in Webster Dictionary, "the totality of the means employed to provide objects necessary for human sustenance and comfort [Webster, 1990]." For example, by standardizing the design plans, tools and databases of all departments, Toyota enabled design work to overlap between stages. Downstream processes can be started while design plans upstream are still being completed [Okino, 1995]. The shortcomings of business operations result from four main sources:

1. *Process stagnation:* Process stagnation examples include tradition (e.g., why fix if it is not broke), business, management, technical, or operational 3Ps—policy, practices and procedures.
2. *Influence of infra-structural factors:* Examples include factors such as culture, mind-set, legacy, and human factors.
3. *Communication roadblocks:* Lack of familiarity, product experience, and training amongst the CE teams are some typical examples of communication roadblocks.

4. ***Organizational roadblocks:*** Lack of management support, confidence, and commitment to apply CE in full force (not haphazardly) throughout an enterprise are some typical examples of organizational roadblocks.

The manufacturing industry is deeply in the paradigm shift, from an *economy of scale*

- To an economy of information.
- To an economy of flexibility (agility).
- To an economy of intelligent manufacturing.

With the emphasis constantly changing in an organization from old systems to new initiatives (such as systems engineering, integrated product development, knowledge-based Engineering, TQM, CALS/EDI), C4 (CAD/CAM/CIM/CAE) tools are also currently in a state of flux. The changes (from the old to the new systems) are putting additional pressure on the C4 tools. These tools are required to provide up-to-date information, at the right place, in the right time, with the right amount, and in the right format. These tools are continuously churning a variety of information at many different places during the product life-cycle, which also needs to be accessed at many more places and applications. To allow the process of product realization to take shape efficiently, tools are being redesigned to reflect an organization's collaborative and competitive posture. Standardization—as in common systems, common methods, and common processes—is becoming more and more important. The quest for standardization is spreading to all disciplines, organizations, and structures like wild fires from a set of *design tools*

- To distributed computing (workstations, mainframe, database)
- To work-group computing (networks, LAN, WAN, etc.)
- To data exchange tools.

By unifying this pattern across all organizational areas—all departments and work groups working on the product—Toyota, for example, intends to reduce the average automobile time-to-market period from 30 months to 18 months by the end of 1986 [Okino, 1995]. In the data exchange area, there is an increased emphasis on the use of new or emerging feature-based standards. PDES/STEP is being implemented in newer CAD tools through a series of application protocols called APs. One of the APs addressed by the initial release of STEP is a configuration controlled design (CCD), formerly designated as AP203. CCD represents the dawning of a new era in digital product data exchange as it specifies how solid models are to be communicated. Using this protocol, one CAD system can now directly exchange solid models in a standardized format with another dissimilar CAD system.

Most research and development (R&D) efforts toward automation for modern manufacturing have been independently developed. For example, faster processors that worked as hardware brains (e.g., silicon graphics) for running high-end graphics applications, in the 1980s, were independently developed. But, these automated applications did not create a panacea, as initially expected, for reducing design and development lead time. In the same decade, design grew more complex, and the amount of time required to prepare the corresponding data (inputs) for each tool to be used in the design process also increased. Figure 8.1 shows the results of a 1994 survey of American manufacturers where the companies were asked about their in house usage of information technologies. The responses were compiled and ranked in nine technological areas: financial accounting, cost accounting and control, CAD, MRP, bar code or data collection, human resource planning, CAM, distribution resource planning and transportation management. The two largest technological uses of information system tools were in non-technical areas: financial accounting and cost accounting and control. In the technical areas: the uses of CAD technology were at the top. Other highly automated areas in manufacturing included computer-aided process planning (CAPP), computer-aided manufacturing (CAM), manufacturing resource planning (MRP), and computer-integrated inspection techniques (CII). With such tools, major functions are performed electronically (using compute power) but the data is nearly always exchanged manually. This is because the interface links between the various tools are either non-existent, weak, or are poorly managed by the vendors of these technological products. There is a recent proliferation of islands of pre- and post-processors generated from using these technologies (see Figure 8.2). They are partly due

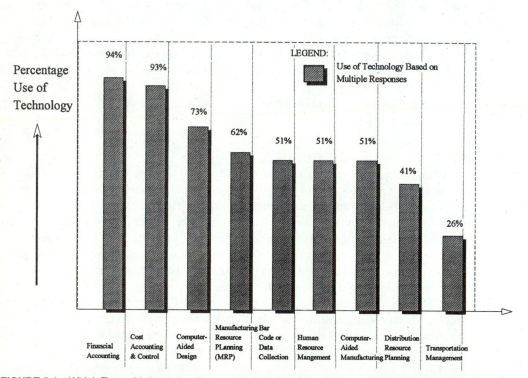

FIGURE 8.1   Which Type of Information Systems do Mid-Size Manufacturers Use? (Data based on Grant Thornton, Survey of American Manufacturers, 1994)

Interface Matrix	Financial Accounting (FA)	Cost Accounting & Control (CA)	Computer-Aided Design (CAD)	Manufacturing Resource Planning (MRP)	Bar Code or Data Collection (BC)	Human Resource Management (HRM)	Computer-Aided Manufacturing (CAM)	Distribution Resource Planning (DRP)	Transportation Management (TM)
Financial Accounting (FA)	→	FA->CA	FA->CAD	FA->MRP	FA->BC	FA->HRM	FA->CAM	FA->DRP	FA->TM
Cost Accounting & Control (CA)	CA->FA	↔	CA->CAD	CA->MRP	CA->BC	CA->HRM	CA->CAM	CA->DRP	BC->TM
Computer-Aided Design (CAD)	CAD->FA	CAD->CA	↔	CAD->MRP	CAD->BC	CAD->HRM	CAD->CAM	CAD->DRP	CAD->BC
Manufacturing Resource Planning (MRP)	MRP->FA	MRP->CA	MRP->CAD	↔	MRP->BC	MRP->HRM	MRP->CAM	MRP->DRP	MRP->TM
Bar Code or Data Collection (BC)	BC->FA	BC->CA	BC->CAD	BC->MRP	↔	BC->HRM	BC->CAM	BC->DRP	BC->TM
Human Resource Management	HRM->FA	HRM->CA	HRM->CAD	HRM->MRP	HRM->BC	↔	HRM->CAM	HRM->DRP	HRM->TM
Computer Aided Manufacturing	CAM->FA	CAM->CA	CAM->CAD	CAM->MRP	CAM->BC	CAM->HRM	↔	CAM->DRP	CAM->TM
Distribution Resource Planning (DRP)	DRP->FA	DRP->CA	DRP->CAD	DRP->MRP	DRP->BC	DRP-HRM	DRP->CAM	↔	DRP->TM
Transportation Management (TM)	TM->FA	TM->CA	TM->CAD	TM->MRP	TM->BC	TM->HRM	TM->CAM	TM->DRP	↔

FIGURE 8.2 Interface Links (Pre- and Post-Processors)

to a piece-wise growth of the technologies themselves and partly due to lack of in-house organizational standards for applying them uniformly across an enterprise. Figure 8.2 shows the number of upstream and downstream interfaces required to communicate among the previous nine-categories of technologies. In each technological category there could be a number of prominent tool vendors supplying computer tools. Each tool for a particular technology in turn may have its own pre- and post-processors for communicating among the tool vendors supporting that technological area. The number of pre- and post-processors needs may grow thus twofolds.

1. Pre- and post-processors for communicating among the various technology ranks (e.g., nine areas shown in Figure 8.1).
2. Pre- and post-processors for communicating among the tools supporting a particular information technology rank themselves.

With increased dependency on computational and logical techniques, the recent emphasis in CE organizations has been on integrating the existing CAD, CAM, CAPP, MRP and CII systems to provide a computer-integrated manufacturing (CIM) environment (see Figure 8.3). Developments in integration area include IGES, design for manufacturing & assembly (DFMA), DBMS, PDES, JIT, MAP and Cell control software [Prasad, 1995a].

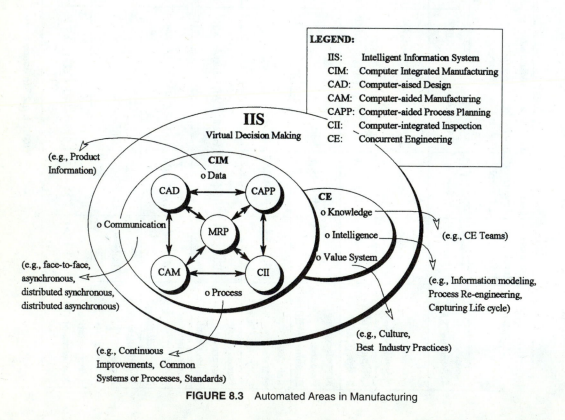

FIGURE 8.3   Automated Areas in Manufacturing

However, they are currently deemed to be independent contributions to improve productivity or efficiency in specific CIM areas or applications.

Today, CIM systems are merely being applied to integration and processing (storage and automation) of data, communication, and processes (common systems and standards) [Alting, 1986]. The communication part of the CIM design is data and process:

1. *Data:* This includes many different categories of product images:
   - CAD data.
   - CAM data.
   - CAPP data.
   - MRP data.
   - CII data.
   - Design specifications.
   - The history of production.
   - Interface information in various forms, including electronic, text, raster images, video, audio, and their mixture, as well as many different types of paper formats and methods.
2. *Process:* This includes methodologies such as continuous process improvements (CPI), QFD, Pugh, Taguchi, TQM, common systems, standards, and so on. Embedded within such a vast information base, there lays a hidden knowledge about the design of product and the processes.

CIM, in the current form, is incapable of dealing with product knowledge or solving problems related to decision and control. Even though, there has been an increasing interest in subjects such as artificial intelligence (AI), distributed blackboards (DBBs), knowledge based systems (KBS), expert systems, and so on. [Lim, 1993]. The latter slew of tools is more cognitive in nature and is quite potent for decision making in CE. More discerning companies that have felt a greater drive to improve the competitiveness urge, are focusing their vision on the combined potential of CE and CIM technologies. CE and Computer Integrated Manufacturing go hand in hand. Together, they are called Intelligent Information System (IIS) [Prasad, 1995b]. CIM plus CE equals IIS (see Figure 8.4).

$$CIM \oplus CE \Rightarrow IIS \qquad (8.1)$$

Most companies consider manufacturing information-intensive. Many more believe that intelligent handling of information through computer techniques can yield a better CIM system since IIS can monitor and correct problems. IIS reduces the need for frequent manual intervention. CE brings forth three missing links of CIM.

1. *Intelligence:* The intelligence comes from the virtual elements of CE teams. This is discussed in Chapters 4 and 5 of volume I.
2. *Knowledge:* The knowledge mainly comes from information modeling and from capturing life-cycle intent. This is discussed in Chapter 7 of volume I and Chapter 6 of volume II.
3. *Value System:* Value system deals with items such as culture, and best industry practices for embedding a procedural discipline in CIM operations and in enter-

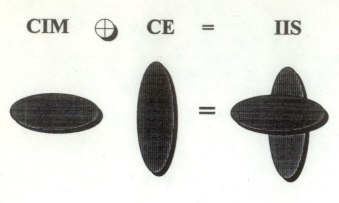

FIGURE 8.4   Intelligent Information System (IIS)

prise-level communications. This is explained in Chapter 6 of volume I and Chapter 3 of volume II.

Concurrent Function Deployment (CFD) is a general definition of many deployment techniques that have been introduced in recent years. They are discussed in Chapter 1 quite extensively. These CFD techniques range from *soft* prototypes, that is, generic templates of product and process design, to so called *smart* prototypes. They are defined as follows:

- **Soft prototypes:** The well-known QFD/House of Quality and CFD/House of Values are examples of *soft* prototypes.
- **Smart prototypes:** CFD techniques for *smart* prototypes include parametric, variational and knowledge-based applications in modeling *physical phenomena*. Current smart CFD techniques include integrated product development, virtual teamwork, collaborative decision making, groupware, virtual mapping, virtual integration, virtual processing, etc. DARPA and the Air Force have mostly expressed interests in these smart techniques as a method of risk reduction. This has allowed new product concepts to be investigated earlier in their design phase by all members of an integrated product development team. Many of these concepts are discussed in other chapters of the *CE Fundamentals* book. The integration of CE and CIM is discussed in this chapter.

The key to the successes of IIS is understanding the obstacles to unifying CE with existing CIM processes and identifying new opportunities for improvement. The identification of improvement opportunities and the implementation of effective product development process control strategies can be facilitated by a systematic collection and monitoring of relevant in-process metrics.

8.1 ENABLING ELEMENTS

The major enabling elements of Intelligent Information System (IIS) applicable to product development are discussed in this section.

Major Enabling Agents	Factors Showing Scope/Range	What typical questions to ask to determine its (scope's) importance or contribution?
Talents	Expertise (competence), Experience, Negotiation ability, Negotiation Power, Intelligent Quotient (IQ), Job skills, Education, Professional Development. Job Training Programs, Technical and Leadership Training, Culture/Attitude	Is team competent to do the job? Is team experience enough? Is team able to come to consensus? Can team resolve its conflict? Does team apply common sense? Has team basic understanding of the engineering concepts? Does culture of the team conducive to cooperation?
Task	Independent, Dependent, Coupled, Size, Complexity, Novelty, Repetitive, Hierarchy, Breakdown Structure, Numbers, Technical Risk, etc.	Are tasks dependent, independent, or coupled? What is the project's size? How complex is an activity? Have the tasks been decomposed enough? Is the task unique? Are tasks repetitive? Do we understand the tasks' hierarchy? How big/small is each decomposed task/hierarchy? What are the probabilities for its successful completion?
Teamwork	Cooperation, Commitment, Motivation, Trust, Morale, Role Balance, Job Rotation, Group Dynamics, Personal Satisfaction, Empowerment, etc.	Are team members cooperative? Are they committed? Is team motivated? Does team members trust each other? Do they respect each other? Are they concern about their personal gains, security? Does the members help each other in needs? Do the teams change hats? Are the teams able to communicate effectively? Do the teams have open and clear channels of communication?
Techniques	QFD, Quality Engineering, CPI, Process Re-engineering, Taguchi Robust Design, Serial, Concurrent Process, Systematic Approach, Decomposition, Integration, Concurrent Function Deployment (CFD), TQM, etc.	Does the team familiar with CE techniques and its usage? Does the team use QFD, quality engineering and CPI principles while doing their job? Does the team re-enginer the process before automation? Does the team understands differences between serial and concurrent development? Does the team able to decompose products into hierarchical structure? Does team understand Concurrent Function Deployment (CFD)? -- understand Total Value Management (TVM)? -- big picture?
Technology	CAD/CAM/CAE/CIM, JIT, Process Planning, NC, DNC, Workstations, Networks, Client Server, Email, Product Tech., Process Tech., Features, Innovation, etc.	Is the team trained in the use of CAD/CAM/CAE/CIM and JIT software systems? Do they understand its usage in product development? Have they done NC and DNC from the same CAD model environment? Are they computer literate? Are they comfortable working on different workstations, distributed networks and client server environment? Does the team have E-mail capability? Do they understand the current design technology, limitation and new product and process features that are coming to the marketplace?
Time	Start Time, Finish Time, Lead Time, Magnitude, Delivery Time Constraints, Productivity, Schedules	What is the lead time for doing an activity? What's the start and finish times, schedules for completing a task or an activity or a phase? How long an activity can take? Can the project be completed on time? Why it takes so long to do an activity? Do the teams working effectively? Are the tools helpful? robust enough?
Tools	Office Tools, Communication Tools, Networking Tools, Project Management, Computer-based Design Modeling Tools, Computer Aids, Product Models, Process Models, Enterprise Models, Codes and Standards	What are the available tools? Do the teams have enough compute power? Do the teams share the compute resources? Are the office tools meeting the teams needs? How the teams communicate? Are the teams networked? Have the teams cooperative Email facility? Are they able to engage in (over-the-net) discussion of parts features, graphics and video transfer. What are the collaborative tools available? Are relevant design codes and standards incorporated?

FIGURE 8.5   Seven Ts—The Enabling Agents

Seven Ts: The seven Ts consist of *talents, tasks, teamwork (or teams), techniques, technology, time and tools* [Prasad, 1995c]. Figure 8.5 identifies factors showing the scope for 7Ts and what typical questions to ask to determine their importance. Teamwork entails equipping manufacturing support personnel with X-abilities talents (expertise) in the product development areas. Teamwork also implies making use of surrogate X-ability tools during the early stages of design rather than only when problems crop up or when the design is set in stone. Tools and technology include a growing set of inter-operable computer aids for geometric design and prototyping networked into a highly extensible environment. Techniques imply analysis and design methods (such as software prototyping, design for X-ability, analysis, simulation, multi-media, virtual reality, etc.) to visualize product and process concepts quickly and in a format that is understandable to all team members. Teams apply these tools and techniques to experiment, in a timely fashion, with a number of product and process options that are available or feasible. For manufactures to become more time competitive teams need tools to match the tasks. In other words, software and hardware tools, techniques and technology must empower concurrent teams at all levels of an organization. This is done either by pro-actively informing teams of tasks that require immediate attention, or by giving teams timely information to make timely decisions before tasks become critical.

Empowerment: This entails empowering multi-disciplinary teams so that they are able to make critical decisions early, preferably during product definition stage. Often, due to lack of empowerment, teams make decisions later in the process, such as during major design reviews, when the cost of engineering changes is much more dramatic.

Requirements Management: This deals with attaining a balance between product and process and managing requirements. The managing process ranges from understanding the requirements, various interfaces, plan of manufacture for an existing design, to extending system support for a new design, if envisioned, while meeting the artifact's major functional goals.

Information Modeling: This entails the use of various models that electronically represent, in convenient forms, information (knowledge, methods and data) about the product, process, enterprise, and the environment in which the product is expected to operate.

Standard Means of Exchange for Data, Methods and Knowledge: This is accomplished by a slew of standardized support systems (computers, networks, tools, database, applications, procedures, etc.) to encourage the sharing of common information among the CE team members. A network of compatible systems quickly relays product design plans, iterations, reviews, and approvals to the CE participants.

Information Sharing Architecture: This includes enabling technologies for concurrent engineering, like multimedia communication, framework integration, enterprise integration and information sharing architecture, in a distributed synchronous setting. Four common types of such interactions are shown in Figure 5.5 of volume II. Standardized means of information sharing foster effective communication among various personnel teams involved in IPD activities. Examples of such activities include recording of design history, processing design plans, common product design and development process, standardized testing, design review, video conferencing, and project management. Mozaic—the Auto-trol Technology's object-oriented CE architecture—is an example of the modularity and architecture that comes from object-oriented technologies. Objects represent the real-world product and process decomposition such as tree structures (PtBS, PsBS, WBS). Their ease of communication across platforms make them ideal for collaborative environments.

Cooperative Problem Solving—Seven Cs: Cooperative problem solving includes seven Cs, *Collaboration, Commitment, Communication, Compromise, Consensus, Continuous Improvement,* and *Coordination.* It means sharing problem-solving information among the teams, so that instead of a single individual, the whole team contributes towards decisions. Functions that focus and facilitate collaborative discussions involve use of sound analytical basis such as recording of design rationale, electronic critiquing of designs, and planning and execution of design changes.

Intelligent Decision Making: Decision making can be viewed as a process of creating an artifact that performs what is expected (specified as goals) in the presence of all sorts of inputs, requirements, constraints and the operating environment that governs its behavior. Depending upon the cognitive knowledge about a product available to a decision maker, design decisions may range from a cognitive to a progressive aspect.

The above enabling elements can be represented in a set form as:

$$\text{Enabling Agents} = \cup \, [\{Seven \, Ts\}, \{Empowerment\}, \{Requirements \, Management\},$$
$$\{Information \, Modeling\}, \{Standard \, Means \, of \, Exchange \, for \, Data, \, Methods$$
$$and \, Knowledge\}, \{Information \, Sharing \, Architecture\}, \{Cooperative \, Problem$$
$$Solving—Seven \, Cs\}, \{Intelligent \, Decision \, Making\}] \qquad (8.2)$$

Although IIS has a major impact on productivity, there are several major barriers that inhibit the above enabling elements from attaining the full potential of manufacturing competitiveness.

8.2   MAJOR BARRIERS

Despite the fact that many of the aspects in design, engineering, prototyping, production, and manufacturing can be assisted and automated by computer aided tools, traditionally each aspect or activity is mostly dealt with in its own domain. Lack of in-house standards for information handling and inflexible organizational, cultural and economic practices impede improvements in efficiency and productivity. Despite the improvement of manufacturing automation protocol (MAP) and developments in PLC to facilitate communication at the equipment level, the problems of inconsistency, translation, scheduling, management and control of data in CIM have yet to be effectively resolved. There are several barriers that hinder the development and processing of a product. In a joint study with DARPA's Initiatives in Concurrent Engineering (DICE) program, many large companies, including General Motors, General Electric, Lockheed Missiles and Space Company, and Boeing Aerospace, have experienced barriers while implementing concurrent engineering concepts. Such classes of barriers can be categorized in the following thirteen ranks (see Figure 8.6).

1. *Organizational Barriers:* These types of barriers are man-made and generally appear due to cultural, social, and behavioral aspects of the teams involved. There are

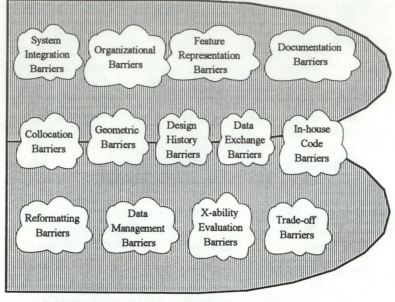

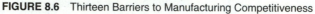

FIGURE 8.6   Thirteen Barriers to Manufacturing Competitiveness

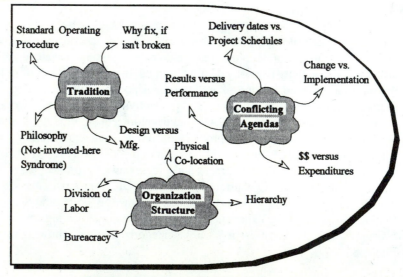

FIGURE 8.7   Subclasses of Organizational Barriers

three sub-classes of organizational barriers as shown in Figure 8.7. They are caused by the presence of the following:

- *Organizational Structure:* Some examples of organizational structures are divisions of labor, teams' cultural roots, procedural bureaucracy, physical co-location, reporting hierarchy, and organizational charts. For example, organizational charts divide work into departments and thus tend to create false boundary lines. Invisible walls develop since most companies evaluate their managers or departments based on the performance of their individual work segments.
- *Conflicting Agenda:* Examples of conflicting agenda include: results versus performance metrics, capital dollars versus expenditures, delivery dates versus project schedules, change plan versus implementation plan, etc.
- T*radition:* Most team-members do not like changes that modify the status-quo that the members have comfortably learned to live with (or cope-up with) over the years. They tend to be uncooperative, not because they love the invisible walls, but because they are innocent victims of those past customary practices. Examples of such traditions or practices include: standard operating procedures, organizational culture, not invented here (NIH) syndrome, "why fix something if it isn't broken?" and so on. Further, since it does not benefit someone individually, many have very little or no personal motivation in wanting to change such practices.

2. **Geometric Barriers:** This results from the differences between underlying mathematical representations of various off-the-self tools or systems and what is required in practice. For example, solid modeling techniques are widely used for defining mechanical components and products. They are represented in various geometrical forms, and often serve as a potential tool to integrate design with manufacturing. Contemporary solid modeling systems provide good support for analytical procedures that can be used for things as interference checking and mass-properties computations such as volumes, etc. However, geometrical forms in those solid models do not explicitly specify the functional requirements and allowable variations of a product (such as dimensional variations) that are important to accurate manufacturing, assembly, and quality evaluations.

3. **Data Exchange Barriers:** Most CAD systems have their own proprietary formats for representing CAD entities. To communicate with another CAD system, design information from a previous CAD system must be converted to a neutral file or use a built-in system-to-system translator. Currently there are several neutral data formats for transferring CAD information from one system to another. These include: IGES, EDIF, DXF formats, and so on. In neutral file format, there are at least three levels of interpretation required.

- *First,* data is written from its proprietary format using the parent CAD system into a neutral file format. In doing so, it uses the best interpretation of what the corresponding geometrical entities mean in the neutral file format.
- *Second,* the neutral file format data is transferred and read-in by the receiving CAD system.
- *Third,* data is then reformatted (again interpreted) from its neutral format to a new proprietary format.

In addition to the above, many new data translation tools are now in use to aid this process such as PDES and STEP. Unfortunately, this process is still not robust. Numerous categories of data loss or errors, such as inaccuracies, incomplete or extraneous information, missing data, ripples, and instabilities, occur during the translation process (see Figure 8.8). Loss of data is defined as the numerical difference between the results of the translated (output data) form from its raw (input data) form. Examples of typical losses are listed in Figure 8.8.

4. *Features Representation Barriers:* Many CAD models include no description of parts functional features and of their interactions with other parts. The lack of modeling facilities for capturing functional requirements creates gaps in integrated product development (IPD) for concurrent engineering [Hummel and Brown, 1989]. For example, the lack of tolerance specification facilities in current CAD environment represents a serious deficiency for the implementation of a process-rich CIM system.

5. *Reformatting Barriers:* Most of the computer programs that are used to support product development require creation of a number of analytical models of the product. Typical product models include: CAD solid models, FEM/FEA models, NC machining models, knowledge-based rules embedded in relational database management

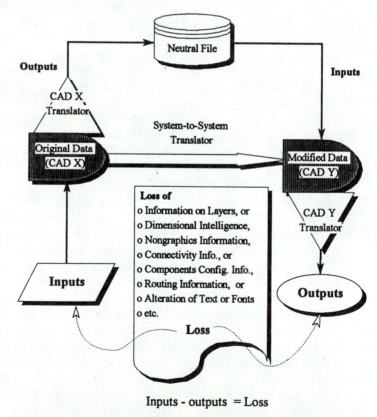

FIGURE 8.8   Types of Output Errors After a Typical Data Translation Process

system (RDBMS), computer-aided software engineering (CASE), graphics user interface (GUI), legacy system, import/export, etc. (see Figure 8.9). Information that is required at various intermediate stages of product development needs to be reformatted to suit the needs of the particular *firmware* application, even though the type of data may remain the same. This is because each application model requires the same data (named entities and the objects) according to his or her *firmware* standards, which is proprietary and unique to the vendor company. Data and information that are suitable for one *firmware* application are often inappropriate for another *firmware* application, requiring development of a large number of pre- and post-processors. This in turn leads to software maintenance problems, data inconsistency, and management of large incompatible data bases. This also makes the control of data tedious and error prone. When manufacturers attempt to integrate incompatible *firmware* (applications) together, translation, scheduling and control of data in a CIM environment become a mammoth problem. Time spent translating, correcting, and rebuilding the data model is costly and adds little value to the core (product design and manufacture) functions.

6. ***Trade-off Barriers:*** One of the major shortcomings of present design and decision-making practices in PD^3 is the lack of adequate analysis and simulation tools for early product trade-off. Due to the crude nature of early product data and stringent time allocations, most trade-off studies at the initial stages of the design cycle are done quickly and hence so many of the important analysis steps cannot be performed. When enough design checks are ignored due to lack of time, it is likely that the design, when passed on for downstream operations, may remain unchecked or incomplete. There are minimal tools to compare design alternatives during early product development. It is well recognized that it is cheaper to solve any problem

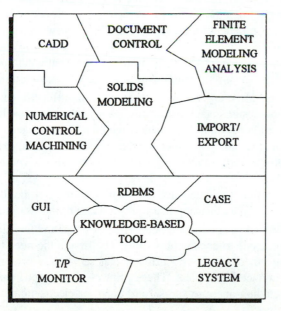

FIGURE 8.9   Reformatting Barriers

early in the design stage rather than in later stages. This is because the cost for introducing changes in later stages of design can be many times higher. Furthermore, early handling of design problems leads to fewer problems at later stages. It is therefore much more cost-effective to carry on more iterations and tradeoffs during an early part of the design cycle rather than in the later part.

7. ***Data Management Barriers:*** In most companies, team members keep their daily work files separate from each other and store them in their own private and local directories. There is a lack of uniformity in work-group filing or network sharing. Even if a product data management (PDM) system is in use, there are problems with maintaining current product versions and revisions. If someone needs to share data, they need to ask for a copy or know the security codes or passwords of the owner to retrieve them. Advances in electronic communication and standards are now being readily used to access information on a network. But incompatibilities of hardware, software, and operators' errors are still very common. Other problems that restrict the smooth flow of communication are the difficulties in properly describing entities in the data base, the speed of transmission and translation of data among different functions, activities, and sub-systems.

8. ***Design History Barriers:*** Tracking the history of a design is normally not an integral part of a PD^3 process. It is typically done after the fact. Thus, there is no way to find out what parameters or values a team member was using, when he or she returns from his or her break, including where they left off and the status at that time. Most team members put down notes they remember at that time. There is no log for tracing out the design history. Notes are subject to errors, which makes such records unreliable and less useful. The problem is somewhat contained when the work group is small. In a small work group, one team member may be responsible for all his or her work and little or no interface is required. When the work group is large and multi-disciplinary, or when the product is complex, a single individual or even a team cannot possibly complete all the assignments. Team co-workers often come in as resources to provide supplementary knowledge or the expertise required for the job at hand. In such a scenario, it is difficult to work without a common design history log and to track progress as a natural part of a teamwork environment.

9. ***In-house Code Barriers:*** Most organizations have made significant investments in commercial analysis and design codes (firmwares). Many have also developed their own niche of customization over and above the commercial codes (such as application-specific integration, pre- and post-processors, macros, or linkages) to help them gain performance or productivity advantages over their competitors. These specialty codes typically remain isolated or loosely integrated into the overall design-to-manufacture process. Frequent maintenance and upgrades are often not considered the teams' job, though they may be using them for product or process optimization.

10. ***Documentation Barriers:*** A user of a Computer-Aided Design and Drafting (CAD) system has often two reasons to document outputs. The first is the need to manage multiple media. Today, there are hybrid printers that serve as a bridge between graphic-plotters and engineering-copiers. These printers can accept inputs from a variety of sources, including aperture cards, hard copy, or electronic devices. The

second need is perception. Even if a design can be viewed on the workstation, designers still want a hard copy in hand, either for themselves or to pass on to someone else. They cannot see everything on a small screen (CRTs). Often they prefer to consult someone or annotate the design on paper. If information is passed on in the form of paper, it is error-prone, subject to delays, and susceptible to traps—not having the right information or forgetting to ask for the right information. Collaborative visual techniques for setting meetings on the network can solve this problem but this has yet to become a reality. There are performance problems in showing all visual effects quickly and inexpensively in a real time setting.

11. *Collocation Barriers:* In a multi-disciplinary organization, it is not possible to collocate all co-workers. People are often geographically distributed by virtue of being responsible for multiple tasks, or being part of different organizational or reporting structures. In a matrix organization, they usually do not report to the same manager. The team, thus, experiences problems in exchanging and sharing of data and information. As electronic communication-based networking becomes more advanced, geographic separation will cease to be a concern when forming work groups.

12. *X-ability Evaluation Barriers:* These come from the teams' inability to use the current CAD system to perform X-ability (Design for Assembly (DFA) or Manufacturability (DFM)) type of evaluations. Most design checks relate to qualitative evaluations relative to Design for X-abilities (DFX). Most commercial CAD systems do not evaluate the degree of difficulty associated with a given mechanical system and assembly (MSA). There are some quantitative methods based on empirical data or index of difficulty, but none are normally tied to a CAD system.

13. *System Integration Barriers:* In designing an information system for an enterprise, one is immediately confronted with inconsistencies between data sets and descriptive forms. Examples where such inconsistencies occur are:
    • Product design data and product family (PtBS).
    • Process engineering data and process hierarchy (PsBS).
    • Work hierarchy (WBS) and the know-how data involved in the manufacturing processes.
    • Production level tasks (programmable logic controller) and data that describe the plant floor.
    • Management steps and business operations.

These inconsistencies may generate the following five types of data sets:

1. A data set for CAD environment (a company is fortunate if there is only one CAD system in use),
2. A second data set for process planning.
3. A third data set for manufacturing processes, MRP, etc.
4. A fourth data set for PLC operations.
5. A fifth data set for business and management operations.

All these data sets are supposed to be homogeneous, that is, the pertinent data forms must exhibit the following properties.

1. Describe the common entities the same way.
2. Use the information that is required without translation.
3. Share a common data representation schema.

In actuality, these data sets are often heterogeneous (describe the same object independently), produce duplicate information, or are incomplete. For large organizations, it is not easy to accomplish a complete level of integration that is based on a single data structure. Most industrial data sets that exist today are partially integrated [Alting, 1986]. Without system integration, it is difficult to improve operational efficiency or be globally competitive. Without integration and automation, it would be impossible to produce complex products such as automobiles or aircraft in a competitive fashion. The above 13 ranks of barriers can be represented in a set as:

$$\text{Barriers} \equiv \cup \, [\{Organizational\ Barriers\}, \{Geometry\ Barriers\}, \{Data\ Exchange\ Barriers\}, \{Features\ Representation\ Barriers\}, \{Reformatting\ Barriers\}, \{Trade\text{-}off\ Barriers\}, \{Data\ Management\ Barriers\}, \{Design\ History\ Barriers\}, \quad (8.3)$$
$$\{In\text{-}house\ Code\ Barriers\} \{Documentation\ Barriers\}, \{Collocation\ Barriers\}, \{X\text{-}ability\ Evaluation\ Barriers\}, \{System\ Integration\ Barriers\}]$$

These barriers act like speed bumps; they tend to slow down the progress for a while. Teams often use stop-gaps, temporary fixes, or alternate means, including 7Cs, to smooth out barriers. However, some barriers are tougher than others. Intelligent KBS for CIM applications have recently emerged from research laboratories to address some of these tough barriers. Details of these intelligent KBS applications are found in fragments in a number of books and research papers. Among these efforts, most tend to be localized— addressing a specific domain and focusing on a particular barrier. Recent domain examples include applications in product configuration, process planning, job-shop scheduling and fixture design [Lim, 1993].

8.3 VISION OF THE FUTURE

Designing a new and complex system requires dividing the work into work breakdown structures (WBS), distributing the tasks into work groups (e.g., design teams, manufacturing teams, etc.), determining the iterations between the computational and data modules, and reordering the sequence of modules. A module in this context does something. It can be, for example, a computer program or a team to minimize iterations. Very few cooperative tools are available to aid the design team in making early decisions regarding proper execution sequences of the design.

   More and more, engineers are looking to software to be a teacher, expert adviser, organizer, problem solver, and specialized librarian in addition to its more traditional function of being a productivity tool. They expect the software to teach them better ways of doing their work. Some are seeking new, better, and more enjoyable work environments (e.g., multi-media, windows, etc.). And a growing number of teams are looking to software to help them do things that they have never done before. Software vendors are also responding with better capabilities, more efficient environments, faster processing, and all-in-one integrated tools.

In the future, an engineer will perform a multitude of highly specialized tasks, each having more than one disciplinary flavor. Intelligent systems that contain the product's specific functional knowledge (algorithmic and heuristic) and the process specific facts pertaining to the product manufacturing operation will be used extensively throughout an enterprise as shown in Figure 8.10. The figure shows most of the manual tasks related to product design, engineering, and manufacturing processes computerized into a series of integrated product and process design modules or knowledge-based tools to support a complete customer-focused manufacturing system. Production planning system (PPS), purchase order management and inventory management provide direct planned order conversion from a MRP/Master Production Schedule (MPS) to manufacturing execution system (MES). Both MRP and MPS can be based on net change or regenerative with respect to *available-to-promise* for forecasted parts. Production program planning targets inventory balancing, smooth production curve, and seasonal product line planning for both new *product roll-outs* and old *product phase-outs*. These are shown (see Figure 8.10) as three vertical parallel rows of functions supporting:

- Product management cycle, including CAE, CAD, CAP, CAM, CAQ, and CAX.
- Process management cycle, including PPS, MRP, MPS, and MES.
- Data management tasks.

Purchase order management can provide a configuration capability from make-to-order to make-to-stock, and to customize sales orders based on model variety or options. The CE approach emphasizes in-process decision-making support in contrast to operational planning emphasis of MRP II and the more recently developed planning systems concepts, such as Enterprise Resource Planning (ERP). Inherent in the notion of in-process decision making or operational support is the need for up-to-the-minute feedback on the execution of the production plan. That is a real time manufacturing execution system (MES) that is fully integrated into the traditional planning, order management, and financial functions. Other CAE, CAD, CAP, CAM, CAQ, and CAX tools assist a work group manager in making early decisions regarding decomposition of a complex design problem (product system) into its constituents: sub-systems, components, parts and features, materials, process, data, and so on. These tools guide the decoupling of decomposed constituents into independent or semi-independent steps. The tools recommend, depending upon the constituent complexity, paralleling steps (concurrent use of a computer program, a team, or a person) that require little or no feedback between themselves. Some tools also recommend those tasks that are unavoidably coupled, thus forcing them to be executed in a serial mode. Other tools allow work groups to visualize the relationships among various tasks or steps. The tools can also be used to reach consensus regarding the design constraints that are common among the steps and where feedbacks are to be allowed. The tools not only provide a framework for analyzing these alternatives but actually show how to alter these

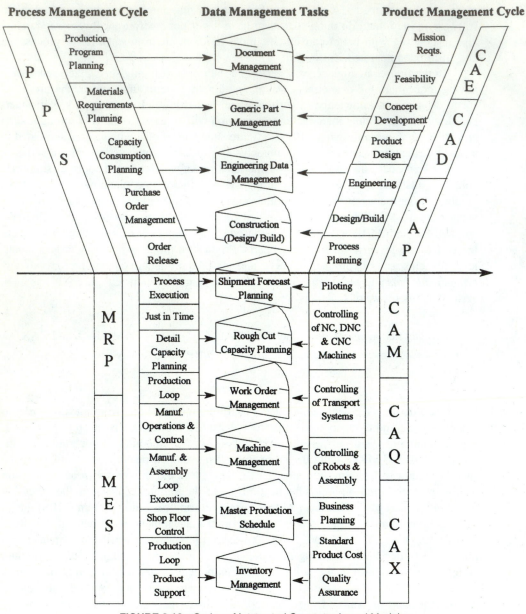

FIGURE 8.10　Series of Integrated Computer-based Modules

processes based on the captured intelligence or knowledge stored in databases as technical memory.

　　Figure 8.11 serves as a road map in pointing out things in relation to automation of an enterprise's CE environment. The tools provide means to synthesize—to assemble a pair of features into a part; a pair of parts into a component; a pair of components into a

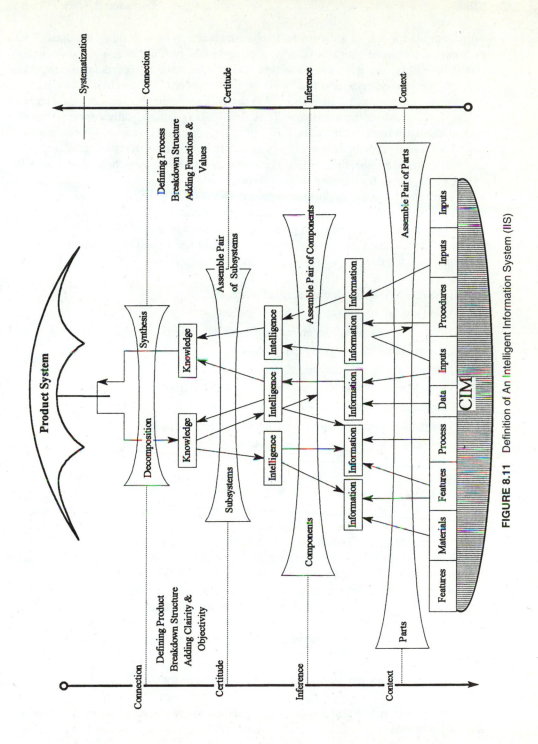

FIGURE 8.11 Definition of An Intelligent Information System (IIS)

subsystem; and a pair of subsystems into a functional system. Joining the activities in pairs can be continued until all of the features, materials, process, data, and so on, are converted into parts. All parts are converted into components, all components are reconstructed into sub-systems, and finally all of the subsystems are synthesized back into a full system. As teams and their tools need to use data in different places, that data must be linked with corresponding translators. They (translators) automatically convert them into more meaningful form, such as new forms of knowledge (see Figure 8.11) or a format required for a firmware application. One way to assure this type of automation is through knowledge-based engineering (KBE). The vision of future contained herein assumes KBE to be a major player for building an intelligent information system.

Perceived functions applicable to each division or work group in the product's life-cycle are formed as virtual agents of the intelligent system as shown in Figure 8.12. These virtual agents aid the CE teams in performing functions of an entire product's life-cycle with accelerated speed and greater accuracy. Typical agents and the functions, they can perform, may include:

- Infrastructure agents for creating infrastructure.
- Standardization agents for establishing standards.
- Need or usage agents for determining the product need or usage.
- Design agents for designing the product.
- Analysis agents for analyzing the product.
- Product development agents for modifying an old design.
- Fabrication agents for manufacturing the parts.
- Production agents for controlling the production operations.
- Distribution agents for sales and marketing.

Some examples of activities that are performed by each agent in a CE office are listed in Figure 8.12. These virtual agents of the intelligent system increase the sharing of heuristics, algorithmic, and derived knowledge, resulting in reduced product lead times, improved accuracy, lower costs, and improved customer satisfaction.

The CE designers, manufacturing engineers, and other teams are able to communicate their ideas early. The IIS environment contains tools and functions for cooperative problem solving. Collaborative discussions can be carried on interactively over the network. Problem solving functions, such as critiquing of ideas, recording of design rationale, and simultaneous planning and execution of design modifications among the CE teams can be done on a real-time basis. Computerized modules constitute a nucleus for speedy communication through a concept known as *soft* and *smart* prototyping discussed earlier.

8.3.1   CE: A Missing Piece of A CIM Architecture

The IIS objective is to transform an initial specification of a product into a description of a physical device or an artifact. A CE system architecture to accomplish this objective is

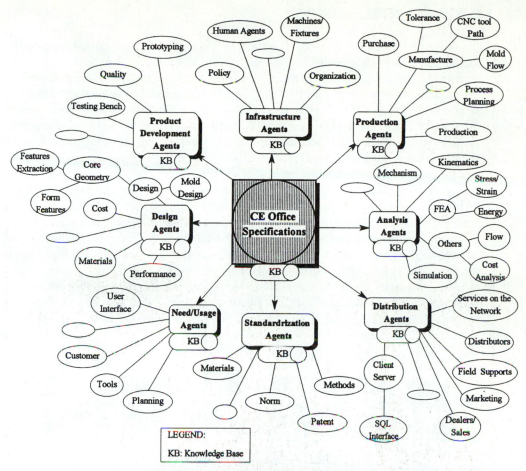

FIGURE 8.12   Population of Agents in a CE office

shown in Figure 5.5, volume I. Four groups of people are shown to be working together: logical teams, personnel teams, technological teams, and virtual teams. The idea is to provide a group problem-solving environment in which human teams (the technology and personnel teams) and life-cycle perspectives (logical and virtual teams) cooperate in the generation of artifacts. Together, they have the opportunity to generate and test design decisions, enabling the simultaneous consideration of all needed perspectives throughout the design process. The competing goals of the teams and the different life-cycle perspectives as well as the interactions between specifications and outputs of a transformation provide many sources of conflict during a product realization process. Consequently, it is necessary to determine dynamically which perspective dominates at what stage of the design process. Specifying a blackboard architecture alone is not sufficient to specify the system's behavior. The human teams' role is to coordinate the activities so that the team members are cooperative and coherent.

8.4   LEVELS OF INTELLIGENCE

There are various types of activities that take place in product design and development. On the one hand, there are repeated or non-creative activities that can be performed by a team member or an individual. Such activities are routine, teams are familiar with them, and they do not require much collaborative effort. Some are middle-of-the-road activities that may require some degree of intelligence for decision making. On the other hand, there are creative activities that require knowledge beyond one's own disciplines or areas of expertise. Depending upon the levels of activities and need for cooperation, the degree of intelligence required varies. This is shown in Figure 8.13 where six levels of techniques or methods required for a class of activity are identified against the degree of creativity and needs for cooperation. The first of such methods is *network-based methods*, which can be performed by an individual team or a team member and where the activities are routine types. This is identified in Figure 8.13 as level 0. The next level is level 1. Over time, team members may have discovered heuristics in performing such tasks—what work the best (best practices)—and what to do in what situation (common systems). Such activities are still routine, though to reduce the lead-time, some level of intelligence, such as spread-sheet logic and documentation-based methods, would be useful. Need for cooperation increases as one moves away from simple problems to family of part—geometry creation (level 2 activities). The use of variable-driven methods (such as parametric, variational or feature-based) are useful for level 2 to alleviate the boredom tasks of recreating the design details repeatedly based on

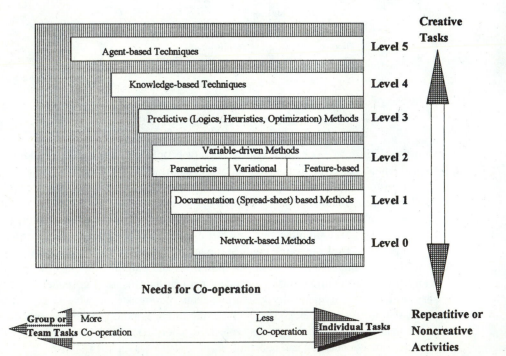

FIGURE 8.13   Levels of Techniques/Methods Driving Cooperation in a CE Office

geometrical compatibility. There are problems beyond geometry whose solutions require non-geometrical knowledge, such as materials substitution, configuration designs, layout designs, knowledge of interaction problems, logics, heuristics, optimization, and so on. These are classified as level 3 or predictive methods. Knowledge-based techniques are more suitable to provide a level of intelligence to deal with such "knowledge-rich" class of problems adequately (level 4). On the other end of the spectrum are the agent-based or *multiple knowledge-based* activities (level 5) that require teams with intelligence, ingenuity, and creativity. An individual team with its own knowledge may not be able to comprehend the magnitude of the decision. Most complex decisions are made during team reviews or quality network circles or from similar collaborative sources. The levels of techniques or methods required to address all these classes of activities are contained in Figure 8.13. There are six levels of techniques identified, one for each level of activities from level 0 to level 5. Level 5 activities are not easily amenable to automation techniques since the possibilities are unlimited. In Figure 8.14, an attempt is made to classify the range of tools by the degree of creativity and degree of uncertainty present. The computerized tools required for creative tasks (levels 4 and 5) are of a very different class than those required to solve routine type activities (level 0).

The above range of tools can be represented in a set as:

$$\text{Range of Tools} \equiv \cup \ [\{\textit{Networking Tools}\}, \{\textit{Work-flow Management Tools}\},$$
$$\{\textit{Modeling \& Analysis Tools}\}, \{ \textit{Predictive Tools} \}, \{\ldots\}, \tag{8.4}$$
$$\{\textit{Knowledge-based Tools}\}, \{\textit{Agent-based Tools}\}]$$

The range of all such tools with product development potential can be classified into the following six difficulty levels.

- *Level 0: Networking Tools:* The types of activities that may fall into this category are document computerization and access facilities for text, graphics, schematics, and distributed data base facilities. Networking tools also include communication tools such as electronic mails, groupware, and multimedia between and across the members of CE teams.

- *Level 1: Work-flow Management Tools:* These control the priority of tasks in a work group, a unit, a department, or in an enterprise setting. Database tools, such as proven systems database, proven components and parts database can be used for this purpose. Other types of tools in this category are word-processing, spreadsheet, schedules, work-flow charting and time management, browsing, graphics/drawing tools, hypertext facilities, intelligent document management, retrieval and version control, quality tools, and so on. The quality tools include an array of conceptual tools, such as cause and effect diagrams, check sheets, histograms, pareto diagrams, control charts, scatter diagrams, matrix charts, SPC, and so on.

- *Level 2: Modeling and Analysis Tools:* Tools of this level should enable the generation, refinement, quantification and prioritization of requirements, such as QFD, Objective tree, and so on. Such tools are the result of modeling engineering activities, for example, geometric modeling tools such as solid modeling, surface modeling, and so on. Tools may also be of product modeling types such as STEP/Express,

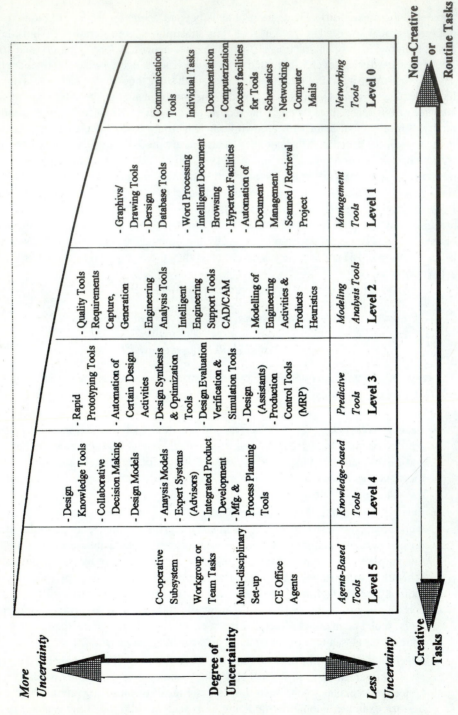

FIGURE 8.14 Automation Levels of Computerized Tools in A CE Office

using feature-based or similar techniques. It also includes engineering analysis and support tools, such as FEA, mechanism analysis, mathematical calculations, and intelligent CAD/CAM, wherein rules of thumbs, heuristics, and parametric rules for model creation are captured.

- *Level 3: Predictive Tools:* These tools are a result of design evaluation, verification and simulation design synthesis and optimization, and automation of design activities based on parametric, simulations, design assistants, advisors or expert types of systems. Tools that are useful for design evaluation and verifications are design for X-ability (reliability, serviceability, assembly, disassembly, manufacturability, testability, safety, etc.), failure mode and effect analysis (FMEA), fault tree analysis, and so on. Tools that are useful for design synthesis are boundary searching, functional analysis, concept selection, feature-based design, design retrieval, materials selections, value engineering, production control tools, and so on.

- *Level 4: Knowledge-based Tools:* These tools help teams to apply manufacturing and engineering intelligence to sort out bad alternative design concepts from good ones. KB tools include design knowledge tools, collaborative decision making tools for coordination, and analysis/design models. This also includes design automation based on optimization techniques, expert systems (advisors), integrated product development, manufacturing and process planning, and so on. The latter includes tools, such as process capability, manufacturing process selection, materials selection, MRP, CAM tools, NC and CNC verification tools, and so on.

- *Level 5: Agent-based Tools:* Such tools are used when constraints are present, when multiple knowledge sources (product and process knowledge) are present as agents, and when conflicts occur requiring trade-off. Agent-based tools belong to distributed AI and cooperative knowledge-base fields, such as cooperative expert system, CE office agents, and so on. Groupware technology replaces the conference room with the *electronic* white-board.

The level of intelligence and degree of cooperation are very much related. Cooperation provides the degree of confidence in the use of knowledge or intelligence. Agent-based tools contain the largest amount of cooperative knowledge or intelligence. The usefulness of tools depends upon the collective creativity of the individual teams participating in applying the tools to problem solving. The teams' dependence on cooperative problem solving decreases as we move to lower level tools (level 3, or level 2, ... 1) requiring less team cooperation and more individual effort. Level 0 tools, for example, do not require any team cooperation. The applicability of a set of tools at a particular automation level depends upon many factors. The important ones are degree of certainty, accuracy and completeness of information, and its integrity in current work environment and procedures. It is not difficult to capture the domain knowledge in most routine tasks with a high degree of confidence. Mining of rules in routine tasks are most common in levels 0 through level 2. Level 2 tools allow teams to build a modeling environment and to capture the domain knowledge before any eventual automation of the design activities can take place. The rest of the levels are more suited for specific applications such as a family of parts' category involving multiple interactions or disciplines. Higher level (levels 4 and 5) tools are useful when a product or a part is frequently redesigned for a variety of specifications. Typical examples include: different bore size and stroke length cylinders for 4-cycle, 6-cycle and 8-cycles engines, and so on.

8.4.1   CE Functions

From any vantage point in the IIS environment, a team should be able to do one or more of the following CE functions:

- Access information about previous product or process designs (past histories) instantly.
- Access information about X-ability considerations for design including manufacturability, reliability, maintainability, safety, cost, quality, performance, etc.
- Access the most current state of the product or process configuration as it exists within a multi-functional PDT unit.

The above information should be available to team members irrespective of the point of origin on the life-cycle, contributing work group that he or she belongs to, or his or her geographical location. In addition, low level automation tools should contain at a minimum the following CE features:

- An outline of the rationale behind recent changes in the design and in the newer versions.
- Capability to notify and record product design changes or manufacturing process changes.
- Major points to support the capture of design and implementation data.

8.5   PRODUCT INTELLIGENCE

The needs of product intelligence in CE depend on several factors. The complexity of the product and the processes, along with the prevalent standards and practices, play a major role in determining how much of the current process can be automated. Such a level of automation is accomplished with the concept of a virtual team. Figure 5.4, volume I, shows the organization of concurrent teams in product development. Virtual team consists of a computer network that integrates all the CE software tools and the needed knowledge bases. The core of the system is a product data management system embedded with product and process knowledge bases. This provides the designer with plenty of intelligence, including design and manufacturing intent, past design histories, lessons learned, and so on.

8.5.1   Knowledge-based Decision Support and Design Control

With product intelligence, the system contains all the features of an expert system. Knowledge-based decision support and design control for all the engineering and non-engineering aspects of a product design are incorporated and artificial intelligence concepts are applied. Such an expert system provides the designer and other members of the CE team with expert advice during the design phase. For example, using the expert system, engineers can determine an optimal mix in terms of performance, manufacturability, and supportability of a new design concept. Networking the expert software and data bases can facilitate more efficient communication among the CE team members. Incorporating CE team's recommendations can make the system self-searching and hence useful for the new design projects. The system can also eliminate inconsistencies or delays associated with lack of expert knowledge when an expert member of the team is not available. One way to design a product system is to base it on the user's input of part geometry, part specification, and recommendations of the production method. The expert system then conducts a check on the capability and limitations of the production methods and other constraints and criteria set up by the CE team to produce the part.

8.6   PROCESS INTELLIGENCE

A virtual team acts as a silent agent (consultant) for the work group and other members of the CE team. Process intelligence is capable of providing members of the CE team with information about every aspect of the product development. This includes different processes that can produce the part, the materials available, the machines in plants and their capacities, the inspection equipment, the market requirements, and so on. In such a system, one must capture the information regarding process limitations that may create difficulties in manufacturing. This can help the designer to tailor the product design so that the requirements of all the departments involved in the production process are satisfied, particularly from the manufacturing point of view. Once such information is captured, the system can allow interaction among members of the CE team.

Process intelligence network, as shown in Figure 8.15, uses an object-oriented database (distributed) to store and share information in a common format (e.g., SQL or RDBMS) with other systems (e.g., CAD, CAM, MRP, Process Planning, DNC Machines, finite scheduling system, remote host computer, etc.). A data import facility helps automate movement of data from an existing mainframe, minicomputer, or any other system to the process intelligence network. The network uses electronic travelers, jobs and queues packets, and operation packets to organize, manage, and deliver work information to the factory floor operators. The electronic traveler provides identification of the operations needed for the manufacture of the part based upon the production method and the desired revision level of the part. Jobs and queue packets contain scheduling and dispatching information associated with jobs in the queue for the specific machine or work center. This is based on the job information from an MRP system and scheduling information

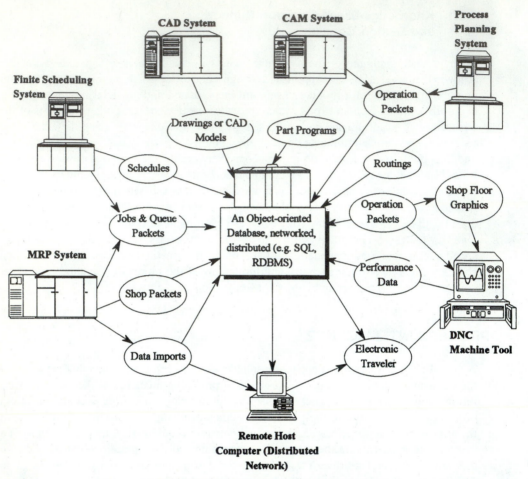

FIGURE 8.15   Process Intelligence Network

from a finite scheduling system or the production scheduler. The operation packets contain the information, documents, part programs, notification services, routing, and so on, needed for each operation on a part at a specific DNC machine tool or work center. The operation packet includes the shop-floor graphics facility. The facilities enable transfer of the needed control information, jobs in the queues, and so on, to the designated machine or work center and display the appropriate graphics and text information to the operator. Individual operators with security privileges can access information associated with the machines and the workstations assigned to their work groups.

CAD/CAM/CAE and other systems approaches are used to streamline the product's design-to-build capabilities. The CAD system is used, for example, to create drawings or CAD solid models. The use of CAM helps ensure manufacturability by generating part-machining programs and tool-path. The tool-path can be used to improve the quality of the finished product. A process planning system provides the routing and operation packets that can be downloaded to a DNC machine tool for further processing.

8.6.1   Process Planning System

A process plan includes a sequence of operations and processes necessary to transform raw materials into a finished product. They are usually broken down into smaller units or sets that represent singular cutting tool operations or *single tool uses*. The latter refers to manufacturing features such as holes, countersunk, and so on. The information can be used as the front end for a knowledge-based numerical control agent (see Figure 8.17). The manufacturing features of a process planning agent provide, in electronic form, *delta-volume* as one of its sub-component. A delta volume of a solid model in turn provides the design information necessary for describing the materials to be removed.

Process Planning Agent: With the distributed data agent, the machining operations, such as Lathe and Milling machines, can be controlled by a computer numerical control (CNC/DNC) agent. This is linked to a remote host computer via a local minicomputer. The tool changer and vision system computers can be networked to the shop floor computers to perform support functions for a DNC machine tool. The data-flow, from the host computer to the shop floor computer and other systems, may consist of operating commands that cause the required machining tasks to be performed. The data flows from other computers, vision system and tool changer computers, back to shop floor computers. Data may consist of sensor data, machining graphics and other status information to help the shop-floor computers direct the machining operations (see Figure 8.16). The machine tool computer provides machine monitoring and notification services through agents. The following items describe the major agents that interact with a process planning agent to render a shop floor intelligence.

- *Machine Monitoring Agent:* The machine monitoring agent provides the necessary software in vision system and machine tool computers to report the collected information such as machine utilization, cycle counts, and so on. In order to employ "*built-in* by design" quality control, it is necessary to monitor and control as many of the important process parameters as possible. In actual practice, it is often difficult to exercise a desired degree of control over all relevant parameters. There are parameters that cannot be easily altered. In such situations, the values of the uncontrolled critical parameters can be monitored and measured. The resulting current state of the process can be compared with the conditions that are necessary for effective operation. The information can be fed into a process control model (see Figure 8.16). This compares the current state with the ideal state and provides an on-line or in-process corrective actions. This process control model may be equipped with entities such as a series of statistical control charts, analytical calculations, signature analysis calculations, or a combination of these and other features. The major benefit of this in-process control is that accurate estimation of part quality is accomplished while the work-piece is still on the fixture.

- *Automated Process Monitoring Agent:* Automated process monitoring agent provides the monitoring data. Monitoring data is collected by real-time spreadsheets that preprocess the data prior to storing it in the native (e.g., SQL) database on the machine tool computer. Interfaces to devices (e.g., SPC/QPC) and in-process qual-

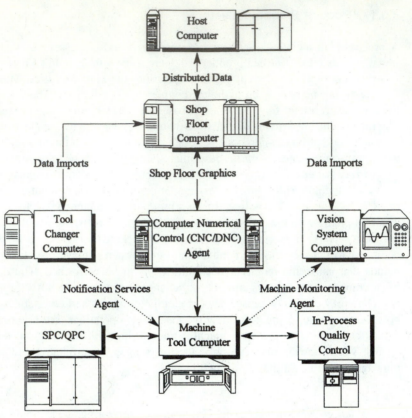

FIGURE 8.16   Block Diagram of a Shop-Floor Process Intelligence

ity control are shown in Figure 8.16. This enables the data to be turned into reduced information ready-to-be-used minimizing storage needs. Besides the optical disk, there are two data sources

- Customers' existing system.
- Standards and control databases.

This *process monitoring agent* eliminates the need to remove the part from the machine tool and perform any in-person inspection. It also minimizes the need for visual inspection at the end of the line when it is often too late. The techniques of measuring quality, however, depend upon the reliability of the process control model and the associated measuring equipment.

- *Notification Service Agent:* The notification service agent provides request for automatic notification of parts and services to a maintenance department, preset toolcrib, NC programming center, MIS support station, and so on. The notification service agent then transmits the electronic response back to the operator requesting such services.

- *Tool-Path (CNC/DNC) Generation Agent:* Tool-path can be employed on a modern machining system, such as lathe or milling, to fabricate the route contour of a particular work-piece. The system can be used to plan the manufacture of a variety of machined parts depending upon a CE work group needs (see Figure 8.17). The result is the creation of right information the first time, which means information gets to the shop floor in a timely fashion, and definitely results into the right part being machined.

- *Acceptance Testing Agent*: The acceptance testing of parts after machining can be based on a variety of product attributes ranging from dimensional features to metallurgical properties. Machining operations and agents only influence the characteristics that are mostly dimensional in nature such as surface finish, and so on. The estimation of a machine's ability to produce a particular part contour can be measured accurately while the work-piece is still secured to the machining fixtures. Specific types of sensors, such as boring bar-mounted touch probe, on-machine camera for optical tool-setting operations, linear variable differential transformer (LVDT), vibration transducers, environmental temperature probes and so on, can be used to monitor in-process feedback information. Furthermore, final product quality can be evaluated later at an inspection station when the work-piece is in an unconstrained condition. However, there is a possibility that such an estimate may not be a good measure for the quality of machining operations. There is a likelihood that some shape uncertainties may crop up in the data due to factors such as residual stresses or inspection fixtures.

To support design-to-build capability, many system control agents and software packages are integrated with management planning, simulation, and machine diagnosis. These developments are able to control production planning and scheduling to facilitate just-in-time delivery schedules. More often, C4 (CAD/CAM/CAE/CIM) and other virtual agents are made an integral part of an organization's Manufacturing Execution System (MES). This is discussed next.

Manufacturing Execution System (MES) Agent: MES is a new concept that is now emerging for linking information with action on the factory floor. Most MRP systems, such as MRP-II, are built around the materials being purchased. MRP's focus is planning, that is, to track starts and stops (when work begins and ends) and to track inventories. Capacity is used to balance demands. In contrast, the point of MES is to manage the information around the factory floor while the product is being manufactured. That is why it is also called plant management system. It allows manufacturing work groups to dance to a new tune by transforming engineering change-order-chaos into a controlled and *agent-managed* procedure. MES fills the communication and information gap between manufacturing planning (e.g., an MRP-II system) and its process control agent. The latter is, what actually happens on the factory floor (a supervisory or physical control of resources, machines, and equipment). MES is a single agent to keep track of all aspects of manufacturing on the factory floor, such as materials, machines, equipments, labor (work force), facilities, and the factory floor environments. It ensures that everyone is looking at the *same sheet of music*. Detailed knowledge of all such aspects is essential to be able to

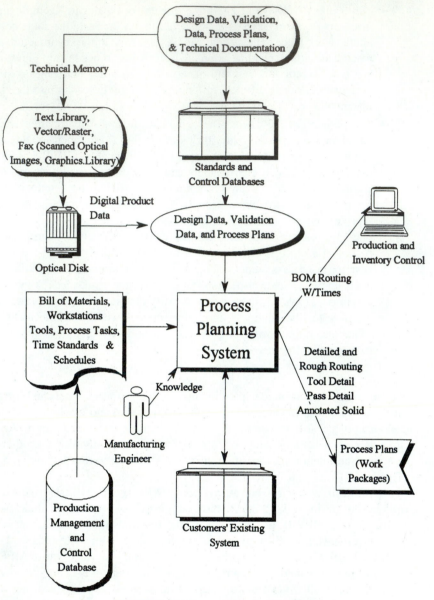

FIGURE 8.17 Process Planning System

ensure good quality of materials, processes, and improved quality of finished goods (see Figure 8.18). Most MES agents require an object-oriented or relational common database and a technical memory to store their sub-agents. A good process planning agent must be capable of assembling detailed technical work procedures using data from many different sources, including workstations, technical memory and databases. It readily co-exists with other computer-based information sources for material, tooling, and scheduling informa-

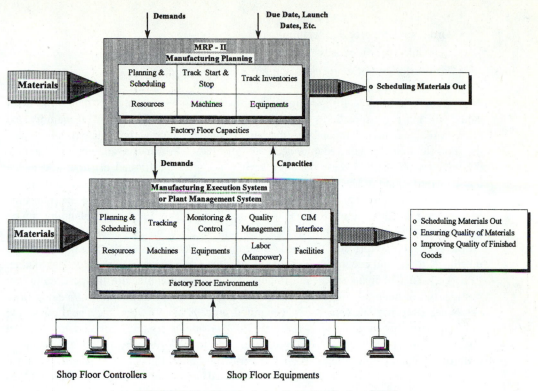

FIGURE 8.18   Manufacturing Execution System (MES)

tion. These sources are organized in Figure 8.17 using the convention described in section 8.5. A technical memory is a repository of design data, validation data, process plan history data, and other technical documentation residing in diverse formats (e.g., text, vector, raster, fax, photographs, and other forms of scanned images). This technical memory resides on an optical jukebox, providing the planner with on-line access to technical information and existing work packages. Another part of the requirements comes from a production management and control database containing more or less the same type of information but in relational forms. Bills of material, workstations, tools, process tasks, and schedules are usually the inputs to a process planning system (PPS) (see Figure 8.17). These come from a production management and a control database. They represent accurate and more up-to-date information, but they appear quite late in the production process. If the planning activities can be postponed, the flexibility and quality of decision making can be improved. However, this tends to move the critical decisions close to the production timeline, and if such decisions are revised so late in the game, this might affect the production schedule. Therefore, it is important to distribute the planning activities according to their timeliness with production schedules. Larsen and Clausen [1992] have proposed an integration approach called concurrent process planning (CPP). In CPP, four types of process planning activities take place concurrently according to their timeliness with production readiness.

- Design-specific process planning.
- Order-specific process planning.
- Schedule-specific process planning.
- Production-specific process planning.

The rationale behind such distribution of process planning activities is that some decisions, which are not specific to production, can be made based on technical memory and other available data. Other decisions that are affected by the real dynamic environment of the shop floor can be deferred until accurate information becomes available. For instance, the analysis phase of process planning activities can be performed off-line, and hence could be performed as early as possible. The selection phase, which requires real dynamic data of the shop floor, can be distributed in the above manner to minimize the total time for planning. By distributing the selection phase, each part of the process planning can be run concurrently with overlap, reducing the "time to planning" significantly. Constraints on the process planning come from the customers' existing system, such as the types of machines, tools, and capacities. The manufacturing work group provides this knowledge. PPS produces and manages all information required to manufacture the part, including routing, bill of material, NC code, and textual and graphical instruction for the shop floor personnel at each work center. The generated process plans (work packages) constitute the output of this system (see Figure 8.17). The output contains detailed information about such things as speeds, feeds, processes, and tools. However, the process planning team or the manufacturing engineer can override any or all the defaults. They can control the acceptable level (automated, assisted or manual) of system automation, and could remain the final authority for approving or altering the knowledge source.

The aforementioned PPS agent can be represented in a set as:

$$\text{PPS Agent} \equiv \cup \, [\textit{\{Machine Monitoring Agent\}, \{Automated Process Monitoring Agent\}, \{Notification Services Agent\}, \{Tool-Path Generation Agent\}, \{Acceptance Testing Agent\}}] \quad (8.5)$$

8.7   TECHNICAL MEMORY

Agents are the computational models of the intelligent information system (IIS). They are *virtual* since they reside on computers in the form of computational models, software, database, or electronic media; such as CD/ROM, technical memory, and so on. PPS and MES are two of its (IIS) big agents. Chapter 5 of volume I describes the concept of teamwork using concurrent teams: technological, logical, personnel and virtual teams. These teams are shown interacting with the virtual agents in Figure 8.19. The compact disk (CD/ROM) is a high-volume low-cost device to store information that is not highly volatile (does not change often). As such, it is an excellent media to store product standards, guidelines, ergonomic data, pictures of illustrations, or other documents. This could be a valuable resource library for both expert and novice product practitioners. Retrieval could be by standard indexing, perusal, key-word searches, or by full-word-indexing or by text retrieval methodology. One of the purposes of technical memory (TM)

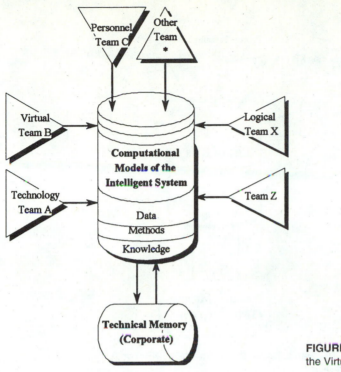

FIGURE 8.19   Teams Interactions with the Virtual Agents

is to increase virtual *contents* in the agents. This can be accomplished by TM infusing knowledge of downstream activities into the upstream process activities so that the balanced designs can be generated rapidly and correctly.

The knowledge-bases that would be required for such an agent-based CE system are shown in Figure 8.20. They are discussed next.

Marketing Technical Memory: This contains information related to product cost, customer demand and other product-specific market information. If this information is made available to the CE teams, it can assist them in the process of product launch and production planning. It could provide the designers with cost restriction, quality dimensions and product appearance information, and could serve as a basis for the conceptual design.

Conceptual Design Technical Memory: The product specifications, from the past histories of conceptual design and lessons learned, can be captured in a knowledge-base together with the on-line design documents. Storing this information in an electronic form facilitates ease of data access as compared to referring via specification manuals. The information can be arranged as rules, commands, or comments that are easy to understand.

Materials Technical Memory: This contains a list of materials, material specifications, properties, actual costs, and other information about materials stocks such as available sizes, purchase costs, and part numbers.

Design for Manufacturability Technical Memory: The determination of the manufacturability of a part requires a wide spectrum of information (particularly dealing with

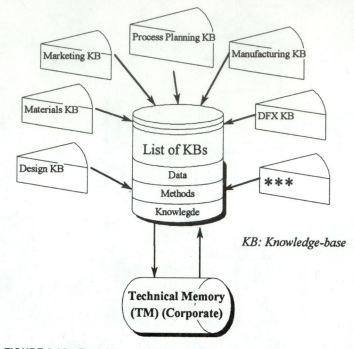

FIGURE 8.20   Partial List of Knowledge-bases (KBs) or Technical Memory for Product Development

data, product, and process knowledge). Data are the facts and knowledge is the guidelines or rules for common manufacturing processes. Some guidelines are found in manufacturing handbooks and many more reside in the minds of manufacturing experts. Such information may be computerized in the form of a data-base for a given manufacturing process, commonly referred to as a knowledge-base. A knowledge-base for the manufacturability of machined parts may contain information on the following.

- *Machinery:* Information about machine tools, NC machines, conventions, including machine type, size, capacity, and quality-related capabilities.
- *Jigs and Fixtures:* Information regarding type, size, conditions, flexibility, and quality-related characteristics of jigs and fixtures.
- *Tools:* Information about tool size, shape, material, and so on.
- *Production Rules:* A set of production rules may be put in place to act as a manufacturability feedback. These rules contain various features and attributes of a part that are related to manufacturability, such as form features, machining features, special tools, inspection and measurement methods.
- *Design for Manufacturing Rules:* As a checklist for manufacturability, a design for manufacturing rules may be created in order to guide a designer through the various stages of design.
- *Design for Assembly Knowledge-Base:* Ease of product assembly is necessary not

only during the production stage but also during maintenance and service stages. Therefore, design for assembly knowledge may contain information about methods of assembly, design for assembly rules, and dealers and users' feedback.

- *Suppliers Data Base:* This data base may contain information on alternative suppliers of raw materials, parts and equipment, pricing, design and production cycle timing, and delivery schedules.

- *Logistics Data base:* Information stored in the database may be derived from material handling, process support, and facility layout areas. The database integrates all of the logistics within a factory. Features of the manufacturing process that must be included in the logistics database are

  - The manufacturing process description—tooling.

  - Estimated time to perform the work process.

  - Production parts used, including all levels of selectivity (color, style, etc.) and options.

The above knowledge-base for DFM can be represented in the set as:

Knowledge-base for DFM ≡ ∪ [{Machinery}, {Jigs and Fixtures}, {Tools}, {Production Rules}, *{Design for Manufacturing Rules}, {Design for Assembly* (8.6) *Knowledge-Base}, {Suppliers Data Base}, {Logistics Data base}*]

Design Technical Memory: The design knowledge-base consists of information needed by a designer during a design stage. It should (a) store design information such as tolerances and standard features (keys and threads, standard parts), and (b) communicate with other knowledge-bases such as materials, design for manufacturability, and supplier parts' base. Design histories of part families of previous successful products can also be stored in addition to mechanical CAD designs of standard components (both in native CAD and neutral data formats).

Process Planning Technical Memory: In many generative process planning systems, major process planning information is stored in a knowledge-base that can be electronically interfaced to other systems. The generative knowledge-base may be transferable to other units or departments within and outside an organization. This potentially offers substantial savings, both monetarily and in the amount of time needed for its implementation in each new production facility.

The technical memory (TM) is a set of its constituents

Technical Memory (TM) ≡ ∪ [*{Marketing Technical Memory}, {Conceptual Design Technical Memory}, {Materials Technical Memory}, {Design for Manufacturability Technical Memory}, {Design Technical Memory},* (8.7) *{Process Planning Technical Memory}*]

Advanced CAD systems exist today that allow designers to create parts as a set of related features and functions. Many newer generation of CAD/CAM packages, like Concentra/ICAD, SDRC/IDEAS, Pro-Engineer, EDS Unigraphics, are adding integrated "design capture and modify" functionality to their systems. In addition to providing the ability to model the geometry of the part as in conventional CAD/CAM systems, they also describe parts as a set of related features or functions including the inherent relationship between the features themselves. It allows the users not to just capture the part geometry parametrically but also to capture the functional intent behind its design. Every part knows its dependencies with other parts and knows the progression of assembly sequences including their orientations and positioning with respect to each other. Compared to a few years ago, integrated design technology is now here to stay and the products we see in the marketplace have a lot to offer.

The next generation of *agent-based* environments will work closely with object-oriented knowledge-bases. The knowledge-bases allow object-oriented systems to retain their identity after the power to the computer is turned-off. The link between design objects and work group objects allows for cooperative designs to take place. The engineering environment will soon include systems that cooperate with each other to produce detailed design in solids or colors while leveraging technical memory or the knowledge of hundreds of individuals around the company. Distributed objects will queue work-plan (responsibility, accountability and schedule) for a product design segment at an engineer's workstation. They will track and maintain logical consistency between that segment and other pieces of design being developed (and simultaneously worked upon) across the various assigned work groups and centers. Specialized knowledge, whenever and wherever it is needed, will be furnished by the object-oriented knowledge-bases distributed across the network.

8.8 FLEXIBLE COMPUTER INTEGRATED MANUFACTURING (FCIM)

The FCIM system is a complete *art-to-part* system starting from feature definitions to automated assembly. FCIM consists of a feature-based product design agent, a computer-aided process planning agent, a shop floor control agent, a CNC machining center, and, finally, a CMM inspection facility. A complete flow of information through a FCIM process is shown in Figure 8.21. The feature-based design agent consists of a set of feature primitives, an engineering feature database, a 3-D CAD system, and a set of engineering analyses or simulation tools. The feature-based solid model of the 3-D CAD system takes the feature definitions, compares them with the existing features on the Features Database and then creates a 3-D solid model. Group technology, fuzzy mathematics and tool systems are used to screen the engineering features knowledge-base to ensure DFX compliance. The 3-D part is further refined to incorporate desirable material, dimensional attributes, and tolerances, and then an appropriate analysis model is constructed. The model just created is passed on to engineering analysis or computer simulation and analyzed or simulated for stress, strain and deformation. Design changes are incorporated to meet the operational requirements and enhance the X-ability (such as design for manufacturability) of the design or of the product. The information is down-loaded to a process planning system. It then determines the manufacturing strategy (machining sequence as well as machining parameters, such as feed, depth of cut, number of passes, surface finish, speed, etc.). The result of this operation is an APT file or a cutter location (CL) data. The work-piece, tools, and fixture details are supplied concurrently to a shop-floor control

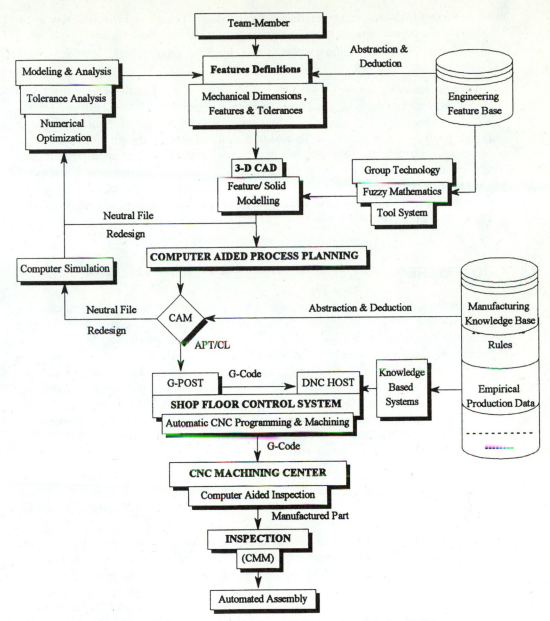

FIGURE 8.21   Flexible Computer Integrated Manufacturing (FCIM)

agent to make the manufacturing plan more realistic and complete. A knowledge-based system that is a part of FCS accesses the manufacturing knowledge-base and draws upon old history files, rules, and empirical production data to create a CNC programming and machining code instructions. Instructions are verified in two ways.

- *First*, a high level cutter simulation is carried out to check and eliminate any possible occurrence of tool gouging, interference, and fixture-tool collision.
- *Second*, by actually cutting a part sample (hardware prototyping).

Upon satisfactory CL data generation, the CNC machines are identified and post-processing instructions are obtained. The Direct Numerical Control (DNC) host built into the network downloads the CAD information to the appropriate CNC machines and the product is machined from raw materials. To ensure a quality aspect of the physical model thus built so far, a coordinate measuring machine (CMM) is used to select appropriate inspection probes and measure the build dimensions. The resulting measurement is compared with the data obtained from the 3-D CAD model for geometry compliance. FCIM is built on standards tools and standard interfaces to ensure compatibility of data, process, and the knowledge. Such standards are discussed in section 5.8 of volume II.

8.9 GROUPWARE

Groupware is the collection of software and best practices that improves the functioning of groups of employees as opposed to most existing computer programs. Most existing computer programs are designed for specific people (one-at-a-time). To be used by the product development community, this would include a tailored spell checker/glossary/thesaurus, a secure, limited e-mail group. A network will directly link all product practitioners to an index of recent and pending multimedia (text, line-art, photographs, audio, and video) documents (in editable format). The network will provide access to scanned product images, (on-line product project journals, personal/group/department/inter-organizational team, calendars, brainstorming aids), outlining aids, and similar applications.
    Groupware tools address two needs.

1. *Facilitate and improve the structured communication processes:* Structured communication processes include smart project management, scheduling, design review, group editing, and so on.
2. *Facilitate and improve the unstructured communication processes:* Unstructured processes include, for example, voice mail, collaborative authoring, document sharing, screen sharing, audio/video conferencing, meeting on the network, and so on. A groupware is often built using some type of electronic mail message handling system that is accessible to all employees through an office network. The combination of a message handling system and access to some real-time communication mechanism such as client/server, provides an electronic transport vehicle to route files, CAD images, forms, designs, drawings, and other objects concurrently.

Groupware tools can be categorized by functionality, modes of communication, or level of teams' interactions. Figure 8.22 shows a categorization based on the dimensions of time and place of group interaction. The interaction matrix has

Type	Same Time	Different Time
Same Place	o Electronic Copy Board o Electronic Decision o Support Technologies	o Team Room Equipment o File storage/Retrival/Reformat o Downloading Machine Instructions
Different Place	o Audioconferencing o Videoconferencing o Screen sharing o Electronic Design Review o Meeting on the Network o Multimedia Sharing	o Word Processing (Glossary, Thesaurus) o Scheduling, Calendaring o Project Management o Voice Mail o Electronic Mail o Nicknames, Phones, Distribution List, Fax o Group Editing o Organization Charts o Employee Directory o Forms, Procedures, Practices, Standards

FIGURE 8.22   Groupware Categorization According to Time and Place Dimensions

Two rows: same place and different place.

Two columns: same time, different time.

The groupware tools corresponding to *same place and same time* are needed to support face-to-face meetings. Tools for *same place and different time* are generally useful to support teams in place. Groupware tools for *different place and same time* can support cross-distance meetings, whereas tools for *different place and different time* can support ongoing coordination. The following are some additional groupware functions that are parts of IIS.

8.9.1   Smart Project Management

Diagramming a Critical Path Method (CPM) or a GANTT Chart for pursuing product development can be a time-consuming, labor-intensive chore for the manager or other core-team participants. A smart project management application could be developed to assist in this effort. This would involve defining sub-sets of a GANTT chart (such as an arrow diagram or an IDEF diagram) and building macros to relate these to overall platform planning, funding, and scheduling tasks. The application could be presented to the work group as an adviser or assistant. As such, the application could serve as one of the series of modules of an overall product intelligent information Management System (IIS). In addition to planning and scheduling, smart project management modules can assist control functions. On-line sign-off on statements of work (or on deliverables) can provide user-wide improved understanding of specific project status and responsibilities.

8.9.2   Common Data Exchange

There are many different computer systems in use in an enterprise. Each of these systems has a number of specific legacy applications that are very reliable in problem solving. Most of these applications are home grown and are not scheduled to be phased out in the immediate future. The applications employed in one functional unit (e.g., a product design unit) should be able to receive data from and transmit data to all of the other remaining units. The following are some common systems through which this data transfer capability (wire frame and surface geometry) ought to be established.

- Strategic CAD/CAM tools.
- CAE tools.
- Data base management system (DBMS) tools.
- Knowledge-based tools (e.g., PTC/Pro-Engineer, Intellicorp/Pro-Kappa, Concentrs/ICAD, etc.).

An important step towards this goal is the creation of the Standard for the Exchange of Product Model Data (STEP), formerly known as ISO 10303. STEP provides a serious attempt toward an unambiguous, computer-interpretable definition of the physical and functional characteristics of a product that design, manufacturing, and logistics support systems can readily use.

8.9.3   Team Tracking System

Early team tracking systems were originally ill-conceived as management reporting systems. Later this was corrected. The current understanding is that they are part of GROUP-WARE tools and constitute an element of a multi-purpose status system for the entire PDT. This type of team tracking system, applied to product design and development ends, can assist in a wide range of work tasks such as

- Dissemination of common goals that meet constancy of purpose.
- Wide understanding of teams' roles and responsibilities.
- Monitoring of emerging product life-cycle plans.
- Monitoring of specific functional steps and their status (including task formulation, staffing, activation, completion, and sign-off).
- Methods to communicate issues to higher management.
- Mechanisms to identify roadblocks and eliminate bottlenecks.
- Processes to develop and implement recovery plans.

- Recording and extracting data to be used as yardsticks for measuring performance (both efficiency and effectiveness) and for developing basic guidelines.

While this team tracking system can certainly help command and control, the tracking part requires both personnel and cooperation from all associated groups.

8.9.4   CE Training

Training is an important part of concurrent engineering. The process of concurrent engineering, no matter how many automated tools and services are in it, will not do much good to an organization until teams of people, who are going to use them, have full faith and they have bought into the basic concepts. Teams ought to be fully committed and be well-versed with the concepts including the philosophies of CE. The CE management should fully support the initiatives with respect to the needed commitment of personnel and resources.

CE should be considered a strategic weapon to tackle lack of productivity, competitiveness, and inefficiency problems. CE has been an accepted IIS management practice in many companies for several years. It is recommended that all product-related analysts, designers, managers, and appropriate vendors be given CE training or a refresher course in CE. Such training should support and emphasize CE teamwork and concurrent product realization strategies. Benefits of using CE and integrating DFX (e.g., DFA and DFM) applications into the product realization cycle should be a central focus.

REFERENCES

ALTING, L. 1986. "Integration of Engineering Functions/Disciplines in CIM." *Annals of the CIRP*, Volume 35, No. 1, pp. 317–320.

CHANG T.C., and R.A. WYSK. 1985. *An Introduction to Automated Process Planning Systems*, Engelwood, NJ: Prentice Hall PTR.

HUMMEL, K.E., and C.W. BROWN. 1989. "The Role of Features in the Implementation of Concurrent Product and Process Design." Proceedings on the Winter Annual Meeting of the ASME, San Francisco, CA, December 10–15, *Concurrent Product and Process Design*, edited by Chao and Lu, DE-Volume 21, PED-Volume 36, pp. 1–8, New York: ASME Press.

LARSEN, N.E., and J. CLAUSEN. 1992. "Applied Methods for Integration of Process Planning and Production Control." Proceedings of the Manufacturing International '92, New York: ASME Press, pp. 349–364.

LIM, B.S. 1993. "ICIMIDES: Intelligent Concurrent Integrated Manufacturing Information and Data Exchange System—A Distributed Blackboard Approach." *Concurrent Engineering: Methodology and Applications*, edited by P. Gu, and A. Kusiak, pp. 135–174. The Netherlands: Elsevier Science Publishers B.V.

PRASAD, B. 1995a. "A Structured Methodology to Implement Judiciously the Right JIT Tactics." *Journal of Production Planning and Control*, Taylor & Francis, Volume 6, No. 6, pp. 564–577.

PRASAD, B., 1995b. "CE Plus CIM Equals IIS: Intelligent Information Systems." Proceedings of the 1995 ASME Mechanical Engineering Congress and Exposition, November 12–17, 1995, San Francisco, CA, edited by R. Rangan, New York, NY: ASME Press,

PRASAD, B. 1995c. "On Influencing Agents of CE." *Concurrent Engineering: Research and Applications—An International Journal*, Volume 3, No. 2 (June 1995) pp. 78–80.

OKINO, S. 1995. "Less is More." *Automotive Industries* (March 1995) Volume 175, No. 3, pp. 81–82.

WEBSTER, A.M. 1990. *Webster's New Collegiate Dictionary*, Springfield, Massachusetts: G. & C. Merriam Company.

TEST PROBLEMS—INTELLIGENT INFORMATION SYSTEM

8.1. What are some primary reasons for the shortcomings of an IPD environment? What are the reasons for the tool-related shortcomings? How would you overcome them?

8.2. What are the four primary reasons for the shortcomings of a business operation? Which of the 7Ts play a heavy part in giving Japanese producers a competitive edge over U.S. companies? Explain why?

8.3. What state are most manufacturing industries in today? What is the type of paradigm this state represents? How has this state changed over the last few years?

8.4. What state are most C4 tools in today? What is being expected from those C4 tools? Today, why are these C4 tools being redesigned?

8.5. Why are standards becoming more and more important for product realization? What types of standardization do you see emerging in areas such as compute platform (e.g., workstations), network (e.g., LAN), and data exchange? What advantages do such standards bring to a parent organization and the associated CE work groups?

8.6. What are the emerging standards in the data exchange area? What are the highly automated areas in manufacturing? Why do some of the most automated applications have grown to be self-assertive? Why do we see an influx or proliferation of *islands of pre- and post-processors* in this area? What impact has such proliferation brought to an organization?

8.7. What is a CIM system? Where are the CIM systems currently being applied and why? What are the two communication parts of a CIM? What are the types of efforts that are required in integrating some existing C4 systems that have steadily grown strong over recent years?

8.8. Why is CIM, in the current form, incapable of dealing with knowledge? What are other companies doing to exploit such PPO (product, process and organization) knowledge? What role does CE play in CIM strategy?

8.9. What is an intelligent information system (IIS)? What are its components? How does the CE get factored in an IIS?

8.10. How do you achieve a better CIM system? What are some of the missing links of a CIM? How do you describe a CIM as it relates to CE?

8.11. What are the two prototype techniques used during virtual modeling? What are the key successes of an IIS?

8.12. Why is some of the activities of a PD^3 process dealt with in its own domain? How do you identify new improvement possibilities?

8.13. What are the eight enabling elements of an IIS for product development? Why is decision making so important so has to become a part of IIS? How do you view the decision making process? What are its ranges?

8.14. What are the thirteen barriers that prevent any organization from regaining a full potential in manufacturing competitiveness? How do you recognize and overcome the barriers of a good intelligent CIM system?

8.15. What are the three sub-classes of organizational barriers? Why is data exchange considered a barrier? Why do we require three levels of interpretations in exchanging data through a neutral file transfer format? What may happen during any of those translation steps? How do you measure loss of data?

8.16. What is a reformatting barrier? Why are some companies attempting to integrate incompatible firmwares? Why is it more cost-effective to do a trade-off study during an early part of a design than later?

8.17. Why is it, in a large or global organization, not possible to collocate all co-workers? Why do most commercial CAD systems not associate degree of difficulties in capturing a design intent? What are some of the system integration (SI) barriers? What types of inconsistencies does a work group encounter in designing an information system? What can work groups do to overcome SI barriers?

8.18. What is involved in designing a new and complex system? Why are there only a few computer tools that are also designed to aid the CE teams? What things do most work group members look for (to do) in a software? What would a work group engineer perform in the future and how?

8.19. What are the manual tasks related to a product design, engineering, and manufacturing processes? How could they be computerized into a parallel row of a vertically aligned series of intelligent functions? What does such an intelligent system contain?

8.20. What is a roadmap to automation? How do the intelligent tools provide product realization means, that is, first do they decompose an artifact into its constituents, and then synthesize or assemble constituents back into a working system?

8.21. What constitutes a virtual agent of an intelligent system? How do these agents aid the CE teams in performing collaborative functions? Give examples of some of those functions.

8.22. List a set of typical agents and functions that an intelligent system performs? What activities does each agent perform in a CE office? What are the results of such sharing?

8.23. What is the nucleus of a CIM architecture? What are the objectives of a soft-prototyping concept? Which of the four teams are more active than others? What are the roles of each team?

8.24. What are some types of activities that take place in a PD^3 process? What are the six levels of techniques or methods that are required to meet all aspects of creativity and cooperation? Give a couple of examples of each level.

8.25. How do you classify tools that provide solutions to the PD^3 tasks or introduce some new techniques? Give examples of tools required for each difficulty level. When would you employ a higher level IIS tool?

8.26. What are the CE functions a work group is expected to perform? What are some of the features you would like to see in some low level automation tools?

8.27. How do you determine how much of the current PD^3 process can be automated? How does the PIM (product information management) system help in knowledge sharing and communication?

8.28. How does intelligence about some PD^3 processes help work groups? How does this contribute in selecting the right process for the design? Why is CAPP more computerized than many other manufacturing elements? List some advantages of CAPP?

8.29. How does the process intelligence network work? What are its components? What facilitates the movement of information between various network services? How is the work information transmitted to a specific machine or a work-center?

8.30. Why has the automation of process planning been the major requirement in many manufacturing industries? What are the different elements of a process planning agent? Who provides the machine monitoring and notification services? What are the major agents that interact with a process planning agent? How is shop floor intelligence rendered?

8.31. What do you understand by a manufacturing execution system (MES)? How do the points of MES differ from a MRP system concept? What role does a MES play? Who ensures that everyone is looking at the *same sheet of music*? What is the foundation on which an MES is built? What are the sources of its intelligence?

8.32. What is a concurrent process planning (CPP) concept? How many activities can take place concurrently in a CPP? What are the rationales behind such distribution of activities? What is the output of this system? Who is the final authority for altering the knowledge source?

8.33. What is the purpose of technical memory (TM)? What is the electronic form in which TM is captured and stored? How do the teams interact with the virtual agents? What types of indexing and key-word searches are used in TM?

8.34. What are the types of knowledge-bases (KBs) that are required for an agent-based system? Describe the types of information make-up the manufacturability (DFM) knowledge-base? What types of KBs make-up the design TM? What types of KBs make-up the process-planning TM?

8.35. What is a flexible computer integrated manufacturing (FCIM) system? What does FCIM contain? Describe a flow of information through a FCIM process? How are the instructions for CNC programming verified? What is the basis on which an FCIM system is built?

8.36. What is *groupware*? Describe examples of a groupware tool that is relevant to a product development community. What are the different types of needs that a groupware tool addresses? How are such tools categorized? Describe a groupware categorization based on the dimensions of time and place. What are some of its functions that are considered important part of a groupware?

CHAPTER 9

LIFE-CYCLE MECHANIZATION

9.0 INTRODUCTION

With everybody jumping on the bandwagon of minimizing waste and designing for value, more and more enterprises are now facing a greater need for advanced productivity tools and efficiency methods. The two main purposes of employing these tools and methods are (a) to cut down on product development lead-time beyond what is feasible through conventional means, and (b) to become and stay competitive in a global market place. This has led to old and new life-cycle mechanization frenzy. Existing and/or new manufacturing systems are being automated (reviewed, modeled, analyzed, synthesized and mechanized). The results of such automation attempts have been largely mixed. Organizations that paid ample attention to re-engineering the process before automation, have done relatively well compared to those that did not. Some organizations have taken the old process and simply applied automation. Some assumed that this process can be captured in a flow chart and have attempted to directly implement an algorithm or a computer system of some sort. Unfortunately, many attempts have been made at applying this approach to design-cycle and most have achieved very little or no success [Phillips and Aase, 1990]. One of the prime reasons for a flowchart based algorithm failure is the fact that a design-cycle, whether serial or concurrent, is not a unique sequence of steps. As discussed in Chapter 9 of volume I, every product realization process follows a taxonomy of transformation driven more by the types of specifications (inputs, requirements and constraints) than by some fixed procedural steps. Another important reason for flowchart based algorithm failure is uncertainty. During a design process, information is continually churning, design is constantly changing, and the level of completeness of data at a particular phase cannot be guaranteed or assured. If a fixed scenario is used for a task, such as an analysis or simulation, it may lead to errors or quality deficiencies in the product design. This, in turn, may require additional iterations or may require building hard prototypes.

As stated in Chapter 2 of volume I, life-cycle mechanization that simply speeds up the phases of a serial process cannot stand up to the test of time. In the global competition that we are facing today, it is not enough to speed up the individual process steps. It is more important to re-engineer the entire life-cycle process before automation. The new reengineered life-cycle process must reflect the changes in major aspects of the seven Ts discussed in Chapter 4 of volume I. There are three key elements to this mechanization.

1. *Common Set of Tools:* The first level of mechanization is to provide an enterprise with a common set of automated tools and techniques for modeling, analysis, simulation, and optimization of modern manufacturing systems. Some examples of these (e.g., 7Cs, 7Ts, 4Ms, etc.) are discussed in Chapter 7. Mechanization happens at various levels of detail and from different angles or viewpoints.

2. *Dynamic Information Models:* The second level of mechanization is the creation of a set of dynamic models (see section 7.4 of Chapter 7, volume I). This can deal effectively with many of the foreseeable changes. These dynamic models are to be shared with all interested parties in the CE work groups. These dynamic models automate routine tasks in the product realization process in a manner that can be reconfigured to suit a wide range of CE information modeling and management needs.

3. *Collaborative Virtual Environment:* The third element of mechanization is the creation of a new collaborative virtual environment that takes advantage of commonality (common processes, common tools, standard protocols, generic tasks, etc.) and exploits concurrency (see Chapter 4, volume II).

Intelligent Information System (IIS) was discussed in Chapter 8. As indicated, IIS can reduce the need and dependency of CE teams and work groups on mutual cooperation. Many believe that mechanization of product development through computer automation can play a major role in reducing this dependency. It can yield a better IIS. However, it has seldom been practical to develop a truly automated IIS (CIM + CE) system. Developing an automated IIS means creating an intelligent concurrent set-up (see Figure 9.1). A set-up such that whenever *a new set of product specifications is fed to the system, a new design of the artifact emerges.* The life-cycle of the product starts from the cradle (e.g., needs, choice of raw materials) and ends in the grave (e.g., absorption, recycle). In actual implementation, this scenario encounters many roadblocks. It is very difficult to remove all the barrier classes (namely, integration barriers, automation barriers, communication barriers, and organizational barriers as discussed in section 8.2) that are usually present in an organization. Those who have tried have only partially succeeded in removing a few barriers. Many others have fallen below expectations. Typical examples of integration barriers are the difficulty in interfacing an array of suppliers' production facilities. Today, in most global-based industries, barriers are the causes of substandard gain in productivity. In Chapter 8 of volume II, IIS was said to be equal to CIM + CE. Life-cycle mecha-

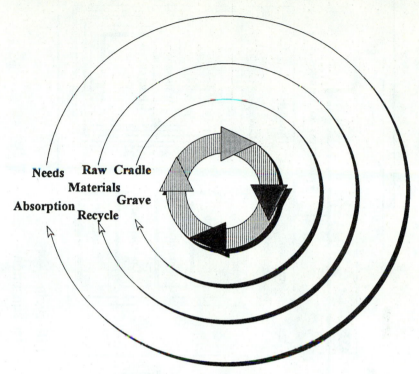

FIGURE 9.1   Life-cycle Mechanization

nization goes one step further. In addition to CIM and CE, life-cycle mechanization includes automation.

$$\text{Life-cycle mechanization} \equiv \underline{C}IM + \underline{A}utomation + \underline{CE}$$

To make this process easier to remember, life-cycle mechanization is arranged under a familiar acronym: CAE, for $\underline{C}IM$, $\underline{A}utomation$, and $\underline{CE}$. Since CAE also equals IIS plus automation, the major benefits of mechanization in CAE come from breaking down and removing the various barriers. The three common barriers are

1. Integration (this is a term taken from CIM).
2. Automation barriers.
3. Cooperation barriers (which is a term taken from CE).

CE provides the decision support element, CIM provides the frameworks and architectures, and both (CE + CIM) provide the information management elements. Figure 9.2 shows the interrelationship between integrated product development (IPD) methodology, frameworks and architectures, intelligent information systems, information management, decision support system, and life-cycle mechanization. *Life-cycle Mechanization* refers to the automation of life-cycle features or functions or creation of computerized modules

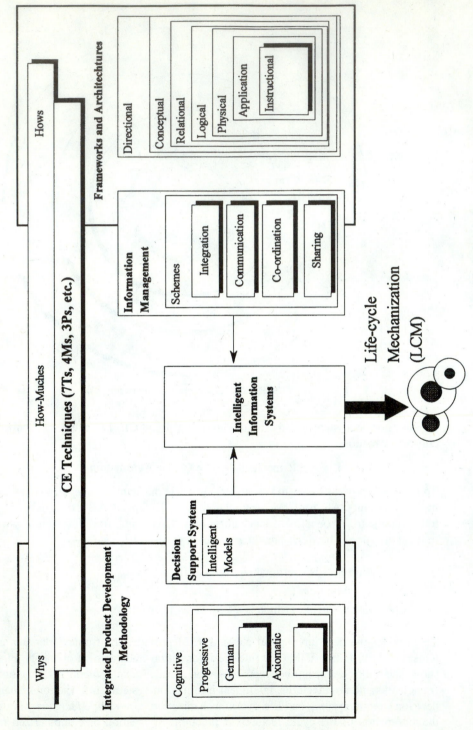

FIGURE 9.2 Bird's Eye View of an LCM and its Interrelationship to Other Areas

408

that are built from each function and share common information with each another. This sharing includes integration and seamless transfer of information (data + method + knowledge) between commercial computer-based engineering tools and product-specific, in-house applications. The creation of such models tends to reduce the dependency of CE teams on communication links (e.g., client/server, UNIX, . . .) and product realization strategies (e.g., decomposition and concatenation). One of the primary factors that drives this mechanization in recent years is the increase in availability of automation tools and technology. Combining the capabilities of surface and 3-D solid modeling, animation software, and powerful computers, the automotive designers can develop vehicles much more cost-effectively than in the past. With newer 3-D integrated CAD tools, work groups will be able to perform the following tasks easily.

- Design work groups can style and immediately view a concept in three-dimension and solids.
- Work group engineers can test every aspect of a vehicle's performance.
- Manufacturing work groups can run pre-production assembly.
- Quality Assurance work groups or teams can fit tests without building a single prototype or clay model.

Recent examples of time and cost savings generated from life-cycle mechanization (such as CAD/CAM/CAE/CIM) include a Ford Explorer rear wiper motor cover. This was designed in 6 weeks, a savings of up to forty-five percent in design costs [Keenan, 1995]. A finite-element mesh rendering of a Volvo side mirror was made in 2 days, and a simulated wind-tunnel testing was accomplished in an additional 48 hours. Ford saved 134 days in designing a light-weight lower suspension control arm with life-cycle mechanization instead of conventional engineering [Keenan, 1995].

9.1   CE MECHANIZED ENVIRONMENT

Concurrent engineering requires a slew of computer tools to facilitate coupling between product and process domains. Figure 9.3 shows an array of underlying tools and technologies required for a mechanized IPD environment. The tools are arranged in five categories. A category indicates a step of the life-cycle mechanization process. Each category identifies a set of tools and technologies that particularly employs a set of CE enablers (e.g., 7Ts, 4Ms, or 3Ps) that are found to be the most relevant. The tools in each of the first four categories, when combined, give the fifth category. The first category contains tools and techniques of networking and the second category has tools for management of information. The third category includes tools for modeling and analysis, and in fourth category are tools for prediction and knowledge based decision making. When all of these categories of tools are combined, it results in the fifth category. This last category of tools and technology (fifth category) includes product and process computer models, automation tools, and training tools. Many of the tools in the first four categories exist today in some form or other. A few

Tools and Technolgy	Description/Remarks
Networking Email (computer mails) PROFS Information Utilities	Internet Windows, X-window LAN, WAN
Management of Information Problem Solving & Culture Word-Processing Project Management Customer Focus marketing Routing & queuing Logistics Data flow diagrams Structured Analysis Process Management	QFD/CAPTURE, Problem-solving Techniques, Branstorming, etc. Word, Word Perfect, Microsoft Office, Word Star, etc. PERT Charts Design for value/value engineering, QFD Accounting & budgeting,, scheduling Spread-sheet IDEF, IDEF0, IDEF1X, IFLOW System Engineering, Yourdon, document improvement TQM, Continuous Improvement, MRP, JIT, Synchronous manufacturing, Systems Engineering
Modeling and Analysis CAD/CAM tools CAE Tools Preprocesing Tools Requirement specs generator Design definitions Knowledge encoding Design for X-ability Activity-based Costing	Parametric, Feature-based, Knowledge-based 3-D CAD modeling FMS, GT, MRP, OPT, JIT, KANBAN, Facilities Layout FEA/FEM, CFD Reliability, Sensitivity PTS, CTS, SSTS, STS Design Assessment, Bill of Materials (BOM) Technical Memory, Knowledge Database DFA/DFM, Taguchi, DOE, FMEA, FTA, FNECA, RMS, MTBF DFA/Cost-drivers, Assembly Motion and Time
Prediction and Decision Making Sensitivity Tools Advisors or Critics Post-processing Tools Optimization Processors	Axiomatic, Design assessment Expert systems Simulation models, Monitoring, Groupware, Conflict Management Product/process optimization, Feature-, parametric-, or knowledge-based optimization, Automation
Virtual Models Product models Process models Automation tools Training tools	Analysis processors, kinematics, mechanism, 3-D geometry, Mechanical Design, assembly design model SQC, Six-Sigma, Simulation, Taguchi, Robust design, DOE, SPC/ SQC, Trade-off, synthesis, continuous improvement, Poka Yoke, Process FMEA Design models, library of parts, receiving, query and tracking Flocharting Software, Training Tools, Computer-based Instruction Software.

FIGURE 9.3   A Palette of Life-cycle Mechanization Tools and Technology

of the powerful tools have not been fully developed or integrated. These slew of tools and technologies is an important part of the productivity equation, but they are only a small part. Not all enterprises can achieve the same level of productivity gains using such tools. It has been observed that there are dramatic differences in productivity gains between two identical firms, even though both may be using the same set of tools and technology. Actually, tools and technology simply establish a set of potential goals for achieving maximum efficiency. Reaching this potential is a function of many factors, such as

- How a process of life-cycle mechanization is planned.
- IIow mechanization is implemented.
- How cooperation between the joining forces is furnished.

A coordinated methodology for the use of tools and techniques is being evolved slowly. The latter will take a center stage in defining a future IPD environment where communication of ideas and exchange of information between 7Ts, 4Ms, and 3Ps (for products and processes) will have to take place simultaneously. Today, companies do not seem to have a good handle on a strategy for attacking problems related to 7Cs (e.g., communications). They do not have the capability for an intellectual synthesis of a mechanized IPD solution. A good and friendly cooperative (and interactive) environment is an important step.

9.1.1   Interactive and Friendly Environment

A friendly user interface is only a first step to providing a mechanized environment. It gives the personnel CE team an interactive environment for product design and development. It gives the CE teams a friendly interface with the global manufacturing world and gives them abilities to do the following:

- Run one or more of their processors in any order the teams desire.
- Define specifications and constraints by running their appropriate processes.
- Select an appropriate configuration from a library of existing configurations for further evaluation.
- Modify parameters at any point during a design session with a click of the mouse, instead of lengthy typed commands or passages.

9.1.2   Human Factors Engineering

Most of the process measures, unfortunately, are driven more by technological advances than by customer or user needs. If such technological advances are blindly implemented in a CE process, the teammates may have to adapt to awkward and inefficient products or processes. Consideration of user capabilities and limitations early in the design process can alleviate this situation. It can result in more useful designs driven by factors that are *human-centered*. It can also enhance inherent product safety and reduce operational costs.

A teammate of the CE organization may be a product assembler, an operator, or a maintainer. The metrics for human factors engineering introduce human factor principles for individuals and teams, and offer consideration for implementing these factors into a group problem-solving environment.

The user interface provides the team with the capability to confirm or override the virtual teams' suggestions at each stage in the logical teams' decision cycle. The system's graphical icons and windows (a Motif-based) enable the team members to move interactively between views, models, and even applications. Next to the user interface is a *virtual team* blackboard that provides an environment for the personnel and technology teams to interact. These tools and advisers, called processors, can be used to model the problem from all aspects of the logical team, from capturing the design intent to obtaining a physical device that meets these requirements. A series of generic processors has been identified in Figure 7.5 of volume II for this purpose. The most popular types of processors often found in an organization are listed in Figure 7.5 of volume II. However, these do not, by any means, represent a full set. Depending upon the level of automation required for the product and the ease of processing, a more refined set of analysis and design toolsets can be generated for a particular family. A family may consist of functions and features to perform the following tasks.

- Model the part and part's geometry.
- Capture the design intent knowledge.
- Define the problem in a set of parameters, constraints, and objective functions.
- Identify conflicts between different design perspectives.
- Analyze the design for structural considerations under a range of field conditions such as stress, heat, shock, vibration, deflection, and other real world situations.
- Provide feedback based on incremental analysis of design as it evolves.
- Generate comments on the design.
- Create a library of critics to provide immediate feedback on the design decisions as they are being made.
- Create a synthesis facility in which humans and automatic synthesis tools, such as optimization, can interact in making design decisions.
- Act as an interface between human synthesis teams and development programs.
- Act as a distributed problem solving environment to support the parallel programs and client/server interface.
- Help designers optimize their product concepts long before constructing physical prototypes when cost associated with a design modification is minuscule.
- Create a library of design agents. Design agents are responsible for the synthesis of an aspect or portion of a design with respect to a chosen set of specifications.

Such a facility utilizes information from many sources such as databases, agents, spreadsheets, application programs, CAD systems, and mechanized processors with inputs from various concurrent work groups.

9.1.3   Modular and Plug-in Modules

These processors are not only a set of sophisticated tools to be used by the personnel and technology CE team members, but also represent an integrated CE environment. The environment may be spanned by a number of specialized end user tools for carving and meeting product developmental needs such as design models, design critics, and so on. They are often built from the ground up using a modern programming language such as C++, or a similar object-oriented technique. The CE environment thus forms a large and integrated set of plug-in design and manufacturing modules that can be quickly configured to meet current and future enterprise needs. User-interface, process, and job-control instructions are *recipe-type* to allow users to define the way they would like to interact with or are most familiar with, as well as the scope and method of required automation. Repetitive and specific processes can be combined and automated through *user-defined* run-stream definitions. The extent to which such automation can be pursued will depend upon the criticality of time reduction and the generality to which they can be programmed. Integrating a variety of life-cycle concerns into a design process through a system based on critics, does reduce the product development time, but it limits their applicability to specialized situations and designs.

9.1.4   Critics

A critic is a conventional analysis or simulation program with an intelligent front end. Design critics are self-activated analysis tools that track the progression of the design, evaluate an aspect of the design, and communicate relevant results to the design agents. Design agents and critics can be set to run in automatic mode or under a human supervision. The purposes of this front-end interface are

- *Pre-processing:* Critics can automatically generate input data for an analysis program from the design representations being considered by high level design work groups.
- *Decision Making:* Critics can follow design decisions of the work groups as they are being made and automatically invoke an analysis program whenever there is a potential for an analysis to provide some useful feedback.
- *Post-processing:* Critics can then make this analysis or feedback available in a form that a designer or a design work group can readily understand and use.

Parametric finite element analysis (FEA) for structural design is a good example of a critic. Although several finite element analysis programs have been available commercially for a number of years, they require a large number of detailed inputs. Such design details are not usually available during a high level design process and are difficult to uncover that early. Besides, FEA operation requires area specialists with extensive training in the field. The program runs in batch mode with a long turn-around time. As such, these analysis programs serve more as checks for evaluating the integrity of completed designs.

They do not actively participate in the concurrent design process. Advanced CAD systems exist today that allow designers to create parts, not as a set of geometry primitives, but as a set of related parametric features. By treating objects as sets of common parametric features, manufacturing constraints are more easily satisfied. Design features become accessible to manufacturing evaluation and manufacturing features become accessible to design evaluation. However, a CAD-based parametric modeling system is not enough by itself to serve as a critic. Just as an FEA software cannot change the geometry of a part, a CAD system by itself cannot tell whether a part will break or hold. Similarly, a shape optimization software by itself cannot vary the topology of the part. What is needed is a synthesis environment to facilitate flow of information between applications, languages, and computer platforms.

With the advent of feature-based parametric CAD systems (e.g., Pro-Engineer, Concentra/ICAD, EDS/Unigraphics, etc.) and parametric or variational FEA (Ansys, Nastran, Mechanica, etc.), optimum design process is becoming much easier to implement. When these systems are linked with a classical optimization theory, they provide a soft-prototyping capability. The same feature-based parametric model that an engineer builds for analysis can be optimized. At a later point, the team can perform what-if experimentation to gain additional knowledge about the optimized design. For the CE work groups, building parametric CAD models, what-if tradeoffs, and software prototyping simply represent an extension of the conventional PD^3 process. Hard rapid prototyping techniques provide another example of such a synthesis environment. In rapid prototyping, such as stereolithography, SLA models are intimately related to CAD models. The precision of CAD model data is one such common parameter controlling the tolerances of the stereolithography model.

9.2 CONCURRENT PRODUCT DEVELOPMENT (CPD)

The decision process for developing a product can be computerized so that major portions of the product can be created with little or no human intervention. This is often the goal of a generative or a variant product design. The product development process can be as simple as a parametric part or as complex as several hundred assemblies of parts. The product may represent an axis-symmetric part, a family of parts, or a complex assembly.

The generative product design system is based on employing an enterprise's expert decision logic as a part of its core intelligent system [Abdalla, and Knight, 1994]. This is referred to in this book as an enterprise knowledge base. The knowledge base contains "histories, why the design looks the way it does" instead of merely capturing the geometry content (static representation of design decisions). Unlike a human mind, a knowledge base can be called by a program over and over again. The knowledge once captured can be preserved and disseminated throughout a CE enterprise. While such a knowledge base can provide information (on a real time basis) about likelihood of occurrence of a design solution, it cannot guarantee the existence of a feasible design. The latter requires a series of checks and balances as outlined in Chapter 4 of this volume.

A variant product design system, on the other hand, can retrieve an existing part with similar features from a historical database and then allow changes to be made to the retrieved design. A generative CE system actually generates the product design based on requirements about the parts, features, and attributes. Supporting database information that is made integral to the generative system includes: geometry, form features, design criteria, heuristics, and best design practices. Non-geometric information includes system analysis tools, materials, assembly sequence, and, of course, the enterprise's knowledge base of design decision logic. System analysis tools include the best of classical analysis, information modeling, and experimentation methods, blended with knowledge-based methodologies and decision logic. The hardest part of a generative system is mining and capturing a complete set of product and process specific decision logics. Adequate support of the organization and commitment of time from the resident experts are critical to its success.

What does it take to change a company that has been manually creating all of its designs to one of generative CE system creation? The generative system needs to know a tremendous amount about parts, processes, and, environments before it can start to generate a new design. The system needs to know the company's specific decision logic. This is where the knowledge engineers teach the product or process specific decision logic steps to the system. The resident experts must agree on the best way to make a part before any team starts computerizing the corresponding logic rules. If a consensus is not reached, one may end-up with contradictory rules and a sub-optimized design.

9.3   CE NETWORK TOOLS AND SERVICES

Integrated Product Development (IPD) encompasses all areas of intelligent programming, methods, tools, and services that capture, integrate, and automate various phases of product design, development and delivery (PD^3) process. Product design consists of a series of cooperating modules building upon each other like objects. Each module contains a subset of knowledge for the total design of the product, describing a set of its own engineering principles and integrating them with the applicable design and manufacturing rules. This evolves into a set of CE network of generative modules. The family of CE network of generative modules is written using a set of object-oriented programming principles, and is based on a parametric, feature, or knowledge-base driven solid modeling technology. The modules encapsulate rules and information that were used in the PD^3 stages. The basic entity understood by the CE network of generative modules is the definition of an object-module. The object-module can be viewed as having a part of inherent product characteristics, and can inherit properties and behaviors from other modules' definitions to become a new type of object-module. The CE network thus comprises a tree structure of generative modules communicating with each other and with the PIM data base. With successive applications, users can automate and integrate the entire product design-through-manufacturing process using a unified set of controlling parameters and a common knowledge-base. Changes, which can occur at any stage in the product development cycle, are reflected automatically in all interfaces and functional deliverables. A particular situation for design improvement can be handled incrementally through a specific set of modules. Since properties of a product are defined as objects in specific modules, there is no need to redefine and re-run the entire set of modules. The proposed changes would propagate automatically through the entire system. CE modules can be employed for manipulating the product design models or for making successive refinements to the product as team moves through its entire life-cycle. Built-in security features prevent unwanted alterations of the master data. Figure 8.12 of volume II shows a context diagram for such a CE office network of generative modules.

Figure 9.4 shows a network of 12 modules that form the infrastructure for the life-cycle mechanization process. Five modules belongs to CIM; four relate to automation; and three modules deal in CE topics. The criteria of mechanization are global in nature (such as 7Ts, 4Ms, and 3Ps) with the overall company goal of making maximum profits and great product. Among the 12 modules (see Figure 9.4) there is marked synergy in that the effectiveness of the whole CE network is far greater than the sum of each of its parts (modules). The extent to which a candidate CE network component resembles these 12 characteristic modules, determines how close the smart regenerative system is to reaching its full mechanized potential. The smart regenerative infrastructure is not an end in itself but a decision support environment intended to aid in achieving the product optimization objectives. For example, standard procedures may exist in multimedia, CAD media (CAE, CAD, CAM and CIM), or human brain-media forms. Multimedia is the presentation and delivery of information in forms other than conventional texts and graphics. Multimedia includes audio, full motion video, animation, high quality still images, in addition to text, line-art, photographs, and various combinations of all these. A slew of requirements, specifications, and standards that need to be integrated into the life-cycle mechanization process is contained in Figure 9.5. The requirements range from personal, local, community, divisional, industry, and corporate to government types. They may appear in various multi-media forms, such as text, line-art, figures, photographs, audio, video, and so on. The requirements may be distributed into a number of computers (e.g., mainframe, workstation, Mac, PC, etc.), operating systems (UNIX, Windows, PC/DOS, etc.), and applications (word-processing, CAD, CAM, etc.).

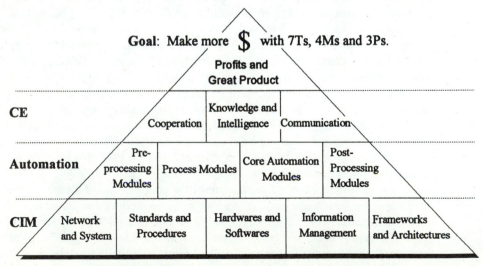

FIGURE 9.4　Modules of Life-cycle Mechanization Infrastructure

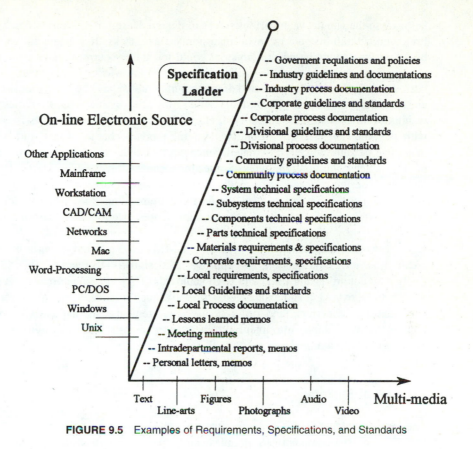

FIGURE 9.5   Examples of Requirements, Specifications, and Standards

9.3.1   Development on the Network

Considering network as a mode in which CE corporating teams want to design and develop a part, a client/server interface can be built. Design Teams feed features, parameters, and attributes of a design through a CE network of generative modules. The generative system then creates appropriate information models and executes a series of generative modules, testing to see if the part can be produced. If it can be produced, the system determines trade-off factors such as manufacturing cost. The manufacturing team now has more information on which to base their decision. The design team could choose to do a number of things depending upon the available design modules, such as introducing a change in inputs or adjusting the design slightly. On screen, a design team sees the spreadsheet of parameters, for example, thickness, materials, geometry, and so on. If there is a change, the program automatically runs the appropriate analysis and alters all the dependent parameter values and the geometry. This allows the trade-off to be completed quickly up-front. This cuts cycle time since the process is chain-activated. One change triggers another change that, in turn, triggers something else, and so on. At the end, the CE teams immediately see the impact of the change on cost and manufacturing. This also helps close the gap between various stages of design. Design for X-ability (such as design for manufacturability, etc.) can be incorporated as checks. If the generative system is based on knowledge-based principles, it could tell the concurrent teams what their specific options are. Additionally, with design history knowledge, the system could bring back a list of other similar parts and drawing numbers that have been previously built. For each similar part found, the system can display the product characteristics (such as weight, volume, materials, actual cost, etc.). If a standard library of parts is constructed with the knowledge-based generative system, product characteristics, like volume or mass, can be automatically calculated. Since both the analysis and design are inter-linked to a product configuration, what-if questions can be answered readily. This allows alternative design concepts to be evaluated and sensitivity analysis to be performed. This decision support information gives the designer tremendous knowledge to design a part with the available resources or to achieve optimal specifications with least manufacturing cost.

A product data management system (PDM) is generally put in place to control the creation, review, modification, and approval of designs. PDES and EXPRESS are relatively new but promising CAD and CAE standards. They allow features and attributes of models to be extracted and exchanged. Currently, such interfaces are unique for different combinations of systems and different user requirements. As standards become more commercially available, true integration between various stages of design and manufacturing will be streamlined.

9.3.2 Pull System

The network of generative modules for accomplishing product design serves as a pull type system. The computerized modules can be pulled and utilized by strategic business units (SBUs) and strategic suppliers. A complete CE network of generative modules for a PD^3 system is made accessible over the network. The use and access of the functions will depend upon the responsibility assigned to the person or the work group or the unit in the SBU product team. In response, computer-based support systems must be re-engineered to support this CE team environment. Most Artificial Intelligence (AI) systems, to date, have focused upon modeling individual expertise [Abdalla, and Knight, 1994].

Figure 9.6 shows (in a schematic form) three categories of modules for building a smart regenerative CE network. The first category of modules is a set of pre-processor units. The second category is a core set of automation units. The third category is a set of post-processing units. The functions corresponding to each module or unit can be executed randomly except in situations where an output from one function is an input to another function. The functions for all modules (or units) in the intelligent PD^3 system are integrated into a common data base and managed by a central object-oriented data management system. The interested units execute only the functions needed to achieve their share of responsibility and electronically pass the updated data base to the next responsible unit. The knowledge-base is thus continuously enriched every time a product function is executed or re-executed.

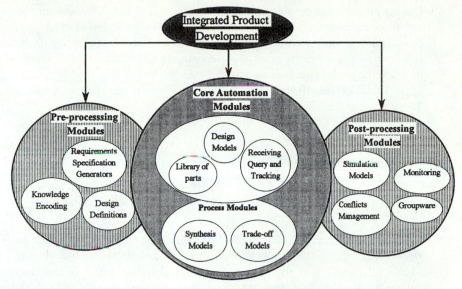

FIGURE 9.6   Categories of Modules for Building a Smart Regenerative CE Network

9.4   IPD AUTOMATION MODULES (PREPROCESSING)

Preprocessors are a network of modules, that:

1. Prepare and condition data to be used by the core set of generative modules, such as translators
2. Define the specifications, such as requirement generators
3. Assign design definitions—methods of capturing the design intent
4. Encode knowledge
5. Forecast estimates

9.4.1   Requirements Specification Generators

CE teams routinely consider and follow requirements from many sources (see Figure 2.22, volume I). There is seldom a record showing what the requirements are, where they originated, to which area of product or process they apply, or why they are important. This limited visibility makes it difficult to verify

- Whether or not the necessary requirements have been considered or applied at the right place.
- Whether or not there are excess requirements in certain areas or omissions in other areas that should be resolved.

Commonly, different levels of requirements are given to the CE work groups, such as customer specifications, industry standards, product feedback, marketing/sales requirements, product development, and so on. Not all of them are required at every design stage. Classification of requirements is an important step in CE that enables meeting these requirements at the lowest cost.

There are four levels of technical specification that are specified for a product. These are

Level I: Assembly Technical Specification (ATS).
Level II: Subassembly Technical Specification (STS).
Level III: Component/Parts Technical Specifications (CTS).
Level IV: Materials/Features Technical Specifications (MTS).

Figure 6.8 of volume I shows a hierarchical breakdown of a car assembly. In this case, *assembly* applies to the vehicle system as a whole. The sub-assembly applies to sub-systems of the vehicle, such as the powertrain, body, chassis, interior or the electrical system. Within a powertrain hierarchy, the engine represents the assembly; the cylinder-block is a subassembly; and cylinders, pistons, and so on, are components or parts of the subassembly. The components can further be broken down into multiple hierarchical layers until the lowest level of parts (bolts, nuts, studs, etc.) is reached. The materials (properties, composites (XMC), SMC, etc.), or the features are the lowest level of technical specifications.

1. *Assembly Technical Specification (ATS) Generator:* There is often a time lag between the customer or QFD data release and the ensuing definition of product specifications (e.g., level I through level IV specifications). The process is manual and time consuming. A level I (ATS) generator computerizes the creation of level I information from customer needs or product QFD data. It provides an expert linkage between customer needs, QFD data, and the corresponding level I specification definition. It may require interfacing with the house of quality (QFD) model.

2. *Subassembly Technical Specification (STS) Generator:* This is aimed at reducing the time lag between level I completion and the definition of level II specifications. A Subassembly Requirements Generator computerizes the level II process and its interfaces. It provides an expert linkage between level I data and the corresponding level II definition. This is coordinated with the level I (ATS) model proposed earlier.

3. *Component Technical Specification (CTS) Generator:* Like the ATS and STS generators, the CTS Generator provides an expert linkage between level II specifications and the corresponding level III definitions. It contains the process knowledge of level II and III experts, thus enabling CTS specification to be derived intelligently from ATS specifications. It is coordinated with the level II (STS) model proposed earlier.

4. *Materials/Features Technical Specifications (MTS) Generator:* The material technical specifications are the most independent of all specifications. However, they are very important since they provide the basis for capturing the design definitions discussed next.

9.4.2   Design Definitions

Design definitions are the common features and parameters associated with a product. They need to be captured so that all subsequent operations and network of modules use those features. They provide a common geometric design model. If a feature is changed during a run of any application, all related applications will automatically be run or tagged for processing to reflect this change on the outcome. These modules thus remove geometry and model barriers.

9.4.2.1   Parametric Product Design System

Depending upon the complexity of the product, automation needs may vary. Such needs are characterized by the amount of effort required to modify or to regenerate a part. The following situations exemplify three cases when development of a parametric program is considered the most appropriate.

1. When a large amount of effort is required in redesigning a part.

2. When most of the design steps have to be repeated for a similar part.

3. When frequent design modifications are required and during each modification cycle, the required steps are repeated.

Parametric design makes the part appear as if the part were designed for the first time. The difference in timing between a serial process and a parametric process indicates the amount of savings achieved by automating the generation of a part. This saving potential applies to all levels of the product's hierarchy from sub-systems and components to parts. Many of these components and sub-systems can be identified as potential candidates for parametric design.

During a parametric product design system, the critical elements of the product system are identified. Rules for the design of those specific elements are captured along with their geometrical rules and heuristics. The structural sub-assembly of a product represents a most widely used candidate for a parametric design. Structural definitions form the major portion of rules for structural subassembly design. Examples of structural definitions are: types and sizes of rods, struts, wide-area-angles, reinforcements, spacers, number of bays or stories including design rules for topology. Feature-based parts or macro objects are intended to make surface or form-feature generation simpler. Examples of macro objects are ribs, pockets, tabs, and so on. Essentially the process is a judicious application of parametric within a parametric representation of part-features, or form-features. Design work groups only have to know where to put form-features and parameters such as the desired sizes. A complex part can be created with a few strokes depending upon the library of form-features available. Such a library is a precursor to creating large parametric systems. Repeated use of form-features followed by successive use of parametric does represent a convenient way of building a CE network of generative system.

9.4.3 Knowledge Encoding

Rules can be gathered and implemented into expert systems [Abdalla, and Knight, 1994]. Parametric engineering can be applied to individual components or sub-systems, and conventional computer automation can be applied to many product procedures. These constitute considerable applications that can be developed.

9.4.3.1 Library of Constituent Rules Knowledge-Base

The rules that support constituent requirements have been identified in section 9.4.1, Requirements Specification Generator. Section 9.4.1 contains high-level reasons for constituent requirements. There is a need to identify the basis for having each constituent requirement. Many product designers and experts possess this type of knowledge. Most of this knowledge is closely associated with individual parts, types, or components of products. It is often characterized as *component reasoning*. The purpose of this library module is to capture the knowledge that supports the reasoning and rationale for having such requirements. Once this is put in a suitable knowledge-base, it can be readily used in the accompanying *library of system knowledge-base* module.

9.4.3.2 Library of System Knowledge-Base

The library of system knowledge-base contains an envelope of system buildability. This knowledge is beyond the capture of component rules and constituent knowledge-base discussed earlier. The envelope of product buildability ties all components reasoning into a total framework for a product system. It also features an on-line reasoning system that provides *intelligent* advice for various stages of design. With this information readily available, a CE team can make sound judgments based on the good of the entire product rather than the sub-optimal good of the component being designed. The above two knowledge-bases, together, can provide product-constituent and product-system interdependency. This also provides a useful means of sharing and preserving knowledge throughout an enterprise.

9.4.3.3 Rule Gathering and Relationship to IFLOW Model

IFLOW modeling provides a systematic network or hierarchy of activities. This network or task hierarchy could then serve for better communication and status reporting. It could also serve as a means to organize the rules collected for life-cycle automation efforts, especially those using knowledge-based engineering (KBE) [Forgionne, 1993] and expert system methodologies [Abdalla, and Knight, 1994]. The decomposition of activity via IFLOW models results in functions small enough to let knowledge engineers implement those rules into a KBE system.

To-be models started during the *as-is* process study often require further refinement. These *to-be* models provide a focus for subsequent development efforts and should be as correct and complete as possible. The models should be reviewed by as many experts in the product development team (PDT) as possible in order to result in an agreed-upon set of diagrams. Chapter 3 of volume 1 describes the mechanics of using this IFLOW methodology.

9.4.3.4   Library of Design Rules

One of the primary benefits of this IFLOW flowchart is that each *activity box* can serve as a focal point for gathering rules that are pertinent to that activity. When dealing with a large design problem, the scope is often so broad that gathering design rules can be quite time-consuming. The individual activity boxes become a pointer for gathering the pertinent design rules and then storing them in a relational database of some form. When dealing with design rules it is easier to think about the activity boxes than referring to a relational database that contains them. The IFLOW format is also useful in controlling piece-wise automation and thus helps team members focus on what can be done in parallel.

9.4.4   Estimating and Forecasting Tools

Estimating and forecasting tools are required during marketing and planning stages of product development. There are a number of places where such tools can be used. Forecasting tools can be used for determining the right time to launch the product and estimating tools for determining costs and its impact on sales. For instance, if sales staff is trying to predict sale figures corrected to environmental adjustment factors, they may need estimating tools.

9.5   IPD AUTOMATION MODULES

Automation modules represent a core set of modules that let work groups model their individual share of the problem, product, process, or the organization. A second set of modules lets work groups analyze the results, and a third set of modules lets the teams optimize the results, if possible. Following are the different set of modules that may be considered as a part of the automation modules.

9.5.1   Process and Product Modeling Systems

The product and process modeling system generally consists of a variety of design and analysis models (see Figure 9.3) built from the basic CAD/CAM system. There are three modes in which a modern CAD/CAM system can be employed to create design models: (a) playback system, (b) parametrically written program specifically for this purpose, and (c) language-based programming.

- *Playback system:* Some CAD/CAM systems allow interactive capture of commands onto a file. These commands can be recorded on a file and later played back, thus providing an instantaneous chain of developments, with facilities to pause or review. Other CAD systems allow interactive modification of the commands on file, thus allowing provision for addition and deletion of information. However, the information is pretty static, that is, it is dimensionally fixed.

- *Parametric-based Programs:* Some programs have provisions for parametrically deriving the geometry through variables whose values are only assigned at run time. Such systems provide a better modeling representation of geometry compared to playback systems and its manipulation if the design is altered.
- *Language-based Programming:* Some CAD/CAM programs contain some type of language based programming environments. This is discussed in great detail in section 6.3 of this volume.

9.5.2   3-D Geometry Models

During an initial modeling process, geometrical attributes, such as wire-frame, surface, and solid entities, are not value-coded. The geometry of the product is captured in a modifiable or a generic form. The related entities belong to an integrated generic family, that is, there is either a one-to-one correspondence between these modifiable entities or they are built from one another. During generic model construction, a team builds the geometry by first selecting the basic construction geometry symbolically and then supplying a set of referencing parameters rather than their actual values. Techniques such as flexible, parametric, features and *spatula* modeling, now common with many modern CAD/CAM software systems, allow quick building and sculpturing of solid geometry. A 3-D representation of a product is created that defines its geometry, topology, and material characteristics precisely and generically. This resulting master geometry model is then used for all downstream engineering activities including mechanical design, assembly design, simulations, analysis, packaging, and numerical control.

9.5.3   Mechanical Design Model

Mechanical design is the next step after geometry modeling. Here the constraints between various geometric entities and constituents of the product are specified. During successive model constructions, work group members decide what portion to parameterize and where to constrain. These steps in a mechanical design model permit the specification of free geometry (2-D lines, surfaces, 3-D solids) with constrained geometry.

9.5.4   Assembly Design Model

The important focus of assembly design is to display the various constituents of the product graphically in an assembled manner. Assembly design can be performed either by exploiting the interactive features found in newly developed CAD/CAM systems directly or by developing a customized assembly program. Through a language-based structure, it is possible to write a custom assembly program that would display an assembly of parts as one united design. The program creates the assembly using the modifiable information held in a series of files. The program first creates the individual library of parts by setting the common set of variables in the assembly design model. In the background, the geometry of the parts corresponding to each node of the part library is created. Then it assembles all the geometric profiles of parts so generated.

The exploitation of the CAD/CAM system in this way, to focus on assembly designs, software renderings, and linking of designs without annotated drawings, is becoming more and more popular.

9.5.5   3-D Machining Models

In a 3-D machining model, a *machining model* or a CAM file for a constituent of a product is created containing the NC machining and process information. The same CAD features and attributes that were used for the mechanical design models can be used as a source of input for the CAM program. Programming language facilities can be used to build this 3-D machining model. Most programming languages for NC include a full array of at least 2 _ axis tool path computations for milling, turning, punching, wire EDM, and other machining operations. Other NC options available include facilities for 2-, 3-, and 5-axis programming. For surfaces with multiple patches, techniques are available for combining the patches to create a continuous offset tool path across all surfaces. Draft surfacing capability can create tool paths specially suited for form and mold die environment. A constant draft profile can be swept along the 3-D curve. As NC machining checks are performed, process information (machining instructions) can be captured and stored in a CAM file. The machining model and the CAM file provide the work groups the ability to view the machined 3-D solid model of the part after each operation, after a series of operations, or to view a completely machined part. The 3-D mechanical design models can be manipulated at any time during the machining operation. The machining model or the CAM file can then be updated immediately to reflect any corrections—to the machining process or changes in design parameters—discovered by viewing the mechanical design models. The steps can be repeated until satisfactory conclusions are reached. The creation of these machining models allows the work group (a) to play *what-if* effects the NC operations will have on design, and (b) to incorporate provisions to change the design geometry or surface quickly. Work groups can alter the machining process if a problem is detected with NC operation, cutter path generation, or tool obstruction.

9.5.6   Structural Design Models

Analysis models are different from design models. An FEA model is an example of a structural analysis model. In an integrated product design approach, many types of structural design models are necessary to satisfy the integrity of parts. These models are used not only for evaluating the integrity of parts, but also for sizing the design to meet a set of specific performance objectives. Examples include models for solving thermal, magnetic, acoustic, and composite analysis, and sizing models for stiffness, strength, and fracture constraints, and so on. In the critical integration from analysis to a CAD-based design, a single source of modifiable geometry and a database management system are essential. Generic mechanical design models could be used as starting template models for obtaining various product configurations. Using some of the advanced analysis tools and languages available with modern tools, such as EAL [Kulkarni, Prasad and Emerson, 1981], or ICAD work groups can write structural design models for appropriate analysis. These design models are built using a CAD/CAM system and taking the same design features and parameters that were used originally in mechanical design models. There is no need to mesh the part, translate the design geometry, or recreate it. From structural design models, compatible FEA meshes can be mapped directly. Work groups can choose to simulate (depending on product analysis needs) the effects of static, dynamic, nonlinear, or thermal behavior on a product design.

9.5.7   Kinematics Design Models

Product subassemblies are subjected to various kinematic movements. Kinematic analysis models often constitute a basis for evaluating a design or its support mechanism. Kinematic design models provide a simulated way of satisfying kinematic constraints (relationships) in a set of predetermined cases. Solid model data is extracted from the mechanical design model database and formatted for use by a kinematic analysis. Kinematic design models employ the appropriate kinematic analysis models to compute the analysis results, such as dynamic interactive simulation and real-time frame-based animation. They are then used to predict design or kinematic (geometrical) changes. The modified mechanical design model geometry is exported back into a CAD system for further (downstream) processing. The kinematic design model eliminates redundant data entry and geometry creation, ensures model data integrity, and automates kinematic model descriptions. Automotive door example applications include: door opening and closing efforts when operated from inside or outside, stability of doors from full open or close cycles, and so on. Kinematic design models for these situations can help accelerate design decisions. The design models once created can be used repeatedly for a number of vehicle lines, when such design situations reoccur.

9.5.8   Acoustic Design Models

Spatial positioning of parts with respect to each other (such as the interior of an automobile) has a large influence on the intensity of noise level. Analysis tools are available to predict noise levels using a wide range of acoustic calculation algorithms. Numerical techniques include the acoustic finite element method suited primarily for modeling interiors or enclosures. When combined with wave envelope elements, exterior radiation patterns can also be predicted. Acoustic design models can be developed integrating the spatial geometry of the product constituents with acoustic analysis. Once the model is generically developed, it can be used repeatedly for solving interior and exterior acoustic design problems.

9.5.9   Linking of Design Models

The focus is on linking a set of product design models to satisfy common performance objectives. Many types of product design models such as structural design, kinematic design, load path design, and acoustic design models have been proposed earlier. They separately address individual performance needs such as structural, kinematic, acoustic, loads, and so on. Linking of design models allows CE work groups to design products that can simultaneously satisfy multi-disciplinary constraints. The efforts complement the generic approach by optimizing the product design for all constraint-types simultaneously as opposed to piece-wise optimization for each constraint-type, one at a time.

9.6   LIBRARY OF PARTS

A library of parts facilitates product design, development and delivery (PD^3) process. It could consist of a complete 3-D database of an existing product and its constituents—sub-assembly, components (inner-body, outer-body, etc.)—and parts. A part library contains data necessary (parametric, non-parametric, vector, or scalar) to define a mathematical representation of the product, sub-systems, and their constituents in an integrated computer model form. It includes a list of typical sections with parametric and non-parametric relationships. Wherever applicable and proven cost-effective or not, parametric and knowledge-based engineering forms are generally used to maintain the family of part's library [Forgionne, 1993]. A library of parts would encompass all interface elements that constitute the product assembly. These interface elements will contain all data necessary to define the mathematical representation of the parts including positioning and orientation, dependencies of parts with each other, and secondary considerations, like packaging, and so on.

9.7   SYNTHESIS MODELS

A synthesis model is shown in Figure 9.7. A synthesis model takes one or more of the design models outlined in section 9.5 and imposes appropriate performance requirements and constraints on them. A synthesis system has built-in processors for sensitivity calculations and optimization, to iterate through the design, based upon a specified set of RCs. The system calls upon the appropriate geometrical processors, analysis processors, sensitivity processors, and design models to compute new constraint values when design variables are perturbed. There are many types of synthesis models. They are discussed in the next sections.

9.7.1   Specification Advisor

Specification advisor is a framework type of development. It would define a user interface and a windowing system to view the various specifications. The four generator modules stated earlier in section 9.4.1, namely, ATS Generator, STS Generator, CTS Generator,

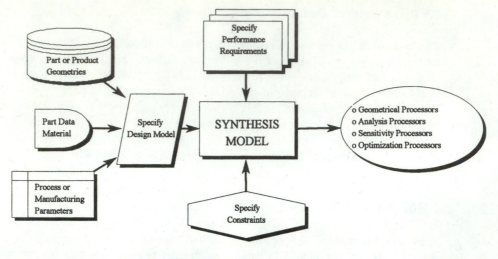

FIGURE 9.7   A Synthesis Model

and MTS Generator, are made integral to a window management system so that changes made in one generator are reflected in other generators.

9.7.2   Planning Advisor

In previous sections we outlined many opportunities to automate manual steps or to assist in the design and development of products. Opportunity exists to extend this approach to include product planning information. Additionally, other AI "assistants" (or expert systems) can be integrated in the following areas.

- Marketing.
- Scheduling.
- Resource Allocation.
- Other similar project management fields.

9.7.3   Assembly Planning Models

Assembly planning models for products are useful to address specific design situations, such as configuration design, addition, deletion or selection of parts. Parts are initially designed in terms of descriptive form features. The assembly product model is then created from the inter-part mating information specified interactively by the design work group. A rule-based procedure is employed to derive an exploded view from the initial assembly model of a product. Assembly planning models can be used to study the impact of the following scenarios.

- Repositioning of parts.
- Addition of parts or reinforcements (for example, brackets, fixtures, welds, and so on).

- Minimizing the number of changes in the direction of assembly.
- Eliminating awkward assembly operations.
- Identifying the number of subassemblies that can be run in parallel plants.
- Minimizing the number of components in a subassembly.

Many analytical tools and models need to be integrated such as Boothroyd and Dewhurst (DFA) [1983], *Vanational Simulation Modeling* (VSM), build variation, and packaging constraints to address one or more of the above issues.

9.7.4 Load Path Synthesis Models

Determinations of forces at various critical positions of the product and kinematic conditions are important considerations in meeting the desired product functions. Load path analyses predict these forces and determine the distribution of loads at various parts of the structure. The configuration of a product normally does not change very much from year to year. Furthermore, since the load carrying member positions and attachments remain fixed and parts are only subject to minor change from one product line to another, it is possible to create a generic load path model. This model can contain pertinent parameters of the products that are subject to minor adjustments, such as hinge locations, size, capacity, impact beam locations, and so on. Creation of such a load path model could be of great assistance in determining the basic loads in parts of the structure, which cannot be easily predicted.

9.8 DECISION SUPPORT TOOLS OR MODELS

A *generative or variant design* is a modifiable form of modeling to capture the configuration intent of a part, product, or structure that enables a specified function to be performed. An optimum design is a decision support tool or a method through which some design aspect, such as weight, shape, manufacturing, cost, or performance, is improved to the greatest extent without compromising the original part's intended function. As part of an IPD suit of models, decision support tools or models facilitate trade-off mediation among the multi-disciplinary work-group members. In the future, these models may use advanced information control methodologies, such as AI, Neural Nets and Fuzzy Logic, to offer appropriate decision advice to the design work group [Poindexter, 1991]. The decision support tools that can assess the optimum situation are disussed in the next sections.

9.8.1 Software Prototyping

Software prototyping is a concept for proving and verifying design options in software (or virtual) mode such as what-if, rather than making any physical (hardware) prototype such as those used in rapid prototyping or stereolithography systems. There are many techniques of software prototyping depending upon the stage of the product development cycle. In order to explain developing designs, annotations are often done in conventional practice. The working teams of production engineers, designers, and draftsmen who are familiar with the CAD/CAM and CAE systems, have realized that it could be a waste of time to annotate developing designs. A CAD/CAM system could be employed instead of annotation to interrogate geometrical profile for any kind, and to view dimensional changes at any time. This has resulted in a concept called software prototyping. Software prototyping is a process of successive refinements of the product on the computer before the product is released for production. This suppresses the urge to annotate details—until the layout has been designed and developed using rendering, scientific visualization, and other techniques—into a viable assembly. Fully associated layout assures that models and display (views) stay in sync. If the parameter of a part is changed in any design view, the corresponding changes are reflected in all design views that are dependent on this part.

9.8.2   Design for X-ability Evaluation

CAD systems can be used for verifying and evaluating design for X-ability (DFX) evaluations. Today this process is mostly manual. Most designers face a multitude of difficulties when attempting to use a CAD system for X-ability. Sturges and Yang [1992] have attempted to use CAD for assembly and encountered ten significant difficulty factors. Based on their work and the decomposition scenario of Chapter 6 of volume I (see Figure 6.8), these difficulty factors can be rearranged in a six-by-three matrix. The matrix has six levels (system, sub-system, component, parts, materials, and process) and three operations (acquisition, positioning and orientations, and an assembly operation). The resulting matrix is shown schematically in Figure 9.8.

9.8.3   Trade-off Advisor

There is a need to develop a more comprehensive computer support system called a Trade-off Advisor. This may require integrating and controling the specific modules or the system (for example, design for X-ability). This may require concentrating on the performance trade-off and overall coordination of the product functionality. This type of support system is particularly suited to the emerging CE environment. When more work is done in parallel, the coordination and control of computer support system becomes increasingly important.

This advisor would be a framework or wrapper type development. It would define an overall computer environment, as well as interfacing standards and protocols for a natural *hand shake* to be performed in implementing specific modules. The list of potential modules that could be a candidate for this trade-off advisor suit are

- Product design models.
- Product simulation models.
- Synthesis models.
- Trade-off models.
- Groupware.

Operations / Levels	Acquisition Operation	Positioning and Orientations	Assembly Operation
Descriptions	Parts are brought from the feeding point to assembly point	Parts are positioned and oriented ready to be assembled	Assembly operation performed
System Level Factors: Factors dependent upon the Subsystem Interactions	*Handling Systems:* Difficulty in handling subsystems and determining the correct orientations to be acquired and brought in.	Difficulty in determining the datum plane, correct orientation and positioning of system. (system stability)	Difficulty in how subsystems are positioned and oriented to assemble into a system.
Subsystem Level factors: Factors pertaining to the individual subsystems; independent of the overall system	*Handling Subsystems* Difficulty in determining the subsystems' shape, size, and weight for handling, rotation, position, and transportation modes.	Difficulty in determining the datum plane, correct orientation and positioning of subsystems.	Difficulty in how components are positioned and oriented to assemble into a subsystem. (clearances, positions and directions)
Components Level factors: Factors pertaining to the individual compoents; independent of the overall subsystem	*Handling Components:* Difficulty in determining the components" shape, size, and weight for handling, rotation, position, and transportation modes.	Difficulty in determining the datum plane, correct orientation and positioning of components.	Difficulty in how parts are positioned and oriented to assemble into a component. (clearances, positions and orientations)
Parts Level factors: Factors pertaining to the individual parts; independent of the overall components	*Handling Parts:* Difficulty in determining the subsystems' shape, size, and weight for handling, rotation, position, and transportation modes.	Difficulty in determining the datum plane, correct orientation and positioning of parts.	Difficulty in how materials are mixed, and features are placed to assemble parts (mixing + positions).
Materials Level factors: Factors pertaining to the individual materials; independent of the overall parts	*Handling Materials:* Difficulty in determining the features' size, and materials properties for the parts to be acquired.	Difficulty in determining what constituent features properties of materials would meet the orientation and positioning needs.	Difficulty in what constituent properties of materials of diiferent parts can be joined, welded, or assembled.
Process Level factors: Factors dependent on the process employed to assemble the mechanical system	*Handling Distance:* Difficulty in bringing a constituent part to the assembly point. Transportation difficulty, speed of movement, types of transportation, wait, etc.	Difficulty associated with fixtures and methods used to orient or restraint a constituent part for assembly (fastening)	Difficulty associated with the path a constituent part would traverse during an assembly process (path).

(Leftmost vertical column label: L E V E L S)

FIGURE 9.8   Assemblability Difficulty Factors (Organized with respect to associated levels and operations)

The above models by themselves can end up into a variety of product-oriented network of tools and services that are simply "islands of automation." The trade-off advisor integrates and controls the models.

9.9 KNOWLEDGE-BASED PRODUCT AND PROCESS MODELS

If we wish to emulate some of the team cognitive functions (learning, remembering, reasoning, intelligence, perceiving), we have to generalize the previous definition of knowledge to include cognitive aspects of modeling. Product and process models must be cognitive or knowledge-based. They must deal with simulation, processing of cognitive information, and fuzzy logic (soft logic) in addition to some hard uncertainty and numerical information [Forgionne, 1993].

Optimization is essential if a product model has to become a predictive tool for design work groups. Knowledge-based feature modeling is often a prerequisite for optimization. Knowledge-based features are named entities that encapsulate how a unit of a product functions using part rules and attributes along with their defining constraints and geometry. For example, a through-hole feature understands the rule that it must pass completely through the unit and will do so no matter how the parts change. An expert relies on accurate product and process modeling to accurately predict any product or process queries [Abdalla, and Knight, 1994]. Knowledge-based optimization involves three things.

1. A knowledge-based feature model.
2. Formulation of design problem as an optimization problem.
3. An optimization algorithm to guide the design to reach a feasible solution.

In a knowledge based expert system [Forgionne, 1993], the teams capture both design and manufacturing intents. This intent information is tied to the past design results, that is, lessons learned. This helps guide work groups to an optimal mix in terms of performance, manufacturability, and supportability of new designs. The optimization process involves automatic looping (searching) through different models of the design to see which of the perturbed design yields the smallest design objective and satisfies most of the design constraints.

9.10 COMPUTER-BASED TRAINING TOOLS

Computer based training tools uplift the confidence of the manufacturing and logistics work groups to use more of the analytical models early in the product development. This is true even when the teams are presented with an incomplete set of information. Computer based training tools capture and codify the manufacturing *black book* expertise required to make the concurrent processes really successful. Underlying these various computer-based training tools are the hidden needs for the information integration technologies. These technologies ensure that the right training information is available at the right time and that all product and process information (data, method, and knowledge) are integrated across the heterogeneous environment.

9.11   COST AND RISK REDUCTION TOOLS

Cost and risk reduction tools enable PDT members to better visualize and understand risk factors associated with selection of product and process alternatives. Tools such as Quality Function Deployment (QFD) and Design Of Experiments (DOE) have been successfully used by Japanese producers to better assess the effect of cost reduction on overall system quality and its impact on critical process control parameters. Rapid and software prototyping are enabling tools for reduction of risks. Risks are reduced through a better understanding of the design and the required processes that make the product.

9.12   IPD AUTOMATION MODULES (POST-PROCESSING)

Post-processing tools are required to ascertain that the core automation modules do function well and that all the necessary operations in the factory proceed as expected. At any time, work groups might be working in a variety of product areas simultaneously: product development, process design, quality and productivity, sales and marketing, and process reengineering. If a problem is detected, irrespective of where it is, it must be solved. In process reengineering, a problem solver (e.g., a process work group) may be experimenting with process parameters to reduce their variability and increase robustness (design of experiments). Good quantitative problem-solving schemes are essential to achieve that goal. They range from simple techniques such as statistical and forecasting techniques for a relatively unknown domain, to more elaborate simulation and mathematical programming methods for relatively well posed problems. For instance, if a quality team detects variability in what it measures, it may need statistical techniques. But if the team is trying to improve a system that is not available for experimentation, or too expensive to experiment on, the work group may need simulation techniques. Similarly, if a production team is trying to analyze the effect of increasing the throughput capability of one machine, the team may solve the problem using simulation. But if the team is trying to reduce the process waste on a certain line, the team may need a simulation tool to predict the amount of waste and a mathematical programming technique to predict the amount of change. These classes of modules are discussed in this section.

9.12.1   Simulation Models

Simulation tools can be used to assess the impact of changes on the business environment. The changes may occur anywhere during the life of a product, could be caused by variations in engineering, manufacturing, or the shop floor.

9.12.1.1 Packaging Simulation Models

Several movable parts of products can interfere with other parts if enough clearances do not exist or adequate real state property is not allocated. Analyses and simulation of packaging envelopes determine the adequacy of the part functions. Design envelopes for the automobile packaging of tailgate cables, door latch release mechanism for glass, or an automobile tailgate panel assembly are examples of packaging simulation. In the case of sliding products, examples include: sizing of the pop-up spring to ensure that the product is pushed out as soon as the product is released; profiles of détentes on the track to ensure that the product swings out completely before sliding, and so on.

Once the constituent parts (e.g., a vehicle's inner-panel, outer panel, and product sub-system assembly) are designed to meet their individual packaging considerations, the entire product assembly must function as a unit. The whole assembly must meet the fits and clearance requirements with respect to the overall fit to the vehicle-body. Simulation of product functions (such as door opening and closing, fits and gaps with A and B pillars and roof-rails) provides the system dependencies of various controlling parameters. Any variation in these parameters ought to be propagated to the sub-system level for their re-design and packaging refinements.

9.12.1.2 Predictive Cost Simulation Models

A predictive tool for determining the life-cycle cost is needed from the very beginning of the PD^3 process. A predictive cost model captures information about the design elements (e.g., components, assembly sequences, manufacturing requirements, quality control, scrap, testing) and estimates the unit cost associated with other processing and field supports. It is based on the actual service data available over the life-cycle history of a similar product. If the product is brand new, the cost associated with doing the design right the first time is much smaller than fixing or correcting the problems later. The cost increases with every bit of delay in identifying the problems. Once the problems are identified, it is far cheaper to correct them sooner in terms of both time and money. After the product gets to the manufacturing floor, major decisions are already made and it is too costly to change them. The predictive cost model at any life-cycle stage computes these costs based on the types of products and the stage of their completion into the life-cycle. Generic profile or a template of a product over its life-cycle can be prepared. New product concepts can then concentrate only on modifications to those bogies. Associated with each part memory, there is a corresponding cost memory. Cost memory can be used as a basis for comparison with other product lines, other model years, and other manufacturing processes.

Unfortunately, many product work groups tend to see parts, but do not see the hidden manufacturing processes behind them. With an appropriate cost model this can be eliminated. All processes, such as manual component preparation, manual assembly, lifting, winding, upside down, centering, masking, and post-wave touch up, add to cost and subtract from quality. Product quality typically drops each time a constituent is handled, screwed, or welded. The greater the number of process steps a product goes through during the life-cycle, the less reliable the resulting product is likely to be. Attention should be directed toward obtaining an integrated cost model that accounts for the life-cycle (design and manufacturing) process steps, time duration, and the associated unit cost.

9.12.1.3   Graphics Animation

Graphics animation is one of the visual attractions that is making the use of simulation tools very popular and acceptable in most industries. Animation of constituents makes simulation easy to understand and intuitive to relate to when analyzing manufacturing systems. Its popularity can be judged by the growing number of simulation programs that are now supported by animation software packages. Simulation is generally considered an add-on (not a replacement) to a traditional sound analysis. For instance, a pretty graphic animation will not do any good to any CE team, if the manufacturing systems are poorly simulated or analyzed.

9.12.2   Reasoning, Query, and Tracking

9.12.2.1   Query: On-line Requirements

The product requirements, as compiled during a product's life-cycle, need to be updated, improved, and maintained over time. Product on-line requirements query system is a menu driven system that provides on-line access to these requirements. Concurrent teams can use and maintain them electronically on a continuing basis. If initially developed using an object-oriented data base technology, conversational query can also be easily addressed.

9.12.2.2   Query: On line Information Flow Diagrams

Every multi-disciplinary organization intends to define the life-cycle phases of a product from its inception to recycling. Automotive industries categorize them as a multi-phase process. An initial set of IFLOW models for each phase in schematic form was defined in Chapter 3, section 3.6 of volume I. Each function-box has associated input, feedback, outputs, constraints, and resources. These are *WHATs* of a product system's multi-phase process. It is often difficult to visualize the relationships and dependencies between one functional box and another, since IFLOW diagrams are hierarchically organized. The information is voluminous and often separated by several pages and embedded in a function later in a chart. A menu driven *Multi-Phase Query System* can provide interactive access to these IFLOW diagrams. Teams can maintain and upgrade them on a continuing basis. If the initial queries were developed using an object-oriented data base approach, conversational-type query can also be furnished.

9.12.2.3   Tracking: Life-cycle Process

When a large program is underway, it is difficult to control the various tasks according to a well-defined multi-phase product process. The access to IFLOW diagram gives an interactive capability to query about work tasks. The multi-phase query system does not provide a useful mechanism to monitor and track the design progress. The multi-phase query system also does not show the status, the required work-force, or identify bottlenecks. Since the output of a task is a resource for a succeeding task, it is important that each task be completed on time by a work group. A *multi-phase tracking system* can provide a way to monitor, track, and control this process systematically, effectively, and on a timely basis. This system can be coordinated with a scheduling system (e.g., a Gnatt chart) so that timing is reflected for each task and sub-type. The *multi-phase query system* can be used as a resource for supporting many of its tracking needs.

9.12.2.4   Reasoning: On-line CAD Guidelines

"On-line guidelines" would be of great value to CAD work groups interested in accessing, for example, "design for X-ability" guidelines at various points along an interactive design session. Examples of such guidelines are

- Metal Stamping.
- Plastic or Composite Fabrication.
- Assembly.

On-line guidelines can be implemented on a stand-alone basis or as an automated checklist mode. In an automated checklist mode, the CAD system can be made to scan an evolving CAD file and reason about whether or not the on-line guidelines have been followed. Additionally, this checklist feature has bearings on the design advisor, the product problem diagnostic system, and the product trade-off analysis system modules discussed earlier.

9.12.2.5   Reasoning for Manufacturing Rules and Guidelines

In section 9.4.3 knowledge encoding describes four categories of knowledge-bases. Three of the knowledge-bases dealt with rule-gathering and the last one dealt with their relationship to an IFLOW model. These libraries of rules are not supported by corresponding *WHYs*. Rules are best understood when they have associated *WHYs*. Most libraries are captured in a knowledge data-base described in section 9.4.3. A *product requirement reasoning system* can interrogate the rules' knowledge base to search for reasons. It can provide traces of why a requirement is or is not important. It can provide a conversational reasoning for having that rule and its implication.

9.12.3   Conflict Management

One notion of a concurrent engineering environment is that it allows a work group to work on his version of a module in parallel with other work groups (see Figure 5.16, volume I). Conflicts may occur if appropriate means are not put in place to manage conflicts. The following techniques have been used in the past to improve communications and reduce such conflicts.

- Electronic or physical conferring.
- Locking items that must not be changed.
- Notifying a user when selected items are changed.
- Allowing users to preview changes being made by others.
- *Parallel release so that any change approved and released in one version of a module is immediately reflected in all remaining work-in progress versions.* This technique can have negative as well as positive consequences.

- *Forcing work groups to merge their design versions at regular predetermined intervals.* This technique has often been used in software engineering projects. In such a project, a project administrator will insist that the latest version of a procedure be included in an overnight build of a system. This way any incompatibilities between that procedure and the modules that call it can be identified as soon as possible.

- *Identifying the amount of interface data that are common and must be shared between work groups.* By dividing each work group module into a portion that can be independently worked upon and an interface portion that needs to be shared, the amount of sharable data is less, and, therefore, conflicts are reduced. In software design, the interface of a module might be the definition of a procedure; in electrical design, the interface might be the external pins of a circuit. In mechanical design, interface is usually some portion of the geometry defining the boundary of a part. In all cases, dividing a module into an interface portion and an independent portion reduces the potential for conflicts because only the interface must be shared. This allows the work group to modify independent modules without affecting other work groups. In other words, the technique emphasizes to the other work groups that changes to the interface modules will have consequences beyond its own scope.

The most drastic method for avoiding conflicts is to insist that only one team work on a module at a time. Even if this step is taken, concurrency is still possible (Figures 4.14 and 4.15, volume I). This is due to the fact that different tools can be applied to a module. If we replace a *design engineer* by a corresponding *tool* in Figure 4.15 of volume I, concurrency can be easily exploited. In this scenario, two tools (tool #1 and tool # 2) edit a design concurrently, while another tool accepts the results of those edits. Resolving conflicts is not an issue because a change to any one of the versions is immediately transmitted to the other versions. However, translating data between the tools is now an issue because the design system may require its data to be reorganized in a different format from its physical layout format. Different tools can share items, but they organize them into different formats, requiring data translation.

9.12.3.1   Identifying Conflicts

Artificial Intelligence (AI) based diagnostic systems have been quite popular in the medical field. They can be applied to mechanical or electromechanical systems such as automotive or aerospace products as well. AI can be used to identify constraints that cause conflicts. A forward chaining in AI can be thought of as similar to a roll-down of *voice of the customer* to key part characteristics (KPCs). Similarly, a backward chaining in AI can be thought of as a roll-up from a set of symptoms to a deficiency in analysis and/or design.

9.12.3.2   Resolving Conflicts

Identifying conflicts is not the same as resolving conflicts. This must be done by the concurrent teams in consultation with the expert (e.g., a human team or an application program). Relying on an applications to resolve conflicts means capturing knowledge that could be used by the teams in resolving constraints. In some cases, the conflicts will occur

because of the different goals, and of different types of work groups. As an aerospace example, in aircraft engine design, the aerodynamic team may want turbine blades to be
long and thin because it enhances efficiency. But the mechanical team may want them to
be short and fat because this makes them structurally stronger. It is a definite advantage
for the CE management teams to concurrently develop modules. This helps identify some
fundamental design conflicts at an earlier stage.

As shown in Figure 4.15 of volume I, the conflicts between the mechanical and aerodynamic team versions can be resolved, if the mechanical team sends a list of changes to the
aerodynamic team as a delta file. The aerodynamic team then applies those changes to their
version to see if they interfere with the original design goals. If they do, then the two teams
may have to iterate until an acceptable compromise on changes is reached. The final
changes are merged into one as shown in Figure 4.15 of volume I or are communicated to
the third team (e.g., a process team) so that all remaining teams can update their versions.

9.12.4 Monitoring and Control

In Chapter 3, section 3.1 of volume II, as a part of total quality management, design-
oriented (here called "off-line") quality monitoring methods were described. They were
of two types: product quality monitoring and process quality monitoring (see Figure 3.4
of volume II). Product quality methods were applicable during product design, such as
QFD, Pugh concept selection matrix, DFX assessments, and so on. Process quality methods were applicable during the process design, such as Taguchi's method, design for experiment, and so on. Deming [1993] and Ishikawa [1985] have emphasized the importance of using such off-line quality control techniques during product realization.

With increasing competitiveness and a desire to excel in quality, there is a growing
need to develop monitoring and control methodology that is based on on-line process
measurements. The method involves tracking a sample of parts moving through a production process and taking measurements at each station. The work groups use statistical
analysis techniques to determine the source and type of variation, transmitted or added.
This is followed by a feedback or a corrective action (see Figure 3.4, volume II). The emphasis of this new process is on the following:

- Achieving *defect prevention* versus *defect inspection.*
- Further maintaining the off-line designed product and process on its target value,
  both in terms of the nominal value and with minimum variation, through a combined use of feedback control and SPC techniques.
- Meeting a target production with minimum variance instead of the design falling
  within tolerances.

On-line corrective monitoring methods contain a feedback element in their configuration.
Monitoring and control tools, such as Statistical Process Control (SPC), have been used
for years during inspection modes. They are being used for the corrective mode as well.
There are also off-line preventive monitoring techniques that control the occurrence of
such variability in the first place.

9.12.5   Software Quality and Productivity

The information technology (IT) department is generally responsible for life-cycle mechanization through software engineering and quality assurance. IT has six layers of responsibility from defining its mission to deploying IT products. Figure 9.9 describes these six layers of responsibility associated with IT software quality and productivity engineering. The six layers are IT mission, IT motivation and objectives, IT tools, methods and metrics, IT services, IT process, and finally IT products. Besides the normal design and build functions, the enterprise functions include the following:

- Operations (e.g., data flow, material handling, set ups, loading and unloading operations)
- Built-in quality control systems (e.g., QFD, QC, error-proofing, SPC, SQC)
- Support systems (e.g., tooling maintenance and cleanup).

IT software quality and productivity functions constitute a part of the CE transformation model.

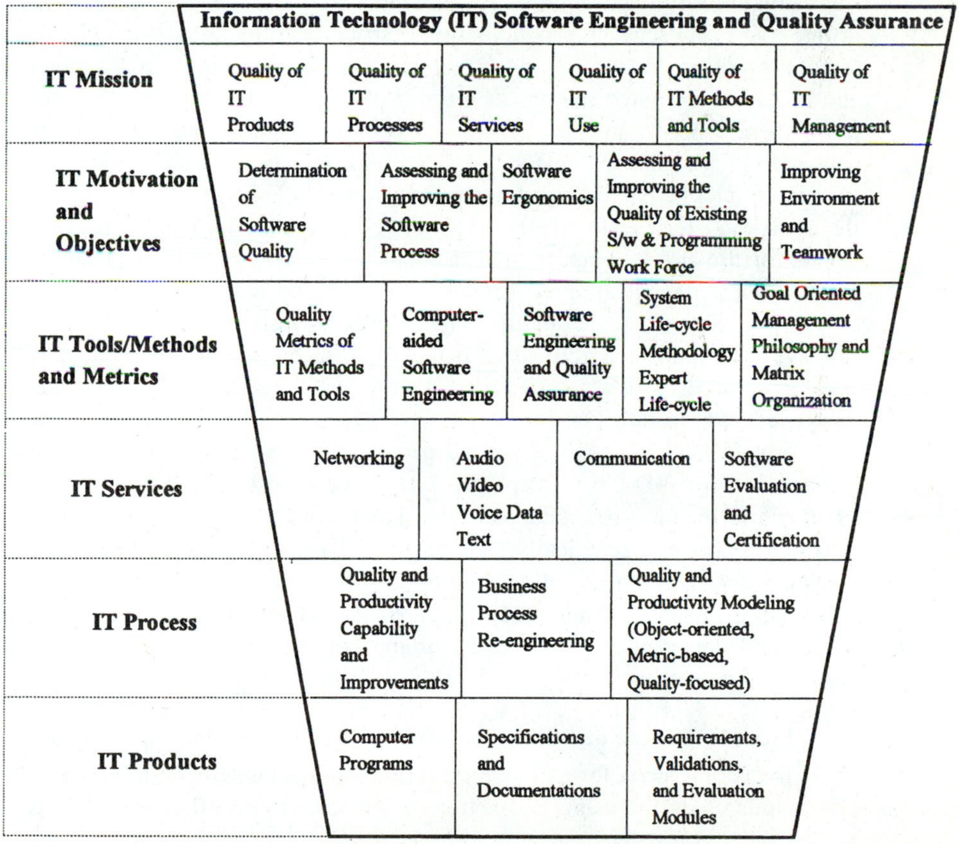

Information Technology (IT) Software Engineering and Quality Assurance

IT Mission	Quality of IT Products	Quality of IT Processes	Quality of IT Services	Quality of IT Use	Quality of IT Methods and Tools	Quality of IT Management
IT Motivation and Objectives	Determination of Software Quality	Assessing and Improving the Software Process	Software Ergonomics	Assessing and Improving the Quality of Existing S/w & Programming Work Force		Improving Environment and Teamwork
IT Tools/Methods and Metrics		Quality Metrics of IT Methods and Tools	Computer-aided Software Engineering	Software Engineering and Quality Assurance	System Life-cycle Methodology Expert Life-cycle	Goal Oriented Management Philosophy and Matrix Organization
IT Services		Networking	Audio Video Voice Data Text	Communication	Software Evaluation and Certification	
IT Process		Quality and Productivity Capability and Improvements	Business Process Re-engineering	Quality and Productivity Modeling (Object-oriented, Metric-based, Quality-focused)		
IT Products		Computer Programs	Specifications and Documentations	Requirements, Validations, and Evaluation Modules		

FIGURE 9.9   Information Technology (IT) Software Quality and Productivity Engineering

9.13   UNIFIED OR SINGLE PPO CONCEPT

One of the most talked-about applications of the life-cycle mechanization is a Single-PPO-concept (SPC). PPO stands for product, process, and organization. The concept (SPC) is based on providing the work groups with a direct access to leading commercial CAE software (Nastran, Patran, Ansys, Rasna, etc.), CAD software packages (such as EDS/Unigraphics, CATIA, SDRC/Master Series, etc.), and CAM software (CAPP, etc.) built on a single consistent PPO representation. Direct access means that CE work groups can eliminate the transfer, translation, or recreation of product features when using SPC. Most of the intelligence in SPC relates to the theoretical foundations for the development of mapping between design support technologies, particularly between intelligent CAD and solid modeling systems. Part functions and possible features are synthesized to align with a desired common product and process mapping (see Figure 7.5 of volume I). Coupling product, process, and organization constraints with one common PPO mapping is critical for SPC. A typical progression of concept from a conventional (serial engineering concept) to an SPC (a 3-D CE concept) is shown in Figure 9.10. Figure 9.10a shows a conventional process discussed in early chapters of volume I several times. Figure 9.10b shows an early version of simultaneous engineering concept, when design and manufacturing functions were carried out simultaneously. The next series of three concepts in Figure 9.10 deals with concurrent engineering (CE). Figure 9.10 also shows the accompanied savings as a result of using such CE concepts. The most popular of these concepts are 1-D, 2-D, and 3-D CE concepts. A unified or single PPO concept (SPC) is another name for a 3-D CE concept. In a 1-D CE concept, the various functional tracks are overlapped linearly as shown in Figure 9.10c. In a 2-D CE concept, the individual tracks are overlapped radially as shown in Figure 9.10d. PIM, which stands for product information management, forms the core of this 2-D CE concept. It serves as a communicating (e.g., 7Cs) block for the radially overlapping tracks. This way, a series of cyclic iterations can take place. Design improvement is radially progressive, and product is refined as it moves from one cyclic iteration to another. This way, to come up with a new product design, reasoning based on the constraints from a combined PPO feature library can be employed by the work groups to search for suitable form features to satisfy a desired part function. Sources of the constraints become immaterial. In a 3-D CE concept, the 2-D cyclic process is repeated along a third-axis as shown in Figure 9.10e. This axis, in 3-D CE concept, represents a process taxonomy dimension, which is generically captured. It is so general that it is applicable across all product lines. The concept of a mapping process between functional models, attributes, symbols, or features is illustrated in Figure 6.11. There are a number of significant outcomes and developments that are becoming commercially available in this area. SPC typically includes

- Product structure tree (PtBS, or an enhanced form of a bill-of-materials).
- Decision criteria for extracting information from external data bases. This may include rules for using geometric design data imported from a C4 (CAD/CAM/CAE/CIM) system.

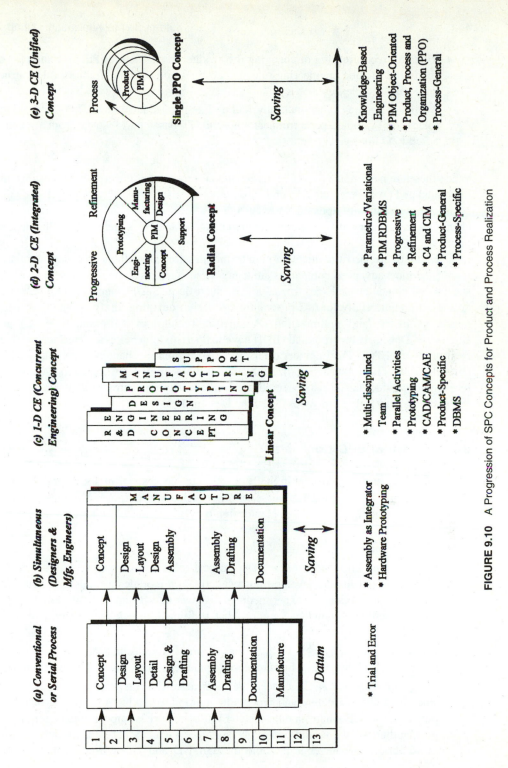

FIGURE 9.10 A Progression of SPC Concepts for Product and Process Realization

441

- Rules for reconfiguring or changing the product tree structure (PtBS). Examples of such rules include engineering rules for design optimization, X-ability rules, structural analysis rules, etc.
- Attributes such as functional, physical and geometric attributes. This may include selection of standard parts from catalogs and GT-based part features from a feature-based PPO library.

When there are new inputs to SPC, interdependent product features (e.g., dimensions and materials) and process features dependent on input specifications (e.g., required loads and costs) are automatically updated. Interdependency between parts of the PPO means that change to one part or a feature automatically changes those parts and features that depend on it.

Furthermore, because intent-capture techniques are extensible, open, and modular, a few of the most advanced companies are using these concepts to tie together many of the specialized computer packages to enable CE. Such tie-ins, as elements of a CE concept, now exist commercially for NC programming, design detailing, drafting, analysis, parametric geometry, and optimization. Parametric Technology Corporation (PTC) and Boothroyd Dewhurst Incorporated (BDI) have teamed up to develop an interface between their products. PTC's Pro/Engineer offers parametric solid mechanical capability for mechanical design. BDI's design for manufacture and assembly software offers capabilities to evaluate product designs for assembly, manufacturing cost, structural efficiency, and optimum manufacturing processes [Machine Design, 1993].

9.13.1   Uses of SPC Concepts

The elements of each CE concept in SPC are software sub-systems or smart modules written for a specific PPO (product, process, or an organization) purpose. Modules automate most of the product life-cycle (e.g., plan/design/build) functions. The various uses of CE concepts are:

- *Discovery of Knowledge Gaps:* The purpose of intent capture techniques in CE concepts is not just to develop the so called smart modules as ends in themselves, but to use the modules as experiments to discover new design domains and to determine if there is a knowledge gap or what additional knowledge, if any, is missing from the SPC set.
- *Elimination of Interfaces:* With a series of smart modules in SPC, the work groups have all critical integrating links in place to practice a successful CE implementation.
- *Range of Outputs:* A CE concept delivers more than its traditional counterparts since it captures structure and process in addition to product features. Typical outputs of a 2-D CE smart module include bill-of-materials, process plans, test strategies, and analysis pre- and post-processors (see Figure 6.11). SPC outputs include, in addition to whatever is possible through 1-D and 2-D CE concepts, the non-

geometrical and process knowledge that can be used to refine the quality of the 1-D and 2-D CE outputs. Figure 6.11 lists most of those popular outputs.

- ***Retention of Technical Memory:*** Similar to expert systems, a CE concept-based system helps retain institutional memory. When a work-group member leaves a company or retires, a portion of the engineering knowledge leaves permanently or is lost. By creating a smart set of CE modules a priori, the knowledge the team members apply to a product realization situation remains with the work group and therefore with the company.

- ***Elimination of Bottlenecks:*** Critical areas of knowledge possessed by relatively few people need to be more widely distributed in the organization. The integration of smart CE modules into a product development framework (discussed in Chapter 4) allows corporate memory to be shared among different CE work groups.

- ***Consistent Results:*** The use of smart CE modules results in PPO designs that are consistent across all product lines.

- ***Improve Task Performance:*** An approach based on a CE concept improves performance in terms of speed, quality, cost, customer satisfaction, and company competitiveness needs.

All three CE concepts in SPC can be used repeatedly by product design work groups or manufacturing work groups to do software prototyping to consistently develop and subsequently redesign and re-engineer a finished manufacturable product. Out of all that, the set of SPC modules facilitates the process of making complex changes to product, process and organization (PPO) designs. By following various taxonomic constructs, work groups are able to perform what if iterations on a PPO design. The approach is *dynamic* and the PPO intent model can be iterated through the various responsible work groups within a concurrent business unit until a set of attributes that provides the best possible (feasible) solution for the PPO's specification is arrived at.

REFERENCES

ABDALLA, H.S., and J. KNIGHT. 1994. "Expert System for Concurrent Product and Process Design of Mechanical Parts." *Journal of Engineering Manufacture*, Volume 208, No. B3, pp. 167–172.

BOOTHROYD, G., and P. DEWHURST. 1983. "Design for Assembly: Manual Assembly." *Machine Design* (November 10) pp. 94–98 and (December 8) pp. 140–145.

DEMING, W.E. 1993. *The New Economics* (November) Cambridge, MA: MIT Center for Advanced Engineering Study.

FORGIONNE, G.A. 1993. "IMDS: A Knowledge-based System to support Concurrent Engineering at Westinghouse." *Expert Systems with Applications*, Volume 6, No. 2 (Apr.–June) pp. 193–202.

ISHIKAWA, K., and D.J. LU. 1985. *What is Total Quality Control? The Japanese Way.* Translated by David J. Lu, Englewood Cliffs, NJ: Prentice Hall.

KEENAN, T. 1995. "Engineering Revolution, CAD/CAE Advancements Changing Vehicle Development." *Ward's Auto World* (March 1995), Volume 31, No. 3, pp. 97–100.

KULKARNI, H.T., B. PRASAD, and J.F. EMERSON. 1981. "Generic Modeling Procedure for Complex Component Design." SAE Paper 811320, Proceedings of the Fourth International Conference on Vehicle Structural Mechanics, Detroit, Warrendale, PA: SAE.

MACHINE DESIGN. 1993. Volume 65, No. 11, p. 77, June 1993.

PHILLIPS, R.E., and J. AASE. 1990. "An Integrated Environment for Concurrent Engineering." Proceedings of the Second National Symposium on Concurrent Engineering, *Emerging Prototypes of Concurrent Engineering*, February 7–9, 1990, Lakeview Resort and Conference Center, Morgantown, WV: Concurrent Engineering Research Center, pp. 487–500.

POINDEXTER, J.W. 1991. "Rapid Prototyping in an Integrated Product Development Environment." SAE Paper No. 911113, Presented at the SAE Aerospace Atlantic Conference, April 22–26, Dayton, Ohio.

STURGES, R.H., and J.-T. YANG. 1992. "Design for Assembly Evaluation of Orientation Difficulty Features." Proceedings of the Winter Annual Meeting of the American Society of Mechanical Engineering, Anaheim, CA, November 8–13, 1992, PED-Volume 59, pp. 221–232, *Concurrent Engineering*, edited by D. Dutta, A.C. Woo, S. Chandrashekhar, S. Bailey, and M. Allen, New York: ASME Press.

TEST PROBLEMS—LIFE-CYCLE MECHANIZATION

9.1. Why are the results of life-cycle automation generally mixed? What should a planning work group do before automation and why? Identify a set of seven selection criteria to help you select the best areas for life cycle mechanization that fits your immediate needs. Why is it not enough to speed up the individual process steps?

9.2. What are the key elements to life-cycle mechanization (LCM)? What are the three components to LCM? What does the CAE acronym stand for?

9.3. Why is it not practical to develop a truly automated intelligent information system (IIS)? What are the two components of IIS? What are some major benefits of LCM?

9.4. What is the meaning of a life-cycle mechanization? What does each LCM component contribute towards time-to-market?

9.5. How do you recognize what tools can do for you, and what they cannot? How are some work groups benefiting from a newer generation of 3-D CAD tools? What is their impact on the time-to-market aspect?

9.6. What array of tools and technologies are required for a mechanized IPD? Describe tools and technology that are parts of a fifth category set (see Figure 9.3). What are those tools in this set? Do the tools create a set of upper limits on which goals could be achieved? What are the factors that affect work groups in reaching this potential?

9.7. What is the first step to providing a mechanized environment? What are the activities this LCM environment should provide to the work groups? What are the advantages of considering human factors engineering in the design process?

9.8. What are the most popular types of processors found in an organization? What are some refined sets of analysis and design functions that can be generated based on these processors?

9.9. What is an extent to which an automation can be pursued? What are some tool types that represent an integrated CE environment? What is a critic? What do you understand by "plug-in"?

9.10. What are some key purposes of creating critics? Give an example of a critic? When do you call a design agent a critic?

9.11. How do you achieve a soft prototyping capability? Why is a CAD-based parametric model not enough by itself to serve as a critic? What are the ten specific actions you can take to eliminate many of the non-creative, routine, labor-intensive, and time consuming tasks?

9.12. How can the decision process for developing a product be computerized? What are the goals of a generative or a variant design? What does it take to change a company that has been manually creating designs to one that is based on a generative CE system?

9.13. What is a CE network of generative modules? What are some twelve modules that form the infrastructure for an LCM process? How many of them belong to CIM, automation, and CE areas? How do you recognize which areas must take precedence in LCM?

9.14. Why is the effectiveness of a whole CE network far greater than sum of its parts? In what form are standard procedures defined?

9.15. Do we need to integrate vendor supplied parts with LCM tools within an enterprise team? If yes, should the assistance be available on-line?

9.16. Are the LCM tools stand-alone tools or a set of cooperative tools? Are the LCM tools centralized? If not, why? Do they serve to a single-discipline team? Why should we integrate tools for each individual team member within a mixed-discipline team?

9.17. What are those sets of requirements, specifications, and standards that need to be integrated into a LCM process? How can a product be designed and developed over a network?

9.18. What are the three categories of modules for building a smart regenerative CE network? How is this transformed into a pull-type of computerized modules? What is meant by a *pull-type* here?

9.19. What are some automation modules that provide pre-processing functions? Can you describe a few that are relevant to you or to your organization in the work you are doing?

9.20. What are some pre-processors that provide design functions? Describe a few that you particularly have found useful for capturing geometry, design-intent, life-cycle intent and for capturing non-geometric properties?

9.21. Summarize a list of automation modules that provide some core functions? What are the three modes in which a modern CAD/CAM system can be employed to create design models? What is a mechanical design model? How does it differ from a geometry model?

9.22. What is the significance of a library of parts? What does a part library contain? What are some techniques used to construct a library of parts?

9.23. What is a synthesis model? What are some different types of synthesis models a product can have? What does a synthesis model contain?

9.24. What is a decision support tool or a method? What are its functions? What are some different types of decision support tools that can assess existence of optimum designs?

9.25. What is a knowledge-based product model? How do you emulate team cognitive functions? When does a knowledge-based model require optimization? What does knowledge-based optimization involve?

9.26. What are the automation tools that provide post-processing functions? Why do work groups need post-processing functions? Give some examples of this class of modules? How do you manage conflicts? What techniques would you use to improve communications and reduce conflicts?

9.27. How do you create the savings from using some advanced concurrent engineering concepts? Describe three such CE concepts that are based on progressively increasing the degree of concurrency. What are the differences between a linear type CE, a planar type CE and a cylindrical type CE concept?

9.28. What is a unified PPO concept? What does a PPO contain? What are the various applicatoins of PPO? Why is PPO considered the most mechanized class of CE concepts?

CHAPTER **10**

IPD DEPLOYMENT METHODOLOGY

10.0 INTRODUCTION

Many work groups understand the concepts and philosophies of integrated product development (IPD), but only a few understand how to undertake an IPD implementation program based on those concepts. Contrary to what anyone may claim, there exists no *cookbook* solution for IPD. Each implementation has to be tailored to a specific environment, company interests, and customer requirements at hand. What exists today in literature is mostly fragmented guidelines built from rudimentary design practices. IPD deployment methodology, the name given to this implementation guideline, is developed around some of the fundamental CE characteristics introduced in the earlier volume and chapters. The notables ones are process re-engineering, system engineering, product realization taxonomy, integrated product development, concurrent function deployment, and total value management. The deployment methodology is presented here in a generic outline form. The implementation details will vary depending upon the type of industries, applications in place, and the environment in question.

10.1 STRATEGIC CE IDEALS

A common implementation mistake committed by a work group is to confuse CE programs with CE Ideals. The concurrent engineering movement is not just a *bunch* of concurrent programs. It is the realization that certain fundamental ideals can have a profound impact on the long-term success of a business [Wesner, Hiatt and Trimble, 1994]. CE programs are the vehicles for implementing these ideals in an organization. They are contained in Table 10.1

TABLE 10.1 Strategic CE Ideals, enablers, and Example CE programs

Influencing Agents	Strategic CE Ideals	CE Enablers	Example CE Programs	Company Pursuing it (Notable Examples)
Talent	Deep Common Understanding	Specialist Experts	Personnel Team Manage Disagreement	Sony
	Motivation and Rewards	Generalist	Education and Training	Motorola
	Technical/Functional Expertise	Consultant	Develop Oneself	Anderson Consulting
	Demonstrate Adaptability		Teamwork	IBM Services
	Act with Integrity			
Tasks	Constancy of Purpose	The Whole System	Deming's 14 Points Policy Deployment	Florida Power & Light
	Structure Definition		Strategic Quality Planning	GM
	Whole-of-the-parts		Integrated System Planning and Scheduling	Hughes
	Parts-of-the-whole		Clear Vision & Goals	
	Class Hierarchy	Work Breakdown Structures (WBS)	Employee Suggestion Programs	Northrop
	Decomposition Hierarchy			
	Integration Hierarchy			
	Cost/Profitability	Capturing Life-cycle Intent	Cost/Profitability Management	Lucky Goldstar
	Functionality		Ease of Use	Hyundai
	Three Ps (Policy, Practices, and Procedures)	Integrated Product Development	Modeling Methodology Transformation Model Integrated Product and Process Design	Boeing Hughes
		Methodology		GE Aircraft Engine
	Taxonomy	Mapping (Orthogonal, the Whole set)	Product-oriented Cycles	
	Promote Innovation	CE Transformation system	Process-oriented Cycles	
		Track & Loop Methodology	Establish Plans	
	Forms of Representation	Information Modeling	CE Process Invariants	McDonnell Douglas
		Physical Model	Specification Models Product Models	
		Conceptual Model	Process Models	
		Analytical Model	Enterprise Models	
			Cognitive Models	
Teams	Total Employee Involvement	Multi-disciplinary Teams Inter-disciplinary Teams Cross-disciplinary Teams	Quality Control Circles	General Motors Ford Motor
	Decision as Low in the Rank	Empowerment	Quality Improvement Team Suggestion Plan	Toyota

Seven Cs (Collaborations, Commitments, Communications, Compromise, Consensus, Continuous Improvements, and Coordination)	Cooperative Work groups Matrix of Teams Foster Open Communications Listen to Others	Process Development Team Reporting Structure Responsibility Management Flattening of Manufacturing Operations	Pratt & Whitney Peugeot
Compositions Team Interactions Value Diversity	CE Teams Personnel Team Technology Team Virtual Team Logical Team	Strategic Business Units Product Development Team (PDT) Project Management	EDS IBM
Performance Collocation	Skill Management	Tiger Team or Self-directed Team Workplace Organization and Visual Control Employee Rotation Program	BMW Porsche VW
Focus on Customer Satisfaction Focus on Customer Needs	Deployment Methodology	Customer Satisfaction Program Customer Action Team	Sony Honda Toyota
Management Style Strategic Partnership	Management Participation and Commitment Management Structure Supply Chain Management	Customer or Supplier Model Partnership Model Strategic Sourcing Supplier Rationalization	Toyota GM Chrysler
Economy of Teamwork and Cooperation Convergence and Collaborative Thinking	Forms of Sharing and Collaboration Team Design	Goal-oriented Project Management Coach & Develop Constancy-of-purpose Management Build Relationships	Federal Express Northrop Sony
Concurrency and Simultaneity Modes of Concurrency	Product Decomposition Product Aggregation Concurrent Resource Scheduling Concurrent Processing Minimal Interfaces Transparent Communications Quick Processing	Fast-to-Market Minimize Product Interfaces Minimize Process Interfaces Minimize Computer Interfaces	Toyota Nissan Honda Motorola
Key Dimensions of a CE Specification Set	Overlapped Pull System Linked System	Product Hierarchy Description (PhD)	HP
Systems Thinking Strategic Thinking System Integration	Systems Engineering Recognize Global Implications	Product Systematization Product Assembly	Dell Compaq

Techniques

449

TABLE 10.1 (continued)

Influencing Agents	Strategic CE Ideals	CE Enablers	Example CE Programs	Company Pursuing it (Notable Examples)
Techniques (cont.)	Focus on Process, Taxonomy, 4Ms, 7Ts, 8Ws, 3Ps, and so on	Product Realization Taxonomy; Structure Descriptions; Concept Descriptions; Frame Descriptions	Reusable Digital Product or Process Models; Integrate Analysis or Simulation Tools with Digital Product Models	Sony; GE; P&G; Nike
	Managing Interactions; Modes of Cooperation	Implementation Guidelines	Quality Function Deployment (QFD)	Apple; Allied Signal
	6Ms; Promote Corporate Citizenship; Value-centered	Total Value Management	TQM; Quality Assurance (QA)	HP; EDS; AT&T
	Focus on Prevention	Design for Quality; Analyze Issues	Six Sigma; Zero Defects	Motorola; Crossby
	Measures of Merits (MOMs); Efficiency; Effectiveness	CE Metrics & Measures	Design for X-ability; Work Efficiently; Quality of Work versus the Quantity	BMW; Mercedes
	Economy of Change Management	Process Re-engineering; Champion Change; Manage Execution	Process Quality Management & Improvement; Benchmarking; Performance Assessment. As-is, To-be Process	AT&T; VW; GE Aerospace; Xerox
	Economy of flexibility or ease	Virtual Prototyping	Agile Organization; Global Manufacturing	General Electric; General Motors
	Decision Making Style; Sound Judgment	Knowledge-based Engineering	Value System; Virtual Organization	EDS ACE; GE Aircraft Engine
Technology	Standards	Mfg. Competitiveness	ISO 9000, ISO QC9000 Certification Programs (Q1)	ISO, FORD, NIST, and so on
	Distributed Computing	Frameworks & Architectures	Communication Networks	EDS; IBM
	Networking	Computers and Networks	Client/Server Distributed System	DEC; HP/Sun
	Inter-operability Communications; Reusability	Decision Support Systems; Common Database Management	Distributed Computing; All-to-one	IBM; AT&T

	Reconfigurability Economy of Seamless Flow of Information Scaleable Size and Structures	Common Process Template Common Representation Forms	One-to-all Virtual Manufacturing Agile Manufacturing	DEC General Motors Gateway Boeing Motorola/ Intel
Time	Promote Simultaneity Integration	Concurrent Function Deployment (CFD) Seamless Integration	QFD, JIT, Synchronous Manufacturing Logistics Integration Horizontal Integration Vertical Integration	Cap Gemini Sogeti Proctor & Gamble Boeing Unilever Douglas Aircraft MD-12
	Time-to-Market Responsiveness	Life-cycle Management	Paralleling of Responsibility	EDS Intel
	Common Model	Product Breakdown Structures (PtBS)	Paralleling of Tasks	Ford Motor Dell
	Common Process	Process Breakdown Structures (PsBS)	Paralleling of Processes	IBM, Honda, Ford General Motors
	Drive Automation Efficiency Effectiveness	Life-cycle Mechanization Effectiveness of Collaboration	Knowledge-based Engineering	Concentra Inc. Intellicorp Boeing
Tools	Commit to Quality Measures of Merits (MOMs)	CE Metrics and Measures	Built-in Quality by Design Statistical Process Control (SPC)	AT&T Anderson Consulting A.T. Kearney
	Types of Knowledge Capture Knowledge	Intelligent Information Systems	Parametric Models Feature-based Models Knowledge-based (Smart) Models	Computer Vision Autodesk Parametric Tech. GM/Delphi
	Productivity Capture Geometry	CAD/CAM	Version Control Change Management	EDS Unigraphics CATIA
	Computer Integration Interface Compatibility Plug-in Compatibility	CIM MRP CII CAPP	X-ability Aspects Manufacturing Execution Planning	Computer Science Corporation General Dynamics Maxtor
	Analysis or Simulation Capture Life-cycle Values	CAE/CAX/KBE/AI, Fuzzy Logic	Design Optimization Sensitivity Analysis	Nastran/ Patran Ansys/ NISA Pratt & Whitney

10.2   TEN COMMANDMENTS OF IPD DEPLOYMENT

The purpose of this section is to offer a set of implementation guidelines for product design, redesign, development, and delivery process through its life-cycle functions. IPD deployment is a multi-plan methodology as shown in Figure 10.1. It consists of a number of activity-plans arranged in increasing order of enrichment. The activity-plans overlap, but provide a structured approach to organizing product ideas and measures for concurrently performing the associated tasks. Concurrency is built in a number of ways (similar to what was discussed in section 4.3, volume I), depending upon the complexity of the process or the system involved. This *ten commandments,* the name given to this deployment methodology, serves to guide the product and process iterative aspects of IPD rather

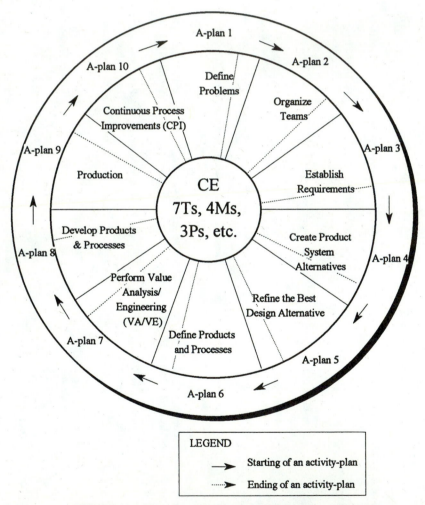

FIGURE 10.1   Ten Commandments of IPD Implementation

than just the work group collaborative aspects of a PD^3 cycle. Examples of collaborative aspects include facility setup, teamwork, organization, management style, education and training, and so on. The CE ideals (7Ts, 4Ms, 3Ps, and so on), in the center of the wheel, ensures that both local, zonal, or global iterative refinements and collaborative refinements take place during this operation. The ten commandments, described in Figure 10.1, are discussed in the following subsections.

10.2.1   Activity-plan 1—Define Problems

In this activity-plan, information related to the product or service is collected. A number of work groups involved in the process (customers, manufacturing personnel, marketing, sales, and product support services, etc.) are interviewed to document the as-is process being followed on the job. The as-is situation may be present in a variety of forms: drawing specifications, part design, analysis, standards, procedures, statements of work, inputs and outputs, required resources, timing, product breakdowns, and so on. The gathering of such information aids in the following aspects.

- Understanding market needs and opportunities.
- Defining objectives, milestones, measurement criteria, and success factors.
- Receiving specific customer needs and expectations.
- Obtaining marketing and technology plans.
- Developing project plans.

If an "as-is" process was already documented in the previous cycle, this activity plan serves as a reminder to update those tasks that have been moved or changed.

10.2.2   Activity-plan 2—Organize Work Groups in a PDT

The second order of business is to establish a cross-functional product development team (PDT) involving all of the product development units that in some way contribute to the PD^3 process. PDT would be the party responsible for establishing a general plan of action. At a minimum, the design and development work groups, manufacturing work groups, and engineering work groups constitute a core of the concurrent PDT. Following the principles of goal-oriented project management, the PDT establishes milestones and constancy of purpose. PDT creates a detailed timeline showing the checkpoints in time when the design would be frozen, approved, released, reviewed, and so on. PDT works backwards from these checkpoints to identify the following:

- The start of the 1-T, 2-T or 3-T concurrent loops.
- The needed expertise of the work groups in PDT.
- A reasonable work schedule (as shown in Figure 9.1, of volume I).

- A set of common check points (such as design freeze point, process freeze point, production release point, and so on) along the planned deployment path.
- Other tasks required for this activity-plan include:
  - Review resource requirements and identify leveraging opportunities.
  - Select and prioritize key projects.
  - Develop a business case and refine project plan, if needed.
  - Designate system manager and form core work groups (CE teams).
  - Identify support personnel.
  - Acquire and allocate resources.
  - Identify a champion: It is not worth investing a lot of time trying to persuade non-believers and critics. It is wise to identify one or two key managers who believe in CE and see the merits of the approach. A successful implementation of a CE project goes a long way toward convincing skeptics and fence sitters. Results tend to speak for themselves.
  - Involve suppliers: With the emergence of value managed supplier relationships and long-term sourcing strategies, it has become essential that the original equipment manufacturer (OEM) and key suppliers work together as a team.
  - Confirm or redirect business strategies.
  - Compute return on investments (ROI). This is a time consuming task, subjective but often needed for project justification. Early adopters of CE seldom relied on detailed ROI analyses to justify the funding. Instead, they focused on cost-avoidance. Continuing to develop products the old way often incurs significant loss in time and cost. Return from investments *"from going out of business"* is zero.

10.2.3   Activity-plan 3—Establish Requirements

Requirements are gathered by a team of multi-disciplinary experts with an overall product development background, a grasp of existing product history, and a knowledge of manufacturing or production methods. Some examples of key requirements (see Figure 2.22 of volume I) are discussed in this section.

Review warranty, field, cost, and capacity data: The team looks at the different company products and their field performance data: failure rates, part recalls, manufacturing defects, maintenance schedules and cost, service history, and so on.

Identify feasible product and process technologies: Identification of best-in-class outputs (products or services) is an ongoing activity for any organization. This may take the form of customer surveys, customer clinics, market research, or other types of trials and evaluations. The steps involved in this activity are:

1. *Background research or investigation:* During this step, the team aims at enumerating the world's best designs from the standpoint of meeting customer needs and expectations. Background research is performed to evaluate current and possible usage of emerging technologies. Investigation is conducted to determine product

feasibility, and to identify and evaluate marketing, tooling, and manufacturing requirements.

2. *Competitive analysis:* PDT looks for fundamental differences in the approach, procedure, and method used internally versus how competitors do. This leads to design improvements that may make the product competitive, more robust, and durable with respect to its overall performance. Customer rating information may be used for this purpose. Such ratings, relative to certain broad characteristics, are available through syndicated research, on-going internal evaluation, and performance analysis.

3. *Competitive tear-down activities:* The team may look into the competitive tear-down information to get an idea of alternative concepts that the competitors might be using. Technical teams often find it interesting to participate in this activity. This could lead to teams' identifying alternative designs that are technically feasible and provide good customer perceived performance.

4. *Benchmark for function, features, cost, weight, and investment:* One of the best ways to convince management of CE is to benchmark a respected competitor's CE process. Benchmark has been found quite useful in studying product functionality, comparing product values, determining material characteristics, and identifying effective manufacturing processes. The activity for competitive benchmark usually starts at the beginning of the concept definition and may continue until the product is engineered. Benchmark is often done in conjunction with competitive tear-down activities. It serves the following purposes:
   - Identify the competitions and their competitive products.
   - Identify internal and corporate expertise.
   - Investigate or analyze product features and technologies.
   - Compare a company product feature with an equivalent feature in a competitor product.
   - Level-set (expand and enhance) team knowledge and an understanding of the product.

   Benchmark studies done by an outside company carry more weight than studies that are conducted internally.

Establish technical performance requirements: Identify the necessary product design life targets, duty cycles, and performance requirements to achieve a set of necessary confidence factors in establishing RMS (reliability, maintainability, and supportability) value sets.

Establish the customer satisfaction requirements: These requirements can come from a variety of sources: engineering or management experience of the teams, technical standards, marketing survey, competitive analysis, field data, VOC, and so on (see Figure 2.22, volume I). Product planning activity flow-charting (see Figure 10.2) is a systematic way of organizing the needs and requirements, and assessing the customer satisfaction levels. Details are discussed in Chapter 2, volume I. During this flow-charting step, the QFD concept is used to identify value characteristics (VCs) necessary to create breakthroughs in customer satisfaction or customer excitement. QFD nomenclatures provide

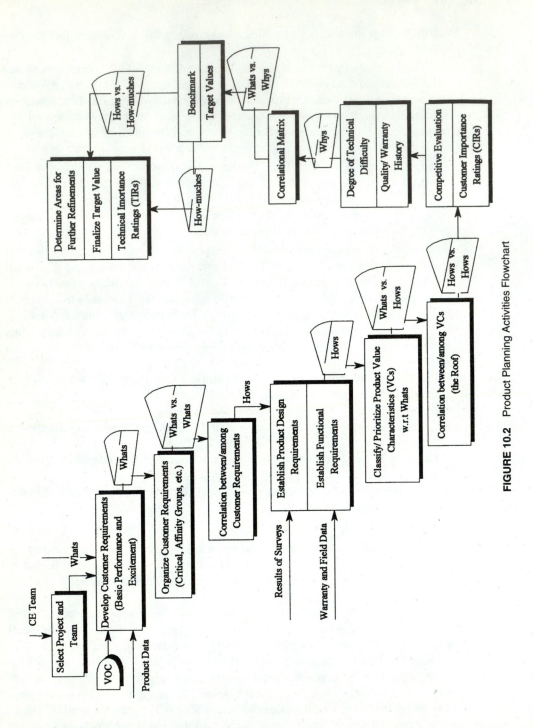

FIGURE 10.2 Product Planning Activities Flowchart

the foundation for prioritizing VCs according to the desired product functions from customer views and determining the target performance levels of the product compared to the competition.

Create consolidated requirements: Merge customer-driven requirements with technical requirements to make up a consolidated list of requirements. The consolidation list of requirements ought to satisfy the expectation of the technical work group community with regard to their feasibility. That is, the consolidated requirements must be in line with the permissible factors such as functionality, reliability, weight, complexity, production process capability, and so on. Consolidated requirements ought to be driven by reality checks, the constraints of the business plan, and must satisfy other product acceptance standards. Examples of the business-plan include existing lines of products, warranty expense and servicing, the life-cycle costs, manufacturing investment constraints, program timing, and so on.

Establish product verification plan: A product verification plan consists of the following four parts.

1. *Evaluate the consolidated requirements against a marketing plan:* It serves no purpose, if the marketing plan is in disagreement with the chosen consolidated requirements.

2. *Establish a list of formal testing and acceptance criteria:* Testing and acceptance criteria are required for checking out an existing design or procedure, and measuring quality levels of a production process.

3. *Establish an accompanying validation procedure:* The validation procedure is a three-step schema:
   - Design verification of parts, components, sub-systems, and finally the product system itself.
   - Production validation of sample parts, components, sub-systems on a pilot basis.
   - A validation procedure for assembling a set of purchased parts and that of the supplier involvement.

4. *Get Customer acceptance of consolidated requirements and constraints (CRs): This consists of the following five activities*
   - Evaluate CRs against a technology plan.
   - Evaluate CRs against a product portfolio or plan.
   - Evaluate CRs against a set of manufacturing capabilities.
   - Evaluate CRs against core competency.
   - Develop execution plan for a functional organization developing the product.

Determine performance gaps: With the benchmark and tear-down information in the work group's hand, it is easier to identify major performance gaps. This is identified by comparing the differences between the current capabilities and the consolidated requirements.

Validate concept requirements: At this activity-plan stage, the current methods of manufacture, advantages and disadvantages of manufacturing processes in-use, and other product design requirements are reviewed. This would include quality requirements as the type of customer-driven design requirements (derived from QFD Phase-1 (product planning) as shown in Figures 2.23 and 2.24 of volume I). Examples of quality requirements include attributes of functional performance (basic function and performance levels), reliability (design life, duty cycle, field testing), manufacturing variation, critical process (PsC*), key process parameters (PsCs), and may even include styling or esthetic features. Besides, there may be other technical or business driven quality requirements, designated by management or CE work groups, that are specific to a company or to a particular business situation.

Establish value characteristics: This consists of broad system level functional requirements to allow constituent-by-constituent benchmarking. Examples of value characteristics (VCs) include: quality (functionality), X-ability (performance), tools and technology, costs, responsiveness, infrastructure, and so on. Quality or functionality (a VC) is often decomposed and expressed in terms of its hierarchical breakdown structures. A breakdown structure for a Quality VC is sub-system, component, part function, materials and common representation, and standards (see Figures 6.3 and 6.4 of volume I and Figure 10.3 for details). The corresponding actions PDT can take are

- Identify systems needs.
- Identify sub-systems needs.
- Identify components needs.
- Identify parts needs.
- Identify materials, attributes, features, or parameters needs.
- Identify common needs (such as standards, process, 4Ms, 7Cs, 3Ps, etc.).

The purpose of the above needs identification is to affect positive system specifications in order to provide a greater likelihood of a product design or redesign success. There are similar needs associated with other VCs. The examples of X-ability needs include simplicity, producibility, reliability, assemblability, and recyclability (see also Table 5.1, volume I). A list of other attributes associated with the remainder of the VCs, tools and technology, costs, responsiveness and infrastructure, is contained in Figure 10.3. The concept is known as CFD—Concurrent Function Deployment. Chapter 1 of volume II contains details on this concept. The following steps are involved in establishing value characteristics.

> ***Generate CFD Product Planning Matrix.*** Here, a house of values, which consists of a series of interconnected matrices, is constructed (see Figures 1.9 through 1.12 of volume II). The first of this series is the product planning matrix. It is obtained through a process of translating consolidated requirements (CRs) into a set of technical *HOWs,* establishing the relationship between the two, identifying technical benchmarks, and setting target values. QFD, a degenerate case of CFD, is used for building a house of quality (see Figure 2.24, Volume I).
>
> ***Determination of VCs Relative Importance***: This step constitutes the roof of a CFD matrix. Normally, each VC from its HOW set will not hold equal importance.

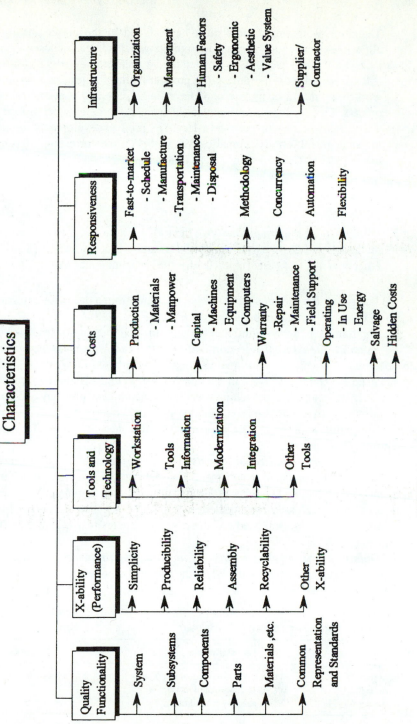

FIGURE 10.3 Taxonomy based on Value Characteristics (VCs)

459

Therefore, weighting factors must be included in the preference formulation to allow CE work groups allocate appropriate weights to each chosen value character-istic (VC). Such weights may need to be changed as new measures of merits (MOMs) are discovered during the product realization process. For instance, a CE work group may need to drop or add a VC at a certain point of realization in the PD^3 process. This leads to a realignment of weights in the new prioritized set of VCs. The weights (w_i/w_j) may also be affected by product development timing or market-ing strategies. If one is comparing VC_i with VC_j, the corresponding sensitivity val-ues, VC_{ij} and VC_{ji} are determined as follows.

1. $VC_{ij} = w_i/w_j = 1.$ (10.1)
2. $VC_{ji} = VC_{ij}$
3. If VC_i is more important than VC_j, then VC_{ij} can be assigned a number based on a methodology known as the Analytic Hierarchy Process (AHP) [Saaty, 1978]. An-other approach is to use Table 10.2 for the computation of VC_{ij}.

In the latter case, the roof of a CFD matrix, resulting from a pair wise comparison of a set of VCs, would be similar to that shown in Figure 10.4.

The following are the additional steps for the completion of activity-plan 3.

Establish the concept selection matrix (Plus and Minus Assessment): The task is to construct a concept selection matrix [Pugh, 1991] based on a set of product require-ments. The method is discussed in Section 2.8.3 of Volume I and the steps are shown in Figure 2.25 of volume I. If the requirement list is incomplete, one may end up approving a sub-optimum design. The common situation in Pugh's matrix is when one design appears better for quality, another for cost, another for weight, and yet another (say the carry-over design) for investment.

Establish functional requirements: Here, the functional requirements for the de-sign are established in quantitative terms. They are expressed in measurable units of prod-uct functions, simulating in-use conditions and degree of compliance in meeting an ex-pected performance under a controlled set of conditions. These functional requirements must agree with the consolidated (customer) requirements that were originally defined in earlier steps.

Obtain management approval of functional requirements and plan: The evalua-tion of the next activity-plan starts when the management has approved the functional re-quirements and plan.

TABLE 10.2   Computation of Sensitivity Values

Value of VC_{ij}	Definition	Remarks
$w_i/w_j = 1$	VC_i and VC_j are of equal importance	
$w_i/w_j = 3$	VC_i is weakly related to VC_j	
$w_i/w_j = 9$	VC_i is strongly related to VC_j	

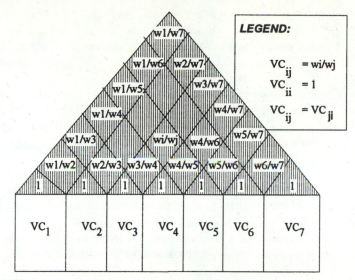

FIGURE 10.4   Determination of VCs Relative Importance (Roof of the CFD Matrix)

10.2.4   Activity-plan 4—Create Product System Alternatives

This Activity-plan focuses on tasks that create a number of product system alternatives. Initially, product ideas may be developed utilizing brainstorming techniques to bring out the experience of the PDT members. The work groups, in most cases, first use Pugh's concept matrix to generate a set of alternatives, then they attempt to do one of the following:

1. Find ways to eliminate the negative points of the strongest design concept, or
2. Find ways to generate a new design idea that is inherently superior to any of the previous designs.

The major steps involved in this activity-plan are described in this section.

Identify systems and components sourcing requirements: Experience has shown that this is usually a difficult step to complete at this point in the PD^3 process since the design alternatives are not yet fully developed and all of the target values are not yet fully known. The task is included at this early stage to challenge the CE work groups to consider sub-systems, components, parts, and material sourcing requirements as a part of the process for evaluating the so-called big picture. The list of other product design requirements will typically include design cost, weight, investment, and timing. It may also include requirements for technology, staff utilization, or budget. A detailed list of RCs is contained in Figure 8.14 of volume I.

Identify candidate sub-systems: Having completed activity-plans 1 through 3, there should be sufficient information to identify a candidate sub-system for further evaluation.

The team's judgment should be reflected in the Pugh's plus or minus assessment (Figure 2.25, volume I), based on the benchmarking activity discussed in activity-plan 3.

Develop preliminary sub-systems specifications: Preliminary sub-system specifications include fits and appearance, functional specs, design for X-ability specs, government specifications, and so on.

Create concept or system alternatives: With the functional requirements established in activity-plan 3, the work groups should be able to study each of the pertinent concepts or system alternatives in terms of their ability to deliver these functional requirements.

Evaluate and analyze concept or system alternatives: Following the construction of a Pugh matrix, the work groups should be able to evaluate concept alternatives and assign plus/minuses depending upon which alternatives have advantages and in what characteristic areas. Certain characteristic areas may not be a candidate for trade-off, while others are. It is often helpful to double check the benefits by looking at a concept alternative with only a few key design requirements in mind. Management should make sure that the work groups fully value the essential considerations in their analysis.

Select the best alternative construction: Two situations are likely to occur.

1. One design is superior in every respect. In this case the choice is simple, select that design.
2. No single design is superior relative to every functional requirement. This situation leads to several alternate concepts. The choice involves determining the following additional steps.
   - Create preliminary sub-system design.
   - Develop packaging or economy criteria.
   - Perform costs versus benefits analysis: Pugh's analysis tends to favor one concept alternative if that alternative has many pluses. However, the work group should perform a cost benefit analysis to be sure of this outcome.
   - Process re-engineering: If the team is unable to develop a single solution, the alternative is to carry out a process re-engineering analysis. Each concept has its own strength and weakness. The needs are evaluated by a system level work group to determine what constitutes the best system concept. If, on the basis of this evaluation, a planned production process is not capable, there are three choices.
     1. *Select the one found the best:* If, one alternative design is clearly superior relative to all of the product and process requirements, concept selection is easy, select this alternative.
     2. *Select an alternative process concept*: If the components work within a system with other components but may not satisfy all requirements, an alternative process concept is selected. The work groups may negotiate the lowest cost and weight system concept with associated project teams.
     3. *Modify the one in consideration:* The work groups conduct design of experiments (parameter design or tolerance design) to reduce the inherent variability of the design by making product more robust for the possible causes of variation [Dika and Begley, 1991].

Generate preliminary parts list: The next step is to compile good solid factual data to support the plus or minus assessments for those design requirements that are most important to the work groups. This is done utilizing an engineering bill-of-materials at the sub-system level. The following are the two additional steps needed.

1. *Estimate weight and cost*: The simplest way to get weight and cost information is through product literature, computing solid volumes, or by weighing current or competitive parts and making adjustments for differences. Weight can also be estimated from a CAD or part drawing details based on the dimensions of the part and the density of the materials.

2. *Estimate design costs:* There are many factors influencing the price of a purchased component that are beyond the control of a design work group. It is important, therefore, to remember that a comparison of inherent design costs is also associated with the complexity of the part, not its price alone. Eventual process costs (manufacturing and assembly) can be larger or smaller depending upon the parts' processing complexity added at a later stage.

Model a preliminary system concept: The information methodology for a product model-class (discussed in Sections 7.7 of volume I) and for a process model-class (discussed in Section 7.8 of volume I) can be used to generically capture a preliminary system concept for further study.

Establish process capability: Process capability identifies how good a process is, in terms of SPC, based on the manufacturing floor's machine capabilities. This includes identifying critical processes or capabilities (PsCs*) and key process characteristics (PsCs). PsCs* is the measure of how well a machining is on target when statistical deviation (from a nominal machining value) of a number of sample parts is measured and plotted on a Six Sigma chart.

Perform System Analysis. A series of system analyses (such as FMEA, DFMA, etc.) is performed to challenge the design relative to its simplicity and ease of manufacture. This is discussed in section 2.33, volume II. The following items outline the major attributes of system analysis.

- Perform FMEA and complete related documentation.
- Identify process requirements.
- Develop material or feature specifications.
- Perform sensitivity and trade-off analysis.
- Perform an alternative manufacturing process analysis.
- Create a list of potential manufacturing facilities.
- Develop a product validation plan.
- Develop a risk management plan.
- Test or validate concept alternatives. DFMA is ordinarily used to reduce the inherent time and effort necessary to make the components and assemble them without

affecting the functionality of the part. During concept development, DFMA is applied very broadly, like FMEA, as a means to challenge the design alternatives.

At this point, the conclusions of the PDT sub-teams are not usually supported by much analytical or experimental data. Certainly, the team would not feel confident in making a commitment to specific design alternative based on the aforementioned approach. Activity-plans 5 through 9 help eliminate that uncertainty.

10.2.5   Activity-plan 5—Refine the Best Design Alternative

In this activity-plan, the team verifies the feasibility of the best design alternative from a set of technical and economical standpoints. As Professor Don Causing of MIT says, we would like "to identify a design that is not immediately vulnerable to being overturned by a superior design" [Clausing and Pugh, 1991]. The team addresses the dependency of design on scarce materials, technological risks, costs of production, and so on. The PDT refines specifications to eliminate wasted staff hours designing a *perfect part* at the expense of excessive product cost.

Draw a process flow chart: The methodology discussed in section 3.6, volume I can be followed to obtain the as-is and to-be process flow charts.

Finalize sub-system specifications: The next step is to finalize the sub-system specifications. This assures that the specifications are complementary and decomposable.

Conduct a sub-system FMEA: FMEA, in this case, is used mainly to identify what could go wrong with the design and process as a system, and to establish corrective steps while the design is still in an early stage. This may involve identifying the expected product reliability levels. These reliability numbers should be added to the bill-of-parts. This was discussed in section 2.5.6, volume II.

Conduct a sub-system DFMA: DFMA is a tool to solve many of the problems discovered during FMEA. DFMA can be used to improve the best design alternative by examining how it will be manufactured. This is where the members from the manufacturing work groups have a direct opportunity to influence the design. The intent is to reduce manufacturing variation, improve reliability, reduce the number of parts (complexity), improve the ease of assembly, and consider the effects of automation.

Conduct CFD process planning analysis: CFD process planning analysis consists of determining a process planning matrix. This matrix is used in identifying the next set of parameters—key process characteristics (PsCs). This is shown in Figure 1.12, volume II.

Build alternative concept replica (software prototype): It may be better to build an electronic replica of an alternative concept through a computer modeling technique such as a software prototype than to check out its functions through some costly hardware experimentation. It then becomes easier to perform tolerance design, parameter design, serviceability analysis, engineering analysis, cost analysis, and so on

Conduct a program team review: Although the collective judgment of a work group is more likely to achieve a better result than an individual, there are still many uncertainties. Management may be reluctant to approve. A program team review is meant to set technical and business direction and to overcome any possible management barriers.

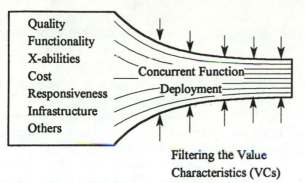

Filtering the Value
Characteristics (VCs)

FIGURE 10.5   Concurrent Function
Deployment Filtering

Concurrent function deployment filtering: Many of the tools used globally during a concept development stage will be required again at the component or part levels. The common purpose of CFD is to filter the value characteristics—improve quality, functionality, design for X-ability, component cost, weight, investment, responsiveness, infrastructure, manufacturing variation, reliability, and so on (Figure 10.5).

Concept alternative selection: The goal during the rest of this activity-plan is to select an alternative concept that not only meets its customer's objectives, but exceeds them.

10.2.6   Activity-plan 6—Define Elements of Products and Processes

It is impossible to be certain that the selected alternative design concept is sound without knowing the capability of the manufacturing method to process the design with low variation.

Finalize system specifications: The first task of this activity-plan is to designate fabrication assembly and component plants and perform a complete preliminary design evaluation. This may involve creating design for demonstration and procuring materials and purchased parts. This consists of the following additional steps.

- *Validate a master concept parts list:* Since a product is a hierarchy, the *bill-of-materials* will follow a hierarchical pattern. The initial *bill-of-materials* will be at the sub-system, component, or assembly level, and the rest will be at a lower level of detail appropriate for the hierarchy, that is at a part level or at a material and feature level. They must relate to a universal *bill-of-materials* system that was approved at the program direction level.

- *Develop a parts list:* A part level bill-of-materials is developed at this step. The engineering bill-of-materials will continue to become more definite until data is confirmed, utilizing production pieces.

- *Develop detailed plans for manufacturing and assembly operations:* Detail will be added continuously until the production stage is complete to the nuts-and-bolts level. This step of completion usually occurs either at the final design release point or at the beginning of the production control work.

- *Confirm resources (7Ts), investments, and schedules:* The 7Ts, investments and schedules are recommitted and rechecked to make sure they fall within the stated budget and finish guidelines.

Conduct a system FMEA: FMEA is used again at the part level to further refine the design and make it more forgiving and reliable in production.

Conduct a system DFMA: DFMA will be used again at the components or parts level to further refine the design, reduce the essential processing cost, and improve quality. During DFMA, the PDT members begin to identify the key process characteristics (PsCs) necessary to achieve the product characteristics (PtCs) identified during the activity-plan 5.

Conduct CFD production planning III analyses: The key production (PnCs) and key process parameters (PsCs) are documented in a CFD Track III for use during production preparation, in developing operator instructions, and shop floor control plans. During this step, production capability (critical processes, PsCs*, and key production control characteristic values, PnCs) are finalized. The work group establishes production quality capabilities for each of the key process characteristics (PsCs) identified in CFD Track II. The work group benchmarks them against industry-wide manufacturing methods. This is shown by a production planning matrix in Figure 1.13 of volume II.

Process Concept Selection Matrix: The work groups construct a process concept selection matrix, similar to the Pugh's alternative selection matrix, to assist in the selection of the best process (see Figure 2.25 of volume I). A work group also develops a process routing.

Finalize engineering and manufacturing criteria: FMEA and PDT design reviews are used to identify the few key process characteristics (PsCs) and target values that are critical to meeting the engineering and manufacturing requirements.

Refine the manufacturing process design proposal: Along with DFMA activity, CE work groups begin to refine the process design proposal. Refinement involves development of more definitive process variable cost data, a roll-up of system costs, and the identification of cost-saving opportunities to offset investments. This activity can be categorized into four definite ranks.

- Refine the process control plan.
- Refine the operation plan.
- Refine the shipping and packaging plan.
- Refine the tooling and equipment design plan.

Confirm manufacturing capability: The work group begins to see where new manufacturing investments may be needed to upgrade existing processes and where quality or cost is at a disadvantage compared to the benchmark processes.

Finalize source (supplier) selection: At this point, outside suppliers submit the bids on the job with a cost figure for a determined quantity and satisfying a production schedule. On the basis of the information supplied and the best match with the overall company goals, a supplier is selected. It is hoped that some degree of supplier participation on the project has been steadily maintained throughout the PD^3 process. Cost reduction efforts may still continue to find the best production alternative and reduce the final production price even further. In cases where the systems in question will be made in-house, this step is not necessary.

Develop a process validation plan: This consists of pursuing the following six subtasks.

- Manufacturing process design.
- Manufacturing specifications including process control plans.
- Manufacturing process validation specifications.
- Tooling and equipment designs.
- Updated FMEA documentation.
- Risk management.

Having done this activity-plan, the risks associated with selecting the best product, hopefully, are greatly reduced. At this point in the program, PDTs feel confident that product and process would meet all of the design objectives including quality, cost, weight, and investment simultaneously. A PDT refines the design alternatives to a bill-of-material level of detail and validates process capability data to support the feasibility of the design. PDTs must also feel confident that they are working with a concept that is the best in the class [Dika and Begley, 1991]. What qualifies a product to be the best in the class is accomplished by performing a series of value analysis (VA) or value engineering (VE) on the concept. This is discussed next.

10.2.7   Activity-plan 7—Perform Value Analysis or Engineering (VA or VE)

The steps involved in this activity-plan are similar to those described for process improvement methodology in Section 3.7 of volume I. The following are the major steps.

Establish a value for each function or service: The value engineering begins with looking at a *candidate* process model of an enterprise. It consists of performing a functional economic analysis in support of process redesign and new investment justification, if any. The major sub-steps follow.

- *List sources of variations:* The work group lists all sources of variation potentially affecting each operation.
- *Identify value-added activities:* Brainstorming and team dynamics are tools often used to identify value-added activities from non-value added activities in the candidate process.
- *Develop Fish-bone diagrams*: This sub-task is typical of most brainstorming sessions. It is equivalent to developing a Fish-bone diagram or a conceptual model for

each operation (see Figure 7.2, volume I for Fish-bone and other examples of conceptual models).

- *Select sources of long lead-time systems or components:* The next step is to identify systems or components that are likely to take a long lead time.
- *Create and refine*: Based upon work group consensus, the concepts are redefined, combined, or subdivided in order to provide a clear identification of value-added activities from non-value added activities. A graphical tool for placing values can be used to aid this process (see Figure 3.21 and Figure 7.11, volume 1).

Seek creative ideas, evaluate or investigate: Creative ideas and graphic pictures obtained during the previous steps are evaluated here. The work group thinks creatively, analyzes cost (such as assigning a dollar sign, $, on tolerances), applies judgment, and timing. The work group also looks at the outgoing characteristics (such as dimensions) affected or changed at each step in the process. A set of manufacturing specifications, including a process control plan, is initiated for each operation. Each outgoing dimension, or characteristic, and its associated description, is recorded. If *gages* are available for measuring process results, they are recorded on a control plan. Most often, performed activities imply a transformation process from inputs to outputs that adds value. Some activities are *means to get to the end,* but are generally not an essential part of the PD^3 process.

- *Evaluate alternatives:* Brainstorming and other tools are used to determine alternative methods of doing the same activities. The work group looks at one activity at a time and studies the measurement system variability. Participants are encouraged to suggest ideas no matter how difficult or costly they may be. The logic is that even inappropriate ideas may lead to good ideas.
- *Redesign an improved process:* The aim of this step is to come up with a redesigned process that has error-free designs and embeds performance monitoring that has provisions for self-correction. During this step, the work group overcomes roadblocks and constructs a characteristic matrix that displays relationships between product requirements and operations. The matrix helps to evaluate the importance of value-added characteristics while showing important upstream relationships. After considering this, an improved process is redesigned. The improved process employs standards and is based on identifying effective employee time, waste elimination or redundancies. Elements of the improved process contain a significant proportion of value-added activities in their flow (*perform* type activities). Output of one activity is normally an input for a subsequent activity in a pipeline mode. The preparation and testing time are kept to a minimum. However, meeting these thresholds consistently at a lower cost is often difficult. Complexity, process variation, and uncontrolled situations, such as dealing with consequences are some of the elements that create difficulties.
- *Define ideal process:* Here the goal is to provide services at the lowest possible cost without affecting performance or diminishing overall quality.
  - Think creatively: This can be a paradigmatic shift in terms of how the concurrent work groups do business. Though the goal is to find the best possible solution

within the limits of organizational and cost barriers, it is possible to cross these barriers, if a new concept provides extraordinary saving and adds values to the company.

* Eliminate waste: Eliminate all the waste inherent in a process. A list of eight wastes is described in section 3.1 (Figure 3.4) of volume I.
* Define an ideal situation: The team looks at all the possible alternatives and recommends an ideal situation that fits the system as a whole.
* Choose an optimal process: This is the last step of this activity-plan. An optimal process is obtained when an essential function is achieved for minimum cost or time. The team determines, from the previous steps, the cost and benefit analysis information along with what is possible to achieve in the short term and what is feasible in the long term. If the elements of the optimal process do not alter the current production plan, it is considered safe and risk-free. Others are looked upon more carefully in the next activity-plan.

10.2.8   Activity-plan 8—Develop Production-intent Products and Processes

The activity-plans described in section 10.2.7 involve a significant amount of process re-engineering effort. This effort should result in development, testing, and releasing of products and processes that are significantly smoother than before. Activity-plan 8 involves the following steps:

Finalize the best design or process objectives: Cost, weight, investment, reliability, and process capability needs are finalized in sufficient sub-component details. The work groups select the design concept that is right for production. Design of experiments is used to further optimize the manufacturing process.

Select manufacturing source: The final production manufacturing source is selected. The work groups identify tooling suppliers, confirm the component configuration, and confirm manufacturing schedules and investment allocations. Work groups build parts, test parts, perform prototype part verification, and identify additional product requirements.

Establish process capability for critical items: If a manufacturing process or an element of the manufacturing process is carried over, the corresponding planning, PsCs*, numbers can be obtained from production. If the intended process is similar to the process being used to manufacture another component elsewhere in the company, those capability values can be shared. If the process is absolutely new, capability data will have to be gathered at the advanced manufacturing tryout or during the process development activities.

Complete validation plan: Considering the process flow-chart and the experience of the work groups involved, it is critical that the complete validation plan, required to produce a quality part, is identified. Therefore, the establishment of engineering, manufacturing, and financial data should be one of the key management tasks in the early part of the production stage.

Validate the manufacturing process design: The work groups validate the manufacturing process design.

Engineering and manufacturing validation: If there is no assurance that the manufacturing process is under control, the alternative design cannot be considered feasible.

Conduct financial validation: The work groups ensure that cost accounting practices are valid both at the design and production levels.

Validate manufacturing facilities: The work groups validate the manufacturing facilities.

Finalize Production Intent Design: The work groups establish the necessary production capability target based from experience or on past policy of the manufacturing discipline. The following are the additional tasks performed during this step.

- Complete first those programs that take long lead times.
- Freeze design and release prototype. Work groups release completed. Design, or re-design the product to manufacturing specs.
- Develop demonstration products.
- Approve production intent design concept.

10.2.9   Activity-plan 9—Production

During this activity-plan, the newly designed, or redesigned, products are actually produced. The production activity-plan includes the following tasks.

- Order production equipment and tooling.
- Qualify production equipment and tooling at the suppliers' facility.
- Order production material and purchased parts.
- Install production equipment and tooling.
- Qualify production equipment and tooling at manufacturing facility.
- Run production schedule.
- Perform production part verification.

10.2.10   Activity-plan 10—Continuous Process
Improvement (CPI)

The process of continuous improvement never ends. This topic is discussed quite elaborately in sections 3.1.1 and 3.8.4 of volume I. CAD/CAM technology works well in a company with a continuous improvement philosophy because the effects of engineering automation are cumulative. As workers become more familiar with CAD tools, they work concurrently. As more programs are written to link CAD tools together and to automate manual tasks, the engineering organization will become more productive. Incremental improvements and functional thinking are not enough. Figure 10.6 shows a cycle of continuous improvement. Let us assume that a current production part is made out of a feature combination—task A, tool B and technology C. Here, three Ts are taken out from a set of seven Ts (see Figure 4.1, volume I). The use of 7Ts in continuous improvement is shown in the bottom second block of the second row of Figure 10.6. In order for companies to

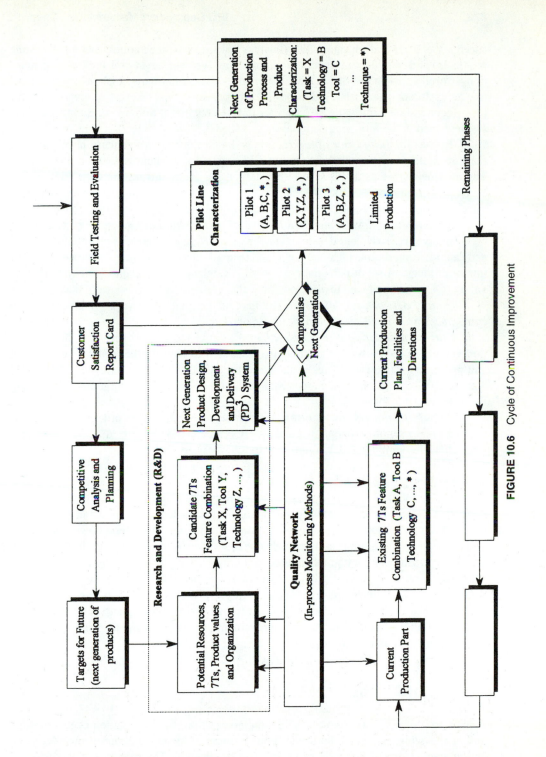

FIGURE 10.6 Cycle of Continuous Improvement

survive the impact of changing 7Ts (talents, tasks, teams, techniques, technology, time and tools) and 3Ps, they must set continuous improvement targets for their next generation of products. Process documents, obtained during earlier activity-plans and competitive analysis, are often employed as basis for improvements. The cycle of continuous improvement is shown in Figure 10.6 by a set of five concurrent rows of parallel tasks. They are described in the following paragraphs.

Targets for future row of tasks: This is the topmost row of concurrent blocks. The purpose of the blocks is to create a set of CPI targets for the next generation of products. The intermediate steps that lead to setting targets for the future are shown in Figure 10.6 by an array of top row blocks. The major ones are field testing and evaluation, customer satisfaction report card, and competitive analysis and planning.

Research and development(R&D) row of tasks: The next sequence of concurrent tasks involves research and development activities. Here a number of potential resources and candidate alternative ideas are combined to arrive at a compromise for the next generation of products. A set of potential resources (7Ts, talents, tasks, teams, etc.) and product-values (QCs) are combined to predict a new value combination that is best of the class.

Quality network row of tasks: The task of the quality network organization is to come up with a limited set of compromising options.

Current production row of tasks: Concurrent to these R&D activities, the production version of the current product is also evaluated to identify the existing key product and process characteristics. This is shown in Figure 10.6 by a fourth layer of horizontal blocks. The R&D recommendation is evaluated in the light of the current production plan, facilities, and directions.

Remaining phases row of tasks: These blocks are required in order to fabricate the product once a set of pertinent QCs and 7Ts (CE ideals) are recognized in the final selection. They are ommitted from Figure 10.6 for clarity purposes.

A limited production run is then performed on each of these options; three are shown in Figure 10.6 as pilot 1 (A, B, C, . . . ,), pilot 2 (X, Y, Z, . . . ,), and Pilot 3 (A, B, Z, . . . ,). On the basis of the results of these production runs, the options are narrowed down to a next generation feature combination. A final selection is shown in Figure 10.6 as (X, B, C). Continuous improvement by itself will not do much good unless there is an *in-process* quality monitoring system that triggers the error-proofing. Waiting to check the quality at the completion of the production cycle does not serve any useful purpose. This means process (3Ps), methods (4Ms), and CE Ideals (7Ts) necessary for ongoing control and improvement must be designed in the overall process. The current layer of quality network provides the methods for an in-process monitoring and control. Such in-process monitoring and continuous improvement are useful tools to guarantee quality from the first manufacturing principle.

10.3   CE CASE HISTORIES

A pilot program is a first step to an IPD implementation. Implementation begins through an initial establishment of pilot needs at one or more of the manufacturing sites. Realistic and attainable sets of common goals are then established. The selection of experienced

Some Case-Histories of CE						Source of Data or Contact Person
Company	Product	Best Development Time			Target of Improvement	
		Before (months)	After (months)	Reduction		
Instron Corp.	Physical Test Equipment	-	-	-	M, T	Stephen Eis
Government Electronic Systems Div, GE Aerospace	Phased Array Antenna Systems (Ground-based)	6-9	3	50-75%	M,T, S	Timothy Fuhr
Thiokol Corp., Strategic Operations	Rocket Motors (Booster Rocket)	24-36	12	50-75%	M, P	Greg Goin
GE Canada's Dominion Engineering Works	Hydroelectric Power Generators and Turbines	24	9-12	50%	A, C, M	Norea Nuon
Litton Guidance & Control Systems	Gyroscopes and Accelerometers	-	-	-	A, C	Douglas Roberts
Abbott Laboratories's Diagnostics	Medical Instruments	-	-	-	A, C	David Mulholland
GE's Fanuc Automation Div.		-	-	-	A, P	
3M Telecom Systems Group	Fiber Optics products	-	-	-	A, P	
Vought Aircraft Company		-	-	-	A, P	
GE's Aeropsace Division		-	-	-	A, P	Welch/ Fullmer
GenCorp Aerojet Div.	Aerospace Equipments	-	-	-	A, P	Bob Culver
Power Team Div. of SPX Corp.	High Pressure Hydraulic Tools and Work Holding Components	-	-	-	A, P	Rick Henderson
United Technologies, Hamilton		-	-	-	A, P	
Telco Systems Fiber Optics Corp.	High Performance Telecommunication Equipment	-	-	-	A, C	Gary Wenger
Westinghouse Electric Corp.'s Electrical Components Div.	Industrial Power Equipments	-	-	-	A, P, S	Bob Beatty
AMP Inc., Automotive Group		-	-	-	A, P	
General Dynamics Corp., Land Sys.		-	-	-	A, P	
Sanborn Manufacturing Co.		-	-	-	A, P	

SOURCES: Machine Design, "How to make Concurrent Engineering Work,"

Part 1: CE: The Need and Enabling Technologies, August 6, 1992;
Part 2: Prerequisites for Successful Implementation, September 10, 1992;
Part 3: Core Technology for Successful Implementation, October 22, 1992;
Part 4: Selling CE to Management, July 23, 1993;
Part 5: Benchmarking Your Company for CE, October 22, 1993;
Part 6: Measuring Success from CE, November 26, 1993

LEGEND:

M: Multi-functional Teams;
C: Computer Integrations,
A: Analytical Methods and Tools;
S: Suppliers in the Project Team
T: Ties between Engineering Model, Drafting and Manufacturing
P: Capturing the Process and Manufacturing Rules

FIGURE 10.7   CE Case Histories

project team leaders are pivotal to program success. Evaluation of pilot projects at the key sites provides the case history and know-how to finalize a company-wide implementation of a full-fledged IPD system. Figure 10.7 lists a collection of successful CE pilot projects tried by eighteen independent companies. Figure 10.7 provides a rich source of case histories in implementing various CE pilot programs. It also lists in a matrix form the product for which CE programs were applied, reduction in development time achieved, target of improvements, and source of data or contact person in each case. Most of the information came from documented successes and from published literatures.

Management participation in the pilot process is even more important than during actual development. This includes selection of the pilot projects, project objectives, project team leaders, and insuring availability of all required resources (7Ts). Versatile highly experienced persons, who have worked as *champions* in multi-disciplinary projects through most of the activity-plans, are ideal candidates for PDT leader positions.

10.4   COMPUTATION OF SAVINGS

There are many ways to quantify CE benefits. CE benefits result not so much from reducing the time in performing the steps as from reorganizing the steps so that the new process is more concurrent. There is a degree of interdependence between the various tasks that can be overlapped, and many of these tasks are automated. Automation means life-cycle mechanization (see Chapter 9), so that each of the reorganized tasks can be run by any concurrent work group in any order depending upon the needs of the organization or the assigned PDT responsibility. The product realization process is captured in a series of computerized modules that share common objects, features, and parameters. These modules are developed in a common collaborative environment so that no exchange of information is necessary to run them. Five categories of information models from specifications to cognitive models (as discussed in Chapter 7 of volume I) are developed for a particular product line. The software modules follow the examples of life-cycle mechanization discussed in Chapter 9 of volume II. The following calculation outlines how one can quantify some of the savings resulting from employing one or more CE principles (outlined in Chapter 4 of volume I). They are all lumped under tangible savings. Intangible savings that cannot be easily quantified, such as increased customer satisfaction, are listed at the end of this section.

10.4.1   Tangible Savings

The following are the different types of tangible savings that can be realized by implementing CE (or IPD) projects through a set of KBE techniques described in Chapters 8 and 9. The accuracy of the computations depends on how one can best estimate the variables in the equations.

10.4.1.1   Productivity Gains

The saving due to productivity gains tends to be the simplest and most straightforward. The most common approach is to get a good estimate from an *as-is* flow charting of a business process that a work group is trying to automate. The enterprise modeling method described in section 3.5 of volume 1 can be used for this purpose. Once the current tasks in the as-is process are identified, the next stage is to determine the amount of time each task may take to complete. Work groups and CE teams then determine what percentage of impact CE can bring in by automating or integrating those functions. It is not necessary to keep the activities, tasks, or the process-flow intact. More savings can be realized by combining or eliminating few of the redundant tasks and rearranging the process or the information flow. The following formula gives a template one can use to compute the savings due to productivity gains:

$$\text{Value of productivity Gain} = [\text{Individual Salary (in \$/Hr.)} \\ \times \text{Estimated Time (in Hours)} \times \text{CE Impact Factor}] \qquad (10.2)$$

There are two types of impacts a work group needs to consider in calculating productivity gains.

1. Impact due to a new set of 7Ts deployment, which will reduce time for each of the individual tasks, assuming they were still done in an "as-is" mode.
2. Impact due to a new set of 4Ms that let this reduced set of tasks be carried out in parallel. This will further reduce the lead time required.

$$\text{Impact Factor} = \text{Impact due to implementation of 7T features} \\ \times \text{Impact Due to Parallel or Simultaneous Processing of Tasks} \qquad (10.3) \\ (\text{e.g., 4Ms, 3Ps, 7Cs, and so on}).$$

10.4.1.2 Technical Memory Retention

A CE application captures life-cycle activities and tasks and stores them in a series of modules (knowledge-based engineering (KBE) software). The captured technical memory be-

TABLE 10.3 Computation of Productivity Gain

Task Identification Number	Task Description	Estimated Time to do the "as-is" Task Identified	Salary of the Person Doing the Task	Percentage Impact due to CE 7Ts Deployment	Percentage Impact due to Paralleling of Tasks, 4Ms, 3Ps, etc.	Value of Productivity Gain
T001.001	Create an Interactive Design of Part X	T_1	S_1	I_{11}	I_{12}	$T_1 \times S_1 \times (I_{11} \times I_{12})$
T001.002	Create an Interactive Design of Part Y	T_2	S_2	I_{21}	I_{22}	$T_2 \times S_2 \times (I_{21} \times I_{22})$
T001.003	Check the Accuracy of Interactive Design	T_3	S_3	I_{31}	I_{32}	$T_3 \times S_3 \times (I_{31} \times I_{32})$
T001.003		T_i	S_i	I_{i1}	I_{i2}	$C_{si} = T_i \times S_i \times (I_{i1} \times I_{i2})$
Total		ΣT_i				ΣC_{si}

comes a corporate asset. It becomes a major source of memory and standardization tool for obtaining consistent designs or redesigns for a future line of similar products. This in turn increases the competitive strength of the company. In the absence of such a KBE/CE application, one will be required to keep resident product and process experts. These experts will be spending a percentage of their time advising the relevant concurrent work groups without regard to problems at hand, diagnosing, troubleshooting, or critiquing, as the case may be. A portion of the time of these experts will also be spent in documenting the expertise in the form of a handbook or a training manual. The time of these experts will also be needed for training the new hires to perform, as well as to provide equivalent functions similar to the KBE applications. Further, if such trained persons or the resident experts quit, the company may lose the entire product or process knowledge. The value of such knowledge is equivalent to how long it will take for a company to redevelop or recoup the lost knowledge. The equivalent cost value of memory retention could be crudely calculated as:

$$
\begin{aligned}
\text{Equivalent Cost Value of the Technical Memory} = \\
(\text{Experts Salary} \times \text{Time used to Advice}) \\
+ (\text{New Hire Training Time} \times \text{New Hire Salary}) \quad (10.4) \\
+ (\text{Competitive Value of the Product or Process} \\
\text{Knowledge Captured}).
\end{aligned}
$$

10.4.1.3 Quality Improvements

The impact of a KBE/CE application on quality improvements (QIs) can be measured in a number of ways. Savings due to quality improvements are really a sum of the following:

$$
\begin{aligned}
\text{Savings due to QIs} \equiv \text{Savings due to reduced Engineering Changes (ECs)} \\
+ \text{Savings due to reduced Prototype Builds (PBs)} \quad (10.5) \\
+ \text{Savings due to reduced scraps.}
\end{aligned}
$$

The computation of indirect savings is as follows:

- *Reduction in Number of Engineering Changes (ECs):* If the number-of-engineering changes$_{before}$ and number-of-engineering-changes$_{after}$ is the number of changes before and after the implementation of CE, and a single engineering change amounts to a loss of C_{ECs} units of dollars, then

$$
\begin{aligned}
\text{Savings due to ECs} \equiv (\text{Number-of-engineering changes}_{before} - \\
\text{Number-of-engineering-changes}_{after}) \times C_{ECs}
\end{aligned} \quad (10.6)
$$

where C_{ECs} is the average cost of a single engineering change.

- *Reduction in the Number of Prototype Builds (PBs):* If the number-of-prototype-built$_{before}$ and number-of-prototype-built$_{after}$ are the number of prototypes before and after the implementation of CE, and a single prototype amounts to a loss of C_{PBs} units of dollars, then

$$
\begin{aligned}
\text{Savings due to reduced PBs} \equiv (\text{Number-of-prototype-built}_{before} - \\
\text{Number-of-prototype-built}_{after}) \times C_{PBs}
\end{aligned} \quad (10.7)
$$

where C_{Pbs} is the cost of a single prototype.

- **Reductions in the Scrap Rate:** If Scrap-rate$_{before}$ and Scrap-rate$_{after}$ are the scrap rates before and after (in parts per millions) the implementation of CE, and Scrap-cost$_{before}$ and Scrap-cost$_{after}$ are the amounts that correspond to cost per part before and after CE implementation, then

$$\text{Savings due to reduced Scraps} = [(\text{Scrap-rate}_{before} \times \text{Scrap-cost}_{before})$$
$$- (\text{Scrap-rate}_{after} \times \text{Scrap-cost}_{after})] \qquad (10.8)$$
$$\times \text{Production-quantity}$$

10.4.1.4 Increased Competitiveness

If CE modules are deployed as KBE applications, one for each family of product line in various product programs, the company would then be able to use them to reduce lead time. Furthermore the company would be able to estimate a better BOM and thereby reduce the unit-cost of the product since work groups have more information available in the knowledge-base. The company would be able to figure out whether they can use a carry-over part or will require the product to be designed afresh. Due to better cost estimates, the company will be able to produce more detailed, accurate, more competitive, and faster proposals. They would also be able to provide a number of options and breakdowns in the proposals due to accurate availability of the historical part data. This will enable the company to win new contracts and grow business in areas where the company has cost-to-market advantage. In its absence, it is assumed that a company will be loosing some of the contracts since it would have difficulty in responding quickly to a competitive bid.

Equivalent Value of Increased Competitiveness $\equiv$
  (Value of winning a new contract for product or services due to CE deployment)   (10.9)
  + (Cost of losing the contract due to inaccurate estimates or poor proposal)

The value of winning a contract can be based on the amount of profits that are built into it. This normally ranges anywhere from 25–75% of the value of the contract itself. The second term in the above equation (10.9) can be calculated on the basis of the time spent in preparing the proposal, writing, gathering the information, performing competitive analysis, and carrying out initial concept-level studies.

10.4.1.5 Use of Standardized parts

The KBE/CE application modules usually contain the library of part information captured as a part of the knowledge-base. If such information is electronically captured, anyone can access databases of standard parts and retrieve the parts. Designers can compare the new specifications of the proposed design with the specifications of parts already in stock to see if there is a match. If the old part can be reused by changing its features, the old part can be used instead of creating a new part number. This could result in a cost savings not only in design and development of parts but also in tooling and manufacturing costs. The average cost savings associated with using an existing part ranges from approximately 10–50 percent of the total product cost. The following formulae can be used to calculate the savings:

Savings due to Standard Parts $\equiv$
  (Number-of-parts-reused $\times$ Operational cost per month per part)   (10.10)
  $\times$ (Design-and-development-cycle time for the part in months).

REFERENCES

CLAUSING, D., and S. PUGH. 1991. *Enhanced QFD Workshop*. Seminar conducted at General Physics Corporation, April 17–18, Troy, MI.

DIKA, R.J., and R.L. BEGLEY. 1991. "Concept Development Through Teamwork—Working for Quality, Cost, Weight and Investment." SAE Paper No. 910212, pp. 1–12, Proceedings of the SAE International Congress and Exposition, SAE, February 25–March 1, 1991, Detroit, MI.

PRASAD, B. 1995a. "A Structured Methodology to Implement Judiciously the Right JIT Tactics." *Journal of Production Planning and Control*, Taylor & Francis, Volume 6, No. 6, pp. 564–577.

PRASAD, B. 1995b. "A Structured Approach to Product and Process Optimization for Manufacturing and Service Industries." *International Journal of Quality and Reliability Management*, Volume 12, No. 9, pp. 123–138, England, U.K.: MCB University Press.

PUGH, S. 1991. *Total Design—Integrating Methods for Successful Product Engineering*, Reading, MA: Addison-Wesley Publishing Company, Inc.

SAATY, T.L. 1978. "Exploring the Interface between Hierarchies, Multiple Objectives and Fuzzy Sets." *Fuzzy Sets and Systems*, Volume 1, pp. 57–68.

WESNER, J.W., J.M. HIATT, and D.C. TRIMBLE. 1994. *Winning with Quality: Applying Quality Principles in Product Development* (September) Reading, MA: Addison-Wesley Publishing Company, Inc.

TEST PROBLEMS—IPD DEPLOYMENT METHODOLOGY

10.1. Since the 1990s, most companies have been down-sizing and cutting back their work-force. Does this mean it will be difficult to implement CE or IPD in late 1990s?

10.2. Saturn Corporation was a major and a radical departure from GM operating procedures. In 1994-95, when Saturn was becoming a more successful business unit, GM decided to undo many of its special privileges. Why?

10.3. Some would argue that the 3Ps at the hub of an IPD implementation wheel in Figure 10.1 belong outside the center so that it will hold the ten activity-plans together. What do you think and why?

10.4. Why are some companies unsuccessful in CE pursuits? What are some characteristics that must be present for a successful implementation of CE?

10.5. What impacts the long-term success of a business: (a) CE influencing agents (b) CE ideals (c) CE enablers (d) CE programs, or (e) all of these?

10.6. Why is there no *cookbook* solution for an IPD deployment? Discuss four important migration considerations when legacy applications and procedures (3Ps) transition to IPD.

10.7. How can you predict your product's success? What are the vehicles for implementing your CE ideals?

10.8. How do you estimate and justify the costs of CE automation tools? Outline seven straightforward methods.

10.9. How do you choose an implementation methodology for rapidly deploying a product design, development and delivery (PD^3) system? Outline six new approaches to improving your existing deployment methodology.

10.10. Why do you or an organization need a deployment methodology? What would the results be if the concurrent work groups do not follow such implementation guidelines?

10.11. What are the seven (7Ts) agents that influence the IPD implementation guidelines? Why do tools and techniques have a bigger influence on deployment than the rest of the 7Ts? How do you prepare for the organizational impact of IPD?

10.12. What are the *ten commandments* for a successful IPD implementation? Do they guide the product and process iterative aspects? If so, how? How do you use IPD techniques to streamline business procedures and reduce cost?

10.13. Why are *defining problems*, *organizing work groups into a PDT*, and *establishing requirements* considered the top three activities of an IPD deployment methodology? Why is it essential to concentrate so heavily in the definition phase? How can you get real cost data up-front?

10.14. Do the three activity plans in exercise 10.13 have equal weight? Do they take an equal amount of time to complete or do they consume equal resources? What would be an ideal distribution of those activity-plans?

10.15. Why does creating a *product system alternative* plan precede a *best design alternative* activity-plan, even though the reliable data for a product system alternative is not available until the best design alternative activity-plan is finalized.

10.16. Describe best practices in integrated product development (IPD). How do you design for quality up-front?

10.17. How do you establish IPD implementation guidelines that suit specific customers and suppliers? How do you test the water before you jump into an IPD implementation?

10.18. Why is senior or top-management involvement essential? How can you get their buy-in?

10.19. How do you achieve best-of-class product definition during an IPD implementation? What are the three guidelines for establishing IPD partnerships with your suppliers and customers?

10.20. What is the basic purpose of benchmarking and QFD study? How do you use CE metrics and measures to provide the most robust implementation?

10.21. How do you measure and identify the need for a mid-course correction while implementing an IPD methodology? How do you avoid the seven most common problems that plague an IPD implementation?

10.22. What are the methods companies have successfully used to prioritize PD^3 improvement efforts?

10.23. What management measures can you use during an IPD Implementation? Define a process for planning and implementing a rapid improvement in reducing a PD^3 cycle-time.

10.24. What are the ten steps of a concurrent PD^3 process? Identify some action steps and tools for meeting (and exceeding) your company product development goals.

10.25. Why do some elements of products and processes have to wait until the *best design alternative* activity-plan is finalized? What are the three most important issues to consider if you are currently deploying an IPD program?

10.26. Following a V-method for product realization shown in Figure 9.19 of volume I, can you determine what activity-plan corresponds to a decomposition phase? What activity-plan corresponds to an aggregation phase?

10.27. Does the beginning of aggregation start with a value *analysis or engineering* activity-plan? What is the ratio of decomposition to aggregation assuming each activity plan takes about equal time for its completion?

10.28. Why is CPI considered a part of an IPD deployment methodology scenario? What are the important differences between a pull-type IPD versus a push-type IPD and how do you decide which is best for your organization?

10.29. Why are a number of concurrent rows of parallel tasks required for a CPI activity-plan? What benefits does such approach bring to IPD? How would you integrate the IPD concept into your overall electronic commerce strategy?

10.30. Why does an organization need a good number of case histories? Does this help in creating a learning organization? How do you apply what you have learned in your IPD implementation plans?

10.31. How do the project leaders justify a CE (or IPD) project? Why do you compute tangible savings? What are the major factors that contribute to tangible savings? Discuss methods to maximize your return on investments (ROI).

10.32. What are the different categories of savings that may result from an IPD deployment? Describe five proven ways to sell IPD to your senior management by demonstrating bottom-line impact. Create a table of savings assuming you can achieve a 10% gain in all categories of tangible savings.

INDEX

484

489